Security and Privacy in Internet of Things (IoTs)

Models, Algorithms, and Implementations

OTHER BOOKS BY FEI HU

Associate Professor
Department of Electrical and Computer Engineering
The University of Alabama

Cognitive Radio Networks
with Yang Xiao
ISBN 978-1-4200-6420-9

Wireless Sensor Networks: Principles and Practice
with Xiaojun Cao
ISBN 978-1-4200-9215-8

Socio-Technical Networks: Science and Engineering Design
with Ali Mostashari and Jiang Xie
ISBN 978-1-4398-0980-8

**Intelligent Sensor Networks: The Integration of Sensor Networks,
Signal Processing and Machine Learning**
with Qi Hao
ISBN 978-1-4398-9281-7

Network Innovation through OpenFlow and SDN: Principles and Design
ISBN 978-1-4665-7209-6

Cyber-Physical Systems: Integrated Computing and Engineering Design
ISBN 978-1-4665-7700-8

**Multimedia over Cognitive Radio Networks: Algorithms, Protocols,
and Experiments**
with Sunil Kumar
ISBN 978-1-4822-1485-7

**Wireless Network Performance Enhancement via Directional Antennas:
Models, Protocols, and Systems**
with John D. Matyjas and Sunil Kumar
ISBN 978-1-4987-0753-4

**Security and Privacy in Internet of Things (IoTs): Models, Algorithms,
and Implementations**
ISBN 978-1-4987-2318-3

Spectrum Sharing in Wireless Networks: Fairness, Efficiency, and Security
with John D. Matyjas and Sunil Kumar
ISBN 978-1-4987-2635-1

Big Data: Storage, Sharing, and Security
ISBN 978-1-4987-3486-8

Opportunities in 5G Networks: A Research and Development Perspective
ISBN 978-1-4987-3954-2

Security and Privacy in Internet of Things (IoTs)

Models, Algorithms, and Implementations

Edited by **Fei Hu**

CRC Press
Taylor & Francis Group
Boca Raton London New York

CRC Press is an imprint of the
Taylor & Francis Group, an **informa** business

CRC Press
Taylor & Francis Group
6000 Broken Sound Parkway NW, Suite 300
Boca Raton, FL 33487-2742

© 2016 by Taylor & Francis Group, LLC
CRC Press is an imprint of Taylor & Francis Group, an Informa business

No claim to original U.S. Government works

Printed on acid-free paper
Version Date: 20160216

International Standard Book Number-13: 978-1-4987-2318-3 (Hardback)

For Fang, Gloria, Edwin, and Edward . . .

Contents

Preface

The Internet of Things (IoT) has attracted strong interest from both academia and industry. The IoT integrates radiofrequency identification (RFID), sensors, smart devices, the Internet, smart grids, cloud computing, vehicle networks, and many other information carriers. Goldman Sachs mentioned that the IoT would bring over 28 billion "things" into the Internet by 2020. Typical "things" include end users, data centers, processing units, smartphones, tablets, Bluetooth, Zig-Bee, IrDA, UWB, cellular networks, Wi-Fi networks, NFC data centers, RFID, their tags, sensors and chips, household machinery, wristwatches, vehicles, house doors, and many other cyberunits. With the growth of nanodevices, smartphones, 5G, tiny sensors, and distributed networks, the IoT is combining the "factual and virtual" anywhere and anytime, and is attracting the attention of both "maker and hacker."

However, interconnecting many "things" also means the possibility of inter-connecting many different threats and attacks. For example, a malware virus can easily propagate through the IoT at an unprecedented rate. In the four design aspects of the IoT system, there may be various threats and attacks: (1) Data perception and collection: In this aspect, typical attacks include data leakage, sovereignty, breach, and authentication. (2) Data storage: The following attacks may occur: denial-of-service attacks (attacks on availability), access control attacks, integrity attacks, impersonation, modification of sensitive data, and so on. (3) Data processing: In this aspect there may exist computational attacks that aim to generate wrong data processing results. (4) Data transmission: Possible attacks include channel attacks, session hijacks, routing attacks, flooding, and so on. Apart from attenuation, theft, loss, breach, and disaster, data can also be fabricated and modified by the compromised sensors.

Therefore, efficient and effective defense mechanisms are of the utmost importance to ensure the security of the IoT. In particular, the U.S. Department of

Energy (DOE) has identified attack resistance to be one of the seven major properties required for the operation of the smart grid, which is an emerging field of the IoT. Then, the question is: how do we use efficient algorithms, models, and implementations to cover the four important aspects of IoT security, that is, confidentiality, authentication, integrity, and availability? Obviously, no single scheme can cover all these four aspects, due to the extreme complexity of IoT attacks.

In this book, we have invited some top IoT security experts from all over the world to contribute their knowledge about different IoT security aspects. We have seamlessly integrated those chapters into a complete book. All chapters have a clear problem statement as well as detailed solutions. More than 100 figures have been provided for graphic understanding.

After reading this book, industrial engineers will have a deep understanding of security and privacy principles in complex IoT systems. They will also be able to launch concrete cryptography schemes based on the detailed algorithms provided in some chapters.

After reading this book, academic researchers will be able to understand all critical issues to be solved in this exciting area. They will get to know some promising solutions to those research problems, and pick up an unsolved, challenging issue for their own research.

After reading this book, policy-makers will have a big picture of IoT security and privacy designs, and get to know the necessary procedures to achieve robust IoT information collection, computation, transmission, and sharing across Internet clouds.

All chapters are written for both researchers and developers. We have tried to avoid much jargon and use plain language to describe profound concepts. In many places, we have also provided step-by-step math models for readers' security test bed implementation purposes.

Overall, this book consists of the following five parts:

Section I. Attacks and Threats: This part introduces all types of IoT attacks and threats. It also demonstrates the principle of countermeasures against those attacks. Moreover, we have given detailed introductions of Sybil attacks, malware propagation, and some other specific attacks.

Section II. Privacy Preservation: Privacy is always one of the top concerns for any network application. The IoT collects data from all the "things" around people. Much data is related to human activities. For example, biomedical data may include patients' health records. How do we distribute those data for Internet sharing while, in the meantime, protecting people's privacy well? In this part, we will discuss privacy preservation issues during data dissemination, participatory sensing, and indoor activities. We will also use smart building as an example to discuss privacy protection solutions.

Section III. Trust and Authentication: The trust model is a critical topic of IoT security design. This part will describe different types of trust models in the

IoT infrastructure. The access control to IoT data is also discussed. A survey of IoT authentication issues is provided in this part.

Section IV. IoT Data Security: This part emphasizes the security issues during IoT data computation. We will introduce computational security issues in IoT data processing, security design in time series data aggregation, key generation for data transmission, as well as concrete security protocols during data access.

Section V. Social Awareness: Any security designs should consider policy and human behavioral features. For example, a security scheme cannot be installed in a real platform without the consent of the users. Many attacks aim to utilize the loopholes of user habits. A security design will have deep impacts on the dissemination of IoT data to each corner of the world. In this part, we will cover social-context-based privacy and trust design in IoT platforms, as well as the policy-based informed consent in the IoT.

We have required each chapter author to provide detailed descriptions of the problems to be solved, the motivations of their proposed solutions, and detailed algorithms and implementations. Our goal is to provide readers with a comprehensive understanding of the security and privacy aspects in the IoT system. A few chapters are written in a survey style. They can be used by beginners to get to know the basic principles of achieving attack-resilient IoT infrastructure.

Due to limitations of time, there may be some points missing in this book. Please contact the publisher if you have any comments for its future improvement.

MATLAB® is a registered trademark of The MathWorks, Inc. For product information, please contact:

The MathWorks, Inc.
3 Apple Hill Drive
Natick, MA 01760-2098 USA
Tel: 508-647-7000
Fax: 508-647-7001
E-mail: info@mathworks.com
Web: www.mathworks.com

Editor

Fei Hu is currently a professor in the Department of Electrical and Computer Engineering at the University of Alabama, Tuscaloosa. He obtained his PhDs at Tongji University (Shanghai, China) in the field of signal processing (in 1999), and at Clarkson University (New York) in electrical and computer engineering (in 2002). He has published over 200 journal/conference papers and books. Dr. Hu's research has been supported by the U.S. National Science Foundation, Cisco, Sprint, and other sources. His research expertise may be summarized as 3S: security, signals, sensors. (1) Security: This concerns how to overcome different cyberattacks in a complex wireless or wired network. Recently he has focused on cyberphysical system security and medical security issues. (2) Signals: This mainly refers to intelligent signal processing, that is, using machine learning algorithms to process sensing signals in a smart way in order to extract patterns (i.e., pattern recognition). (3) Sensors: This includes microsensor design and wireless sensor networking issues.

Contributors

Krishnashree Achuthan
Center for Cybersecurity Systems and
   Networks
Amrita Vishwa Vidyapeetham
Kerala, India

Kemal Akkaya
Department of Electrical and Computer
   Engineering
Florida International University
Miami, Florida

Antonio Marcos Alberti
Instituto Nacional de Telecomunicações
Santa Rita do Sapucaí, Brazil

Gianmarco Baldini
European Commission
Joint Research Centre
Ispra, Italy

Haiyong Bao
School of Electrical and Electronic
   Engineering
Nanyang Technological University
Singapore

Jorge Bernal Bernabe
Department of Information and
   Communications Engineering
University of Murcia
Murcia, Spain

Fang Bingxing
School of Computer Science
Beijing University of Posts and
   Telecommunications and Institute of
   Information Engineering
Chinese Academy of Sciences
Beijing, China

Abdur Rahim Biswas
CREATE-NET
Trento, Italy

Andreas Brauchli
Department of Information and
   Computer Sciences
University of Hawaii at Manoa
Honolulu, Hawaii

Kwang-Cheng Chen
Graduate Institute of Communication
   Engineering
National Taiwan University
Taipei, Taiwan

Pin-Yu Chen
Department of Electrical Engineering
and Computer Science
University of Michigan
Ann Arbor, Michigan

Xiang Chen
School of Information Science and
Technology
Sun Yat-sen University
Guangzhou, China

Shin-Ming Cheng
Department of Computer Science and
Information Engineering
National Taiwan University of Science
and Technology
Taipei, Taiwan

Bertrand Copigneaux
Inno-Group
Sophia Antipolis, France

Pablo Cortijo Castilla
OSNA Cyber Security Research Group
National University of Ireland
Galway, Ireland

Edielson Prevato Frigieri
Instituto Nacional de Telecomunicações
Santa Rita do Sapucaí, Brazil

Romeo Giuliano
Department for Innovation Technologies
and Processes
Guglielmo Marconi University
Rome, Italy

Jose Luis Hernandez
Department of Information and
Communications Engineering
University of Murcia
Murcia, Spain

Cheng Huang
School of Electrical and Electronic
Engineering
Nanyang Technological University
Singapore

Xumin Huang
School of Automation
Guangdong University of
Technology
Guangzhou, China

Bharat Jayaraman
Department of Computer Science and
Engineering
University at Buffalo (SUNY)
Buffalo, New York

Na. Jeyanthi
School of Information Technology and
Engineering
Vellore Institute of Technology
University
Vellore, India

Roger Piqueras Jover
AT & T Security Research Center
New York, New York

Jiawen Kang
School of Automation
Guangdong University of
Technology
Guangzhou, China

Jinesh M. Kannimoola
Center for Cybersecurity Systems and
Networks
Amrita Vishwa Vidyapeetham
Kerala, India

Chin-Fu Kuo
Department of Computer Science and
Information Engineering
National University of Kaohsiung
Kaohsiung, Taiwan

Depeng Li
Department of Information and
    Computer Sciences
University of Hawaii at Manoa
Honolulu, Hawaii

Liu Licai
School of Computer Science
Beijing University of Posts and
    Telecommunications and Institute of
    Information Engineering
Chinese Academy of Sciences
Beijing, China

Yin Lihua
Institute of Information Engineering
Chinese Academy of Sciences
Beijing, China

Xiaodong Lin
Faculty of Business and Information
    Technology
University of Ontario Institute of
    Technology
Oshawa, Canada

Hong Liu
Department of Electrical and Computer
    Engineering
University of Massachusetts Dartmouth
North Dartmouth, Massachusetts

Pavel Loskot
College of Engineering
Swansea University
Swansea, United Kingdom

Rongxing Lu
School of Electrical and Electronic
    Engineering
Nanyang Technological University
Singapore

Yung-Feng Lu
Department of Computer Science and
    Information Engineering
National Taichung University of Science
    and Technology
Taichung, Taiwan

Liangli Ma
School of Electronic Engineering
Naval University of Engineering
Wuhan, China

Franco Mazzenga
Department of Enterprise
    Engineering
University of Rome Tor Vergata
Rome, Italy

Hugh Melvin
OSNA Cyber Security Research Group
National University of Ireland
Galway, Ireland

Klaus Moessner
Centre for Communications Systems
    Research
University of Surrey
Surrey, United Kingdom

Mara Victoria Moreno
Department of Information and
    Communications Engineering
University of Murcia
Murcia, Spain

Michele Nati
Centre for Communications Systems
    Research
University of Surrey
Surrey, United Kingdom

Ricardo Neisse
European Commission
Joint Research Centre
Ispra, Italy

Alessandro Neri
Department of Engineering
Roma Tre University
Rome, Italy

Jason M. O'Kane
Department of Computer Science and
Engineering
University of South Carolina
Columbia, South Carolina

Niklas Palaghias
Centre for Communications Systems
Research
University of Surrey
Surrey, United Kingdom

Ranga Rao Venkatesha Prasad
Mathematics and Computer Science
Department
Delft University
Delft, the Netherlands

Wei Ren
School of Computer Science
China University of Geosciences
Wuhan, China

Yi Ren
Department of Computer Science
National Chiao Tung University
Hsinchu, Taiwan

Rodrigo da Rosa Righi
Interdisciplinary Program
Universidade do Vale do Rio dos Sinos
São Leopoldo, Brazil

Nico Saputro
Department of Electrical and Computer
Engineering
Florida International University
Miami, Florida

Michael Schukat
OSNA Cyber Security Research Group
National University of Ireland
Galway, Ireland

Antonio Skarmeta
Department of Information and
Communications Engineering
University of Murcia
Murcia, Spain

Arif Selcuk Uluagac
Department of Electrical and Computer
Engineering
Florida International University
Miami, Florida

Anna Maria Vegni
Department of Engineering
Roma Tre University
Rome, Italy

Miao Xu
Department of Computer Science and
Engineering
University of South Carolina
South Carolina, Columbia

Wenyuan Xu
Department of Computer Science and
Engineering
University of South Carolina
Columbia, South Carolina

Rong Yu
School of Automation
Guangdong University of
Technology
Guangzhou, China

Guo Yunchuan
Institute of Information Engineering
Chinese Academy of Sciences
Beijing, China

Ali Ihsan Yurekli
Department of Electrical and Computer
Engineering
Florida International University
Miami, Florida

THREATS AND ATTACKS

I

Chapter 1

Internet of Things (IoT) as Interconnection of Threats (IoT)

N. Jeyanthi

CONTENTS

1.1 Introduction

People worldwide are now ready to enjoy the benefits of the Internet of Things (IoT). The IoT incorporates everything from the body sensor to the recent cloud computing. It comprises major types of networks, such as distributed, grid, ubiquitous, and vehicular; these have conquered the world of IT over a decade. From parking vehicles to tracking vehicles, from entering patient details to observing postsurgery, from child care to elder care, from smart cards to near field cards, sensors are making their presence felt. Sensors play a vital role in the IoT as well. The IoT works across heterogeneous networks and standards. Exceptionally, no network is free from security threats and vulnerabilities. Each of the IoT layers is exposed to different types of threats. This chapter focuses on possible threats to be addressed and mitigated to achieve secure communication over the IoT.

The concept of the IoT was proposed in 1999 by the Auto-ID laboratory of the Massachusetts Institute of Technology (MIT). ITU released it in 2005, beginning in China. The IoT can be defined as "data and devices continually available through the Internet." Interconnection of things (objects) that can be addressed

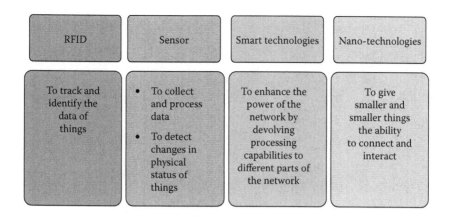

Figure 1.1: IoT underlying technologies.

unambiguously and heterogeneous networks constitute the IoT. Radiofrequency identification (RFID), sensors, smart technologies, and nanotechnologies are the major contributors to the IoT for a variety of services, as shown in Figure 1.1. Goldman Sachs quoted that there are 28 billion reasons to care about the IoT. They also added that in the 1990s, the fixed Internet could connect one billion end users, while in the 2000s, the mobile Internet could connect another two billion. With this growth rate, the IoT will bring as many as 28 billion "things" to the Internet by 2020. With the drastic reduction in the cost of things, sensors, bandwidth, processing, smartphones, and the migration toward IPv6, 5G could make the IoT easier to adopt than expected. Every "thing" comes under one umbrella encompassing all the things.

The IoT also views everything as the same, not even discriminating between humans and machines. Things include end users, data centers (DCs), processing units, smartphones, tablets, Bluetooth, ZigBee, the Infrared Data Association (IrDA), ultra-wideband (UWB), cellular networks, Wi-Fi networks, near field communication (NFC) DCs, RFID and their tags, sensors and chips, household equipment, wristwatches, vehicles, and house doors; in other words, IoT combines "factual and virtual" anywhere and anytime, attracting the attention of both "maker and hacker." Inevitably, leaving devices without human intervention for a long period could lead to theft. IoT incorporates many such things. Protection was a major issue when just two devices were coupled. Protection for the IoT would be unimaginably complex.

1.2 Phases of IoT System

The IoT requires five phases, from data collection to data delivery to the end users on or off demand, as shown in Figure 1.2.

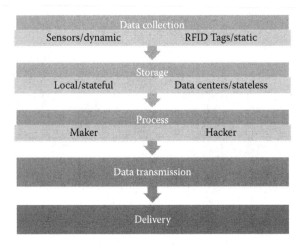

Figure 1.2: Phases of IoT system.

1.2.1   Phase I: Data collection, acquisition, perception

Be it a telemedicine application or vehicle tracking system, the foremost step is to collect or acquire data from the devices or things. Based on the characteristics of the thing, different types of data collectors are used. The thing may be a static body (body sensors or RFID tags) or a dynamic vehicle (sensors and chips).

1.2.2   Phase II: Storage

The data collected in phase I should be stored. If the thing has its own local memory, data can be stored. Generally, IoT components are installed with low memory and low processing capabilities. The cloud takes over the responsibility for storing the data in the case of stateless devices.

1.2.3   Phase III: Intelligent processing

The IoT analyzes the data stored in the cloud DCs and provides intelligent services for work and life in hard real time. As well as analyzing and responding to queries, the IoT also controls things. There is no discrimination between a boot and a bot; the IoT offers intelligent processing and control services to all things equally.

1.2.4   Phase IV: Data transmission

Data transmission occurs in all phases:

- From sensors, RFID tags, or chips to DCs

- From DCs to processing units

- From processors to controllers, devices, or end users

1.2.5 Phase V: Delivery

Delivery of processed data to things on time without errors or alteration is a sensitive task that must always be carried out.

1.3 Internet of Things as Interconnections of Threats (IoT vs. IoT)

In the future, maybe around the year 2020 with IPv6 and the 5G network, millions of heterogeneous things will be part of the IoT. Privacy and security will be the major factors of concern at that time. The IoT can be viewed in different dimensions by the different sections of academia and industry; whatever the viewpoint, the IoT has not yet reached maturity and is vulnerable to all sorts of threats and attacks. The prevention or recovery systems used in the traditional network and Internet cannot be used in the IoT due to its connectivity.

Change is the only thing that is constant, and end users strive to develop technology to suit their needs. The evolution of threats has caused an increase in the security measures that need to be taken into consideration. This chapter presents security issues in three dimensions, based on phase, architecture, and components. Figures 1.3 through 1.6 show all possible types of attacks in these three different views, thus depicting the IoT as the Interconnection of Threats.

1.3.1 Phase attacks

Figure 1.3 demonstrates the variety of attacks on the five phases of IoT. Data leakage, sovereignty, breach, and authentication are the major concerns in the data perception phase.

1.3.1.1 Data leakage or breach

Data leakage can be internal or external, intentional or unintentional, authorized or malicious, involving hardware or software. Export of unauthorized data or information to an unintended destination is data leakage. Generally, this is done by a dishonest or dissatisfied employee of an organization. Data leakage is a serious threat to reliability. As the cloud data move from one tenant to several other tenants of the cloud, there is a serious risk of data leakage. The severity of data leakage can be reduced by the use of DLP (data leakage prevention).

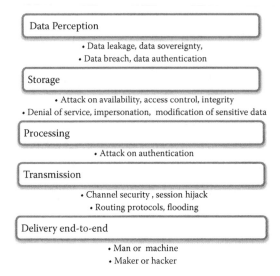

Data Perception
- Data leakage, data sovereignty,
- Data breach, data authentication

Storage
- Attack on availability, access control, integrity
- Denial of service, impersonation, modification of sensitive data

Processing
- Attack on authentication

Transmission
- Channel security , session hijack
- Routing protocols, flooding

Delivery end-to-end
- Man or machine
- Maker or hacker

Figure 1.3: Attacks on phases.

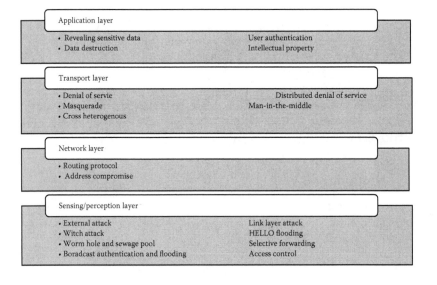

Application layer
- Revealing sensitive data
- Data destruction

User authentication
Intellectual property

Transport layer
- Denial of servie
- Masquerade
- Cross heterogenous

Distributed denial of service
Man-in-the-middle

Network layer
- Routing protocol
- Address compromise

Sensing/perception layer
- External attack
- Witch attack
- Worm hole and sewage pool
- Boradcast authentication and flooding

Link layer attack
HELLO flooding
Selective forwarding
Access control

Figure 1.4: Possible attacks based on architecture.

1.3.1.2 Data sovereignty

Data sovereignty means that information stored in digital form is subject to the laws of the country. The IoT encompasses all things across the globe and is hence liable to sovereignty.

1.3.1.3 Data loss

Data loss differs from data leakage in that the latter is a sort of revenge-taking activity on the employer or administrator. Data loss is losing the work accidentally due to hardware or software failure and natural disasters.

1.3.1.4 Data authentication

Data can be perceived from any device at any time. They can be forged by intruders. It must be ensured that perceived data are received from intended or legitimate users only. Also, it is mandatory to verify that the data have not been altered during transit. Data authentication could provide integrity and originality.

1.3.1.5 Attack on availability

Availability is one of the primary securities for the intended clients. Distributed denial of service (DDoS) is an overload condition that is caused by a huge number of distributed attackers. But this not the only overload condition that makes the DCs unavailable to their intended clients. The varieties of overload threat occurrence that cause DCs to freeze at malicious traffic are analyzed here:

- Flooding by attackers

- Flooding by legitimates (flash crowd)

- Flooding by spoofing

- Flooding by aggressive legitimates

1.3.1.5.1 Flooding by attackers

DDoS is flooding of malicious or incompatible packets by attackers toward the DCs. This kind of overload threat can be easily detected by Matchboard Profiler. If the attacker characteristic is found, the user can be filtered at the firewall.

1.3.1.5.2 Flooding by legitimates (flash crowd)

Flash crowd is an overload condition caused by huge numbers of legitimate users requesting the DC resources simultaneously. This can be solved by buffering an excess number of requests so that this overload condition remains live only for a certain period of time.

1.3.1.5.3 Flooding by spoofing attackers

This is caused by impersonation which can be detected by acknowledging each request and by maintaining the sequence number of the requests and requesters' Internet protocol (IP) address.

1.3.1.5.4 Flooding by aggressive legitimates

Aggressive legitimates are users who are restless and repeatedly initiate similar requests within a short time span. This leads to an overload condition, where the legitimate users flood the server with requests that slow down the DC performance. These attacks are difficult to detect because of their legitimate characteristics. By analyzing the inter-arrival time between data packets as well as the values of the back-off timers, those attacks can be detected.

1.3.1.6 Modification of sensitive data

During transit from sensors, the data can be captured, modified, and forwarded to the intended node. Complete data need not be modified; part of the message is sufficient to fulfill the intention.

Modification takes place in three ways: (1) content modification, in which part of the information has been altered; (2) sequence modification, in which the data delivery has been disordered, making the message meaningless; and (3) time modification, which could result in replay attack.

For example, if an ECG report has been altered during a telemedicine diagnosis, the patient may lose his or her life. Similarly, in road traffic, if the congestion or accident has not been notified to following traffic, it could result in another disaster.

1.3.2 Attacks as per architecture

The IoT has not yet been confined to a particular architecture. Different vendors and applications adopt their own layers. In general, the IoT is assumed to have four layers: the lowest-level perception layer or sensing layer, the network layer, the transmission layer, and the application layer. Figure 1.4 depicts the layers and the possible threats to each layer.

1.3.2.1 External attack

In order to make full use of the benefits of the IoT, security issues need to be addressed first. Trustworthiness of the cloud service provider is the key concern. Organizations deliberately offload both sensitive and insensitive data to obtain the services. But they are unaware of the location where their data will be processed or stored. It is possible that the provider may share this information with others, or the provider itself may use it for malicious actions.

1.3.2.2 Wormhole attack

Wormhole attack is very popular in ad hoc networks. IoT connects both stationary and dynamic objects, ranging from wristwatches and refrigerators to vehicles. The link that binds these objects is also heterogeneous, may be wired or wireless, and depends on the geographical location. Here, the intruder need not compromise any hosts in the network. The intruder just captures the data, forwards them to another node, and retransmits them from that node. Wormhole attack is very strange and difficult to identify.

1.3.2.3 Selective forwarding attack

Malicious nodes choose the packets and drop them out; that is, they selectively filter certain packets and allow the rest. Dropped packets may carry necessary sensitive data for further processing.

1.3.2.4 Sinkhole attack

Sensors, which are left unattended in the network for long periods, are mainly susceptible to sinkhole attack. The compromised node attracts the information from all the surrounding nodes. Thereby, the intruder posts other attacks, such as selective forward, fabrication, and modification.

1.3.2.5 Sewage pool attack

In a sewage pool attack, the malicious user's objective is to attract all the messages of a selected region toward it and then interchange the base station node in order to make selective attacks less effective.

1.3.2.6 Witch attack

The malicious node takes advantage of failure of a legitimate node. When the legitimate node fails, the factual link takes a diversion through the malicious node for all its future communication, resulting in data loss.

1.3.2.7 HELLO flood attacks

In HELLO flood news attacks, every object will introduce itself with HELLO messages to all the neighbors that are reachable at its frequency level. A malicious node will cover a wide frequency area, and hence it becomes a neighbor to all the nodes in the network. Subsequently, this malicious node will also broadcast a HELLO message to all it neighbors, affecting the availability. Flooding attacks cause nonavailability of resources to legitimate users by distributing a huge number of nonsense requests to a certain service.

1.3.2.8 Addressing all things in IoT

Spoofing the IP address of virtual machines (VMs) is another serious security challenge. Malicious users obtain the IP address of the VMs and implant malicious machines to attack the users of these VMs. This enables hacking, and the attackers can access users' confidential data and use it for malicious purposes.

Since the cloud provides on-demand service and supports multitenancy, it is also more prone to DDoS attack. As the attacker goes on flooding the target, the target will invest more and more resources into processing the flood request. After a certain time, the provider will run out of resources and will be unable to service even legitimate users. Unless DLP agents are embedded in the cloud, due to multitenancy and the movement of data from users' control into the cloud environment, the problem of data leakage will also exist.

The Internet has been expanding since its inception, and with it, threats to users and service providers. Security has been a major aspect of the Internet. Many organizations provide services through the Internet that involve banking transactions, registrations, and so on. As a consequence, these websites need to be protected from malicious attacks.

1.3.2.9 Distributed denial of service (DDoS)

DDoS, an attack initiated and continued by some hundreds or even thousands of attackers, starts by populating unwanted traffic packets with enormous size in order to capture and completely deplete memory resources. At the same time, the traffic disallows legitimate requests from reaching the DC and also depletes the bandwidth of the DC. This eventually leads to unresponsiveness to legitimate requests. A denial of service (DoS) or DDoS attack can overwhelm the target's resources, so that authorized users are unable to access the normal services of the cloud. This attack is a cause of failure of availability. Table 1.1 shows the various types of DDoS attacks, the tools used, and the year of origination.

1.3.2.10 Flash crowd

A flash crowd is basically a sudden increase in the overall traffic to any specific web page or website on the Internet and the sudden occurrence of any event that triggers that particular massive traffic of people accessing that web page or website.

Less robust sites are unable to cope with the huge increase in traffic and become unavailable. Common causes of flash crowd are lack of sufficient data bandwidth, servers that fail to cope with the high number of requests, and traffic quotas.

1.3.2.11 IP spoof attack

Spoofing is a type of attack in which the attacker pretends to be someone else in order to gain access to restricted resources or steal information. This type of attack can take a variety of different forms; for instance, an attacker can impersonate the IP address of a legitimate user to get into their accounts. IP address spoofing, or IP spoofing, refers to the creation of IP packets with a forged source IP address, called spoofing, with the purpose of concealing the identity of the sender or impersonating another computing system.

Table 1.1   **Origin of DDoS attacks**

DDoS Tool	Possible Attacks	Year
Fapi	UDP, TCP (SYN and ACK), and ICMP floods	June 1998
Trinoo	Distributed SYN DoS attack	June 1999
Tribe Flood Network (TFN)	ICMP flood, SYN flood, UDP flood, and SMURF-style attacks	August 1999
Stacheldraht	ICMP flood, SYN flood, UDP flood, and SMURF attacks	Late summer of 1999
Shaft	Packet flooding attacks	November 1999
Mstream	TCP ACK Flood attacks	April 2000
Trinity	UDP, fragment, SYN, RST, ACK, and other flood attacks	August 2000
Tribe Flood Network 2K (TFN2K)	UDP, TCP, and ICMP Teardrop and LAND attacks	December 2000
Ramen	Uses back chaining model for automatic propagation of attack	January 2001
Code Red and Code Red II	TCP SYN Attacks	July and August 2001
Knight	SYN attacks, UDP flood attacks	July 2001
Nimda	Attacks through e-mail attachments and SMB networking and backdoors attacks	September 2001
SQL slammer	SQL code injection attack	January 2003
DDOSIM (version 0.2)	TCP-based connection attacks	November 2010
Loris	Slowloris attack and its variants, viz. Pyloris	June 2009
Qslowloris	Attacks the websites, e.g., IRC bots, botnets	June 2009
L4D2	Propagation attacks	2009
XerXeS	WikiLeaks attacks, QR code attacks	2010
Saladin	Webservers attacks, Tweet attacks	November 2011
Apachekiller	Apache server attacks, scripting attacks	August 2011
Tor's Hammer	http POST attacks	2011
Anonymous LOIC tool	—	2013

IP spoofing is most frequently used in DoS attacks. In such attacks, the goal is to flood the victim with overwhelming amounts of traffic, and the attacker does not care about receiving responses to the attack packets. They have additional advantages for this purpose—they are more difficult to filter, since each spoofed packet appears to come from a different address, and they hide the true source of the attack.

There are three different types of spoof attacks: impersonation, hiding attack, and reflection attack. Congestion is a threat in any network if the number of incoming packets exceeds the maximum capacity. The factor that is affected at the time of congestion is throughput.

1.3.2.12 Types of spoof attacks

Among the several types of spoofing attacks, the following attacks are addressed, as they are launched on behalf of clients and destroy the DC's resources.

Type I, Hiding attack: Attackers simultaneously send a large number of spoofed packets with random IP address. This creates chaos at the DC regarding which specific packets should be processed as legitimate packets, shown in Figure 1.5.

Type II, Reflection attack: Attackers send spoof packets with the source IP address of the victim to any unknown user. This causes unwanted responses to reach the victim from unknown users and increases the flood rate, shown in Figure 1.6.

Type III, Impersonation attack: Attackers send spoof packets with the source IP address of any unknown legitimate user and acting as a legitimate user. This is equivalent to a man-in-the-middle attack. The spoof attacker receives requests from clients, spoofs IP, and forwards the requests to the DC, acting as a legitimate user. The responses of the DC are again processed intermediately and sent to the clients. This leads to confidentiality issues and data theft or loss at the DC, as shown in Figure 1.7.

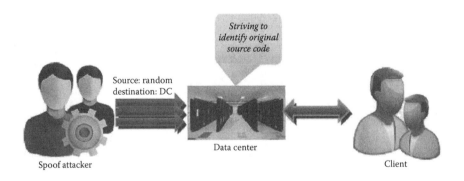

Figure 1.5: Hiding attack.

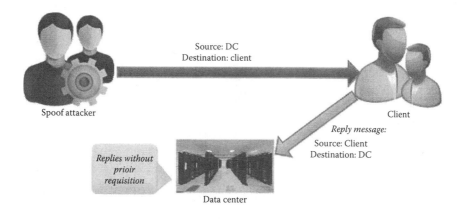

Figure 1.6: Reflection attack.

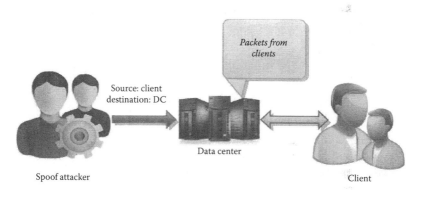

Figure 1.7: Impersonation attack.

If a proper spoof detection mechanism is not in place, the DC could respond badly, leading to a partial shutdown of services.

- In network-level DDoS, the attackers will try to send invalid requests with the aim of flooding the cloud service provider (CSP); for example, requests for a half-open connection.

- In service-level DDoS, the attacker will be sending requests that seem to be legitimate. Their content will be similar to a request made by a legitimate user. Only their intention is malicious.

1.3.2.13 *Goodput*

Goodput is the application-level throughput, that is, the number of useful information bits, delivered by the network to a certain destination, per unit of time.

The amount of data considered excludes protocol overhead bits as well as retransmitted data packets. The goodput is a ratio between the amount of information delivered and the total delivery time. This delivery time includes interpacket time gaps, overhead in transmission delay, packet queuing delay, packet retransmission time, delayed acknowledge, and processing delay.

1.3.2.14   Data centers (DCs)

A DC is a centralized repository, either physical or virtual, for the storage, management, and dissemination of data and information organized around a particular body of knowledge or pertaining to a particular business.

A DC is a facility used to house computer systems and associated components and huge storage systems. The main purpose of a DC is to run the applications that handle the core business and operational data of the organization. Such systems may be proprietary and developed in house by the organization, or bought from enterprise software vendors. Often, these applications will be composed of multiple hosts, each running a single component. Common components of such applications are databases, file servers, application servers, middleware, and various others.

1.3.2.15   Botnet

A *botnet* is a collection of Internet-connected computers whose security defenses have been breached and control ceded to a malicious party. Each such compromised device, known as a "bot," is created when a computer is penetrated by software from a malware distribution, otherwise known as malicious software. The controller of a botnet is able to direct the activities of these compromised computers through communication channels formed by standards-based network protocols such as Internet Relay Chat (IRC) and hypertext transfer protocol (http).

In DDoS attacks, multiple systems submit as many requests as possible to a single Internet computer or service, overloading it and preventing it from servicing legitimate requests. An example is an attack on a victim's phone number. The victim is bombarded with phone calls by the bots, attempting to connect to the Internet.

1.3.2.16   Confidentiality

All the clients' data are to be transacted in a network channel with greater visibility regarding assurance for the intended clients that data are tamperproof.

1.3.2.17   Physical security

Hardware involved in serving clients must be continuously audited with a safe checkpoint for the sake of hysteresis identification of threats.

1.3.2.18 Software security

Corruption or modification of application software by threats could affect several clients who depend on that particular application programming interface (API) and related software interfaces.

1.3.2.19 Network security

Bandwidth attacks such as DoS and DDoS can cause severe congestion the network and also affect normal operations, resulting in communication failure.

1.3.2.20 Legal service-level agreement (SLA) issues

SLAs between customer and service provider must satisfy legal requirement, as the cyber laws vary for different countries. Incompatibilities may lead to compliance issues.

1.3.2.21 Eavesdropping

Eavesdropping is an interception of network traffic to gain unauthorized access. It can result in failure of confidentiality. The *man in the middle attack* is also a category of eavesdropping.

The attack sets up a connection with both victims involved in a conversation, making them believe that they are talking directly but infecting the conversation between them.

1.3.2.22 Replay attack

The attacker intercepts and saves old messages and then sends them later as one of the participants to gain access to unauthorized resources.

1.3.2.23 Back door

The attacker gains access to the network through bypassing the control mechanisms using a "back door," such as a modem and asynchronous external connection.

1.3.2.24 Sybil attack

Impersonation is a threat in which a malicious node modifies the data flow route and lures the nodes to wrong positions. In *Sybil attack*, a malicious user pretends to be a distinct user after acquiring multiple identities and tries to create a relationship with an honest user. If the malicious user is successful in compromising one of the honest users, the attacker gains unauthorized privileges that help in the attacking process.

1.3.2.25   Byzantine failure

Byzantine failure is a malicious activity that compromises a server or a set of servers to degrade the performance of the cloud.

1.3.2.26   Data protection

Data Protection It is difficult for the cloud customer to efficiently check the behavior of the cloud supplier, and as a result, the customer is confident that data is handled in a legal way. But practically, various data transformations intensify the job of data protection.

1.3.2.27   Incomplete data deletion

Incomplete Data Deletion Accurate data deletion is not possible, because copies of data are stored in the nearest replica but are not available.

1.3.3   Attacks based on components

The IoT connects "everything" through the Internet. These things are heterogeneous in nature, communicating sensitive data over a distance. Apart from attenuation, theft, loss, breach, and disaster, data can also be fabricated and modified by compromised sensors. Figure 1.8 shows the possible types of attacks at the component level.

Verification of the end user at the entry level is mandatory; distinguishing between humans and machines is extremely important. Different types of Completely Automated Public Turing test to tell Computers and Humans

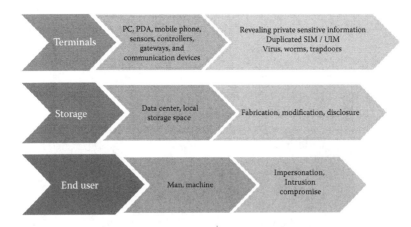

Figure 1.8: Possible attacks based on components.

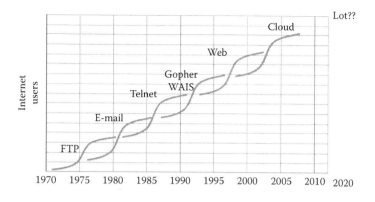

Figure 1.9: Growth of IoT. (*Courtesy of Forrester.*)

Apart (CAPTCHA) help in this fundamental discrimination. With its exponential growth, the IoT will soon dominate the IT industry, as shown in Figure 1.9.

Bibliography

[1] Chuankun, Wu. A preliminary investigation on the security architecture of the Internet of Things. *Strategy and Policy Decision Research*, 2010, 25(4): 411–419.

[2] Goldman Sachs. *IoT Primer, The Internet of Things: Making Sense of the Next Mega-Trend.* September 3, 2014.

[3] International Telecommunication Union. ITU Internet reports 2005: The Internet of Things. 2005.

[4] Ibrahim Mashal, Osama Alsaryrah, Tein-Yaw Chung, Cheng-Zen Yang, Wen-Hsing Kuo, Dharma P. Agrawal. Choices for interaction with things on internet and underlying issues. *Ad Hoc Networks*, 2015, 28: 68–90.

[5] Jeyanthi, N., N.Ch.S.N. Iyengar. Escape-on-sight: An efficient and scalable mechanism for escaping DDoS attacks in cloud computing environment. *Cybernetics and Information Technologies*, 2013, 13(1): 46–60.

[6] Kang Kai, Pang Zhi-bo, Wang Cong. Security and privacy mechanism for health Internet of Things. *The Journal of China Universities of Posts and Telecommunications*, 2013, 20(Suppl. 2): 64–68.

[7] Kim Thuat Nguyen, Maryline Laurent, Nouha Oualha. Survey on secure communication protocols for the Internet of Things. *Ad Hoc Networks*, 2015, 32: 17–31.

[8] Lan Li. Study on security architecture in the Internet of Things. *Measurement, International Conference on Information and Control (MIC)*, 2012, pp. 374–377.

[9] Peng, Xi, Zheng Wu, Debao Xiao, Yang Yu. Study on security management architecture for sensor network based on intrusion detection. *2009 International Conference on Networks Security, Wireless Communications and Trusted Computing*, IEEE, New York.

[10] Prabadevi, B., N. Jeyanthi. Distributed denial of service attacks and its effects on cloud environment: A survey. *The 2014 International Symposium on Networks, Computer and Communications*, June 17–19, 2014, Hammamet, Tunisia, IEEE.

[11] Qazi Mamoon Ashraf, Mohamed Hadi Habaebi. Autonomic schemes for threat mitigation in Internet of Things. *Journal of Network and Computer Applications*, 2015, 49: 112–127.

[12] Qinglin, Cao. Review of research on the Internet of Things. *Software Guide*, 2010, 9(5): 6–7.

[13] Rodrigo Roman, Jianying Zhou, Javier Lopez. On the features and challenges of security and privacy in distributed Internet of Things. *Computer Networks*, 2013, 57: 2266–2279.

[14] Rolf H. Weber. Internet of Things—New security and privacy challenges. *Computer Law and Security Review*, 2010, 26: 23–30.

[15] Sicari, S., Rizzardi, A., Grieco, L.A., Coen-Porisini, A. Security, privacy and trust in Internet of Things: The road ahead. *Computer Networks*, 2015, 76: 146–164.

[16] Wang, Y.F., Lin, W.M., Zhang, T., Ma, Y.Y. Research on application and security protection of Internet of Things in smart grid, *Information IET International Conference on Science and Control Engineering 2012 (ICISCE 2012)*, 2012, pp. 1–5, Shenzhen, China.

[17] Xingmei, Xu, Zhou Jing, Wang He. Research on the basic characteristics, the key technologies, the network architecture and security problems of the Internet of Things. *3rd International Conference on Computer Science and Network Technology (ICCSNT)*, 2013, pp. 825–828.

[18] Yang Guang, Geng Guining, Du Jing, Liu Zhaohui, Han He. Security threats and measures for the Internet of Things. *Tsinghua University (Science and Technology)*, 2011, 51(10): 19–25.

[19] Yang Yongzhi, Gao Jianhua. A study on the "Internet of Things" and its scientific development in China. *China's Circulation Economy*, 2010, 2: 46–49.

[20] Yang Geng, Xu Jian, Chen Wei, Qi Zheng-hua, Wang Hai-yong. Security characteristic and technology in the Internet of Things. *Journal of Nanjing University of Posts and Telecommunications (Natural Science)*, 2010, 30(4): 21–28.

[21] Zhang Fu-Sheng. *Internet of Things: Open a New Life of Intelligent Era.* ShanXi People's Publishing House. 2010, pp. 175–184.

Chapter 2

Attack, Defense, and Network Robustness of Internet of Things

Pin-Yu Chen

CONTENTS

2.1 Introduction

The Internet of Things (IoT) [3] enables ubiquitous communication among different devices. However, the functionality and operations of the IoT heavily depend on the underlying network connectivity structure. Despite the fact that the IoT features ubiquitous communication among all kinds of electronic devices, it inevitably raises security concerns due to seamless penetration and automated integration among all sorts of applications. For example, an adversary may leverage the interconnected devices for malware propagation [7, 16–19]. Therefore, efficient and effective defense mechanisms are of the utmost importance to ensure the reliability of the IoT [9, 12]. In particular, the U.S. Department of Energy (DOE) has identified attack resistance to be one of the seven major properties required for the operation of the smart grid [1], which is an emerging field of the IoT.

By representing the intricate connections of the IoT as a graph, we can investigate the network vulnerability of the IoT to various attack schemes. Three defense schemes are investigated to counter fatal attacks: the intrinsic topological defense scheme, the fusion-based defense scheme, and the sequential defense scheme. Furthermore, by formulating the interplay between an adversary and a defender as a two-player zero-sum game, in which they aim to maximize their own payoffs in terms of network connectivity, we can use the game equilibrium to evaluate network robustness. A sequential defense scheme is also introduced to defend against fatal attacks in the IoT. The results are demonstrated via real-world network data.

Throughout this chapter, we use the undirected and unweighted graph $\mathcal{G} = (\mathcal{V}, \mathcal{E})$ to characterize the network connectivity structure of the IoT, where $\mathcal{V}$ is the set of nodes (devices) with size n, and $\mathcal{E}$ is the set of edges (connections) with size m. Equivalently, the graph can be represented by an n-by-n binary symmetric adjacency matrix $\mathbf{A}$, where $\mathbf{A}_{ij} = 1$ if there is an edge between nodes i and j; otherwise, $\mathbf{A}_{ij} = 0$. For the following sections, we use the fraction of the largest connected graph as a measure of network resilience to node or edge removals in the IoT. Node or edge removals can be viewed as temporal device or connection failures or targeted attacks in the IoT setting. For instance, node or edge removals in a graph can be caused by denial of service (DoS) or jamming attacks, or by natural occurrences.

2.2 Centrality Attacks, Network Resilience, and Topological Defense Scheme

2.2.1 Centrality attacks

A node centrality measure is a quantity that measures the level of importance of a node in a network. The utility of centrality measures is that they can break the combinatorial bottleneck of searching through all the possible permutations and combinations of nodes that might reduce largest component size. An attack that removes nodes according to a measure of centrality will be referred to as a *centrality attack* [14]. For example, the authors of [2, 6, 11, 28] study the effectiveness of degree centrality attacks, that is, removing the largest hub nodes, as a way to reduce the size of the largest component of the network. However, it has been shown in [13] that node degree is not the most effective centrality measure for minimizing largest component size. For different network topologies, investigating resilience of network connectivity to centrality attacks provides a unified metric for evaluating network vulnerabilities.

Let $\mathcal{N}_i$ denote the set of nodes connecting to node i (i.e., the set of neighbors of node i), and let $|\mathcal{N}_i|$ denote the set size. The degree of node i is the number of edges connected to it, that is, $d_i = \sum_{j=1}^{|\mathcal{V}|} \mathbf{A}_{ij} = |\mathcal{N}_i|$. The degree matrix $\mathbf{D}$ is defined as $\mathbf{D} = \mathrm{diag}\left(d_1, d_2, \ldots, d_{|\mathcal{V}|}\right)$, where $\mathbf{D}$ is a diagonal matrix with degree information on its main diagonal, the rest of the entries being 0. The graph Laplacian matrix $\mathbf{L}$ is defined as $\mathbf{L} = \mathbf{D} - \mathbf{A}$, and therefore it encodes degree information and connectivity structure of a graph. $\mathbf{L}$ is a positive semidefinite matrix, all its eigenvalues are nonnegative, and $\mathrm{trace}(\mathbf{L}) = 2|\mathcal{E}|$, where $\mathrm{trace}(\mathbf{L})$ is the sum of eigenvalues of $\mathbf{L}$, and $|\mathcal{E}|$ is the number of edges in $\mathcal{G}$. Moreover, the smallest eigenvalue of $\mathbf{L}$ is always 0, and the eigenvector of the smallest eigenvalue is a constant vector. The second smallest eigenvalue of $\mathbf{L}$, denoted by $\mu(\mathbf{L})$, is also known as the algebraic connectivity [21]. It has been proved in [21] that $\mu(\mathbf{L})$ is a lower bound on node and edge connectivity for any noncomplete graph. That is, algebraic connectivity ≤ node connectivity ≤ edge connectivity.

The centrality of a node is a measure of the node's importance to the network. Centrality measures can be classified into two categories: *global* and *local* measures. Global centrality measures require complete topological information for their computation, whereas local centrality measures require only partial topological information from neighboring nodes. For instance, acquiring shortest path information between every node pair is a global method required for the betweenness centrality measure, and acquiring degree information of every node is a local method. Some commonly used centrality measures are

■ *Betweenness* [22]: Betweenness is the fraction of shortest paths passing through a node relative to the total number of shortest paths in the network. Specifically, it is a global measure defined as betweenness $(i) =$

$$\sum_{k \neq i} \sum_{j \neq i, j > k} \frac{\sigma_{kj}(i)}{\sigma_{kj}},$$

where σ_{kj} is the total number of shortest paths from k to j, and $\sigma_{kj}(i)$ is the number of such shortest paths passing through i.

■ *Closeness* [25]: Closeness is a global measure of shortest path distance of a node to all other nodes. A node is said to have higher closeness if the sum of its shortest path distance to all other nodes is smaller. Let $\rho(i,j)$ denote the shortest path distance between node i and node j in a connected graph; then $\text{closeness}(i) = 1/\sum_{j \in \mathcal{V}, j \neq i} \rho(i,j)$.

■ *Eigenvector centrality* (eigen centrality): Eigenvector centrality depends on the ith entry of the eigenvector associated with the largest eigenvalue of the adjacency matrix $\mathbf{A}$. It is defined as $\text{eigen}(i) = \lambda_{\max}^{-1} \sum_{j \in \mathcal{V}} \mathbf{A}_{ij} \xi_j$, where $\lambda_{\max}$ is the largest eigenvalue of $\mathbf{A}$, and ξ is the eigenvector associated with $\lambda_{\max}$. It is a global measure, since the eigenvalue decomposition of $\mathbf{A}$ requires complete topological information of the entire network.

■ *Degree* (d_i): Degree is the simplest local centrality measure, which is simply the number of neighboring nodes.

■ *Ego centrality* [20]: Consider the $(d_i + 1)$-by-$(d_i + 1)$ local adjacency matrix of node i, denoted by $\mathbf{A}(i)$, and let $\mathbf{I}$ be an identity matrix. Ego centrality can be viewed as a local version of betweenness that computes the shortest paths between its neighboring nodes. Since $[\mathbf{A}^2(i)]_{kj}$ is the number of two-hop walks between k and j, and $\left[\mathbf{A}^2(i) \circ (\mathbf{I} - \mathbf{A}(i))\right]_{kj}$ is the total number of two-hop shortest paths between k and j for all $k \neq j$, where $\circ$ denotes entrywise matrix product, ego centrality is defined as $\text{ego}(i) = \sum_k \sum_{j>k} 1/\left[\mathbf{A}^2(i) \circ (\mathbf{I} - \mathbf{A}(i))\right]_{kj}$.

■ *Local Fiedler Vector Centrality* (LFVC) [15]: LFVC is a measure that characterizes vulnerability to node removals. A node with higher LFVC is more important for network connectivity structure. Let $\mathbf{y}$ (the Fiedler vector) denote the eigenvector associated with the second smallest eigenvalue $\mu(L)$ of the graph Laplacian matrix $\mathbf{L}$. LFVC is defined as $\text{LFVC}(i) = \sum_{j \in \mathcal{N}_i} (y_i - y_j)^2$. Although LFVC is a global centrality measure, it can be accurately approximated by local computations and message passing using the distributed power iteration method of [5] to compute the Fiedler vector $\mathbf{y}$.

Note that the edge centrality measure can be defined in a similar fashion.

2.2.2 Network resilience

When evaluating network resilience to different centrality attacks, we often compare the number of node removals needed by a centrality attack to reduce the largest component size to a certain amount, say, the number of nodes required to reduce the largest component size to 10% of its original size. For illustration, Figure 2.1 shows the network resilience of the Europe Internet backbone network topology (GTS-CE dataset) [23]. This network contains 149 nodes

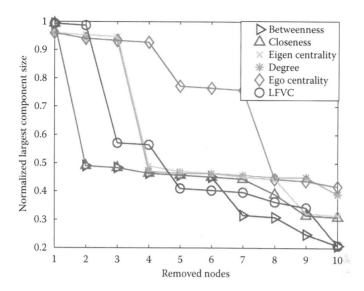

Figure 2.1: Resilience of network connectivity to different centrality attacks on the Europe Internet backbone network topology (GTS-CE dataset). The largest component size can be reduced to 20% of its original size by removing 10 nodes based on LFVC or betweenness attacks. (Data from S. Knight, H.X. Nguyen, N. Falkner, R. Bowden, and M. Roughan. The Internet topology zoo. *IEEE J. Sel. Areas Commun.,* **29(9), 1765–1775, 2011.)**

(routers) and 193 edges (physical connections). In this network, betweenness and LFVC attacks have comparable performance that results in 20% reduction of the largest component size by removing 10 nodes from the network. The topological information needed to compute the centrality measures are updated when a node is removed from the graph (i.e., a greedy removal approach). The network resilience of a Western U.S. power grid can be found in [14].

2.2.3 Topological defense scheme

A topological defense scheme allows change of network topology to enhance network resilience. It has been found in [14] that by swapping a small number of edges in the network topology, one is able to greatly improve network resilience without including additional edges. As shown in Figure 2.2, the Europe Internet backbone network can be secured by swapping 20 edges, such that the rewired network is more robust to centrality attacks. Moreover, the proposed edge rewiring method in [14] can be implemented in a distributed fashion, which is particularly preferable for the IoT due to scalability.

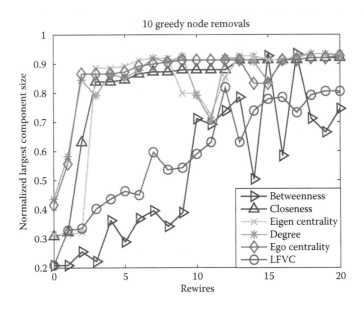

Figure 2.2: Network connectivity of the edge rewiring method when restricted to 10 greedy node removals on the Europe Internet backbone network topology (GTS-CE dataset) [23]. The edge rewiring method can greatly improve network resilience without introducing additional edges into the network. (Data from S. Knight, H.X. Nguyen, N. Falkner, R. Bowden, and M. Roughan. The Internet topology zoo. *IEEE J. Sel. Areas Commun.***, 29(9):17651775, 2011; edge rewiring method proposed by Pin-Yu Chen and Alfred O. Hero. Assessing and safeguarding network resilience to nodal attacks.** *IEEE Commun. Mag.***, 52(11):138–143, 2014.)**

2.3 Game-Theoretic Analysis of Network Robustness and Fusion-Based Defense Scheme

In many cases, edge rewire is not permitted in the IoT due to circumstances such as protocol confinement, geolocation constraint, and so on. In this scenario, one seeks to use the nodal detectability to infer the presence of an attack [6, 8, 11]. A fusion-based defense mechanism is proposed [6, 8, 11] to infer the presence of an attack based on the feedbacks from each node. The feedback information can be as simple as a binary status report reflecting that each node is, or is not, under attack, based on the node-level detection capabilities. Then, a network-level attack inference scheme is carried out at the fusion center.

An illustration of the attack and fusion-based defense model for the IoT is shown in Figure 2.3. A two-player game between the defender (the fusion center) and the attacker is naturally formed, given the critical value of network resilience (e.g., the largest component can be no less than 50% of its original size) and the node-level detection configurations. Intuitively, from the adversary's perspective,

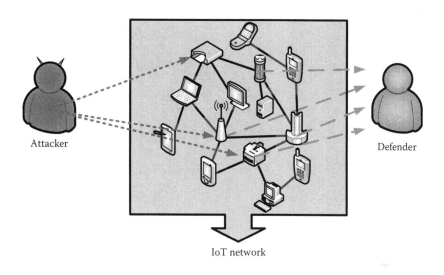

IoT network

Figure 2.3: Illustration of the attack and fusion-based defense model for the IoT. The adversary attacks a subset of nodes, as indicated by the red dotted arrows. The defender performs attack inference based on the attack status feedbacks from another subset of nodes, as indicated by blue dashed arrows.

too few node removals cause hardly any harm to the network connectivity, while too many node removals are prone to be detected by the fusion center, which means that the attack is eventually in vain. From the defender's perspective, inferring attacks using all feedbacks might treat the topological attack as a false alarm, since only a small subset of nodes are targeted. On the other hand, inferring attacks using only a few feedbacks might suffer from information insufficiency and therefore fail to detect the presence of attacks. Consequently, there exists a balance point at which both attacker and defender are satisfied with their own strategies, which is exactly the notion of Nash equilibrium in game theory [24]. At game equilibrium, no player's payoff can be increased by unilaterally changing strategy. As a result, the game payoff at game equilibrium can be used to study the robustness of a network.

As an illustration, we evaluate the network robustness of the Internet router-level topology [2] and the EU power grid [26] in terms of the payoff of the defender at the game equilibrium in Figure 2.4. The parameter P_D (P_F) denotes the probability of declaring an attack when the attack is actually present (absent). It is observed that the EU power grid is more robust to the Internet router-level topology given the same parameters P_D and P_F, and the network robustness approaches 1 as the detection capability increases, which suggests that the adversary gradually loses its advantage in disrupting the network, and the damage caused by malicious attacks can be alleviated by the fusion-based defense mechanism.

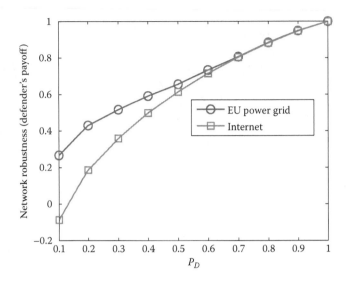

Figure 2.4: Network robustness of the Internet router-level topology and the EU power grid under degree attack when $P_F = 0.01$. The topological map of the Internet contains 6,209 nodes and 12,200 edges, and the EU power grid contains 2,783 nodes and 3,762 edges. (The empirical data are the network parameters collected by Réka Albert, Hawoong Jeong, and Albert-Laszlo Barabási. Error and attack tolerance of complex networks. *Nature*, 406(6794):378–382, 2000; Ricard V. Solé, Martí Rosas-Casals, Bernat Corominas-Murtra, and Sergi Valverde. Robustness of the European power grids under intentional attack. *Phys. Rev. E*, 77:026102, 2008.)

These results suggest that in addition to topological defense approaches (e.g., the edge rewiring method), one can improve network resilience of the IoT by implementing network-level defense mechanisms. However, one main disadvantage of fusion-based defense is the acquisition of feedbacks from all nodes, which may not be applicable to the IoT due to its enormous number of devices. Nonetheless, fusion-based defense can be used in a hierarchical manner for multilayer defense.

2.4 Sequential Defense Scheme

A sequential defense scheme is proposed by [10] that sequentially collects feedbacks from high degree nodes for attack inference. The advantage of sequential defense is that there is no need to acquire feedbacks from all nodes, and it terminates the collection process once sufficient feedbacks have been collected for attack inference. The enormous network size (e.g., Internet routers or sensors in the IoT) renders simultaneous data transmissions infeasible, especially for wireless networks with scarce radio resources. Moreover, due to the large network size and limited computational power, analyzing the collected information from all nodes incurs tremendous computation overheads, and it may fail to provide timely defense.

It is worth mentioning that the sequential defense scheme is quite distinct from the traditional data fusion scheme [27] due to the fact that the attack may not be a common event to all the nodes in the network. In other words, an intelligent adversary can target some crucial nodes instead of launching attacks on the entire network to efficiently disrupt the network and reduce the risks of being detected, which therefore hinders the precision of attack inference and poses severe threats to the network robustness.

It is proved in [10] that a relatively small fraction of feedbacks is sufficient to detect fatal attacks on the network prior to network disruption. We compare the number of node removals required for a network to break down and the number of feedbacks needed for the sequential defense scheme to detect the attack under three different real-world networks: the webpage links in the World Wide Web (WWW) [4], the Internet router-level topology [2], and the EU power grid [26]. Figure 2.5 shows the number of feedbacks needed for sequential defense under different parameters P_D and P_F. It can be observed that there is a surge in the number of required observations when P_F is large and P_D is small, as intuitively one needs more observations to verify the presence of an attack in the circumstances of low detection capability and high false alarm rate. Comparing the

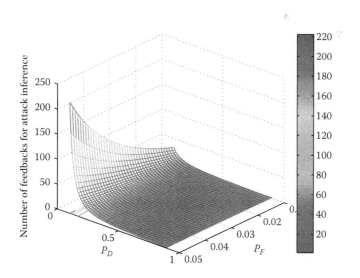

Figure 2.5: Expected number of feedbacks required for the sequential defense scheme to detect a degree attack. The critical values for the WWW, the Internet, and the EU power grid to break down are 21,824, 187, and 766, respectively.

critical number of node removals for network breakdowns, the required numbers of feedbacks for these three networks are less than the critical value for moderate P_D and P_F. These results suggest that sequential defense can effectively detect an attack prior to network breakdown by acquiring only a small number of feedbacks.

2.5 Conclusion

This chapter introduces several centrality attacks that aim to maximally disrupt the connectivity of an IoT network, and three defense schemes to counter these fatal attacks are investigated. The first one is the topological defense scheme, which allows edge swapping to enhance intrinsic network resilience. The second one is the fusion-based defense mechanism and the game-theoretic perspective of network robustness. The third one is the sequential defense scheme, which enables efficient attack inference with only a few feedbacks from the network.

Acknowledgment

The author would like to thank Dr. Alfred Hero at the Department of Electrical Engineering and Computer Science, University of Michigan, Ann Arbor, United States, Dr. Shin-Ming Cheng at the Department of Computer Science and Information Engineering, National Taiwan University of Science and Technology, Taiwan, and Dr. Kwang-Cheng Chen at the Department of Electrical Engineering, National Taiwan University, Taiwan, for their valuable discussion and collaboration.

Bibliography

[1] U.S. Department of Energy (DOE). *A System View of the Modern Grid.* National Energy Technology Laboratory (NETL), U.S. Department of Energy (DOE), 2007.

[2] Réka Albert, Hawoong Jeong, and Albert-Laszlo Barabási. Error and attack tolerance of complex networks. *Nature*, 406(6794):378–382, 2000.

[3] Luigi Atzori, Antonio Iera, and Giacomo Morabito. The internet of things: A survey. *Computer Networks*, 54(15):2787–2805, 2010.

[4] Albert-Laszlo Barabási and Réka Albert. Emergence of scaling in random networks. *Science*, 286(5439):509–512, October 1999.

[5] Alexander Bertrand and Marc Moonen. Distributed computation of the Fiedler vector with application to topology inference in ad hoc networks. *Signal Processing*, 93(5):1106–1117, 2013.

[6] Pin-Yu Chen, Shin-Ming Cheng, and Kwang-Cheng Chen. Smart attacks in smart grid communication networks. *IEEE Commun. Mag.*, 50(8): 24–29, August 2012.

[7] Pin-Yu Chen and Kwang-Cheng Chen. Information epidemics in complex networks with opportunistic links and dynamic topology. In *Proceedings of IEEE Global Telecommunications Conference (GLOBECOM)*, pages 1–6, December 2010.

[8] Pin-Yu Chen and Kwang-Cheng Chen. Intentional attack and fusion-based defense strategy in complex networks. In *Proc. IEEE Global Telecommunications Conference (GLOBECOM)*, pages 1–5, December 2011.

[9] Pin-Yu Chen and Kwang-Cheng Chen. Optimal control of epidemic information dissemination in mobile ad hoc networks. In *IEEE Global Telecommunications Conference (GLOBECOM)*, pages 1–5, December 2011.

[10] Pin-Yu Chen and Shin-Ming Cheng. Sequential defense against random and intentional attacks in complex networks. *Phys. Rev. E*, 91:022805, February 2015.

[11] Pin-Yu Chen, Shin-Ming Cheng, and Kwang-Cheng Chen. Information fusion to defend intentional attack in internet of things. *IEEE IoT-J.*, 1(4):337–348, August 2014.

[12] Pin-Yu Chen, Shin-Ming Cheng, and Kwang-Cheng Chen. Optimal control of epidemic information dissemination over networks. *IEEE Trans. Cybern.*, 44(12):2316–2328, December 2014.

[13] Pin-Yu Chen and Alfred O. Hero. Node removal vulnerability of the largest component of a network. In *Proceedings of IEEE GlobalSIP*, 2013.

[14] Pin-Yu Chen and Alfred O. Hero. Assessing and safeguarding network resilience to nodal attacks. *IEEE Commun. Mag.*, 52(11):138–143, November 2014.

[15] Pin-Yu Chen and Alfred O. Hero. Local Fiedler vector centrality for detection of deep and overlapping communities in networks. In *IEEE International Conference on Acoustics, Speech and Signal Processing (ICASSP)*, pages 1120–1124, 2014.

[16] Pin-Yu Chen, Han-Feng Lin, Ko-Hsuan Hsu, and Shin-Ming Cheng. Modeling dynamics of malware with incubation period from the view of individual. In *79th IEEE Vehicular Technology Conference (VTC Spring)*, pages 1–5, May 2014.

[17] Shin-Ming Cheng, Weng Chon Ao, Pin-Yu Chen, and Kwang-Cheng Chen. On modeling malware propagation in generalized social networks. *IEEE Commun. Lett.*, 15(1):25–27, January 2011.

[18] Shin-Ming Cheng, Pin-Yu Chen, and Kwang-Cheng Chen. Ecology of cognitive radio ad hoc networks. *IEEE Commun. Lett.*, 15(7):764–766, July 2011.

[19] Shin-Ming Cheng, Vasileios Karyotis, Pin-Yu Chen, Kwang-Cheng Chen, and Symeon Papavassiliou. Diffusion models for information dissemination dynamics in wireless complex communication networks. *Journal of Complex Systems*, Article ID 972352, 2013.

[20] Martin Everett and Stephen P. Borgatti. Ego network betweenness. *Social Networks*, 27(1):31–38, 2005.

[21] Miroslav Fiedler. Algebraic connectivity of graphs. *Czech. Math. J.*, 23(98): 298–305, 1973.

[22] Linton C. Freeman. A set of measures of centrality based on betweenness. *Sociometry*, 40:35–41, 1977.

[23] Simon Knight, Hung X. Nguyen, Nickolas Falkner, Rhys Bowden, and Matthew Roughan. The Internet topology zoo. *IEEE J. Sel. Areas Commun.*, 29(9):1765–1775, October 2011.

[24] Martin Osborne and Ariel Rubinstein. *A Course in Game Theory*. MIT, Cambridge, MA, 1999.

[25] Gert Sabidussi. The centrality index of a graph. *Psychometrika*, 31(4): 581–603, 1966.

[26] Ricard V. Solé, MartíRosas-Casals, Bernat Corominas-Murtra, and Sergi Valverde. Robustness of the European power grids under intentional attack. *Phys. Rev. E*, 77:026102, February 2008.

[27] Pramod K. Varshney. *Distributed Detection and Data Fusion*. Springer, New York, 1996.

[28] Shi Xiao, Gaoxi Xiao, and Tee Hiang Cheng. Tolerance of intentional attacks in complex communication networks. *IEEE Commun. Mag.*, 45(1): 146–152, February 2008.

Chapter 3

Sybil Attack Detection in Vehicular Networks

Bharat Jayaraman

Jinesh M. Kannimoola

Krishnashree Achuthan

CONTENTS

In this chapter, we consider safety issues arising in vehicular ad hoc networks (VANETs) (Figure 3.1). Although vehicular networks originated in the infotainment domain, today they are also used in many safety-critical systems such as in an emergency vehicle grid. Due to the open nature of vehicular networks,

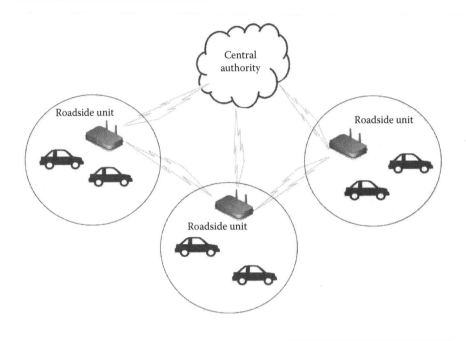

Figure 3.1: Architecture of vehicular ad hoc network

they are more amenable to malicious attacks; and, due to their high mobility and dynamic topology, the detection and prevention of such attacks is also more difficult. We consider one such attack in this chapter, the Sybil attack, in which an attacker tries to violate the unique vehicular ID property by forging or fabricating it and presenting multiple identities. A Sybil attack is a serious threat because it can result in large-scale denial of service or other security risks in the network. This chapter presents a new method to prevent Sybil attacks in a vehicular network based on the traditional cryptographic techniques, as well as the unique features of the network. A key feature of the methodology is the use of fixed roadside units and a central authority. This chapter presents a formal model of the system using the Promela language and shows how the safety property can be verified using the SPIN model checker.

3.1 Introduction

The automobile today has evolved from a complex electromechanical system to a "computer system on wheels" and vehicular networks are pushing the frontier of the internet of things (IoT) to include the large class of highly mobile entities; namely, vehicles. With the inclusion of vehicles and communication between vehicles, as well as between vehicles and the infrastructure, the "internet of vehicles" can potentially provide real-time connectivity between vehicles around the globe. By further providing connectivity with entities such as traffic lights and RFID devices, we move closer toward the goal of a safe and efficient traffic environment. A vehicle has potentially has more storage, communication, and computing capacity compared to other embedded and mobile devices, and hence, vehicular networks can act as core infrastructure to connect various things.

The vehicular ad hoc network (VANET) facilitates communication between vehicles in the network by sharing road conditions and safety information. The network is especially useful in dense urban regions in promoting greater road safety and efficient traffic control. In contrast with a mobile ad hoc network, a vehicular ad hoc network has a highly dynamic network topology owing to the rapid movement of vehicles, with frequent disconnections in the network and more resource constraints [13]. It uses a combination of networking technologies such as Wi-Fi IEEE 802.11p, WAVE IEEE 1609, WiMAX IEEE 802.16, Bluetooth, IRA, and ZigBee.

There are two types of communication in a vehicular network: (i) vehicle-to-vehicle and (ii) vehicle-to-network-infrastructure. The open nature of VANET communication makes it much more amenable to malicious attacks [11, 18], and the dynamic nature of vehicular movement makes it difficult to protect against these. In this chapter, we consider one such attack, the Sybil attack, in which a single entity can gain control over a substantial fraction of the system by presenting multiple identities [4]. There are mainly two types of Sybil attacks: (i) a single node presents multiple identities; and (ii) a Sybil node uses the identity of another node. Sybil attacks violate the fundamental assumption of one-to-one correspondence of a node with its identity. There are several adverse effects that result from a Sybil attack in a VANET environment [1, 14]:

■ Routing: The Sybil attack affects the performance of geographical routing and leads to large-scale denial of service.

■ Tampering with voting and reputation systems: Reputation and trust management system crucially depend upon the unique ID and authenticity of the node. A Sybil attack violates this assumption and results in erroneous computation of reputation values.

■ Fair resource allocation: A node with multiple identities can exploit the network to its advantage by using more bandwidth and network time.

■ Data aggregation: Wireless sensor networks typically aggregate the values from sensor nodes rather than sending individual values. A Sybil node can manipulate these values, resulting in misleading aggregate values.

Our motivation in this chapter is to present an effective approach for Sybil attack detection in the setting a highly dynamic vehicular ad hoc network. Basically, a Sybil attack can be prevented by using public key certificates issued by a central authority (CA) [4]. Such an approach is not scalable because the CA can become a bottleneck in communication. Although methods have been proposed to prevent a Sybil attack in a VANET [2, 10, 15], they fail to capture the dynamic characteristics of the network. Our method makes use of the roadside unit (RSU) along with a cryptographic certificate scheme with position verification to capture the dynamic context of a vehicle in the network. Essentially, in our approach, the RSU acts as an authority to verify the authenticity of a vehicle node by using the information in nearby RSUs. The idea is that an RSU can contact nearby RSUs more quickly compared with the CA.

Thus, the contribution of our work is an effective detection mechanism for Sybil attacks, using a semicentralized approach, by taking advantage of the presence of RSUs in addition to the CA. Essentially, we distribute the function performed by the CA through the RSUs to capture the dynamic nature of the network. A real vehicular network typically contains thousands of vehicular nodes and hundreds of RSUs. Before deploying the system in a real environment, it is desirable to model the key aspects of the technique at an abstract level and check the correctness of the proposed protocol. We therefore develop a formal model of our approach and verify its key properties using a model-checking approach [3], since it supports reasoning over all possible paths of execution.

We develop a specification of the vehicular network using Promela (Process Meta Language) and check its correctness using the open-source model checker SPIN (Simple Promela Interpreter) [6]. Vehicles, RSUs, and the CA are modeled as Promela processes, and the communication between them is represented by Promela channels. Promela supports the dynamic creation of processes as well as channels, the latter being a crucial capability for modeling the mobility of vehicles from one RSU to another. Attack detection is also modeled as a process that continuously observes the network for any violation of the key system properties, including the property that only one vehicle uses a given ID for communication.

The remainder of this chapter is organized as follows: Section 2 presents closely related approaches for Sybil attack detection and their limitations; Section 3 presents the overall design of our Sybil attack detection method; Section 4 gives a formal specification and verification of our method using Promela/SPIN; and Section 5 presents conclusions and areas of further work. The full Promela model is given in the appendices.

3.2   Related Work

A Sybil attack can cause harm to various layers of communication [7, 16]. In this section, we discuss the methods that have been proposed for different network scenarios.

Newsome et al. [8] propose a detection scheme based on resource testing for wireless sensor networks, assuming that each entity has limited resources. According to this approach, communication capability is used for resource testing. The principle here is that radio is incapable of simultaneously sending or receiving on more than one channel. If a node wants to verify its neighboring nodes, it will assign each of its neighbors to a different channel to broadcast messages. The verifier node then randomly selects a channel to listen to. If it receives the message on the assigned channel, then it is a legitimate node; otherwise it is a Sybil node. But an attacker can, in practice, use unlimited resources or radios to launch an attack.

Douceur [4] notes that a Sybil attack can be effectively prevented using public key certificates issued by a CA. However, due to the dynamic nature of VANET, it is impractical to communicate with a CA each time. Also, in this method, an attacker can easily use a stolen certificate for communication because there is no certificate binding with a unique physical identification.

Zhou et al. [19] proposed a scheme to preserve privacy based on pseudonyms. Here, each vehicle has a set of pseudonyms issued by the Department of Motor Vehicles (DMV). For each communication, a vehicle uses one of its pseudonyms rather than its real ID. Pseudonyms in vehicles are hashed to a unique value, and hence cannot be used to launch a Sybil attack. This scheme needs a lot of communication with the central authority for pseudonym verification, making it less practical in the highly dynamic context of a vehicular network.

Park et al. [10] suggest a scheme based on the time-series approach. Here, each vehicle-to-vehicle communication contains a unique time series certificate certified by the RSU. This method is based on the basic assumption that it is not possible for two vehicles to pass through the same RSU at same time. A vehicle can detect a Sybil attack when it receives a similar certificate from a different vehicle. This method can identify a Sybil attack to some extent, but here, attack detection occurs at the vehicle level. This scheme is based on the dense deployment of RSUs. This method is applicable only if both the Sybil node and the actual node are within the range of the same RSU.

The position verification scheme [12] is an another approach to detect a legitimate node. It is based on the assumption that a vehicle can be present at only one position at a particular time. Yan et al. [17] presented an approach for position verification using onboard radar at a node to verify the location of a neighboring vehicle. Here, each vehicle sends a message with location information. A vehicle can cross-check the presented location information using onboard radar, but location verification is limited by its range.

According to Guette et al. [5], secure hardware built on a trusted platform (TPM) can be used to prevent Sybil attacks in a VANET. Secure information is stored in the TPM; hence, forging and fabrication of data is impossible. Credentials are trusted by car manufacturers and the communication between two TPMs are protected from attack.

3.3   Location Certificate-Based Scheme

Our proposed approach for Sybil attack detection is based on a traditional public key certificate together with position verification. In this approach, the whole network is viewed as a tree-like structure rooted at the CA, which maintains information about all vehicles in the VANET. At the second level, or layer, from the root is the set of all RSUs, which effectively constitute a fixed infrastructure. Unlike a normal tree structure, there are links between RSUs. The third (and last) level from the root contains the mobile nodes (vehicles). Each vehicle has a unique ID and certificate registered with the CA.

The main properties of the proposed design are as follows:

■ *No dependence on specialized hardware*: This scheme does not need any special type of hardware. It makes use of existing infrastructure for the detection of attacks.

■ *The CA and RSUs both participate in detection*: This approach avoids a central bottleneck in communication, and attack detection happens at both the CA and RSU levels. The support of other vehicles in the network is not needed.

■ *Node authentication depends on geolocation information*: The claimed location of nodes is verified using the strength of received signals and also the geographic location of nodes.

■ *Support for high vehicular mobility*: Our proposed approach supports a high mobility of vehicles between RSUs. The overhead associated with attack detection does not affect the performance of the VANET.

■ *Sybil nodes are isolated from the network*: The Sybil node will be automatically removed from the network and will be prevented from engaging in any further communication.

The fundamental assumptions for the proposed scheme are as follows: (i) Each RSU must know its geographical location. (ii) The RSUs are connected to adjacent RSUs and the CA with a high-speed back end. (iii) RSUs are considered as trusted entities. (iv) Each vehicle is registered with the CA with a unique ID and public key certificate. (v) Each vehicle has a GPS device to acquire its geolocation.

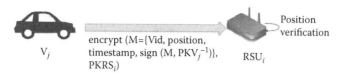

Figure 3.2: Communication from vehicle to RSU

3.3.1   Sybil node detection scheme

Our proposed scheme is founded on the concept of a location certificate issued by a RSU for communication with other vehicles under the same RSU. For each vehicle j, the CA stores the vehicle ID with the corresponding public key (PKVj). Each RSU continuously broadcasts its public key (PKRS) using the beacon signal. Before we describe the major steps in this scheme, we first clarify the common notations used in this scheme in the table below.

Notation	Meaning
(PKCA,PKCA $^{-1}$)	Public and private key of CA
(PKRS $_i$,PKRS $_i$ $^{-1}$)	Public and private key of ith RSU unit
(PKV $_j$,PKV $_j$ $^{-1}$)	Public and private key of jth vehicle

1. Suppose the jth vehicle enters the ith RSU's range (Figure 3.2). This step is a one-time process for each session and occurs only if the vehicle does not have a valid location certificate. The vehicle creates a location certificate request in the following format: {$vehicle\_ID$, $position$, $timestamp$}. Here, $position$ is taken from the GPS sensor. For communication security, a message is signed using the vehicle's private key, PKV$_j$ $^{-1}$, and encrypted by the ith RSU's public key.

2. When obtaining a location request from the jth vehicle, the RSU first verifies the claimed position using the received signal strength (RSS), since it is possible to calculate the distance from a node using the RSS [9]. If it is valid, the RSU forwards the encrypted request to the CA using PKCA (Figure 3.3). If the claim of the vehicle is invalid, the RSU notifies the vehicle ID to adjacent RSUs.

3. The CA verifies the request using PKV$_j$ and checks if the jth vehicle is registered anywhere in the network. If it is not, it registers the vehicle location with the RSU and notifies the corresponding RSU using the vehicle's PKV$_j$ (Figure 3.4). The CA knows the public key of all RSUs and hence can securely communicate with RSUs.

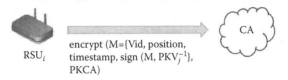

Figure 3.3: Communication from RSU to CA

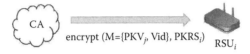

Figure 3.4: Communication from CA to RSU

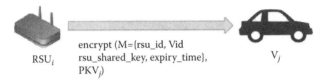

Figure 3.5: Communication from RSU to vehicle

4. After obtaining the confirmation from the CA, the RSU issues a location certificate with {rsu_ID, rsu_shared_key, vehicle_ID, expiry_time} which is encrypted with the vehicle's public key (Figure 3.5). If the CA detects a Sybil attack, it will inform the RSU concerned, which in turn will not issue a certificate to the vehicle.

5. A particular vehicle communicates with other vehicles using an *rsu_shared_key*. Each vehicle continuously checks the *expiry_time* of the location certificate and sends a location certificate request before the expiration of the previously issued certificate. The valid location certificate acts as a key for vehicle-to-vehicle communication.

6. If a vehicle enters the range of the next RSU, it again sends a location certificate request to the RSU, but it includes the position certificate from previous RSU (Figure 3.6). When the kth RSU gets a request with a position certificate from the ith RSU, it checks the validity of the certificate from the ith RSU and acquires the public key of the corresponding vehicle. The kth RSU then issues the certificate and notifies the CA of the vehicle ID and *rsu_ID*. Subsequently, the ith RSU removes the corresponding vehicle from its storage.

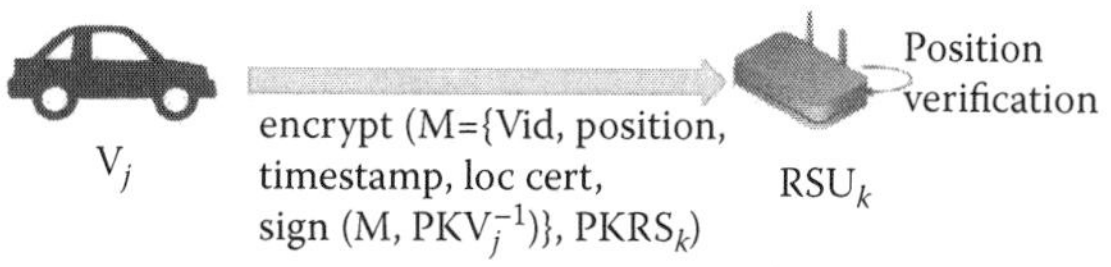

Figure 3.6: Communication from vehicle to RSU with previous location certificate

Evaluation. In this scheme, the Sybil node detection happens at two levels. Each RSU can verify the node (vehicle) based on location information, and the CA can check whether the node registration occurred anywhere in the network using a unique ID. An attacker cannot send a legitimate request to the CA, since the CA can check the validity of the message using the vehicle's public key. Each RSU requires less storage space because it stores information only of those vehicles that are within its range—an RSU erases a vehicle's details after it moves to the next RSU. Without a location certificate, vehicles cannot communicate with other vehicles and this prevents a Sybil node from taking part in further communication. If an RSU or the CA detects a Sybil attack, it informs nearby RSUs, which in turn can reject a vehicle request without going through the remaining process.

3.4   Formal Modeling and Verification

Formal verification is a method to check various system properties such as liveness, deadlock, and design errors. SPIN is a powerful tool to conduct formal verification of concurrent systems using specifications in Promela, a process specification language. We model the location certificate distribution method without considering the cryptographic processes involved with it. Here the CA, RSUs, and vehicles are modeled as a Promela proctype. The system has multiple instances of RSUs and vehicles. Communication between these processes occurs through Promela channels. Each RSU maintains, for vehicle communication, a certificate request channel (veh_rsu_chan) and a certificate response channel (rsu_veh_chan). These channels are asynchronous and defined as

```
chan veh_rsu_chan[NO_OF_RSU+1]=[0] of {CER_REQ};
chan rsu_veh_chan[NO_OF_RSU+1]=[0] of {CER_RES};
```

The types CER_REQ and CER_RES represent, respectively, the certification request from a vehicle and the response from the RSU. A location certificate request contains the vehicle identity (veh_ID), vehicle location (veh_loc), request time, and available location certificate. The initial RSU ID is supplied by the initialization process. Vehicle movement is achieved by changing RSU IDs and locations. These structures are defined through Promela typedefs (Figure 3.7).

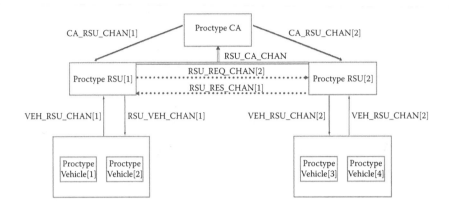

Figure 3.7: Promela channels and process types

```
typedef CER_REQ {byte veh_ID; byte veh_loc; byte rsu_ID;
                 int loc_cert; int time;
} typedef CER_RES {byte veh_ID;  byte rsu_ID;  int loc_cert;}
```

The type loc_cert contains a positive value for a valid certificate. After obtaining a valid certificate from the RSU, the vehicle process increments the RSU ID by one and tries to associate with it using the current location certificate. The vehicle proctype is detailed in Appendix 3A.1. The RSU uses two other channels to communicate with the CA: one for communication from the RSU to the CA (rsu_ca_chan) and the other from the CA to the RSU (ca_rsu_chan).

```
chan rsu_ca_chan=[0] of {RSU_REQ};
chan ca_rsu_chan[NO_OF_RSU+1]=[0] of {CO_RES}
```

According to the proposed scheme, the RSS (received signal strength) method is used at each RSU for location verification. It is difficult to model such an environment in SPIN. Therefore, here we are using a simpler method for location verification. The RSU checks that the location is within a 4 km range of the RSU. If this is so and the request contains an invalid certificate, then the RSU forwards it to the CA using the RSU_REQ data structure.

```
typedef RSU_REQ {byte veh_ID;  byte rsu_ID;  bit update;}
```

The update field in RSU_REQ is set to zero for a request with an invalid certificate. If a request from a vehicle contains a valid certificate, the request is forwarded to the RSU which issued the current certificate. To communicate with nearby RSUs, each RSU maintains a request and response channel.

```
chan rsu_req_chan[NO_OF_RSU+1]=[0] of {CER_REQ,byte}
chan rsu_res_chan[NO_OF_RSU+1]=[0] of {CO_RES}
```

Here, CO_RES is the common response format of an RSU:

```
typedef CO_RES {byte veh_ID;  bit status;}
```

After receiving a positive request from a nearby RSU or the CA, the RSU issues the new certificate through the cer_res_chan channel. To update the certificate, the RSU informs the CA of the new rsu_ID. Here, we again use the CA_REQ structure with an enabled update bit. The CA maintains a database with veh_ID and rsu_ID information, which effectively maps the vehicle identity to the RSU identity under which it is present.

```
typedef VEHID_STORE {byte veh_ID;  byte rsu_ID;}
```

The CA can check whether a vehicle is registered in any other RSU using the database. If it is already registered, the CA informs the RSU about a possible Sybil attack attempt. In this situation, the RSU updates its local storage with the invalid certificate for the particular vehicle. This will prevent the vehicle from obtaining a valid certificate on a subsequent request. The complete RSU and CA proctypes are detailed in Appendices 3A.2 and 3A.3, respectively.

Verification. The verification process ensures that no vehicle has a valid location certificate from two different RSUs at the same time. Each RSU maintains a copy of the currently active location certificate within its range. We use an observer process to verify this property, by having it scan different RSUs and ensuring that each vehicle has only one valid location certificate. In the specification below, the assert clause fails if two RSUs have a valid certificate for the same vehicle.

```
active proctype Observer(){
  int i;
  int j;
  do
   :: for (i : 1 .. NO_OF_RSU) {
       atomic{
         for(j:1 .. NO_OF_RSU){
          if
            :: (i!=j && rsu_pids[i] > 0 && rsu_pids[j] > 0) ->
               for(k:1 .. NO_OF_VEH){
                  assert(!(RSU[rsu_pids[i]]:loc_cert[k] > 0 &&
                           RSU[rsu_pids[j]]:loc_cert[k] > 0));
               }
            :: else -> skip;
          fi
       }}}
  od
}
```

Figure 3.8 shows the kind of output produced by SPIN. We briefly explain this output:

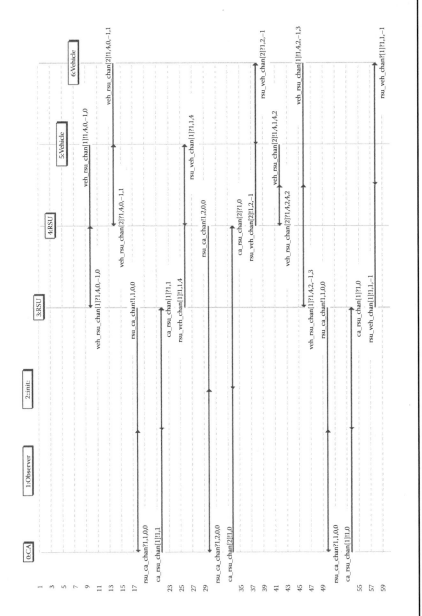

Figure 3.8: SPIN sequence diagram. The malicious Vehicle 6 is a Sybil node and tries to impersonate Vehicle 5 but keeps obtaining invalid certificates.

- 5:Vehicle is a legitimate node with identity 1, and 6:Vehicle is a malicious (Sybil) node that is using the same ID as 5:Vehicle.

- The input received by veh_rsu_chan[1]?1,4,0,-1,0 represents the location certificate requested by 5:Vehicle through RSU 3:RSU. This RSU consults with the CA and assigns a valid certificate through rsu_veh_chan[1]?1,1,4.

- The Sybil node, 6:Vehicle, tries to associate with 4:RSU but obtains an invalid certificate from the RSU via rsu_veh_chan[2]?1,2,-1).

- The Sybil node repeatedly tries to obtain a valid certificate, but does not succeed.

3.5 Conclusion

This chapter presents a novel approach to Sybil attack detection in a vehicular network based upon both cryptographic and location verification. Such an approach helps avoid exclusive dependence on a CA, which can be a bottleneck in communication. Our approach leads to a semicentralized architecture wherein RSUs also participate in the detection process. The accuracy of the system is modeled using the Promela specification language and verified using the SPIN model checker.

One of the shortcomings of our scheme is that it does not consider the hand off between RSUs. At the point of hand off, the vehicle will fail to produce a valid certificate for vehicle-to-vehicle communication. Also, the Promela model presented in this chapter covers only vehicle-to-infrastructure communication. By incorporating the cryptographic modeling along with the vehicle-to-vehicle communication we obtain a more complete coverage of VANET communication. The SPIN model checker does not have the notion of time as a quantitative measure. This is an important factor in security, and the introduction of time into the SPIN model would produce a more powerful verification method of security protocols.

Finally, it may be noted that the Sybil attack is one among many types of attacks on vehicular networks. Our goal is to integrate mitigation techniques for different types of attacks into a single abstract model so as to achieve a more secure and effective vehicular network.

3A Appendices

3A.1 Vehicle proctype

```
proctype Vehicle(byte veh_ID; byte rsu_ID;byte loc){

CER_REQ cer_req;
CER_RES cer_res;
```

```
byte new_rsu_ID;
veh_pids[veh_ID]=_pid;
/*Setup inital request*/
new_rsu_ID=rsu_ID;
cer_req.veh_ID=veh_ID;
cer_req.veh_loc=loc;
cer_req.loc_cert=-1; //Invalid certificate -1
do
  ::true->
  cer_req.time=g_curr_time;
  veh_rsu_chan[new_rsu_ID]!cer_req;        //send request to RSU
  g_curr_time=g_curr_time+1;               //update time
  rsu_veh_chan[new_rsu_ID]?cer_res;
  cer_req.loc_cert=cer_res.loc_cert;       //Get response from RSU
  cer_req.rsu_ID=cer_res.rsu_ID;           //Setup new request
  new_rsu_ID=(cer_res.rsu_ID%NO_OF_RSU)+1;//Increment RSU_ID
od
}
```

3A.2   RSU proctype

```
/* Input : rsu_ID and location of RSU*/
proctype RSU(byte rsu_ID;byte loc) {
int loc_cert[NO_OF_VEH+1];
int i;
for(i:1 .. NO_OF_VEH){
 loc_cert[i]=-1;
}
rsu_pids[rsu_ID]=_pid;
CER_REQ  cer_req;
RSU_REQ rsu_req;
CO_RES co_res;
CER_RES cer_res;
int loc_cert_temp;
int loc_cert_old;
byte req_rsu;
do
/*Listener for cer_req request from vehicle*/
::veh_rsu_chan[rsu_ID]?cer_req->
if
  ::cer_req.veh_loc-loc<=4 ->
  cer_res.veh_ID=cer_req.veh_ID;   //Set location response veh_ID
  cer_res.rsu_ID=rsu_ID;   //Set location response rsu_ID
  loc_cert_old=cer_req.loc_cert;   //Taking location certificate
  rsu_req.veh_ID=cer_req.veh_ID;
  rsu_req.rsu_ID=rsu_ID;
  if
  /*For 1st request(without a valid loc_cert)*/
    ::loc_cert_old==-1->
      rsu_req.update=0;   //Update 0 means vehicle is not registered in any RSU
      rsu_ca_chan!rsu_req;   //send RSU request to CA
      ca_rsu_chan[rsu_ID]?co_res;   //Get response from CA
    ::loc_cert_old>=0->
      rsu_req_chan[cer_req.rsu_ID]!cer_req,rsu_ID;
rsu_res_chan[rsu_ID]?co_res;
::else->skip;
```

```
fi

if
 ::co_res.status==1-> //
 loc_cert_temp=cer_req.time+cer_req.veh_loc;
 loc_cert[cer_req.veh_ID]=loc_cert_temp;
 cer_res.loc_cert=loc_cert_temp;
 ::else->cer_res.loc_cert=-1;
 ::else -> cer_res.loc_cert=-1;
  fi
    rsu_veh_chan[rsu_ID]!cer_res;
    if
    /*if request for loc_cert update initamte CA with new RSU_ID*/
::loc_cert_old>=0->
rsu_req.update=1;
rsu_req.old_rsu_ID=cer_req.rsu_ID;
rsu_ca_chan!rsu_req;
ca_rsu_chan[rsu_ID]?co_res;
if
 ::co_res.status==0->loc_cert[co_res.veh_ID]=-1;
 ::else->skip;
fi
 ::else -> skip;
     fi
      /*listener for near by RSU request to check validity
      of certificate */
 ::rsu_req_chan[rsu_ID]?cer_req,req_rsu->
 co_res.veh_ID=cer_req.veh_ID;
 if
 ::loc_cert[cer_req.veh_ID]==cer_req.loc_cert->
 co_res.status=1;
 loc_cert[cer_req.veh_ID]=-1;
 ::else ->
 co_res.status=0;
fi
   rsu_res_chan[req_rsu]!co_res;
 od
}
```

3A.3   CA proctype

```
active proctype CA(){

 RSU_REQ rsu_req;
 CO_RES co_res;

 do
 /*listener for request from RSU*/
 ::rsu_ca_chan?rsu_req ->
  co_res.status=0;
  co_res.veh_ID=rsu_req.veh_ID;
  if
    ::rsu_req.update==0&&vehid_store[rsu_req.veh_ID]==0->
 vehid_store[rsu_req.veh_ID]=rsu_req.rsu_ID;
 co_res.status=1;
    ::rsu_req.update==1->
```

```
    if
        ::rsu_req.old_rsu_ID==vehid_store[rsu_req.veh_ID]->
    vehid_store[rsu_req.veh_ID]=rsu_req.rsu_ID;
    co_res.status=1;
    ::else->co_res.status=0;
    fi
    ::else->co_res.status=0;
        fi
        ca_rsu_chan[rsu_req.rsu_ID]!co_res;
    od
}
```

Bibliography

[1] Nitish Balachandran and Sugata Sanyal. A review of techniques to mitigate Sybil attacks. *Int. J. Adv. Netw. Applic.*, 4:1514–1518, 2012.

[2] Brijesh Kumar Chaurasia and Shekhar Verma. Infrastructure based authentication in VANETs. *Int. J. Multimed. Ubiquitous Eng.*, 6(2):41–54, 2011.

[3] Edmund M. Clarke, Orna Grumberg, and Doron Peled. *Model Checking*, MIT Press, Cambridge, MA 1999.

[4] John R. Douceur. The Sybil attack. In *Peer-to-Peer Systems*, pages 251–260, Vol. 2429 of Lecture Notes in Computer Sciences. Springer, 2002.

[5] Gilles Guette and Ciarán Bryce. Using TPMS to secure behicular ad-hoc networks (VANETs). In *Information Security Theory and Practices. Smart Devices, Convergence and Next Generation Networks*, pages 106–116. Springer, 2008.

[6] Gerard J. Holzmann. The model checker SPIN. *IEEE Trans. Softw. Eng.*, 23(5):279–295, 1997.

[7] Gökhan Korkmaz, Eylem Ekici, Füsun Özgüner, and Ümit Özgüner. Urban multi-hop broadcast protocol for inter-vehicle communication systems. In *Proceedings of the 1st ACM International. Workshop on Vehicular Ad-hoc Networks*, pages 76–85. ACM, 2004.

[8] James Newsome, Elaine Shi, Dawn Song, and Adrian Perrig. The Sybil attack in sensor networks: Analysis & defenses. In *Proceedings of the 3rd International Symposium on Information Processing in Sensor Networks*, pages 259–268. ACM, 2004.

[9] Charalampos Papamanthou, Franco P. Preparata, and Roberto Tamassia. Algorithms for location estimation based on RSSI sampling. In *Algorithmic Aspects of Wireless Sensor Networks*, pages 72–86. Springer, 2008.

[10] Soyoung Park, Baber Aslam, Damla Turgut, and Cliff C. Zou. Defense against Sybil attack in vehicular ad-hoc network based on roadside unit support. In *Military Communications Conference, 2009. MILCOM 2009. IEEE*, pages 1–7. IEEE, 2009.

[11] Bryan Parno and Adrian Perrig. Challenges in securing vehicular networks. In *Workshop on Hot Topics in Networks (HotNets-IV)*, pages 1–6. ACM, 2005.

[12] Ali Akbar Pouyan and Mahdiyeh Alimohammadi. Sybil attack detection in vehicular networks. *Computer Science and Information Technology*, 2(4):197–202, 2014.

[13] Prabhakar Ranjan and Kamal Kant Ahirwar. Comparative study of VANET and MANET routing protocols. In *Proceedings of the International Conference on Advanced Computing and Communication Technologies (ACCT 2011)*, pages 517–523.

[14] Mukul Saini, Kaushal Kumar, and Kumar Vaibhav Bhatnagar. Efficient and feasible methods to detect Sybil attack in VANET. *Int. J. Eng.*, 6(4):431–440, 2013.

[15] Amol Vasudeva and Manu Sood. Sybil attack on lowest ID clustering algorithm in the mobile ad-hoc network. *Int. J. Netw. Security Applic*, 4(5):135, 2012.

[16] Ravi M. Yadumurthy, Mohan Sadashivaiah, Ranga Makanaboyina, et al. Reliable MAC broadcast protocol in directional and omni-directional transmissions for vehicular ad-hoc networks. In *Proceedings of the 2nd ACM International Workshop on Vehicular Ad-Hoc Networks*, pages 10–19. ACM, 2005.

[17] Gongjun Yan, Stephan Olariu, and Michele C Weigle. Providing VANET security through active position detection. *Comput. Commun.*, 31(12): 2883–2897, 2008.

[18] Jie Zhang. Trust management for VANETs: Challenges, desired properties and future directions. *Int. J. Distrib. Sys. Technol. (IJDST)*, 3(1):48–62, 2012.

[19] Tong Zhou, Romit Roy Choudhury, Peng Ning, and Krishnendu Chakrabarty. Privacy-preserving detection of Sybil attacks in vehicular ad-hoc Networks. In *Mobile and Ubiquitous Systems: Networking & Services, 2007. MobiQuitous 2007. Fourth Annual International Conference*, pages 1–8. IEEE, 2007.

Chapter 4

Malware Propagation and Control in Internet of Things

Shin-Ming Cheng

Pin-Yu Chen

Kwang-Cheng Chen

CONTENTS

4.1 Introduction

Cyberphysical systems (CPSs) integrate computing and physical processes; embedded computers monitor and control the physical process. The system consists of a set of nodes with various communication capabilities, including sensors, actuators, a processor or a control unit, and a communication device. The nodes constitute a network and communicate with each person to support everyday life in a smart way, which is known as the *Internet of Things (IoT)*. The "smartness" in IoT applications, such as smart home, smart factory, smart grid, and smart transportation, implies that nodes could automatically sense the environment, collect data, communicate with each other, and perform corresponding actions with minimal human involvement [44]. Some interesting features of the IoT are listed as follows:

■ Numerous objects: The IoT evolves into a large number of objects that collectively move toward a state of pervasiveness.

■ Autonomous functioning: With minimal human intervention, objects in the IoT will perform data collection, processing, collaborating with each other, and decision-making in an autonomous fashion [44].

■ Heterogeneous communication and computation capabilities: Objects in the IoT might support different wireless communication technologies (such as Bluetooth low energy [BLE], Global System for Mobile Communications [GSM], near field communication [NFC], Wi-Fi, and

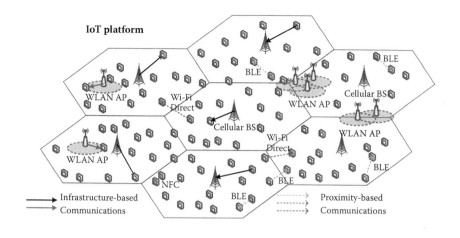

Figure 4.1: IoT platform with infrastructure-based and proximity-based communications.

Zigbee) and computing power. As a result, objects might play different roles in different IoT scenarios [36].

■ Interdependency between the cyber and the physical world: For example, in one well-known IoT, the smart grid, the physical world cooperates with the cyber network [12].

■ Complex network structure: With various radio interfaces, objects can communicate with each other in more complicated ways, forming a complex [53]. For example, an object may communicate with another object via a GSM interface over cellular networks, while also communicating with a different object in the geographic vicinity via proximity-based communication technologies using BLE or Wi-Fi Direct.

Figure 4.1 shows the network architecture of an IoT platform. The security issue in the IoT has received much attention [23]. Obviously, the growing popularity of objects with rich wireless communication capabilities has made the IoT attractive to digital viruses and malicious content. Moreover, the mobility and novel proximity-based communication technologies increase the possibility of spreading malware [14, 16, 17]. In the following, we summarize vulnerabilities to malware due to the unique features in IoT.

■ Weakness of objects with limited computing power: Due to the nature of the limitations of computing capability and energy, the algorithm and mechanism applied to the object are relatively simple. Moreover, conventional security mechanisms such as real-time antivirus scanning cannot be used for the IoT platform due to the unaffordable overhead. As a result, attackers can spend much less resource to break in, and thus, the object becomes a target of malicious users. Another good example is the limited logging, which makes the identification of intrusion harder.

■ Identity hinding in a complex environment: The great number of objects with various, heterogeneous actions and behaviors facilitates the fabrication of identity. Moreover, an intelligent adversary will start infecting some crucial nodes first, instead of launching attacks on the entire network simultaneously, to efficiently disrupt the network and reduce the risks of being detected, thereby posing severe threats to the network robustness.

■ Various infecting patterns under rich wireless communication capabilities: Being capable of infrastructure-based and proximity-based communication technologies, the malware propagates more rapidly, therefore causing more severe results [36].

Typically, after the nodes are infected by the malware, the adversary can control those nodes to launch other attacks. We summarize the impacts of infected nodes on IoT platforms below.

■ Availability of precious network resource: When a large number of infected nodes access the wireless resource simultaneously, the service might be disrupted. Moreover, disruption attacks aim to paralyze IoT operations by launching denial-of-service attacks to jam the entire system. Such destructive consequences for the entire network have a negative impact on the public acceptance and adoption of the IoT, and thus might forestall the widespread deployment of the IoT platform.

■ Safety of human lives and environment: An attack might be launched from the physical world or a cyber network and might impact both domains. In the case of a smart grid, the consequences of cyberattacks could have a severe impact on human lives and the environment [12]. U.S. Executive Order 13636 [1] and Presidential Policy Directive 21 [2] state that proactive and coordinated efforts are necessary to strengthen and maintain secure, functioning, and resilient critical infrastructure and include interdependent functions and systems in both the physical space and cyberspace.

Due to the above vulnerabilities and negative feedback, modeling the behavior of malware propagation in the current world, with its explosive growth in adoption of IoT objects, is an interesting issue that is receiving lots of attention [40]. This chapter aims to provide a theoretic framework for evaluating malware propagation dynamics and to establish a parametric plug-in model for malware propagation control in an IoT network. In particular, we will investigate malware propagation from the viewpoint of both whole networks and individual objects. Understanding the propagation characteristics of malware in both macroscopic and microscopic fashion could aid in estimation of the damage caused by the malware and the development of detection processes.

4.2   Malware Schemes in IoT

Typically, IoT malware can propagate via infrastructure-based communication technologies such as GSM/General Packet Radio Service (GPRS)/Universal Mobile Telecommunications System (UMTS)/Long-Term Evolution (LTE) and wireless local area network (WLAN). The other approach is to exploit proximity-based wireless media, such as BLE, Wi-Fi direct, and NFC, to infect the objects in the vicinity [59], as shown in Figure 4.1. With two kinds of infection path, the malware propagation dynamics might significantly change; Figure 4.2 illustrates an example. As a result, an analytical model is necessary to examine the complicated malware dynamics so that malware mitigation schemes can be proposed accordingly.

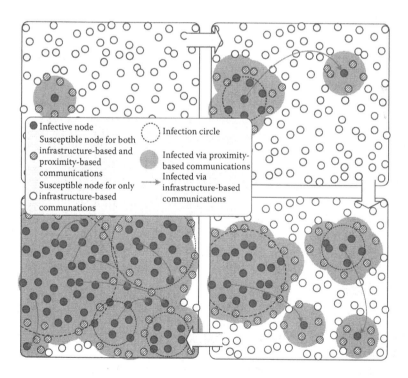

Figure 4.2: The spreading phenomenon of IoT malware.

4.2.1   *Modeling from the view of individuals*

Darabi Sahneh and Scoglio proposed using continuous-time Markov process to build the model [43], and Karyotis proposed a model for malware propagation using Markov random field (MRF) [29], which are both based on a stochastic model. Szongott et al. proposed a spatial-temporal model [48]. It seems that there is enough knowledge on malware propagation modeling; however, all these studies are from the network viewpoint; that is, they regard the nodes as smartphones and the edges as the contact of smartphones in a graph, and implicitly assume that all nodes should possess the same infection rate, unlike our models, which from the viewpoint of the individual. In the real world, every smartphone should have a different reaction when facing spreading malware. Thus, the network view is not suitable to solve this problem [52], because the identities are actually lost when we consider the issue from this viewpoint.

4.2.2   *Modeling from the viewpoint of whole networks*

Since the spread of epidemics among people is similar to the spread of malware over the IoT platform, we typically adopt ideas from epidemiological models [3, 18, 26] to build the models for malwares [10, 15, 17, 19]. The current propagation dynamics of malware can be classified into categories: deterministic models, stochastic models, and spatial-temporal models [40]. Deterministic models use differential equations to describe the spread of infectious malware from the network's point of view, including susceptible-infection (SI) models [16, 28], susceptible-infection-susceptible (SIS) models [7, 8, 25, 35, 39], and susceptible-infection-recover (SIR) models [31, 33]. The authors of [33] further considered the concept of an incubation period from the perspective of the whole network.

Malicious codes such as Internet worms may leverage the inherently fixed topology to sabotage network operations [22, 46] due to the complicated interactions and immense size of communication networks. In [30, 47], the authors find that the spread of Internet worms is similar to the spreading patterns of epidemics and poses severe threats to system security. In [6], Castellano and Pastor-Satorras show that an epidemic will break out if the infection rate exceeds a certain threshold in a network with fixed topology, and the threshold tends to vanish when the network has a skewed degree distribution [24], such as the Internet [20]. In [9], Chen and Carley propose countermeasure competing strategies based on the idea that computer viruses and countermeasures spread through two separate but interlinked complex networks.

Investigations into the dynamics of Internet worm propagation show that the damage caused by Internet worms can be greatly mitigated with efficacious detection techniques or defense at the imminent stages [15, 45, 50, 54, 56–58]. Hu et al. also show that a tightly interconnected proximity network can be exploited as a substrate for spreading malware to launch massive fraudulent attacks [27]. Moreover, in the case of mobile environments, malware can still propagate in such intermittently connected networks by taking advantage of opportunistic encounters [49]. Wang et al. studied spreading patterns of mobile phone viruses, which may traverse through multimedia messaging services (MMS) or Bluetooth, using simulations [51]. In [16], Cheng et al. further modeled malware propagation in generalized social networks consisting of delocalized and localized links. The results show that the contamination by malware speeds up drastically if the malware is able to propagate through heterogeneous links.

4.2.3   Control of malware propagation

In the following, we are going to explore the immunity mechanisms via epidemiology, as well as direct mapping to control of malware propagation. Two schemes are considered, as follows.

- ■ *Self-healing* scheme: On the expiration of the global timer, the infected nodes delete the data, and therefore the nodes transit from the infected state to the recovered state.

- ■ *Vaccine-spreading* scheme: A recovered node participates in vaccinating the susceptible nodes against the malware. In this case, a susceptible node becomes a vaccinee and is therefore immune to the epidemic. The probability that a susceptible node becomes a vaccinee is denoted by κ.

Throughout this chapter, we will investigate the engineering interpretations and the effects of these two immunity schemes on control of malware propagation. To the best of the authors' knowledge, the trade-offs between the time-dependent control capability and the resulting malware propagation dynamics still remain open [21], and the task is further complicated in IoT networks with heterogeneous links.

Traditionally, most research implicitly assumes that the control capability (i.e., the ability to recover from infection) takes effect immediately after the malware propagation. However, this assumption may not be viable in IoT networks, especially for the execution of real-time applications such as antivirus processes [21, 38], since the control signals (e.g., security patches or system updates) are usually not available when a new malware emerges. Alternatively, we consider a more realistic scenario: that the control capability is a function of its distribution time.

4.2.4   Optimal control of malware propagation

How to solve the optimal control signal distribution time is an important issue to mitigate the effects from malware [11, 13]. We first formulate the problem via optimal control theory [34] with the aim of minimizing the accumulated cost, which relates not only to the damage caused by malware but also to the number of replicated data packets in relay-assisted networks. However, optimal control theory assumes full manipulation of the control function, and therefore its solution is inadequate for determining the optimal control signal distribution time. Considering time-dependent control capability, dynamic programming [4] is proposed to obtain the optimal control signal distribution time in real time with respect to the information dissemination process. We also provide early-stage analysis [56] to obtain closed-form expressions of such an SIR model. Using the proposed techniques, we show that the accumulated cost for information dissemination in mobile networks and generalized social networks can be greatly reduced via the proposed approach. Furthermore, the controllability of a network is illustrated by the phase diagram to study the relations between control capability and infection rate.

4.3 Modeling Malware Dynamics from the Individual Viewpoint

4.3.1 Impulse-free model (IFM)

We first consider the simple condition, regardless of incubation period, that the dynamics of malware for an individual due to contacts with infected individuals and the infection rate of the malware are a homogeneous Poisson process with exposure rate λ (contacts/unit time), and the recovery dynamics for an individual due to firewall or antivirus software is exponentially distributed with mean $\frac{1}{\mu}$ unit time. Thus, we model the dynamics of malware without an incubation period with the aid of continuous-time Markov chain (CTMC) $\{X(t), t \geq 0\}$ with the states representing the level of such malware quantized by N degrees, and hence we have $N + 1$ states in total. The CTMC is ergodic with finite states, and the state transition rate diagram is shown in Figure 4.3.

However, the CTMC described in Figure 4.3 is not suitable to describe malware that possesses the property of an incubation period. To make our model more realistic, we define the incubation period T as the time from state 0 to some threshold δ for an individual, and the probability $P_{0\delta}(t)$ that an individual

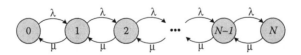

Figure 4.3: Continuous-time Markov chain without incubation period.

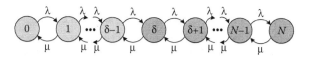

Figure 4.4: Continuous-time Markov chain with incubation period.

is initially safe and will eventually be infected at time t. This interpretation is more practical, since the spread of malware depends on both the exposure rate and the self-immune ability, which is due to firewall or antivirus software, of an individual, and the threshold has meaning not only from the viewpoint of the mobile network but also from the viewpoint of an individual.

We are interested in the status of the expected level of the malware $E[X(t)]$, the incubation period T, the remaining lifetime R, the probability $P_{ij}(t)$ that the status of an individual changes from safe to infected at time t, and the steady-state probability P_n to evaluate the characteristics of the malware. As a consequence, we remodel the CTMC as in Figure 4.4.

4.3.1.1 Expected malware level $E[X(t)]$

We assume that N is relatively large, in the sense that the fatal level is very difficult to achieve (i.e., the malware is fatal if it reaches state N). so that we may view this CTMC as an unbounded CTMC. For a small interval h, given $X(t)$ and $X(0) = i$, we have

$$X[t+h|X(t)] = \begin{cases} X(t)+1, & \lambda h + o(h), \\ X(t)-1, & \text{with prob.} \quad \mu h + o(h), \\ X(t), & 1-(\lambda+\mu)h + o(h) \end{cases}$$

Thus, we have

$$E[E[X(t+h)|X(t)]] = E[(X(t)+1)(\lambda h + o(h)) + (X(t)-1)(\mu h + o(h))$$
$$+ X(t)(1-(\lambda+\mu)h + o(h))]$$
$$= E[X(t)] + (\lambda-\mu)hE[X(t)] + o(h) = E[X(t+h)]$$

Denote $M(t) = E[X(t)]$, then

$$M'(t) = \lim_{h \to 0} \frac{M(t+h) - M(t)}{h} = (\lambda-\mu)M(t)$$

and

$$M(t) = E[X(t)] = \begin{cases} (\lambda-\mu)t + i, & \text{if } \lambda \neq \mu, \\ i, & \text{if } \lambda = \mu \end{cases}$$

4.3.1.2 Incubation period T and remaining lifetime R

We define the incubation period T as the time from state 0 to threshold δ for an individual. For a birth and death process with constant parameters λ and μ, the time taken to leave state x for state $x+1$ is denoted as Z_x, and hence the expected time and variance for Z_x are

$$E[Z_x] = \begin{cases} \dfrac{1 - \left(\frac{\mu}{\lambda}\right)^{x+1}}{\lambda - \mu}, & \text{if } \lambda \neq \mu, \\[2ex] \dfrac{x+1}{\lambda}, & \text{if } \lambda = \mu \end{cases}$$

and

$$Var(Z_x) = \frac{1}{\lambda(\lambda+\mu)} + \frac{\mu}{\lambda} Var(Z_{x-1}) + \frac{\mu}{\lambda+\mu}(E[Z_{x-1}] + E[Z_x])^2$$

where $E[Z_0] = \frac{1}{\lambda}$ and $Var(Z_0) = 1/\lambda^2$.

The expected time to go from state k to state j is

$$E\left[\sum_{x=k}^{j-1} Z_x\right] = \sum_{x=k}^{j-1} E[Z_x] = \sum_{x=k}^{j-1} \frac{1 - \left(\frac{\mu}{\lambda}\right)^{x+1}}{\lambda - \mu}$$

$$= \begin{cases} \dfrac{1}{\lambda - \mu}\left[j - k - \dfrac{\left(\frac{\mu}{\lambda}\right)^{k+1} - \left(\frac{\mu}{\lambda}\right)^{j+1}}{1 - \frac{\mu}{\lambda}}\right], & \text{if } \lambda \neq \mu, \\[3ex] \dfrac{j(j+1) - k(k+1)}{2\lambda}, & \text{if } \lambda = \mu \end{cases}$$

Hence,

$$T = E\left[\sum_{x=0}^{\delta-1} Z_x\right] = \begin{cases} \dfrac{1}{\lambda - \mu}\left[\delta - \dfrac{\left(\frac{\mu}{\lambda}\right) - \left(\frac{\mu}{\lambda}\right)^{\delta+1}}{1 - \frac{\mu}{\lambda}}\right], & \text{if } \lambda \neq \mu, \\[3ex] \dfrac{\delta(\delta+1)}{2\lambda}, & \text{if } \lambda = \mu \end{cases} \tag{4.1}$$

and $Var(T) = \sum_{x=0}^{\delta-1} Var(Z_x)$.

If we assume that malware reaching the fatal level N is the cause of system breakdown for an individual, then the remaining lifetime is defined as the time from the emergence of the malware to state N. Hence,

$$R = E\left[\sum_{x=\delta}^{N-1} Z_x\right] = \begin{cases} \dfrac{1}{\lambda - \mu}\left[N - \delta - \dfrac{\left(\frac{\mu}{\lambda}\right)^{\delta+1} - \left(\frac{\mu}{\lambda}\right)^{N+1}}{1 - \frac{\mu}{\lambda}}\right], & \text{if } \lambda \neq \mu, \\[3ex] \dfrac{N(N+1) - \delta(\delta+1)}{2\lambda}, & \text{if } \lambda = \mu \end{cases}$$

and $Var(R) = \sum_{x=\delta}^{N-1} Var(Z_x)$.

4.3.1.3 Transition probability $P_{ij}(t)$

The transition probability $P_{ij}(t)$ is defined as $P_{ij}(t) = \{X(t) = j|X(0) = i\}$, since we care about the probability $P\{X(t) = j, j \geq \delta | X(0) = i, i < \delta\}$ that an individual is initially in safe status and will eventually be infected at time t, where we set the observation time 0 to mean the time when an individual undergoes inspection or diagnosis for a certain malware by firewall or antivirus software. We can rewrite the Kolmogorov forward equation as $P'(t) = P(t)\mathbf{R}$, where $P(t)$ is the transition probability matrix with elements $P_{ij}(t)$ and $\mathbf{R}$ is the rate transition matrix with elements

$$
r_{ij} = \begin{cases} q_{ij}, & \text{if } i \neq j, \\ -v_i, & \text{if } i = j \end{cases}
$$

So we can write the rate transition matrix $\mathbf{R}$ for the CTMC model in Figure 4.4 as

$$
\mathbf{R} = \begin{pmatrix} -\lambda & \lambda & 0 & 0 & 0 \\ \mu & -\lambda - \mu & \lambda & 0 & 0 \\ 0 & \mu & -\lambda - \mu & \lambda & 0 \\ 0 & 0 & \mu & -\lambda - \mu & \lambda \end{pmatrix} \tag{4.2}
$$

The transition probability matrix $P(t)$ has the solution $P(t) = e^{\mathbf{R}t}$, and we can apply an approximation method in $[X]$ to obtain the result by $P(t) = \lim_{n \to \infty} (\mathbf{I} + \mathbf{R}\frac{t}{n})^n$, where $\mathbf{I}$ is the identity matrix and $e^{\mathbf{R}t}$ is defined as $e^{\mathbf{R}t} = \sum_{n=0}^{\infty} \mathbf{R}^n \frac{t^n}{n!}$. Thus, we may obtain the transition probability $P_{ij}(t)$ and have more information about the probability $P\{X(t) = j, j \geq \delta | X(0) = i, i < \delta\}$ that an individual is initially in safe status and will eventually be infected at time t.

4.3.1.4 Steady-state probability P_n

The steady-state probability P_n of a birth and death process with constant parameters λ, μ, and finite states is a truncated M/M/1/N queue with

$$
P_n = \begin{cases} \dfrac{(1-\rho)\rho^n}{\sum_{x=0}^{N}(1-\rho)\rho^x}, & \text{if } \lambda \neq \mu, \\[2ex] \dfrac{1}{N+1}, & \text{if } \lambda = \mu \end{cases} \qquad 0 \leq n \leq N, \quad \rho = \frac{\lambda}{\mu} \tag{4.3}
$$

4.3.2 Impulse-reaction model (IRM)

The CTMC model shown in Figure 4.4 assumes that the exposure rate and the mean self-immunity period at state n are the same for all states and implicitly indicates that the exposure rate never decreases. With the modeling experiences above, we propose a general CTMC model that aims at capturing the dynamics in a much more general way in Figure 4.5.

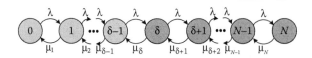

Figure 4.5: A general continuous-time Markov chain model.

To examine the convenience of the general model for the dynamics of malware with an incubation period, we consider a practical case in which an individual looks for help on the emergence of the malware with the aid of a computer engineer and use a method which reduces malware mobility. Moreover, the impulse reaction of the self-immune elements, such as firewall and antivirus software, contributes to enhanced recovery rate as well. Hence, one practical assumption is

$$\lambda_n = \begin{cases} \lambda, & 0 \leq n \leq \delta - 1, \\ \alpha\lambda, & \delta \leq n \leq N - 1 \end{cases} \tag{4.4}$$

and

$$\mu_n = \begin{cases} \mu, & 1 \leq n \leq \delta - 1, \\ \beta\mu, & \delta \leq n \leq N \end{cases} \tag{4.5}$$

where $0 \leq \alpha < 1$ and $\beta > 1$. From Equations 4.4 and 4.5, if we denote $\lambda' = \alpha\lambda$ and $\mu' = \beta\mu$, then we can rewrite the incubation period T, remaining lifetime R, transition probability $P_{ij}(t)$, and steady-state probability P_n.

4.3.2.1 Incubation period T and remaining lifetime R

From Equation 4.1, the incubation period T is unchanged:

$$T = E\left[\sum_{x=0}^{\delta-1} Z_x\right] = \begin{cases} \frac{1}{\lambda-\mu}\left[\delta - \frac{(\frac{\mu}{\lambda})-(\frac{\mu}{\lambda})^{\delta+1}}{1-(\frac{\mu}{\lambda})}\right], & \text{if } \lambda \neq \mu, \\ \frac{\delta(\delta+1)}{2\lambda}, & \text{if } \lambda = \mu \end{cases}$$

and $Var(T) = \sum_{x=0}^{\delta-1} Var(Z_x)$.

The remaining lifetime R from Equation 4.2 is

$$R = E\left[\sum_{x=\delta}^{N-1} Z_x\right] = \begin{cases} \frac{1}{\lambda'-\mu'}\left[N - \delta - \frac{(\frac{\mu'}{\lambda'})^{\delta+1}-(\frac{\mu'}{\lambda'})^{N+1}}{1-(\frac{\mu'}{\lambda'})}\right], & \text{if } \lambda' \neq \mu', \\ \frac{N(N+1)-\delta(\delta+1)}{2\lambda'}, & \text{if } \lambda' = \mu' \end{cases}$$

and $Var(R) = \sum_{x=\delta}^{N-1} Var(Z_x)$.

where:

$$E[Z_x] = \begin{cases} \dfrac{1 - (\frac{\mu}{\lambda})^{x+1}}{\lambda - \mu}, & \text{if } \lambda \neq \mu \\[2ex] \dfrac{x+1}{\lambda}, & \text{if } \lambda = \mu \end{cases} \quad \text{for } 0 \leq x \leq \delta - 1$$

and

$$E[Z_x] = \begin{cases} \dfrac{1 - (\frac{\mu'}{\lambda'})^{x+1}}{\lambda' - \mu'}, & \text{if } \lambda' \neq \mu' \\[2ex] \dfrac{x+1}{\lambda'}, & \text{if } \lambda' = \mu' \end{cases} \quad \text{for } \delta \leq x \leq N - 1$$

$$Var[Z_x] = \begin{cases} \dfrac{1}{\lambda(\lambda + \mu)} + \dfrac{\mu}{\lambda}Var(Z_{x-1}) + \dfrac{\mu}{\lambda + \mu}(E[Z_{x-1}] + E[Z_x])^2, & 0 \leq x \leq \delta - 1, \\[2ex] \dfrac{1}{\lambda'(\lambda' + \mu')} + \dfrac{\mu'}{\lambda'}Var(Z_{x-1}) + \dfrac{\mu'}{\lambda' + \mu'}(E[Z_{x-1}] + E[Z_x])^2, & \delta \leq x \leq N - 1 \end{cases}$$

4.3.2.2 Transition probability $P_{ij}(t)$

The rate transition matrix **R** from Equation 4.2 is shown in Equation 4.6.

$$\mathbf{R} = \begin{array}{c} \\ 0 \\ 1 \\ 2 \\ \vdots \\ \delta \\ \vdots \\ N-1 \\ N \end{array} \begin{pmatrix} -\lambda & \lambda & 0 & & & & & & \\ \mu & -\lambda - \mu & \lambda & \cdots & & & & & \\ 0 & \mu & -\lambda - \mu & \cdots & & & & & \\ 0 & 0 & \vdots & \vdots & & & & & \\ 0 & \cdots & & \beta\mu & -\alpha\lambda - \beta\mu & \alpha\lambda & \cdots & & \\ 0 & & \cdots & 0 & \beta\mu & -\alpha\lambda - \beta\mu & \cdots & & \\ 0 & & \cdots & & & \mu & -\alpha\lambda - \beta\mu & \lambda \\ 0 & & \cdots & & & & \beta\mu & -\beta\mu \end{pmatrix}$$

And the transition probability matrix can be obtained by the same procedure described above for $P(t)$.

4.3.2.3 Steady-state probability P_n

The steady-state probability P_n for the general birth and death model in Figure 4.5 is

$$P_n = \left[1 + \sum_{n=1}^{N} \frac{\prod_{x=0}^{n-1} \lambda_x}{\prod_{x=1}^{n} \mu_x} \right]^{-1} \frac{\prod_{x=0}^{n-1} \lambda_x}{\prod_{x=1}^{n} \mu_x}$$

For the case with parameters in Equation 4.3 with $\rho = \frac{\lambda}{\mu}$, if $\rho \neq 1$, we have

$$
P_n = \begin{cases}
C(1-\rho)\rho^n, & \text{if } 0 \leq n \leq \delta - 1, \\
C(1-\rho)\rho^n \frac{\alpha^{n-\delta}}{\beta^{n-\delta+1}}, & \text{if } \delta \leq n \leq N
\end{cases}
$$

where C is defined as

$$
\left[\sum_{x=0}^{\delta-1} (1-\rho)\rho^x + \sum_{y=\delta}^{N} (1-\rho)\rho^y \frac{\alpha^{y-\delta}}{\beta^{y-\delta+1}} \right]^{-1}.
$$

In the case of $\rho = 1$, we have

$$
P_n = \begin{cases}
\left[\delta + \sum_{y=\delta}^{N} \frac{\alpha^{y-\delta}}{\beta^{y-\delta+1}} \right]^{-1}, & \text{if } 0 \leq n \leq \delta - 1, \\
\left[\delta + \sum_{y=\delta}^{N} \frac{\alpha^{y-\delta}}{\beta^{y-\delta+1}} \right]^{-1} \frac{\alpha^{n-\delta}}{\beta^{n-\delta+1}}, & \text{if } \delta \leq n \leq N
\end{cases}
$$

4.3.3 Numerical results

We denote the model described in Figure 4.4 as IFM and the model described in Figure 4.5 and Equations 4.4 and 4.5 as IRM. Without loss of generality, we further assume that the recovery rate before the threshold, μ, equals one. We will show the incubation period T, the remaining lifetime R, the transition probability $P_{ij}(t)$, and the steady-state probability P_n for both models and provide intuitive explanations for the results. For the parameters of numerical results, we set $N = 100$, $\delta = 20$, $\alpha = 0.6$, and $\beta = 1.5$.

We show the numerical results of incubation period T and remaining lifetime R in feasible regions for both IFM and IRM in Figure 4.6. The results indicate that the incubation period is unchanged, since both models have the same CTMC parameters before the threshold. However, due to the impulse reaction, the reduced infection ratio directly gives rise to higher remaining lifetime, which is quite plausible, since an individual may recover from illness by self-immunity or with the aid of an engineer.

We are also interested in the transition probability that an individual is originally safe (initial state is 0) and eventually comes to the threshold δ, which evolves with time, that is, $P_{0\delta}(t)$. Moreover, we are also interested in the accumulated probability that an individual is originally safe (initial state is 0) and will eventually be infected (reach states above δ), which also evolves with time, that is, $F(t) = \sum_{k \geq \delta} P_{0k}(t)$.

We show the two dynamics in Figure 4.7 and Figure 4.8, respectively. In Figure 4.7, the peak of transition probability $P_{0\delta}(t)$ emerges earlier if the infection ratio ρ is larger, which is plausible, since we have a higher exposure rate for

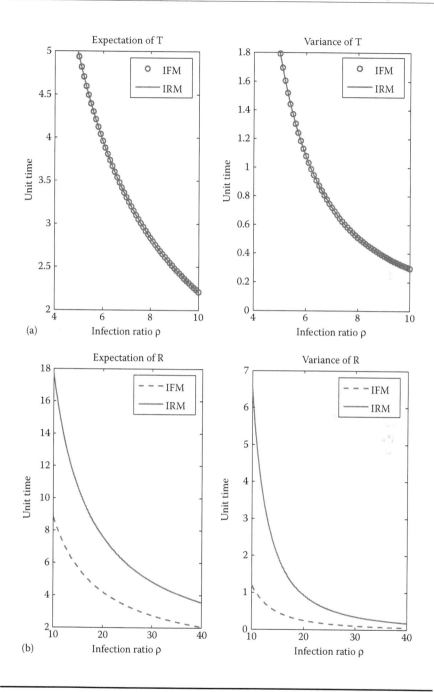

Figure 4.6: Numerical results of incubation period T and remaining lifetime R in feasible regions. (a) Expectation and variance of T. (b) Expectation and variance of R.

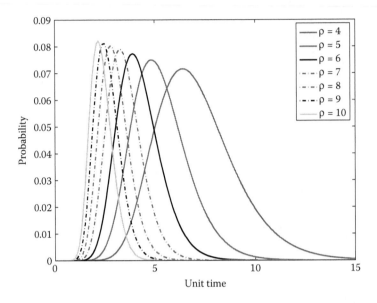

Figure 4.7: Transition probabilities $P_{0\delta}(t)$.

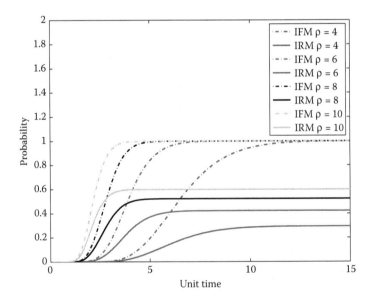

Figure 4.8: Infection transition probability $F(t)$.

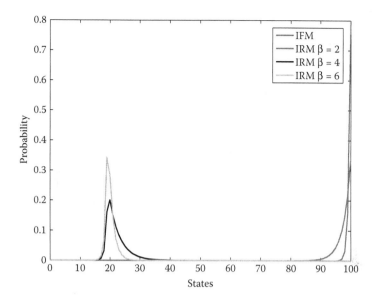

Figure 4.9: Steady-state probabilities corresponding to β.

a larger ratio. To emphasize the effect of the impulse reaction, we set $\alpha = 0.6$ and $\beta = 8$ in Figure. 4.8. Intuitively, the infection transition probability $F(t)$ is greatly reduced if we take the IRM model into consideration.

Finally, we present the steady states in Figure 4.9. For IRM, we fix $\alpha = 0.6$ and observe the distribution of steady-state probabilities with different β compared with that of IFM. The tendency of steady-state probabilities shows that higher β may better refine the distribution of steady-state probability around the threshold or at least reduce the steady-state probability at extremely high levels. This result illustrates well the effect of the impulse reaction.

4.3.4 Summary

In this section, we modeled the dynamics of malware with an incubation period by CTMC, concerning the exposure and recovery rates of an individual and the generalized birth and death process, along with the fact that an individual is infected once the malware level overpasses some defined threshold, providing better insight into the dynamics of malware. We proposed two models, IFM and IRM, and derived the analytic solutions for the malware level, incubation period, remaining lifetime, transition probability, and steady-state probability. Furthermore, we provide numerical results to show that while the expected incubation period is the same in both cases, IRM has higher remaining lifetime and tends to reduce both accumulated infection probability and steady-state probability at fatal levels and refine the malware level around the threshold in the feasible regions of infection ratio.

4.4 Modeling Malware Dynamics from the Network Viewpoint

From the perspective of the whole IoT network, the population is regarded as the total number of nodes N, which are assumed to be stationary and uniformly distributed in an $L \times L$ square area with population density $\rho = (N/(L_2))$. We assume that all objects have both infrastructure-based and proximity-based communications capabilities to maintain the homogeneous mixing property. Denote the subpopulation function $I(t) = I_{pro}(t) + I_{inf}(t)$ as the total number of compromised handsets at time t, where $I_{pro}(t)$ and $I_{inf}(t)$ are those that have been infected via proximity-based and infrastructure-based communication channels at time t, respectively. Likewise, $S(t)$ denotes the set of susceptible nodes at time t.

We assume that all nodes are with identical proximity-based transmission range δ. The average number of proximity-based communication contacts is denoted by $\eta_{pro} = \rho\pi\delta^2$. We further assume that every node randomly selects η_{inf} nodes as its infrastructure-based communication contacts, which are distinct from proximity-based communication contacts. Note that the results are still valid if we treat η_{inf} and η_{pro} as random variables and apply their means to our model. The pairwise infection rates on an infrastructure-based communication link and a proximity-based communication link are respectively denoted as λ_{inf} and λ_{pro}.

4.4.1 Malware dynamics: SI model

In this subsection, we look into the SI model, in which a susceptible node acquires infection and never becomes susceptible again. This is due to users' lack of concern about the threat of malware and the limited capability of current antiviral software. Obviously, we have

$$I(t) + S(t) = I_{pro}(t) + I_{inf}(t) + S(t) = 1 \tag{4.6}$$

and

$$I(t) = \dot{I}_{pro}(t) + \dot{I}_{inf}(t) \tag{4.7}$$

Without loss of generality, we assume that only one handset is infected at the initial stage, that is, $I(0) = I_{inf}(0) = 1$ and $I_{pro}(0) = 0$. Malware is propagated through proximity-based and infrastructure-based communication links. The control signal distribution resembles the malware propagation in the sense that it is distributed through these heterogeneous links to alleviate network cost. The state equation of infrastructure-based infection is

$$\dot{I}_{inf}(t) = \lambda_{inf}(\eta_{inf} - 1)S(t)I(t) \tag{4.8}$$

where $\eta_{inf} - 1$ accounts for the fact that a node being infected implies that at least one of its neighbors is infected [10].

On the other hand, due to the interdependency of proximity-based and infrastructure-based infections, the proximity-based infection stretches out from the infected source nodes generated by infrastructure-based infections, as shown in Figure 4.2. The proximity-based infection spreads out like a ripple centered at the infected source node, and grows with time. In other words, the spatial spreading of the epidemics through proximity-based communications is only contributed by the wavefronts of infection circles, while the infected nodes located in the interior of the infection circles are not engaged in further spatial infections. For a single ripple with radius $r(t)$, $\rho\pi r^2(t) = N \cdot I_{pro}(t)$, and the infected population in the peripheral circular strip of width δ is $\rho\pi r^2(t) - \rho\pi(r(t) - \delta)^2$. We have

$$\Upsilon_{S \to I_{pro}}(t) = \frac{1}{N}\lambda_{pro}\frac{1}{2}\eta_{pro}S(t)\left[\rho\pi r^2(t) - \rho\pi(r(t) - \delta)^2\right]$$

$$= \frac{1}{N}\lambda_{pro}\frac{1}{2}\eta_{pro}S(t)\left[2\rho\pi\delta r(t) - \rho\pi\delta^2\right]$$

$$= \frac{1}{N}\lambda_{pro}\eta_{pro}S(t)\left[\delta\sqrt{\rho\pi N I_{pro}(t)} - \frac{1}{2}\rho\pi\delta^2\right]$$

$$\cong \frac{1}{N}\sigma\lambda_{pro}\eta_{pro}S(t)\sqrt{N I_{pro}(t)} \tag{4.9}$$

where $\sigma = \delta\sqrt{\rho\pi}$ and $\frac{1}{2}\rho\pi\delta^2$ is usually negligible compared with N [37]. Please note that $\Upsilon_{X \to Y}(t)$ is the expected population transition rate from state X to state Y at time t. $\frac{1}{2}\eta_{pro}$ accounts for the average number of proximity-based communication contacts that are located outside of the peripheral circular strip. Since infrastructure-based infection creates multiple infected source nodes over time, we denote the incremental spatially infected population of a ripple that is generated at time z and keeps stretching for s time units by

$$\dot{W}(z,s) \triangleq \frac{dW(z,s)}{ds} = \sigma\lambda_{pro}\eta_{pro}S(z+s)\sqrt{W(z,s)} \tag{4.10}$$

where $W(z,0) = 1$. The state equation of the aggregated proximity-based infection can be characterized as

$$\dot{I}_{pro}(t) = \frac{1}{N}\int_0^t \dot{I}_{inf}(\tau)\dot{W}(\tau, t - \tau)d\tau \tag{4.11}$$

This means that $\dot{I}_{inf}(t)d\tau$ infected source nodes are generated at time τ, and each contributes to $\dot{W}(\tau, t - \tau)$ incremental spatial infection at time t. The overall state equation of $I(t)$ becomes

$$\dot{I}(t) = \lambda_{inf}(\eta_{inf} - 1)S(t)I(t) + \frac{1}{N}\int_0^t \dot{I}_{inf}(\tau)\dot{W}(\tau, t - \tau)d\tau \qquad (4.12)$$

4.4.1.1 Numerical results

Figure 4.10 illustrates the analytical and simulation plots depicting the propagation dynamics of a hybrid malware spreading via only proximity-based communications, only infrastructure-based communications, and both among 2000 nodes uniformly deployed in a 50×50 plane under $\rho = 0.8$. We consider the impact of η_{pro} on the propagation process in terms of speed and reachability. The parameter setups are $\lambda_{pro} = \lambda_{inf} = 0.05$ and $\eta_{inf} = 6$ (follow the data sheet in [51]). We observe that the curves of propagation dynamics closely match our analytical model; the limited discrepancy that exists is mainly due to the fact that the hybrid malware may propagate to objects that have already been infected, and uncertain boundary conditions could not be considered in the analysis.

This figure also shows that propagation via only proximity-based communication is relatively slow compared with that via only infrastructure-based communications due to spatial spreading characteristics. We also observe the same phenomenon for the hybrid malware with much faster propagation speed, where the rapid invasion via infrastructure-based communications dominates the propagation dynamics. When η_{pro} increases from 2 to 3, our model indicates a significant increase in the propagation speed in the early stages of the spreading process. This is in accordance with the fact that a larger η_{pro} results in a larger infected subpopulation, which could exploit both proximity-based and infrastructure-based communications to spread, increasing propagation severity.

Note that the agent-based emulation model [5] and simulation [51] try to characterize behaviors of the N nodes and all interactions among them, which requires huge computational power. In contrast, our model aggregates the N nodes into two states and only tracks the behavior of these two states and the interactions between them, such that our model can be more computationally effective.

4.4.1.2 Summary

Compared with the existing agent-based model or simulation with its computational burden, our analytical model based on differential equations works more efficiently and could act as a quick reference to gather approximate knowledge of propagation speed and severity of hybrid malware with various settings of infection rates and average node degrees in IoT networks. The security assessment could adopt such results to develop detection and containment strategies and processes so as to avoid a major outbreak.

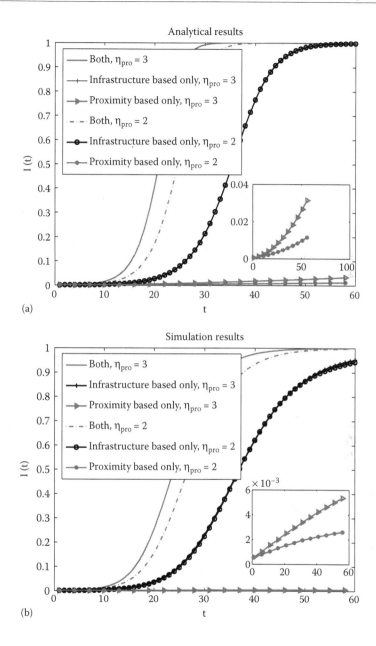

Figure 4.10: Infected population in IoT networks. $N = 2000$, $L = 50$, $I_0 = 1/N$, $\lambda_{inf} = \lambda_{pro} = 0.05$, $\eta_d = 6$, $\eta_{pro} = 3$ and 2.

4.4.2 Malware dynamics under malware control: SIR model

When the control mechanisms for malware (such as self-healing and vaccine) are considered, the "recovery" state is involved in explaining immunity. Analogously to epidemiology, a node is in the infected state if it receives the malware and turns itself into an infectious node. A node that recovers from the epidemic or becomes a vaccinee against the epidemic is said to be in the recovered (immune) state. Please note that a node transits from the infected state to the recovered state in the former case, while a node transits from the susceptible state to the recovered state in the latter case. Only susceptible nodes are vulnerable to the epidemic, and recovered nodes are immune to the epidemic for good. Throughout this chapter, such state transitions are referred to as the SIR model, where $S(t), I(t)$, and $R(t)$ are the normalized susceptible, infected, and recovered population at time t, respectively, that is, $S(t) + I(t) + R(t) = 1$. Considering the immunity schemes and the time-dependent control capability, let $u(t)$ be the recovery probability of the self-healing scheme, where

$$u(t) = \begin{cases} 0, & t < T_D, \\ f(T_D), & t \geq T_D \end{cases} \tag{4.13}$$

By substituting the equation $S(t) = 1 - I(t) - R(t)$ and relaxing the states to be continuous and nonnegative valued, we have, for a small interval Δt,

$$I(t + \Delta t) = I(t) + \Upsilon_{S \to I}(t)\Delta t - \Upsilon_{I \to R}(t)\Delta t \tag{4.14}$$

Please note again that $\Upsilon_{X \to Y}(t)$ is the expected population transition rate from state X to state Y at time t. We obtain the first-order ordinary differential equation (ODE) (state equation)

$$\dot{I}(t) = \lim_{\Delta t \to 0} \frac{I(t + \Delta t) - I(t)}{\Delta t} = \Upsilon_{S \to I}(t) - \Upsilon_{I \to R}(t) \triangleq G_I(I(t), R(t), u(t)) \tag{4.15}$$

Similarly, let $\phi(t)$ be the recovery probability of the vaccine-spreading scheme; the ODE of the recovered population is

$$\dot{R}(t) = \Upsilon_{I \to R}(t) + \Upsilon_{S \to R}(t) \triangleq G_R(I(t), R(t), u(t), \phi(t)) \tag{4.16}$$

where

$$\phi(t) = \begin{cases} 0, t < T_D, \\ \kappa, t \geq T_D \end{cases} \tag{4.17}$$

When $\kappa = 0$, the fluid model degenerates to a noncooperative network in which no nodes participate in vaccine spreading. Without loss of generality, we use the state equations of vaccine spreading to obtain the optimal control signal distribution time T_D^*, since self-healing is a special case of vaccine spreading when there is no cooperation (i.e., $\kappa = 0$).

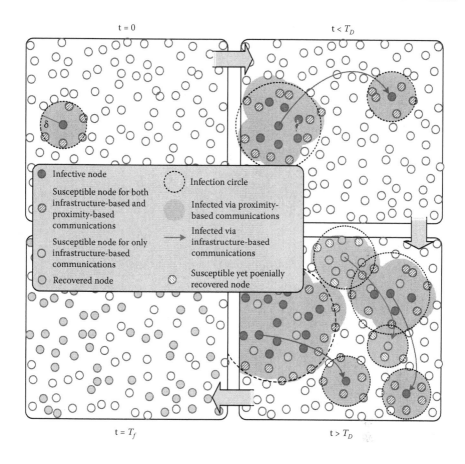

t = 0

t < T_D

● Infective node	$\left(\begin{smallmatrix} \\ \end{smallmatrix}\right)$ Infection circle
⊘ Susceptible node for both infrastructure-based and proximity-based communications	Infected via proximity-based communications
○ Susceptible node for only infrastructure-based communications	→ Infected via infrastructure-based communications
○ Recovered node	⊘ Susceptible yet poenially recovered node

t = T_f

t > T_D

Figure 4.11: Illustration of malware propagation and control signal distribution in IoT networks. Proximity-based and infrastructure-based communication links are exploited for malware propagation. A node is infected at *t* = 0. *T_D* denotes the control signal distribution time, and *T_f* denotes the time instance for the eradication of the epidemic.

The dynamics of malware and control signal distribution are illustrated in Figure 4.11. Malware is propagated through proximity-based and infrastructure-based communication links. The control signal distribution resembles malware propagation in the sense that it is distributed through these heterogeneous links to alleviate network cost. The state equation of infrastructure-based infection is

$$\dot{I}_{inf}(t) = \lambda_{inf}(\eta_{inf} - 1)S(t)I(t) - u(t)I_{inf}(t) \tag{4.18}$$

Compared with Equation 4.8, $u(t)I_{inf}(t)$ in this equation is related to recovery state. On the other hand, the state equation of the aggregated proximity-based infection can be characterized as

$$\dot{I}_{pro}(t) = \frac{1}{N} \int_0^t \dot{I}_{inf}(\tau) \dot{W}(\tau, t - \tau) d\tau - u(t) I_{pro}(t) \qquad (4.19)$$

Compared with Equation 4.11, $u(t)I_{pro}(t)$ in this equation is related to recovery state. The overall state equation of $I(t)$ becomes

$$\dot{I}(t) = [\lambda_{inf}(\eta_{inf} - 1)S(t) - u(t)]I(t) + \frac{1}{N} \int_0^t \dot{I}_{inf}(\tau) \dot{W}(\tau, t - \tau) d\tau \qquad (4.20)$$

Similarly, the immunity scheme can also leverage the proximity-based and infrastructure-based communication links to eradicate the epidemic. The state equation of recovery via infrastructure-based communications is

$$\dot{R}_d(t) = u(t)I_{inf}(t) + \phi(t)(\eta_{inf} - 1)S(t)R(t) \qquad (4.21)$$

The incremental spatial recovery process is characterized by

$$\dot{Q}(z,s) = \sigma\phi(t)\eta_{pro}S(z+s)\sqrt{Q(z,s)} \qquad (4.22)$$

with $Q(z,0) = 1$. The state equation of proximity-based recovery is

$$\dot{R}_{pro}(t) = \frac{1}{N} \int_0^t \dot{R}_{inf}(\tau) \dot{Q}(\tau, t - \tau) d\tau + u(t) I_{pro}(t) \qquad (4.23)$$

The overall state equation of $R(t)$ becomes

$$\dot{R}(t) = u(t)I(t) + \frac{1}{N} \int_0^t \dot{R}_{inf}(\tau) \dot{Q}(\tau, t - \tau) d\tau + \phi(t)(\eta_{inf} - 1)S(t)R(t) \qquad (4.24)$$

4.4.3 Performance evaluation

To demonstrate the trade-offs between control signal distribution and the resulting impacts on malware propagation, we set the function $f(T_D)$ in Equation 4.13 to be $f(T_D) = \min\{1, c \cdot T_D^{\alpha}\}$, where α is a nonnegative value that accounts for the effectiveness of the control signal, and c is a positive constant. The effect of control signal has a power-law growth with respect to the control signal distribution time. This power-law growth model is a general parametric model, and it can be used to investigate the trade-offs between control capability and control signal distribution timeliness.

The exponent α is associated with the effectiveness of the control capability. $\alpha = 0$ degenerates to the scenario that the control capability is irrelevant to its distribution time. For the simulation setup, N nodes are traversing in the square

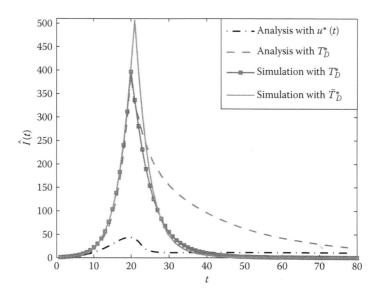

Figure 4.12: Infected population under self-healing scheme in IoT networks. $N = 2000$, $L = 50$, $I_0 = 1/N$, $\delta = 1.1$, $\lambda_{inf} = \lambda_{pro} = 0.05$, $\eta_d = 6$, $\eta_{pro} = 3$, $\alpha = 2$, $\beta = 1$, $\kappa = 0$, $T_f = 200$, $M = 1000$, $\Lambda_I(0) = 200$, $\Lambda_R(0) = 100$, $t' = 1$, and $c = 10^{-3}$ over 300 simulations.

area in wrap-around condition via the Lèvy walk mobility model [42], where the step length and the pause time follow a power-law distribution with negative exponent, respectively. We set the length exponent $l = 1.5$ and the pause time exponent $\varphi = 1.38$, which fit the trace-based data of human mobility patterns collected in the University of California, San Diego and Dartmouth [33]. The simulation setup is the same as that in the previous subsection, except that we fix $\eta_{pro} = 3$ ($\delta \approx 1.1$), since in general, the proximity-based communication range is limited.

The infected population under the self-healing scheme is shown in Figure 4.12. Prior to the control signal distribution, our SIR model captures the simulation results of malware propagation in IoT networks. Although both infrastructure-based and proximity-based communication pairwise infection rates are quite low ($\lambda_{inf} = \lambda_{pro} = 0.05$), the infection spreads rapidly, since the malware propagation benefits from these heterogeneous links. After control signal distribution, the analytical infected population decreases at a slower speed compared with the simulation results, due to the fact that recovery actually disrupts the spread of proximity-based infection, and the ripples are likely to coincide with other ripples as time evolves, which leads to overestimation of malware propagation. In addition, early-stage analysis suggests early distribution, and hence the infection has a slow decaying curve. The infection curve via

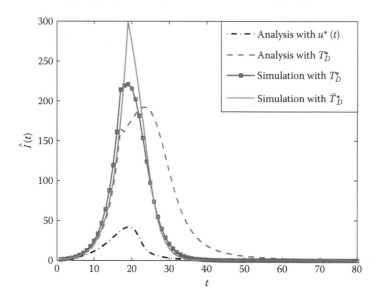

Figure 4.13: Infected population under vaccine-spreading scheme in IoT networks. $N = 2000$, $L = 50$, $I_0 = 1/N$, $\delta = 1.1$, $\lambda_{inf} = \lambda_{pro} = 0.05$, $\eta_{inf} = 6$, $\eta_{pro} = 3$, $\alpha = 2$, $\beta = 1$, $\kappa = 0.1$, $T_f = 200$, $M = 1000$, $\Lambda_I(0) = 200$, $\Lambda_R(0) = 100$, $t' = 1$, and $c = 10^{-3}$ over 300 simulations.

optimal control theory also implies that we can have better control of the malware propagation if we can have full manipulation of the control capability.

Similar results can be found in Figure 4.13 for malware propagation under the vaccine-spreading scheme. With the help of vaccine spreading through infrastructure-based and proximity-based links, we can further mitigate the infection compared with the self-healing scheme. Since susceptible nodes are likely to become vaccinees under the vaccine-spreading scheme, the immune nodes may hinder the growth of the proximity-based infection ripple and thereby decelerate the infection, which again leads to overestimation of the SIR model after the control signal distribution. In Figures 4.12 and 4.13, $u^*(t)$ from optimal control theory elucidates the discrepancy of taking the time-dependent control capability $f(T_D)$ into consideration. Time-dependent control capability inevitably incurs more network cost than optimal control function.

4.5 Optimal Control of Malware

The ultimate goal of this section is to determine the optimal distribution time T_D^* such that the accumulated cost caused by the epidemic is minimized. Via optimal control theory [34], we aim to solve the optimization problem.

$$\text{Minimize} \quad J = \int_{T_0}^{T_f} [NI(t)]^\beta + v \cdot u^2(t) \, dt$$

$$\text{Subject to} \quad \dot{I}(t) = G_I(I(t), R(t), u(t)),$$

$$\dot{R}(t) = G_R(I(t), R(t), u(t), \phi(t)),$$

$$S(t) + I(t) + R(t) = 1,$$

$$S(t) \geq 0, \; I(t) \geq 0, \; R(t) \geq 0 \tag{4.25}$$

where $\beta > 0$ represents the severity of the epidemic, T_0 is the initial time, which is set to be 0, and T_f is the completion time, which is assumed to be free. v is the coefficient representing the cost of control signal distribution with respect to the malware propagation process, and for simplicity it is normalized to $v = \frac{1}{2}$. If $v = 0$, then the cost of control signal distribution is irrelevant to the malware propagation process. The performance measure J represents the accumulated cost caused by the epidemic, and it takes its quadratic form for the control function $u(t)$, such that it is jointly convex in $I(t)$ and $u(t)$. The physical interpretation of J is that it is proportional to the accumulated infected population, which relates to the number of nodes that have received the malware over time. Moreover, when $\beta = 1$, it accounts for the accumulated infected population from T_0 to T_f, which coincides with the performance measure in various networks of interest to us [19, 32, 55].

With Equation 4.25, we aim to find the optimal control signal distribution time T_D^* such that $T_D^* = \arg\min_{T_D} J$. By Pontryagin's minimum principle [41], if $G_I(I(t), R(t), u(t))$ and $G_R(I(t), R(t), u(t), \phi(t))$ are jointly concave in $I(t)$, $R(t)$, $u(t)$, and $\phi(t)$, the optimal control function $u^*(t)$ can be obtained by minimizing the Hamiltonian (Lagrangian dual function) with costate variables $\Lambda_I(t)$ and $\Lambda_R(t)$, where

$$\mathcal{H}(I(t), R(t), u(t), \phi(t), \Lambda_I(t), \Lambda_R(t)) = J(I(t), u(t)) + \Lambda_I(t) G_I(I(t), R(t), u(t))$$
$$+ \Lambda_R(t) G_R(I(t), R(t), u(t), \phi(t))$$

The costate variables are updated by the costate equations

$$\dot{\Lambda}_I(t) = -\frac{\partial \mathcal{H}}{\partial I}; \; \dot{\Lambda}_R(t) = -\frac{\partial \mathcal{H}}{\partial R} \tag{4.26}$$

where $\dot{\Lambda}_I(t) \geq 0$ and $\dot{\Lambda}_R(t) \geq 0$ with boundary conditions $\Lambda_I(T_f) = \Lambda_R(T_f) = 0$. Note that during the update process, the negative state values are truncated to zero, such that the nonnegativity state constraints ($S(t), I(t), R(t) \geq 0$) are satisfied.

The solution of optimal control theory resides in the fact that there is no inherent restriction on the control function $u(t)$. However, it is worth noting that when the control capability is associated with T_D, the solution of optimal control theory only provides the trends of the system outputs and may fail to be a feasible operation for control signal distribution. Despite its impracticality, the results obtained from Pontryagin's minimum principle provide performance comparisons to our proposed approach. To compensate the insufficiency of optimal control theory, we adopt dynamic programming [4] to solve the optimal control signal distribution time. By discretizing the time into M intervals with length $\Delta t = T_f/M$, we define the cost C_m as a function of the infected population at the mth period and the newly infected population between the mth and $m+1$th stage, $0 \le m \le M-1$, where

$$C_m = [NI(m\Delta t) + NG_I(I(m\Delta t), R(m\Delta t), u(m\Delta t)) \cdot \Delta t]^\beta = [NI((m+1)\Delta t)]^\beta \tag{4.27}$$

Let $V_m(I(m\Delta t), R((m\Delta t)), u(m\Delta t))$ denote the accumulated cost from the mth stage with terminal condition $V_M(I(M\Delta t), R(M\Delta t), u(M\Delta t)) = 0$ (i.e., the entire system is in its stable stage); the optimal distribution time can be obtained by solving the optimality equation

$$V_m = \min_{a_m \in \{0,1\}} \{C_m + V_{m+1}\}, \, 0 \le m \le M-1 \tag{4.28}$$

where $a_m = 1$ means that the control signal is distributed, and the immunity mechanisms take effect from the mth stage. That is, $T_D^* = m\Delta t$ and $f(m\Delta t) = f(n\Delta t)$, $\forall\, n \ge m$. V_0 represents the minimum accumulated cost, which is equivalent to the performance measure J in Equation 4.25. Equation 4.28 is equivalent to finding an optimal one-time switch from 0 to 1 among all possible one-time switch paths of the M stages to minimize the accumulated cost, and it can be solved via Bellman–Ford algorithm [4] with $O(2^M)$ complexity. In other words, incorporating the malware propagation process and the time-dependent control capability, the optimal control signal distribution time can be obtained via dynamic programming in Equation 4.28 in real time to minimize the accumulated network cost.

With the state equations, the corresponding Hamiltonian is obtained by plugging the parameters in Equations 4.13, 4.17, 4.20, 4.24, and 4.25 into Equation 4.26:

$$\mathcal{H} = [NI(t)]^\beta$$

$$+ \frac{1}{2}u^2(t) + \Lambda_I(t)\left[\lambda_{inf}(\eta_{inf}-1)S(t)I(t) + \frac{1}{N}\int_0^t \dot{I}_{inf}(\tau)\dot{W}(\tau, t-\tau)d\tau - u(t)I(t)\right]$$

$$+ \Lambda_R(t)\left[u(t)I(t) + \frac{1}{N}\int_0^t \dot{R}_{inf}(\tau)\dot{Q}(\tau, t-\tau)d\tau + \phi(t)(\eta_{inf}-1)S(t)R(t)\right] \tag{4.29}$$

from which the costate equations are $\dot{\Lambda}_I(t) = -\partial \mathcal{H}/\partial I$ and $\dot{\Lambda}_R(t) = -\partial \mathcal{H}/\partial R$. With the switching function $\theta^*(t) = [\Lambda_I^*(t) - \Lambda_R^*(t)] I^*(t)$, the constrained optimal control function $u^*(t)$ that minimizes J is the saturation function

$$
u^*(t) = \begin{cases} 0, & \theta^*(t) \leq 0, \\ \theta^*(t), & \theta^*(t) \in (0,1), \\ 1, & \theta^*(t) \geq 1 \end{cases} \tag{4.30}
$$

Considering the time-dependent control capability, the optimal control signal distribution time T_D^* can be obtained by solving the dynamic programming in Equation 4.28. Similarly, the saturation function in Equation 4.30 only provides an attainable lower bound on control of malware propagation with time-dependent control capability.

4.5.1 Early-stage analysis

With the approximation that $S(t) \approx 1$ at early stages and the initial condition $W(z,0) = 1$, from Equation 4.10, we have the approximation of incremental spatial infection

$$
W(z,s) = \left(\frac{\sigma \lambda_{pro} \eta_{pro}}{2} s + 1 \right)^2 \tag{4.31}
$$

Moreover, we also have the approximation that $I(t) \approx I_{inf}(t)$, since at early stages $I_{inf}(t) \propto I(t)$, while $I_{pro}(t) \propto \sqrt{I(t)}$. That is, the malware propagates at a faster speed through infrastructure-based links than through proximity-based links [16, 51]. At some early stage t',

$$
S(t') = 1 - I_0 - I_0 u(t) \left[t' + \frac{\phi(t)(\eta_{inf} - 1)}{2} t'^2 \right],
$$

and we have the first-order ODE

$$
\dot{I}(t) = [\lambda_{inf}(\eta_{inf} - 1) S(t') - u(t)] I(t) + \frac{1}{N} \int_0^t I(\tau) \dot{W}(\tau, t - \tau) d\tau \tag{4.32}
$$

Using the subgradient of $u(t)$ at $t = T_D$ to define the subderivative $\dot{u}(T_D) = 0$, and differentiating Equation 4.32 with respect to t at both sides, we have the second-order ODE (neglecting the second-order term of $W(z,s)$)

$$
\ddot{I}(t) = [\lambda_{inf}(\eta_{inf} - 1) S(t') + \sigma \lambda_{pro} \eta_{pro} N^{-1} - u(t)] \dot{I}(t) \triangleq [K_1 - K_2 \phi(t) - u(t)] \dot{I}(t) \tag{4.33}
$$

where $K_1 = \lambda_{inf}(\eta_{inf} - 1)[1 - I_0 - I_0 u(t) t'] + \sigma \lambda_{pro} \eta_{pro} N^{-1}$ and

$$
K_2 = I_0 u(t) \frac{\eta_{inf} - 1}{2} t'^2.
$$

With the initial values $I(0) = I_0$ and $\dot{I}(0) = \lambda_{inf}(\eta_{inf} - 1)(1 - I_0)I_0 \triangleq K_3$, we obtain

$$I(t) = \frac{K_3}{K_1 - K_2\phi(t) - u(t)} \exp\{[K_1 - K_2\phi(t) - u(t)]t\} + I_0 - \frac{K_3}{K_1 - K_2\phi(t) - u(t)}$$

$$= \begin{cases} \frac{K_3}{K_1} \exp\{K_1 t\} + I_0 - \frac{K_3}{K_1}, & t < T_D, \\ \frac{K_3}{K_1 - K_2\kappa - f(T_D)} \exp\{[K_1 - K_2\kappa - f(T_D)]t\} + I_0 - \frac{K_3}{K_1 - K_2\kappa - f(T_D)}, & t \geq T_D \end{cases}$$

(4.34)

The performance measure J in Equation 4.25 can be evaluated as

$$J = \int_0^{T_D} [NI(t)]^\beta \, dt + \int_{T_D}^{T_f} [NI(t)]^\beta + \frac{1}{2} f^2(T_D) \, dt$$

$$= \frac{\left(\frac{NK_3}{K_1}\right)^\beta}{K_1 \beta} (\exp\{K_1 \beta T_D\} - 1) + \left(I_0 - \frac{K_3}{K_1}\right) T_D$$

$$+ \frac{\left(\frac{NK_3}{K_1 - K_2\kappa - f(T_D)}\right)^\beta}{[K_1 - K_2\kappa - f(T_D)]\beta} \times \left(\exp\{[K_1 - K_2\kappa - f(T_D)]\beta T_f\} \right.$$

$$\left. - \exp\{[K_1 - K_2\kappa - f(T_D)]\beta T_D\} \right)$$

$$+ \left(I_0 - \frac{K_3}{K_1 - K_2\kappa - f(T_D)} + \frac{1}{2} f^2(T_D) \right) (T_f - T_D) \qquad (4.35)$$

For early-stage analysis, the optimal control signal distribution time can be obtained by $\widehat{T_D^*} = \arg\min_{T_D} J$.

4.5.2 Performance evaluation

The parameter setup of the simulation is the same as in the previous section. When dynamic programming is applied to determine the optimal distribution time, severe epidemics (large β) contribute to early distribution to minimize the accumulated cost, as shown in Figure 4.14. Moreover, both optimal control and early-stage analysis suggest early distribution as the effectiveness of signal (α) increases, as shown in Figure 4.15. The relative difference of these two approaches is plotted in Figure 4.16. Compared with early-stage analysis, optimal control via dynamic programming prefers early distribution when α is small, while it prefers late distribution as α increases.

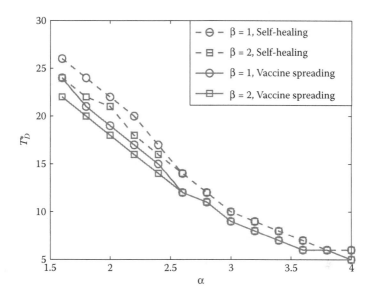

Figure 4.14: Optimal control signal distribution time via dynamic programming under different (α, β) configurations in IoT networks. $N = 2000$, $L = 50$, $I_0 = 1/N$, $\delta = 1.1$, $\lambda_{\text{inf}} = \lambda_{\text{pro}} = 0.05$, $\eta_{\text{inf}} = 6$, $\eta_{\text{pro}} = 3$, $\kappa = 0.1$, $T_f = 200$, $M = 1000$, $t' = 1$, and $c = 10^{-3}$.

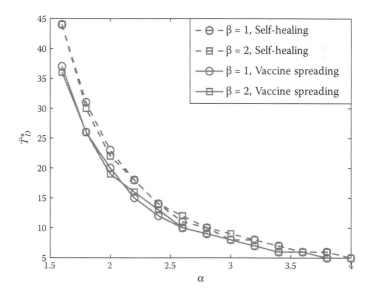

Figure 4.15: Optimal control signal distribution time via early-stage analysis under different (α, β) configurations in IoT networks. $N = 2000$, $L = 50$, $I_0 = 1/N$, $\delta = 1.1$, $\lambda_{\text{inf}} = \lambda_{\text{pro}} = 0.05$, $\eta_{\text{inf}} = 6$, $\eta_{\text{pro}} = 3$, $\kappa = 0.1$, $T_f = 200$, $t' = 1$, and $c = 10^{-3}$.

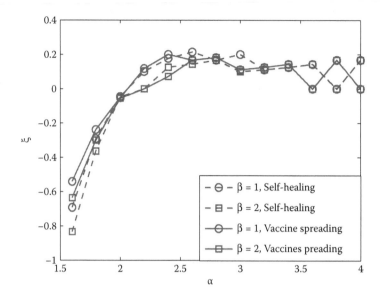

Figure 4.16: Relative difference of optimal control signal distribution time under different (α, β) configurations in IoT networks. $N = 2000$, $L = 50$, $I_0 = 1/N$, $\delta = 1.1$, $\lambda_{inf} = \lambda_{pro} = 0.05$, $\eta_{inf} = 6$, $\eta_{pro} = 3$, $\kappa = 0.1$, $T_f = 200$, $M = 1000$, $t' = 1$, and $c = 10^{-3}$.

4.5.3 Summary

The contributions of this section are twofold. First, with the aid of epidemic modeling, we provide an analytically tractable parametric plug-in model for malware propagation control regarding the time-dependent control capability, with the aim of determining the optimal control signal distribution time to minimize the accumulated network cost in real time via dynamic programming. Second, we demonstrate how to use our developed tools to control malware propagation in IoT networks. Compared with the self-healing scheme, we show that vaccine spreading further mitigates the accumulated cost when the immune nodes participate in forwarding control signal. Consequently, this section provides novel mathematical tools for malware propagation with and without control over IoT networks.

4.6 Conclusion

This chapter introduces malware propagation and control models from different aspects of IoT architecture that involves heterogeneous communication capabilities. We investigate malware propagation from the microscopic view of an individual device as well as the macroscopic view of the entire system. Optimal control approaches are proposed to alleviate malware propagation and enhance system reliability.

Bibliography

[1] President Barack Obama. Improving critical infrastructure cybersecurity. Executive Order, Office of the Press Secretary. 12 February 2013.

[2] President Barack Obama. Presidential policy directive 21: Critical infrastructure security and resilience. Washington, DC, 2013.

[3] Roy Malcolm Anderson and Roy Malcolm May. Directly transmitted infections diseases: Control by vaccination. *Science*, 215(4536):1053–1060, May 1982.

[4] Dimitri P. Bertsekas. *Dynamic Programming and Optimal Control (2 Vol Set)*. Athena Scientific, 3rd edition, 2007.

[5] Abhijit Bose and Kang G. Shin. On capturing malware dynamics in mobile power-law networks. in *Proceedings of the 4th International Conference on Security and Privacy in Communications Networks (SecureComm '08)*, New York, number 12, September 2008.

[6] Claudio Castellano and Romualdo Pastor-Satorras. Thresholds for epidemic spreading in networks. *Phys. Rev. Lett.*, 105(21):218701, November 2010.

[7] Eric Cator and Piet Van Mieghem. Second-order mean-field susceptible-infected-susceptible epidemic threshold. *Phys. Rev. E*, 85(5):056111, May 2012.

[8] Eric Cator and Piet Van Mieghem. Susceptible-infected-susceptible epidemics on the complete graph and the star graph: Exact analysis. *Phys. Rev. E*, 87(1):012811, January 2013.

[9] Li-Chiou Chen and Kathleen M. Carley. The impact of countermeasure propagation on the prevalence of computer viruses. *IEEE Trans. Syst. Man., Cybern. B*, 34(2):823–833, April 2004.

[10] Pin-Yu Chen and Kwang-Cheng Chen. Information epidemics in complex networks with opportunistic links and dynamic topology. In *Proceedings of the Global Telecommunications Conference, GLOBECOM 2010*, Miami, FL, 6–10 December 2010, pp. 1–6.

[11] Pin-Yu Chen and Kwang-Cheng Chen. Optimal control of epidemic information dissemination in mobile ad hoc networks. In *Proceedings of the Global Telecommunications Conference, GLOBECOM 2011*, Houston, TX, 5–9 December 2011, pp. 1–5.

[12] Pin-Yu Chen, Shin-Ming Cheng, and Kwang-Cheng Chen. Smart attacks in smart grid communication networks. *IEEE Commun. Mag.*, 50(8):24–29, August 2012.

[13] Pin-Yu Chen, Shin-Ming Cheng, and Kwang-Cheng Chen. Optimal control of epidemic information dissemination over networks. *IEEE Trans. Cybern.*, 44(12):2316–2328, December 2014.

[14] Pin-Yu Chen, Han-Feng Lin, Ko-Hsuan Hsu, and Shin-Ming Cheng. Modeling dynamics of malware with incubation period from the view of individual. In *Proceedings of the Vehicular Technology Conference (VTC Spring)*, 2014 IEEE 79th, Seoul, 18–21 May 2014, pp. 1–5.

[15] Thomas M. Chen and Jean-Marc Robert. Worm epidemics in high-speed networks. *IEEE Computer*, 37(6):48–53, June 2004.

[16] Shin-Ming Cheng, Weng Chon Ao, Pin-Yu Chen, and Kwang-Cheng Chen. On modeling malware propagation in generalized social networks. *IEEE Commun. Lett.*, 15(1):25–27, January 2011.

[17] Shin-Ming Cheng, Vasileios Karyotis, Pin-Yu Chen, Kwang-Cheng Chen, and Symeon Papavassiliou. Diffusion models for information dissemination dynamics in wireless complex communication networks. *Journal of Complex Systems*, vol. 2013, pp.1–13.

[18] Daryl J. Daley and Joseph Gani. *Epidemic Modelling: An Introduction.* Cambridge University Press, 2001.

[19] Patrick T. Eugster, Rachid. Guerraoui, A.-M. Kermarrec, and L. Massoulie. Epidemic information dissemination in distributed systems. *IEEE Computer*, 37(5):60–67, May 2004.

[20] Michalis Faloutsos, Petros Faloutsos, and Christos Faloutsos. On power-law relationships of the Internet topology. In *Proc. ACM SIGCOMM 1999*, pages 251–262, October.

[21] Eric Filiol, Marko Helenius, and Stefano Zanero. Open problems in computer virology. *J. Comput. Virol.*, 1(3):55–66, February 2006.

[22] A. Ganesh, L. Massoulie, and D. Towsley. The effect of network topology on the spread of epidemics. In *Proceedings of IEEE Infocom 2005*, volume 2, 13–17 March 2005, pp. 1455–1466.

[23] Jorge Granjal, Edmundo Monteiro, and Jorge Sá Silva. Security for the Internet of Things: A survey of existing protocols and open research issues. *IEEE Commun. Surv. Tut.*, 17:1294–1312, January 2015.

[24] Christopher Griffin and Richard Brooks. A note on the spread of worms in scale-free networks. *IEEE Trans. Syst. Man. Cybern. B*, 36(1):198–202, February 2006.

[25] Chang-Rui Guo, ShaoHong Cai, HaiPing Zhou, and DaMin Zhang. Susceptible-infected-susceptible virus spread model in 2-dimension regular network under local area control. In *Proc. ICNDS 2009*, volume 1, Guiyang, Guizhou, 30–31 May 2009, pp. 97–100.

[26] Herbert W. Hethcote. The mathematics of infectious diseases. *SIAM Rev.*, 42:599–653, December 2000.

[27] Hao Hu, Steven Myers, Vittoria Colizza, and Alessandro Vespignani. WiFi networks and malware epidemiology. *Proc. Natl. Acad. Sci. USA*, 106(5):1318–1323, February 2009.

[28] Jennifer T. Jackson and Sadie Creese. Virus propagation in heterogeneous bluetooth networks with human behaviors. *IEEE TDSC*, 9(6):930–943, November 2012.

[29] Vasileios Karyotis. Markov random fields for malware propagation: The case of chain networks. *IEEE Commun. Lett.*, 14(9):875–877, September 2010.

[30] Jeffrey O. Kephart and Steve R. White. Directed-graph epidemiological models of computer viruses. In *Proc. IEEE Computer Society Symposium on Research in Security and Privacy*, Oakland, CA, 20–22 May 1991, pp. 343–359.

[31] William Ogilvy Kermack and Anderson Gray McKendrick. Contributions to the mathematical theory of epidemics. Part I. *Proc. R. Soc. A*, 115(5):700–721, August 1927.

[32] Mohammad Hossein Rezaei Khouzani, Eitan Altman, and Saswati Sarkar. Optimal quarantining of wireless malware through reception gain control. *IEEE Trans. Autom. Control*, 57(1):49–61, January 2012.

[33] Seong-Woo Kim, Jong-Ho Park, Eun-Dong Lee, Mid-Eum Choi, and In Proc. IEEE VTC 2010, Taipei, 16–19 May 2010, pp. 1–5.

[34] Donald E. Kirk. *Optimal Control Theory: An Introduction*. Dover Publications, Mineola, NY, 2004.

[35] Cong Li, Ruud van de Bovenkamp, and Piet Van Mieghem. Susceptible-infected-susceptible model: A comparison of N-intertwined and heterogeneous mean-field approximations. *Phys. Rev. E*, 86(2), September 2012.

[36] Yong Li, Pan Hui, Depeng Jin, Li Su, and Lieguang Zeng. Optimal distributed malware defense in mobile networks with heterogeneous devices. *IEEE Trans. Mobile Comput.*, 13(2):377–391, February 2014.

[37] Yao Liu, Peng Ning, and Michael K. Reiter. False data injection attacks against state estimation in electric power grids. In *Proc. ACM Conf. Comput. Commun. Security*, pages 21–32, November 2009.

[38] Alun L. Lloyd and Robert M. May. How viruses spread among computers and people. *Science*, 292(5520):1316–1317, May 2001.

[39] Piet Van Mieghem. The N-intertwined SIS epidemic network model. *Computing*, 93(2–4):147–169, October 2011.

[40] Sancheng Peng, Shui Yu, and Aimin Yang. Smartphone malware and its propagation modeling: A survey. *IEEE Commun. Surv. Tut.*, 16(2): 952–941, April 2014.

[41] Lev Semyonovich Pontryagin, Vladimir Grigorevich Boltyanskii, Revaz Valerianovich Gamkrelidze, and E. Mishchenko. *The Mathematical Theory of Optimal Processes (International Series of Monographs in Pure and Applied Mathematics)*. Interscience, New York, 1962.

[42] Injong Rhee, Minsu Shin, Seongik Hong, Kyunghan Lee, Seong Joon Kim, and Song Chong. On the Levy-walk nature of human mobility. *IEEE/ACM Trans. Netw.*, 19(3):630–643, June 2011.

[43] Faryad Darabi Sahneh and Caterina Scoglio. Epidemic spread in human networks. In *Proc. IEEE CDC-ECC 2011*, Orlando, FL, 12–15 December 2011, pp. 3008–3013.

[44] Chayan Sarkar, Akshay Uttama Nambi S. N., R. Venkatesha Prasad, Abdur Rahim, Ricardo Neisse, and Gianmarco Baldini. DIAT: A scalable distributed architecture for IoT. *IEEE Internet Things J.*, 2:230–239, June 2015.

[45] Sarah H. Sellke, Ness B. Shroff, and Saurabh Bagchi. Modeling and automated containment of worms. *IEEE TDSC*, 5(2):71–86, April-June 2008.

[46] Daniel Smilkov and Ljupco Kocarev. Influence of the network topology on epidemic spreading. *Phys. Rev. E*, 85:016114, January 2012.

[47] Stuart Staniford, Vern Paxson, and Nicholas Weaver. How to own the Internet in your spare time. In *Proc. USENIX Security 2002*, San Francisco, August 5–9, 2002, pp. 149–167.

[48] Christian Szongott, Benjamin Henne, and Matthew Smith. Evaluating the threat of epidemic mobile malware. In *Proc. IEEE WiMob 2012*, Barcelona, 8–10 October 2012, pp. 443–450.

[49] Sapon Tanachaiwiwat and Ahmed Helmy. Encounter-based worms: Analysis and defense. *Ad Hoc Netw.*, 7(7):1414–1430, September 2009.

[50] Richard Thommes and Mark Coates. Epidemiological modelling of peer-to-peer viruses and pollution. In *Proc. IEEE Infocom 2006*, Barcelona, Spain, April 2006, pp. 1–12.

[51] Pu Wang, Marta C. Gonzalez, Cesar A. Hidalgo, and Albert-Laszlo Barabasi. Understanding the spreading patterns of mobile phone viruses. *Science*, 324(5930):1071–1075, May 2009.

[52] Mina Youssef and Caterina Scoglio. An individual-based approach to SIR epidemics in contact networks. *J. Theor. Biol.*, 283(1):136–144, August 2011.

[53] Shui Yu, Guofei Gu, Ahmed Barnawi, Song Guo, and Ivan Stojmenovic. Malware propagation in large-scale networks. *IEEE TDSC*, 27:170–179, January 2015.

[54] Wei Yu, Xun Wang, P. Calyam, Dong Xuan, and Wei Zhao. Modeling and detection of camouflaging worm. *IEEE TDSC*, 8(3):377–390, May–June 2011.

[55] Xiaolan Zhang, Giovanni Negli, Jim Kurose, and Don Towsley. Performance modeling of epidemic routing. *Comput. Netw.*, 51(8):2867–2891, July 2007.

[56] Cliff C. Zou, Weibo Gong, Don Towsley, and Lixin Gao. The monitoring and early detection of Internet worm. *IEEE/ACM Trans. Netw.*, 13(5): 961–974, October 2005.

[57] Cliff C. Zou, Don Towsley, and Weibo Gong. On the performance of Internet worm scanning strategies. *Perform. Eval.*, 63:700–723, July 2006.

[58] Cliff C. Zou, Don Towsley, and Weibo Gong. Modeling and simulation study of the propagation and defense of Internet e-mail worms. *IEEE TDSC*, 4(2):105–118, April–June 2007.

[59] Gjergji Zyba, Geoffrey M. Voelker, Michael Liljenstam, A. Mehes, and Per Johansson. Defending mobile phones from proximity malware. In *Proc. IEEE Infocom 2009*, Rio de Janeiro, 19–25 April 2009, pp 1503–1511.

Chapter 5

A Solution-Based Analysis of Attack Vectors on Smart Home Systems

Andreas Brauchli

Depeng Li

CONTENTS

Abstract

The development and wider adoption of smart home technology has also created an increased requirement for safe and secure smart home environments with guaranteed privacy constraints. We first present a short survey of privacy and security in the more broad smart world context. The main contribution is then to analyze and rank attack vectors or entry points into a smart home system and propose solutions to remedy or diminish the risk of compromised security or privacy. Further, we evaluate the usability impacts resulting from the proposed solutions. The smart home system used for the analysis in this chapter is a digital-STROM installation, a home-automation solution that is quickly gaining popularity in central Europe. The findings, however, aim to be as solution independent as possible.

5.1   Introduction

As welfare increases and technological gadgets become ubiquitous, we lighten our daily lives by automating trivial and common tasks. The last few years have shown a clear trend of automation technology usage within both personal homes and commercial buildings. The increasing adoption of smart home systems (SHS) leads to the need for not only more functionality but also for a safe, secure, and functional environment. The ongoing battle for smart grid security [1] includes smart homes [2] and, especially when one technology becomes particularly widespread, it automatically creates a high-reward target type. One specific area that is seeing a particular technological increase, and is thus at higher risk of becoming such a target, is home automation for personal use. Several companies offer products on the market to automate lighting, shades, heating, cooling, and the like. Among the many systems that feature different wired or wireless topologies is digitalSTROM (dS) with its powerline-based bus and embedded central server. This research is dedicated to finding security and privacy weaknesses in SHS using the example of a dS system. Wherever possible, we try to approach the problem in a generic way that can also be applied to other systems.

This works is organized as follows: We begin with the introduction and proceed with a review of how smart homes fit into the broader smart world context and present related work. In the fourth section, the dS environment is covered before listing possible attack vectors on SHS in the fifth section along with two example attacks on the dS infrastructure. In Section 5.6, solutions to prevent or diminish those attack vectors are proposed and discussed. In Section 5.7, we analyze the proposed solutions, which are followed by the conclusions.

5.1.1 Smart world

With our world growing "smarter" than ever, there are different ways of integrating smart homes into the broader context of smart services, smart grids, and even smart cities. Researchers of different fields have been studying this ongoing trend and have come up with interesting and useful applications. We briefly present some of those applications to emphasize the security and privacy needs of a modern SHS—especially in the light of a majority of consumers being agnostic of technology and not necessarily trained in computer and network security. They may thus not be fully aware of what privacy invasions must be expected when certain sensory data are leaked or revealed from their smart environment.

Ref. [4] defines smart communities as interconnected sets of colocated homes that share certain common processing infrastructure. The authors give an example of a distributed intrusion detection/aversion scheme based on surveillance data from multiple homes that is processed centrally in the community and an example of smart healthcare where neighbors are alerted when a critical health situation is detected. The paper also projects a call center responsible for multiple smart communities for emergencies or further assistance. While the authors propose the centralized processing of data for privacy reasons, it is likely that not every smart community will want to maintain a data center on its premises. The next paper [6], though unrelated, predicts a trend toward more artificial intelligence and thus processing power in future smart homes. The paper foresees that with the increasing number of sensors and readings, a single smart home might not be able to process all the data and thus it processes them in a cloud environment. Privacy is listed as a potential issue. In a similar light, [5] proposes a framework to integrate smart homes into platform as a service clouds. Data privacy is supposedly managed by the user but the decision on which data to use and process seems to take place post transmission in the cloud. The cloud interface provides additional services or virtual smart home devices provided by third parties. Further improvements in the broad context of lifestyle are presented in [7], where ambient intelligence is mentioned. The paper predicts how smart devices will carry an individual's preferences who will then experience personalized results in places such as museums and other public places.

Overall, the trend is clearly geared toward a highly interconnected smart world where data are processed in a distributed fashion and the line between private data sharing, such as highly sensitive and individual medical records, and beneficial services is at risk of becoming increasingly blurred. In fact, both might not even remain separable due to design, marketing, or infrastructural decisions.

The problem is further amplified when individuals may no longer have a choice as to what part of the collected bulk sensor data is shared or even transmitted over a potentially insecure network. It is thus crucial to set the bar high for both security and privacy. Even when individuals do explicitly consent to data sharing, the actual transmission protocol must always be open and reviewable for potential leaks.[1] Accountability is key to gain the user's trust and, once obtained, can only be beneficial to the product's success. Since smart home installations have a comparatively long lifetime of several years or decades and process sensitive sensory information, interested parties will likely take their time to evaluate and research their options. Open-source approaches are, in general, very favorable toward trustworthiness and, possibly, also toward the longevity of a product when the modifications and extensions can be installed by the owner/user without requiring specific tools or requiring digital signatures. Unfortunately, the shift toward more open protocols is slow and customers might not always see the benefits of open solutions. A change is only expected to happen when demanded by a majority of customers or with comparably successful open solutions.

5.2 Related Work

This section lists related security research in the smart home context and explains the differences to this work. We conclude that there has been—to the best of our knowledge—no previous security assessment of this kind on smart home environments with a wired powerline bus type and, particularly, not for the dS architecture. The journal article in [8] surveys the available SHS technology but only briefly lists potential attack vectors on the SHS control infrastructure (DDoS). It also details personal security, that is, not software system–related security, automation logic proposals such as notifying emergency services when a fire is detected, unusual user behavior detection using neural networks, and a privacy guard to protect against sensitive information leakage. The paper in [9] covers the *detect and prevent* approach to several security issues in wireless sensor networks in the SHS context. Several attack vectors that compromise confidentiality, integrity, and availability are shared in this paper. In contrast, we analyze security issues on the example of dS products, which uses a wired bus system with non-factory default and optional wireless connectivity. Ref. [10] proposes a meter-reporting system based on public key encryption that does not reveal specific power usage to the utility company. The system is based on signed readings by a trusted reader. The processing then directly applies the matching price tariff to those readings, resulting in a fully verifiable bill without specific usage information. This paper creates a good solution to verifiably aggregate metering data but requires a trusted meter by the utility company, which the dS

---

[1]Ideally, the protocols are independently audited and published with unredacted raw data to the general public.

environment does not target or provide. Ref. [11] proposes a framework for evaluating security risks associated with technologies used at home. The paper also associates high-level attacker goals such as extortion or blackmail to low-level attacks compromising the infrastructure. We focus solely on low-level security issues and leave out inferring the potential consequences. Finally, [3] presents a deep literature review of smart homes and provides a prediction of future development going toward integrated healthcare systems. Due to the amount of time that people spend in their homes, there is a large economic potential for integrated services. Additionally, the paper includes a section of papers dedicated to security. dS does not appear in any of the papers; however, some wired systems such as KNX are listed.

5.3   The digitalSTROM Environment

The dS environment is SHS designed primarily for personal home use. It can also be simultaneously used in multiple apartments of a building, whereas each apartment has its own installation. The installation consists of one (optional[2]) digitalSTROM server (dSS), usually one digitalSTROM meter (dSM) and one digitalSTROM filter (dSF) per circuit, and numerous terminal blocks (small clamps) with a digitalSTROM chip (dSC) for each device. The dSF is responsible for filtering out dS messages on the power bus and prevent them from reaching the outside world. This is technically required when multiple dS installations are present nearby to prevent cross talk. Each dSM can handle up to 128 clamps and communicates with the other dSM and the dSS using the ds485 two-wire protocol.[3] The ds485 bus can span up to 100 m but is usually confined within the cabinet (dashed line in Figure 5.1). dSC are conventionally integrated in a terminal block ("clamp") that, in turn, is connected directly to a power switch or an appliance. The dSC can also be integrated directly into an appliance, into a power socket, or onto a socket list by a licensed manufacturer. The appliances communicate over the power wire using a proprietary closed protocol (dash-dotted line in Figure 5.1). The bandwidth available to dS devices is very limited with 100 bauds (dSM → dSC) / 400 bauds (dSC → dSM) [12]. The reaction time for events is between 250 and 750 ms. Figure 5.1 shows a simplified SHS consisting of three separate power circuits (one per floor), two dS appliances (TV, light on the dash-dotted line), and a non-dS charging electric vehicle on an outdoor plug. The dSM are interconnected (dashed lines) with the dSS by the two-wire bus. The dSS is connected to the home network, symbolized by the wireless router, by a Cat.5 cable or, optionally, by a supported wireless universal serial bus (USB) dongle.

---

[2]Although the basic configuration can be made without a dSS, more complex events such as timer-based ones are only possible with a dSS installed.
[3]The name ds485 is an analogy for the serial RS485 bus protocol.

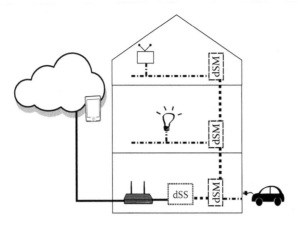

Figure 5.1: A sample digitalSTROM SHS.

A control device (typically a smartphone or tablet) is connected to the home network with the wireless network. The dSS provides a web interface for configuration and an AJAX/JSON application programmable interface (API) for control.

5.4 Attack Vectors on SHS

We grouped the possible SHS attack vectors into five vulnerability categories, which are detailed in this section: wired SHS commonly use (1) a server for state management and to provide a control interface or API, (2) a bus for communication with the appliances, and (3) a small clamp or control device for switching individual appliances. This system is ultimately controlled by the user with (4) a control device such as a smartphone. Additionally, (5) remote third-party services may be contracted to extend the system's core functionality. The categories and their communicative interaction are visualized in Figure 5.2.

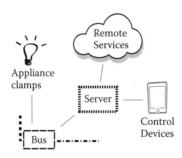

Figure 5.2: The five risk categories.

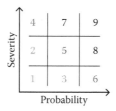

Figure 5.3: The nine risk categories.

We divided the attacks into nine relative and perceived risk categories: low, medium, and high in each of the two dimensions, severity and probability, shown in Figure 5.3. The risk is based on how likely and how severe a given attack is. We note that more probable attacks are assigned higher risk ratings than more severe ones.

5.4.1 Central digitalSTROM server

This first subsection elaborates on the possibilities of gaining access to the central dS server as a means to compromise the entire SHS. The central server has total access privileges to the SHS: it can switch appliances, read out metering values, manage API connections on the home network, and run virtually permanently. The server is thus the most crucial component to secure within the SHS. Due to the many interfaces, it is also the most exposed part. This server role is assumed by the dSS component and is located in the cabinet. In dS systems, the location is dictated by the proximity to the dSM circuit meters. The dSS is an embedded Linux platform with 400 Mhz ARM9 CPU, 64 Mb ram, 1 Gb flash memory, two USB ports, and an RJ45 100 Mbit Ethernet port. It features an onboard RS-232 serial port for recovery purposes [13]. The first possibility to attack the dSS is by gaining physical access and compromising the root system password. This can be done using the debug ports to gain access to the serial console and thus the (uBoot) boot loader. Earlier versions of the dSS featured only 256 Mb flash storage but used an SD card as the main storage drive, which added the possibility of maliciously switching SD cards to one with added or modified credentials. Due to the high impact but local constraint (physical access required), this attack is rated at risk level 4. The second possibility is to gain access to the local wired or, if available, wireless network and (1) exploit a system vulnerability (e.g., TCP/IP vulnerability in the Linux IP stack or network driver both local area network (LAN) or wireless local area network (WLAN) if a WLAN dongle is plugged into the dSS); or (2) exploit a service vulnerability of a service running with system privileges, for example, an SSH server (Dropbear), if enabled. We note at this point that the dSS process handling dS events does not run with elevated privileges. Alternatively, an attacker can (3) exploit an API vulnerability within the dSS process. This attack is generally locally bound to the home network and wireless range but weak router/firewall rules may directly expose the dSS to the Internet and thus pose a major potential flaw. Since home automation systems are long-term systems with expected run times of 10–15 years, the software is highly likely to become outdated and unmaintained during its life cycle, thus greatly increasing the risk. Due to the high potential severity, we assign these two vectors the risk rating 7. Third, an attacker may target the server via the dS485 bus interconnecting the dSM by (1) directly gaining wire access or (2) indirectly by a rogue dSC that injects events that trigger a given message by the dSM on this bus. This attack is judged as having medium impact due to the ability to control the whole static SHS, that is, the functionality of the SHS available when no dSS is installed, with low probability. Besides the impact on the powerline bus, it is questionable whether such an attack would be able to compromise the dSS integrity and would have to be determined by a code analysis of the dS485 bus handler process. We thus assign this attack vector the risk level 2. The fourth attack possibility is to redirect or abuse the app store to (1) inject rogue updates with open backdoors, which is possible because updates are not digitally veri- fied; or (2) rogue apps may be installed either by mistake or by misguiding the user into installing them. As dS apps do not have system privileges because they are run from within the dSS process and are restricted to a JavaScript sandbox, the main threat is to privacy, as all events can be triggered and registered. Both rogue updates and apps can be installed when the attacker has control over the local network and can intercept and modify the home network traffic from and to the dSS as the updates are served through an unencrypted hypertext transfer protocol (HTTP) connection. Without local access, it is very hard to manipu- late network traffic; however, due to the high impact of a compromised update, this attack vector is assigned the risk level 4. When considering rogue apps, we increase the risk to level 5 due to the higher probability of such an attack but lower severity: tricking a user into installing a rogue app is possible but greatly depends on the victim.

5.4.2 Smart control devices

This subsection describes how a compromised smart control device (SCD) such as a smartphone or a control station leads to a compromised SHS.

Besides the wall switches in rooms, control of the SHS is generally dele- gated to trusted or authenticated control devices or both such as smartphones or control terminals. In the dS case, the JSON-API is only accessible by a secure HTTPS connection and requires a token that is obtained after success- ful authentication. If, however, a control device such as an Android or iPhone smartphone is compromised, the control of the whole system, as far as API sup- port reaches, is consequently compromised until the token is revoked or expires, in case the device does not store the actual credentials. dS does not currently feature specific (usually wall-mounted) control terminals, thus this scenario is omitted. dS published both an iOS and an Android app. Since smartphones are mostly connected to the Internet, they are exposed to many third-party apps and, possibly, viruses and worms. Additionally, the device usually has full access to the home network. These facts lead to a high-risk attack vector with risk category nine.

5.4.3 Smart home communication bus

In this subsection, we analyze the risks of a compromised communication bus, the implications of which directly lead to a largely compromised SHS. dS uses a proprietary but unencrypted protocol for its communication on the power wiring (powerline) [12]. As the protocol uses neither encryption nor authentication, any received messages are assumed to be valid. This opens the possibility for (1) injecting control signals to directly control appliances or disrupt the system or (2) injecting invalid power readings to falsify the reporting system on power consumption. When falsifying consumption readings, this only falsifies the reading of individual single devices as the dSM is aware of the total sub-circuit consumption independently of any attached dSC. Having access to the communication bus allows easy jamming of the SHS, thereby creating a denial of service (DoS)-type attack. The low bus bandwidth makes this attack particularly effective. The attacker has the choice of jamming only the subcircuit with the attached rogue sender device or the whole system by continuously sending system-wide events, such as alarms, which are then broadcast by the dSM into the adjacent subcircuits. As all dS appliances have access to the powerline bus and thus have full control of the bus within their subcircuit, an attacker may attach a rogue appliance anywhere in the system. If the attacker does not have physical access, he or she may still trick someone who does have access into plugging in an appliance for him or her, for instance, by gifting or lending such a prepared appliance. dS appliances can be anything from a lamp to a TV or a computer. As dSC clamps are relatively small and only draw minimal power, they are easily hidden inside an appliance case. An alternative and limited attack consists of connecting unmodified original dS clamps to the system, which automatically registers and adds the device, an automatic plug-n-play (PnP) procedure that takes less than 10 min. Once registered, the device is ready for use, for example, a clamp with a yellow color code[4] switches all room lights in the room that it is plugged in. A generic panic button will trigger the panic procedure, which defaults to turning on all lights and opening all shades and blinds in the entire installation. With the locally limited exposure of the powerline bus, generally secure premises (except for outdoor plugs) but with high control level, this attack

---

[4]dS clamps are color coded according to their functionality.

vector is rated a risk category 4 (private home without outdoor plugs) or 7 (with an easily reachable outdoor plug or when the SHS is a semipublic environment such as an office space). An alternative point of entry is the ds485 bus interconnecting the dSM and dSS. The implications are the same as compromising the powerline bus with an additional small but unverified possibility of exploiting the dSS's process by buffer overflow. This attack does not seem very likely or attractive as dSM are usually located next to the more rewarding dSS. We thus assign the risk category 4.

5.4.4 Remote third-party services

This subsection analyzes the trust implications of connecting third-party services with the SHS. A third-party service provides additional functionality to the SHS. Those services can be classified into two categories: (1) monitoring services and (2) control delegation services. A service can also be classified in both categories simultaneously. The monitoring services accept consumption statistics, system events, or other collected data and provide a suggested or analytical service based on the data interpretation. As such, this type of service imposes solely a privacy risk as identifying events such as home presence and activities may be leaked [18]. We rank this attack vector at risk level 3 but the actual danger could greatly vary depending on the nature of the leaked information and the danger that such a leak could go unnoticed for a very long time. The second category of services requires control permissions and thus API access by a token, which may be revoked individually [18]. Such services may, for instance, provide an alternative Internet-based user interface. As a consequence, a compromised third-party service directly implicates a compromised SHS and carries an elevated risk rated at 7 or 9, depending on how secure and trustworthy the third-party service is. dS offers such a service called *mein.digitalSTROM* [19] using a dS app that allows installation via remote control. It also allows temporary control delegation with a time-expiring link and backs up local configuration and metering data. Based on these facts, it is inevitable that all third-party services be trusted with private data and system control, respectively.

5.4.5 Two attack scenarios

In this subsection, we elaborate on two theoretical attack scenarios based on our previous analysis.

The first attack uses the dS Android smartphone app [14] as the entry vector and switches lights on at night when the homeowners are sleeping. The second attack uploads power readings to a remote server, allowing the attacker to know when the home is empty or is likely to be empty. The first attack is created by installing a rogue app on the homeowner's Android smartphone. This app poses as a totally unrelated app to the SHS. Once the app is installed on the SHS owner's smartphone, it launches a background service that sends an Android intent [16], a cross-app message using the dS app's public interface, to the dS app sometime during the night. The unmodified and unknowing dS app then performs the action using the stored credentials. The malicious app does not need to know any connection details or the API token. While the attack may sound banal, more frightening scenarios can be envisaged. In the second attack, the dS app is installed by the user on the dSS using the official dS app store. Once installed, the app collects consumption data from all connected dSM and periodically uploads them to a remote location. The attacker uses this collected data to establish when the residence is likely to be empty. We do note that third-party apps will likely have to pass a code review before being entered into the dS app store. There are enough legitimate uses for sending private data and the app should thus pass a code inspection based on different expectations by the reviewer and the app's user, especially if the documentation is ambiguous, suggestive, or simply missing.

5.5   SHS Hardening

This section is modeled from Chapter 4. It is organized into the central dS server, SCDs, smart home communication bus, and third-party services. In an effort to harden SHS against the attacks described in the previous section, we recommend adopting proven strategies from other domains. In addition to providing security-enhancing suggestions, we reflect on the usability impact of the proposed solutions.

5.5.1   *Central digitalSTROM server*

This subsection reiterates the crucial role of the central dS server in the overall system security. Because of its central role and exposure to different interfaces in the SHS, a physical server breach is rated at both the highest severity and highest probability. To protect against physical server breaches, the easiest and, at the same time, most effective method is, arguably, to lock the cabinet if it is located in a (semi)public space. This should be recommended to every customer through the installation documentation. This solution has a low usability impact and leaves the choice and risk assessment to the customer. Within private spaces, the risk of a physically compromised dSS is rated low. If additional security is desired, one could make use of a tamper-evident case, which may avert certain attackers. This change requires a customer to be aware of how to check the integrity seal, which could possibly be done remotely, but still requires a lock-secured cabinet. A tamper-evident case incurs a high usability impact due to the need for additional training. To protect against network-based attacks on the dSS, it is important to make the user change the default access password, preferably during the initial setup. A default access password together with an open network results in a very high probability and high severity risk. Usability is only minimally impacted by requesting the user to set a password on setup. The initial setup could be streamlined by a setup wizard, which would cover this step. To prevent man-in-the-middle (MitM) attacks, such as modifying the system or app updates, dS update servers should default to an encrypted HTTPS connection with a valid SSL certificate. Such a secure connection is transparent to the user and thus does not incur any usability changes. To reduce the risk of a totally compromised SHS, the introduction of a permission-based access control system for the API is suggested. Possible permissions include reading out meter values, controllable dSMs/rooms such that an application may be restricted to controlling appliances in one subcircuit or even individual appliances, the events that can be triggered, and the events that one can register with. This list is not exhaustive and further permissions may be applicable. There is a certain trade-off between usability and permission configurability as analyzed by [15]; however, the impact could be lessened by allowing full permissions by default and leaving the specific constraining to knowledgeable users within the "advanced settings" menu option.

5.5.2 Smart control devices

SCD have full control over the SHS. Thus, it is crucial to educate all users that a compromised SCD implies a compromised SHS. The dS app for Android provides other apps on the smartphone with the possibility to send intents (Android control messages) that the app will then react on. Thus, any app on such a smartphone can control the SHS. We propose adding a white list of registered apps, managed by the user, to the Android dS app to verify that a certain app is allowed to control the SHS. The list would be updated on the fly upon first request as to impact usability only minimally. Users may also feel more secure when they know which apps can, or are trying to, control their SHS.

5.5.3 Smart home communication bus

dS uses a proprietary protocol for communication between a dSC and a dSM. The technology does not permit inter-dSC communication without going through a dSM first due to separate up- and downstream channels. If one were to reverse engineer the communication protocol and implement a device speaking the protocol—or reverse engineering a dSC's interface/firmware—an attacker could easily inject messages or jam the circuit and installation and create a DoS attack. We thus strongly recommend investigating adding an encryption layer such as targeting low power and very low overhead settings [17]. An encryption layer may incur a moderate overhead in usability if keys have to be set up by the user. We further suggest adding an option to disable the PnP functionality for automatically registering new devices, especially in semiprivate environments such as offices where power plugs are readily available to anyone having physical access. For ease of use, we do not suggest disabling the PnP by default, but when the auto-registration function has been disabled, we suggest adding a timer-based enable function——analogous to how Bluetooth pairing works—that allows auto-registering appliances that are plugged in during a short time frame. The usability impact of such a feature is minimal, resulting in only one more option that could be placed within the advance configuration mode.

5.5.4   Remote third-party services

Remote services provide additional functionality to the SHS by either providing remote access to the dSS or by analyzing and reporting on collected data. To harden the system against privacy leaks, we suggest implementing configurable time-resolution limit permissions to the already proposed permission system. Such a resolution limit would, for example, not allow access to resolutions below a 15 min aggregation in order to maximize privacy. As such a restriction is optional, the usability impact remains small while giving the user a much greater sense of privacy. To harden against compromised third-party services, a restricted set of permissions should be applied to remote-controlled API accesses; additionally, all API accesses and transactions should be logged for a future audit. As the user is responsible for checking the logs, he or she does incur a great usability impairment unless combined with a method of automatically checking logs for irregularities. A third-party app should only be accepted into the dS app store when sufficient, clear, and unambiguous documentation is available as to what data are being processed and sent off remotely and what control events are raised by the app. The code reviewers are responsible for checking the code paths against the documentation and asking for corrections before accepting it. Before installing an app, a user should have the possibility to accept or reject the requested functionality. There is a minimal usability overhead to display the app documentation, which has to be manually accepted or rejected by the user.

5.6   Solution Analysis

We now look back on the sample attacks in the light of the suggested improvements and find that the attacks would no longer be possible. We do note that all proposed solutions are theoretical improvements based on research and experience in related fields. The physical experimentation of the suggested solutions in this exact context is left as a future work item.

The first attack scenario uses the dS Android app to stealthily inject control events into the SHS. With a white list of apps that are allowed to send control events through the dS Android app, any app on the smartphone would have to request permission before being granted access, thus thwarting a stealthy attack. A visual clue should make it apparent that the said app, which has nothing to do with the SHS, pursues a malicious purpose when it seeks to access the SHS via the exposed Android intent.

The second app that sends consumption events to a remote server would have to declare its intent to send readings to a remote service in the documentation and request those specific permissions during the installation. If this is against the purpose of the app, the user should recognize the threat and choose not to install the app.

After implementing our proposed solutions, both sample attacks would thus no longer be possible.

5.7 Conclusion

We conclude this chapter by reiterating that homes are very intimate places where people expect and deserve a high level of privacy and security; this level is currently not being satisfactorily offered by the feature-driven industry. We have elaborated different attack vectors on a dS SHS, which range from physical breaches to networked attacks all the way to third-party remote issues. We have demonstrated the actual abuse of two of those attack vectors and suggested various improvements to all of the identified attack vectors along with possible usability impairments resulting from the solutions. We hope that this research will lead to an increase in openness and security awareness from the early development process on in generic SHS products and particularly to an improved dS system.

Bibliography

[1] European Union Agency for Network and Information Security ENISA, Smart Grid Security Recommendations, 2012.

[2] National Institute of Standards and Technology NIST, NISTIR 7628 guidelines for smart grid cyber security, 2010.

[3] Alam, M.R., Reaz, M.B.I., Ali, M.A.M., A review of smart homes— Past, present, and future, *Systems, Man, and Cybernetics, Part C: Applications and Reviews, IEEE Transactions on*, vol. 42, no. 6, pp. 1190, 1203, 2012.

[4] Li, X., Lu, R.X., Liang, X.H., Shen, X.M., Chen, J.M., Lin, X.D., Smart community: An Internet of Things application, *IEEE Communications Magazine* vol. 49, pp. 68, 75, 2011.

[5] Eom, B., Lee, C., Yoon, C., Lee, H., Ryu, W., A platform as a service for smart home, *International Journal of Future Computer and Communication* vol. 2, no. 3, pp. 253, 257, 2013.

[6] Cook, D., How smart is your home? *Science* vol. 335, no. 6076, pp. 1579, 1581, March 2012.

[7] O'Grady, M., O'Hare, G., How smart is your city? *Science* vol. 335, no. 6076, pp. 1581, 1582, 2012.

[8] Robles, R., Kim, T., A review on security in smart home development, *International Journal of Advanced Science and Technology* vol. 15, pp. 13–22, 2010.

[9] Islam, K., Sheng, W., Wang, X., Security and privacy considerations for wireless sensor networks in smart home environments. In *Computer Supported Cooperative Work in Design (CSCWD), 2012 IEEE 16th International Conference on*, pp. 622–633. Wuhan, China, 2012.

[10] Rial, A., Danezis, G., Privacy-preserving smart metering. In *Proceedings of the 10th Annual ACM Workshop on Privacy in the Electronic Society (ACM WPES11)*, pp. 49–60. ACM, Chicago, IL, 2011.

[11] Denning, T., Kohno, H.M., Leving, T., Computer security and the modern home. *Communications of the ACM*, vol. 56, no. 1, pp. 94–103, 2013.

[12] Aizo AG (12/2013). digitalSTROM FAQ [Online] digitalSTROM, Schlieren, Switzerland. Available: http://www.digitalstrom.com/documents/A0818D044V005_FAQ.pdf

[13] Aizo AG (12/2013). dSS 11 Produktinformation [Online]. Available: http://www.digitalstrom.com/documents/digitalSTROMServerdSS11 ProduktinformationV1.0.pdf.

[14] Google Playstore, Aizo AG (12/2013). dS Home Control [Online]. Available: https://play.google.com/store/apps/details?id=com.aizo.digitalstrom.control.

[15] Kim, T. H.-J., Bauer, L., Newsome, J., Perrig, A., Walker J., Challenges in access right assignment for secure home networks. In *Proceeding Hot-Sec'10 Proceedings of the 5th USENIX Conference on Hot Topics in Security*, Washington, DC, 2010.

[16] Google Ltd. (12/2013). Android API Reference [Online]. Available: http://developer.android.com/reference/android/content/Intent.html

[17] Luk, M., Mezzour, G., Perrig, A., Gligor, V., MiniSec: A secure sensor network communication architecture, In *Proceedings of the 6th International Conference on Information Processing in Sensor Networks*, pp. 479–488. Cambridge, MA 2007.

[18] Rouf, I., Mustafa, H., Xu, M., Xu, W., Miller, R., Gruteser, M., Neighborhood watch: Security and privacy analysis of automatic meter reading systems. *Proceedings of the 2012 ACM Conference on Computer and Communications Security* (ACM CCS12), pp. 462–473. ACM, 2012.

[19] Aizo AG (11/2013). digitalSTROM Installation Manual [Online]. Available: http://www.aizo.com/de/support/documents/html/digitalSTROM_Installationshandbuch_A1121D002V010_EN_2013-11-12/index.html#page/digitalSTROM%2520Installationshandbuch/digitalSTROM%2520Installationshandbuch_A1121D002V010_EN_12-11-2013_Final.1.56.html

PRIVACY
PRESERVATION

Chapter 6

Privacy Preservation Data Dissemination

Miao Xu

Wenyuan Xu

Jason M. O'Kane

CONTENTS

6.1 Introduction

With the advances of wireless communication technologies, wireless sensor networks (WSNs) have been widely deployed to monitor the surroundings. As those WSNs scale in size, the large volume of sensed data and the required energy of collecting them have led to data-centric sensor networks (DCSNs) [9, 21]. In DCSNs, sensed data are stored among a few dedicated storage nodes in the network, and a mobile sink will visit the network occasionally to collect the stored data. Unlike its previous counterpart, the sink-based sensor network, where one sink is used to collect and store sensed data, a DCSN is efficient and robust, since it does not require every sensor node to deliver data to the sink, which may be far away and may also become a single point of failure.

Once deployed, possibly in a remote environment, DCSNs are typically left unattended, with occasional human visits, and can create vast quantities of information. The characteristic of little physical protection combined with their low-cost nature makes DCSNs vulnerable to a wide variety of network dynamics and attacks, including node capture, node compromise, node failure, packet injections, jamming attacks, and so on. As a result, an adversary may breach data privacy by acquiring sensitive data stored in the network through compromising nodes, or may affect data availability by removing data permanently via disabling network nodes. For instance, in a DCSN that is deployed in a forest for monitoring and tracking endangered animals, obtaining the stored data will reveal location information about targets, which may create life-threatening risks.

To overcome these problems, many cryptography-based methods [3, 18, 21] have been designed to ensure data integrity, confidentiality, and access control for sensor networks. Although those cryptography-based strategies are essential in protecting WSNs against various attacks, they can only partially address the threats against data privacy and data availability. For instance, they cannot cope with information leakage caused by node compromise or communication disturbances caused by jamming attacks. Additionally, most cryptography-based strategies rely on robust key management schemes, which will impose extra storage costs and complicate network deployment as well as its operations. Therefore, in this research, we are interested in whether we can mitigate threats against data privacy and data availability by *non-cryptography-based* methods that only exploit the sensor location diversity exhibited in the typical wireless sensor network.

Addressing data privacy issues together with data availability is problematic. To increase data availability against node failure, it is natural to replicate data to many nodes. However, this replication introduces the risk of data privacy leakage due to node compromise. The requirement of energy efficiency further complicates the solution. To strike a balance among these three goals, in this research we construct a graph called the spatial privacy graph (SPG) to guide data dissemination and ensure that the scheme can achieve a higher level of data privacy and data availability at lower energy cost compared with other data dissemination schemes.

6.2   Problem Overview

Since cryptography-based strategies cannot address all the threats against data privacy and data availability, we study the *noncryptography* schemes that can achieve the goal. We first overview the problem by examining the network model and threat model. We summarize the notations that are used in Table 6.1.

6.2.1   Network model

This work focuses on a data-centric sensor network that is deployed for tracking targets. Specifically, the sensing application first utilizes trusted data collectors to collect messages generated by every sensor, and then derives the location information of the target from the messages. The network consists of sensor nodes, storage nodes, and mobile sinks, as shown in Figure 6.1.

Table 6.1 Frequently used notations

Notation	Explanation	Notation	Explanation
S_n	The set of sensor nodes	n_n	The total number of sensor nodes
S_s	The set of storage nodes	n_s	The total number of storage nodes
x_i	A sensor node, where $i \in \{1,\ldots,n_n\}$	y_i	A storage node, where $i \in \{1,\ldots,n_s\}$
r_s	The sensing radius of sensor nodes	r_c	The communication radius of sensor nodes
$\eta_i(t)$	The I-state of storage node i at time t	$\eta^*(t)$	The master I-state, $\eta^*(t) = \bigcap_{i \in S_s} \eta_i(t)$
p	Duplication probability	$V(\eta(t))$	The area of I-state $\eta(t)$
P	I-state based privacy measure	A	I-state based availability measure
E	Energy cost		

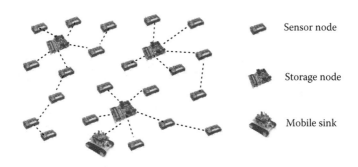

Figure 6.1: An illustration of a data-centric sensor network (DCSN).

6.2.1.1 Sensor nodes

A network of n_n static sensor nodes S_n are deployed through a planar environment W at positions $x_1, x_2, \ldots, x_{n_n}$ and $S_n = \{x_i\}_{i \in [1\ldots n_n]}$. Each sensor node continually senses its surroundings, and sends an event message to storage nodes whenever it senses an event of interest. Sensor nodes are identical, with the same sensing range r_s and the same communication range r_c. The sensor nodes do not store data because of their lack of sufficient memory to store data for months or years and the prohibitive number of nodes from which a mobile sink needs to offload data. Instead, they always forward data to storage nodes.

Additionally, the network consists of low-cost sensors capable of *coarse sensing*. That is, each sensor is equipped with a long-range proximity sensor that can detect the target whenever $\|q(t) - x_{n_i}\| \leq r_s$, where $q(t)$ is the position of a target at time t. This sensing is Boolean, in the sense that the node knows only whether or not the target has been detected, but no other information. Thus, the reported measurement will be a circle with radius r_s. Moreover, the value of r_s is sufficiently large that the capture one message does not breach the privacy requirements.

Finally, each sensor node is aware of the relative location of its neighbors. Such information can be obtained by wireless localization algorithms [19].

6.2.1.2 Storage nodes

A collection of n_s storage nodes S_s are deployed across the environment W at position $y_1, y_2, \ldots, y_{n_s}$, where $n_s \ll n_n$, and $S_s = \{y_i\}_{i \in [1 \ldots n_s]}$. Storage nodes have larger memory and larger battery capacity. They are in charge of storing data before mobile sinks offload the data. To prevent malicious users from overflowing the storage nodes by injecting faulty packets, each storage node will perform data filtering to sterilize the data. Thus, no matter whether the data are encrypted or not during message deliveries, storage nodes are required to access the plaintext of each packet.

6.2.1.3 Mobile sinks

From time to time, one or more mobile sinks will visit the network, and they will get close to each storage node to offload data. Because of their relatively small number, mobile sinks are equipped with tamperproof hardware, or guarded by humans. Thus, mobile sinks cannot be compromised by any adversary or followed by a jammer that may interfere with their communication. In summary, mobile sinks are reliable and trustworthy.

6.2.2 Threat model

Both unintentional and malicious threats that breach data privacy and harm data availability are considered here. In particular, we make the following assumptions about the damage that adversaries or network dynamics can cause:

Nodes can be compromised. Since both sensor nodes and storage nodes are left in the field unattended and are prone to compromise, we assume both of them to be untrustworthy. However, an adversary can only compromise up to g storage nodes, sensor nodes, or any combination of them. As a starting point, we assume that $g = 1$ and adversaries are only interested in capturing storage nodes due to the higher payoff of compromising a storage node than a sensor node. When a node is compromised, adversaries can obtain all stored data including secret keys and sensed data. Moreover, we assume that adversaries do *not* have a global view of the network and are unaware of all the locations of sensor nodes as well as storage nodes.

Nodes can fail or be jammed. We assume both that sensor nodes and storage nodes can fail during the lifetime of the network. They can experience hardware problems, causing permanent data loss, or their communication channel can suffer from severe radio interference, resulting in an inability to receive or send data. In either case, the data that are stored or scheduled to be stored on the affected storage nodes will not be available to mobile sinks.

In summary, data can be leaked to adversaries or can be unavailable to mobile sinks due to various reasons, breaching data privacy and harming data availability.

6.3 Problem Formulation

6.3.1 Privacy scope

Data privacy of a network includes *content* privacy and *context* privacy [10]. This study focuses on content privacy breaches that are caused by node compromise, node failure, or even DoS attacks. We refer readers to other research [4, 10] that deals with preserving context privacy; for example, where the communication has occurred and who has participated in the communication. We note that those two problems are complementary: our *content*-aware data dissemination problem focuses on *which* storage node to deliver while *context*-aware routing problems deal with *how* to deliver data.

6.3.2 Motivation for privacy and availability definition

Preserving privacy is normally considered as the guarantee that data is observable only by those who are supposed to access it. However, such a definition does not capture the fact that privacy is closely linked to its resolution of uncertainty. Taking location privacy, for example, we generally do not want to reveal where we are. Here, the definition of *where we are* determines the boundary of the tolerance level of privacy, and it can be quite different in various cases. As an example, Alice might be willing to reveal her location information if the granularity of location is at the level of city, while she is unwilling to reveal her current street address. Similarly, a granularity of no less than 250 m may be acceptable for protecting endangered animals, but not less than 25 m. Thus, the definition of privacy should quantify the level of information *uncertainty*. Similarly, the goal of data availability is not necessarily to guarantee that all data records are accessible, but to ensure the available data set produces enough information about the target with acceptable levels resolution, that is, uncertainty.

Before quantifying information uncertainty, it is important to clarify the relationship between information and messages in sensor networks. Since the

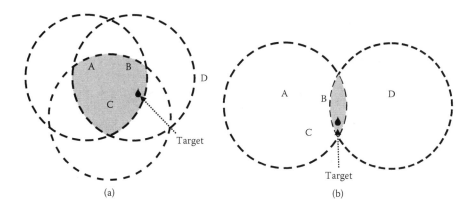

Figure 6.2: Illustration that a combination of two potential related nodes provides more valuable information than three nodes which possess similar information.

message generated by each node only provides a portion of the global location information that the sensing application has, one naive method to quantify information uncertainty is to count the number of messages. For instance, breaching data privacy can be quantified by the number of messages obtained by adversaries, and data availability can be defined as the number of available messages.

However, with regard to privacy and availability, the *content* of messages is more important than the quantity of messages. Figure 6.2 provides a simple illustration of the idea in the context of target-tracking applications, where the content refers to the location of the target. In the figure, nodes A, B, C, and D detect the target using their proximity sensors, and each generates a message reporting the possible region of the target as a circle centered at itself. The location information of the target provided by a set of messages is the intersection of corresponding disks. Combining three messages from nodes A, B, and C results in an intersection region much larger than the intersection of nodes A and D's sensing ranges. Thus, leaking three messages does not necessarily map to a worse privacy breach than leaking two messages, and the definition of data privacy and data availability should be *content-aware* rather than just counting the messages.

6.3.3 Uncertainty and information states

6.3.3.1 Modeling the uncertainty

We employ the concept of *information states* (I-states) [6, 17] to capture the tolerance level of uncertainty on both privacy and availability associated with a set of messages. I-states are used in robotics to reason about uncertainty and explicitly encode the uncertainty about the target. More precisely, the term *state* refers to an instantaneous description of this target at a given time. In target tracking, I-states are the set of possible states that are consistent with the measurements provided by sensors, for example, the possible locations of the target that can produce such measurements; I-states are calculated according to the content of messages. The main advantage of using the concept of I-states is that no prior knowledge of the target but the message contents is required. In comparison, *entropy* has been used to define privacy [5, 20], but it is only applicable to limited scopes because its calculation requires prior knowledge of the probability distribution for the targets' movements.

Formally, in a network that tracks the motion of a target through a planar environment W using proximity sensors, let us suppose that prior to some time t_f sensor nodes have measured m samples that map to m messages,

$$\{(O_1,t_1),\ldots,(O_m,t_m)\} \tag{6.1}$$

in which O_i is a circle known to contain the true state, and t_i is a timestamp at which this information was known to be valid. Then, a target position $\hat{q}$ is *consistent* with those messages if and only if there exists a continuous trajectory $q : [0,t_f] \to W$ such that

1. $dq/dt \leq v_{max}$ for all $t \in [0,t_f]$, where v_{max} is the target's maximum speed.

2. $q(t_i) \in O_i$ for all $i \in [1,m]$.

3. $q(t_f) = \hat{q}$.

The I-state $\eta(t)$ at time t is the set of target positions consistent with the messages with timestamps prior to time t. $V(\eta(t))$ denotes the area of the I-state $\eta(t)$, which quantifies the level of uncertainty. A larger value for $V(\eta(t))$ means that the target can be anywhere inside a larger area, corresponding to a higher level of uncertainty.

Consider the example illustrated in Figure 6.2a, and let us assume at time $t = 0$ nodes A, B, and C generate three messages. The I-state $\eta(0)$ associated with all three messages is the points inside the intersection of those three disks centered at nodes A, B, and C, respectively; and $V(\eta(t))$ is the area of that intersecting region, denoted by the shaded region in Figure 6.2a.

6.3.3.2 Computing the information state

Figure 6.3 illustrates the calculation of the I-state. It starts with an initial state $\eta(0) = W$, and is updated after time passes or new messages are received:

■ When the time passes from t_1 to t_2 without any messages being received, $\eta(t_2)$ is computed from $\eta(t_1)$ by performing a Minkowski sum of $\eta(t_1)$ with a ball of radius $(t_2 - t_1)v_{max}$. Informally, this "expands" the I-state to

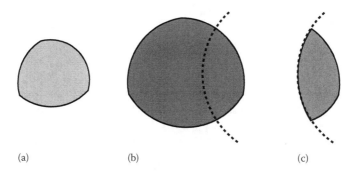

(a)   (b)   (c)

Figure 6.3: Computing the I-state: (a) an initial information state; (b) expansion to account for the passage of time, and intersection with received message disks; (c) the resulting updated I-state.

reflect the fact that the state may have changed since the previous message was received. The resulting region is retained as $\eta(t_2)$.

■ When a message (O, t) is received, the existing I-state is updated to the correct $\eta(t)$ by intersecting the current I-state with O. This takes the information provided by the message into account.

6.3.3.3 Information states in the network

For a network with n_s storage nodes, each storage node y_j will calculate its I-state $\eta_j(t)$ based on its received messages. Additionally, there exists a "master" I-state $\eta^*(t)$ derived from all the messages received across all storage nodes, and $\eta^*(t) = \eta_1(t) \cap \cdots \cap \eta_{n_s}(t)$. Thus, there exist $n_s + 1$ I-states in the network in total.

In a normal scenario, without any attacks or hardware failures, the mobile sink is able to collect all data stored at each storage node and to obtain $\eta^*(t)$, while in practice, some storage nodes may fail and prevent the mobile sink from obtaining $\eta^*(t)$, reducing the amount of information available to the mobile sink. Moreover, it is possible that an adversary compromises one storage node y_j and acquires its I-state $\eta_j(t)$, breaching network privacy.

6.3.4 Evaluation criteria

We target to design an energy-efficient data dissemination scheme that can enhance both privacy and availability. Thus, we define three evaluation metrics.

6.3.4.1 Privacy

Consider the case that the adversary is able to compromise one storage node i. We define the levels of this privacy breach as the size ratio between $\eta_i(t)$,[1] which the adversary can access, and $\eta^*(t)$, which is the knowledge of the entire network. This ratio is a measure of the quantity of information that is protected in spite of the compromise. Of course, compromising different storage nodes may lead to a different level of payoff. In light of the fact that security is typically determined by the weakest point in the system, we define privacy by considering the worst case across all possible compromised storage node:

$$P = 1 - \frac{V(\eta^*(t))}{\min_{i \in S_s} V(\eta_i(t))} \tag{6.2}$$

for the privacy level at time t. The interpretation of this metric is that when $P = 0$, a single storage node has access to the full knowledge of the network, and privacy cannot be preserved against that storage node being compromised. Similarly, $P = 1$ would indicate "perfect" privacy, but this clearly cannot be achieved, since it would require the network to retain information that is not stored at any of its storage nodes.

6.3.4.2 Availability

Similarly to the definition of privacy, to define network availability, we consider the area of the I-state available to the entire network, in comparison to the area that is stored at each individual storage node. If a storage node fails, then the knowledge that can be reconstructed from the remaining $n_s - 1$ storage nodes is simply the intersection of their I-states. As a result, we can define availability by considering the worst case across all possible storage node failures:

$$A = \frac{V(\eta^*(t))}{\max_{i \in S_s} V\left(\bigcap_{j \in S_s - \{i\}} \eta_j(t)\right)} \tag{6.3}$$

To interpret this metric, we observe that if all of the messages are sent only to a single storage node, then we obtain $A = 0$, the worst availability, since the network then has a single point of failure. In contrast, if each message is sent to at least two distinct storage nodes, then $A = 1$, the "perfect" availability, because no single failure can result in data loss. Realistic, energy-efficient protocols fall somewhere between these two extremes.

6.3.4.3 Energy

Because the energy available to each wireless sensor node is generally limited by battery capacity, one important objective is to minimize the amount of energy

---

[1] Since adversaries do not possess the global information of the network, we do not consider the privacy breaches caused by the absence of sensed data at storage nodes; for example, node A did not detect a target.

consumed by delivering messages per unit of time. Let $E(i)$ denote the number of messages forwarded or generated by the sensor node i between $t = 0$ and $t = T$. The system seeks to keep E as small as possible:

$$E = \frac{1}{T} \sum_{i=1}^{n_n} E(i) \tag{6.4}$$

We note that this energy representation is sufficient to model energy used both at the sending and at the receiving end, since we can scale E up by multiplying by a coefficient α. This coefficient may include the energy consumed both as the sender transmits the message and as its neighbors overhear and process the message.

6.3.5   Problem definition

The goal of the proposed data dissemination scheme is to let sensor nodes determine to which storage node they should deliver their observations, so that the overall privacy P and availability A are both good while the energy consumption E is small. As such, the data dissemination protocol can be modeled as a color assignment function. Each storage node is labeled with a unique color ID, for instance, the same as the storage node ID; and assigned colors to each sensor to indicate to which storage nodes to deliver its data. We define the color assignment C as a function mapping each sensor node x_i to one or multiple storage nodes in S_s, that is,

$$C : S_n \rightarrow 2^{S_s}$$

where 2^{S_s} is the power set of S_s. The problem of preserving privacy and availability is equivalent to finding a color assignment function C that maximizes the privacy and availability of the network at minimum energy cost.

Solving this nonlinear multiobjective optimization problem is challenging, since these three evaluation criteria, P, A, and E are at least partially in conflict with one another: Intuition suggests—and our experiments confirm—that increasing A generally reduces P and increases E. To tackle the problem, we first analyze a few baseline data dissemination technologies to gain insights.

6.3.6   Baseline data dissemination

Essentially, the data dissemination protocols are designed with inspiration from secret-splitting algorithms [22]. Each sensor is capable of observing a coarse measurement of the target, similar to the concept of small pieces of the secret. Storage nodes combine multiple messages, analogous to gaining larger portions

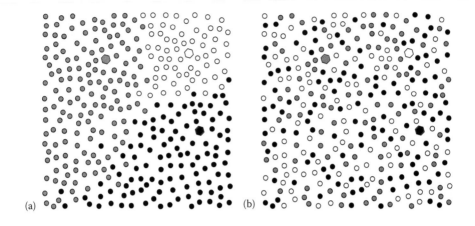

Figure 6.4: Illustration of (a) shortest path coloring and (b) random coloring.

of the secret. Finally, the trusted data collector can obtain $\eta^*(t)$ by combining all messages and can pinpoint the location of the target, corresponding to obtaining the secret.

Intuitively, the data dissemination protocol should guide the messages to be distributed across several storage nodes, and thus split the secret evenly among storage nodes. To illustrate this intuition, we analyze two baseline data dissemination protocols:

6.3.6.1 Shortest path

The shortest path coloring algorithm represents general data dissemination schemes [14] that aim at reducing energy consumption without considering data privacy or data availability. It involves a sensor node choosing the closest storage node to store its data. Figure 6.4a depicts an example of such a coloring scheme with three storage nodes, in which each sensor node transmits to the closest storage node, measured by hop counts in the network, that is, $C(x_i) = \arg\min_{y_j \in S_s} \mathrm{h}(x_i, y_j)$, where $\mathrm{h}()$ returns the hop count between x_i and y_j. Although such a shortest-hop-count-based coloring scheme consumes the smallest amount of energy, it will not provide good privacy and availability. For instance, if we imagine that a target is moving in the white region (upper-right corner), the I-state stored at the white storage node $\eta_w(t)$ equals $\eta^*(t)$. If the white storage node happens to be compromised, the adversary can obtain the same location information about the target as the trusted data collector. Moreover, if the white storage node is unavailable due to hardware failure, then no target movement information will be available.

Table 6.2 **Comparison of the shortest path coloring and the random coloring schemes in a network of three storage nodes**

Scheme	Shortest Path	Random Color
P	0.30	0.49
A	0.02	0.28
E	36	61

6.3.6.2 *Random coloring*

A naive technique to improve the data distribution across the network is to randomly assign each sensor node a color, corresponding to a storage node. That is, the function C is randomly selected, and only one color is assigned to each sensor node. Figure 6.4b gives an example of random coloring under the same network deployment as Figure 6.4a.

To evaluate the performance of the shortest path and the random coloring schemes, we simulated a network with 325 identical sensor nodes spread across a 2000 m × 2000 m network field. A single target moved through the field and each sensor node detected the target whenever it was within the sensor's 250 m range. The results, which are listed in Table 6.2, confirm that the shortest path scheme achieves a low availability A and privacy P but consumes a small amount of energy E. In comparison, the random coloring scheme consumes almost twice the amount of energy as the shortest path, but achieves a higher level of data privacy and data availability.

6.4 SPG-based Data Dissemination

6.4.1 *Spatial privacy graph*

The random coloring scheme improves privacy and availability by simply distributing equal numbers of messages to each storage node. However, equal distribution of messages is not sufficient. Take Figure 6.2, for example; the combination of A's and D's information states $\eta_A(t) \cap \eta_D(t)$ is more "valuable" compared to $\eta_A(t) \cap \eta_B(t) \cap \eta_C(t)$. Thus, nodes A and D must transmit their observations to different storage nodes to improve privacy and availability. In contrast, it is relatively harmless for the three nodes A, B, and C to transmit to the same storage node, because the sensors for these nodes will provide very similar information. This observation motivates us to construct an SPG that identifies those pairs of sensor nodes that, in combination, can determine the position of the target within a small region.

Formally, a set of sensor nodes S forms an SPG $G_P = (S, E_P)$ where a pair of nodes (x_i, x_j) are connected by an edge e_{ij} if and only if they form a *privacy*

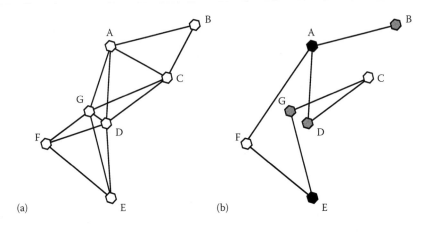

Figure 6.5: Construction of spatial privacy graph: (a) communication topology; (b) spatial privacy graph.

pair. Given a scalar parameter *privacy factor a*, a pair of nodes is a privacy pair, if their distance $d \in [2r_s - a, 2r_s]$. Intuition tells us that these privacy pairs are nodes whose sensing regions have small, nonzero intersections. Figure 6.5 illustrates this process. Figure 6.5a presents a simple network scenario with seven nodes, where the edges represent communication links. Figure 6.5b depicts the resulting spatial privacy graph, where the edges link privacy pairs. Although nodes G and D are within each other's communication range, they are too close to have an overlapping sensing range that is small enough to be considered as a privacy pair. Thus, G and D are not connected in the spatial privacy graph. Assuming that $2r_s > r_c$, then the distance between node pair (A, F) is larger than their communication range r_c but smaller than $2r_s$. As a result, nodes A and F are connected in the spatial privacy graph.

6.4.2 Enhancing privacy via a distributed coloring algorithm

The SPG identifies the privacy pairs that should select different storage nodes to save their data. Thus, to enhance data privacy, each sensor node can determine its storage node by executing a distributed graph coloring scheme. Given an n-vertex SPG with $G_P = (S, E_P)$, the output of the distributed coloring scheme is a colored graph $G_c = (S, E_P, C)$. Without loss of generality, we assign one color to each sensor node, and denote the color assignments C as $C = \{\mathbf{c}_{x_i} | \mathbf{c}_{x_i} = C(x_i)\}_{\forall x_i \in S}$. Ideally, G_c should satisfy two requirements: *valid* and *feasible*. Here, valid means that for every edge $e_{ij} \in E_P$, its vertices x_i and x_j have different colors, that is, $\mathbf{c}_{x_i} \neq \mathbf{c}_{x_j}$, and feasible means that the color of every vertex should be one of the storage nodes' colors. A valid and feasible coloring can guide the network to disseminate messages that belong to the same privacy pairs to different storage nodes and thus achieve high privacy. However, for any SPG and given number of storage nodes, it is not always possible to obtain a valid yet feasible colored graph. For instance, if there are only two storage nodes available to color the SPG shown in Figure 6.5b, then it is impossible to obtain a valid coloring among nodes A, C, and D. To address this issue, our distributed coloring algorithm will first generate a valid coloring and then change infeasible colors into feasible ones.

6.4.2.1 Algorithm walk-through

The distributed coloring algorithm is motivated by Linial's coloring scheme [13], which starts with a valid colored graph with a large number of colors and then reduces the total number of colors iteratively. However, Linial's coloring scheme cannot be simply applied to this problem because it does not consider the factor of energy consumption, which is crucial to sensor networks.

The distributed algorithm works in the following way. Prior to coloring sensor nodes, we map each storage node to a unique color numbered from 1 to n_s. Then, each sensor node assigns its color purely based on its neighbors' colors by executing `Distributed_Coloring`, which is shown in Algorithm 6.1, in parallel. Here, we call a pair of nodes neighbors if they are connected in the SPG, which is different from the concept of neighbors defined according to communication abilities. Each sensor x_i initializes its color to a unique infeasible one; for example, adding its own ID I_{x_i} to n_s. As such, we prevent any sensor node from preassigning itself a feasible color. Then, each sensor node participates in iterative coloring updating until no color is updated between two consecutive iterations.

At the beginning of each iteration, node x_j announces its current color with its ID I_{x_j} to all its neighbors by broadcasting a message $(I_{x_j}, \mathbf{c}_{x_j})$, where $\mathbf{c}_{x_j}$ is its current color. At the same time, it records its neighbors' current colors $\{\mathbf{c}_{x_i}\}_{x_i \in \mathbf{Nbr}}$. In each iteration, only a sensor node that satisfies the following conditions is allowed to update its color:

1. It has not been assigned a feasible color yet.

2. Its color is larger than those of all its neighbors.

Function `UpdateColor()` first tries to find a new color that satisfies all conditions listed below.

1. Feasible: The new color should be one of the storage nodes' colors, $\mathbf{c}'_{x_j} \in \{1, \ldots, n_s\}$.

2. Valid: None of its neighbors has chosen this color, $\mathbf{c}'_{x_j} \notin \{\mathbf{c}_{x_i}\}_{x_i \in \mathbf{Nbr}}$.

3. Nearest: Among all valid and feasible colors, it chooses the storage node that is separated by the fewest hop counts from itself.

Sometimes it is possible that no feasible and valid color is available, as shown in Figure 6.5b. In those cases UpdateColor() returns $-|\mathbf{c}_{x_i}|$. The algorithm terminates when none of the nodes can update its color further, and the following lemma holds.

Algorithm 6.1: Distributed_Coloring

Require: INPUT:
  Nbr: neighbor set
  I_o: local sensor ID
  PROCEDURES:
1: $\mathbf{c}_o = I_o + n_s$;
2: **repeat**
3:   Announce($I_o, \mathbf{c}_o$);
4:   $\{\mathbf{c}_{x_i}\}_{x_i \in \mathbf{Nbr}}$ = ReceiveAnnounce();
5:   **if** $\mathbf{c}_o > n_s$ **and** $\mathbf{c}_o > \max\{\mathbf{c}_{x_i}\}_{x_i \in \mathbf{Nbr}}$ **then**
6:     $\mathbf{c}_o$ = UpdateColor($\{\mathbf{c}_{x_i}\}_{x_i \in \mathbf{Nbr}}$);
7:   **end if**
8: **until** NoChange($\mathbf{c}_o$) **and** NoChange($\{\mathbf{c}_{x_i}\}_{x_i \in \mathbf{Nbr}}$)

Lemma 6.1
Algorithm 6.1 always terminates after $|S|$ iterations and terminates with a valid (but not necessarily feasible) colored graph $G_c = (S, E_P, C)$.

Proof. Termination: In each iteration, a node that can update its color must have a color that is larger than n_s. Meanwhile, a node can only update its color either to the number between 1 and n_s, or to its negative node ID. Thus, each node $x_i \in S$ will only update its color at most once. The algorithm terminates when none of the nodes can update its color, and the total number of iterations $I \leq |S|$.

Validity: We prove validity by induction on k. Let $G_c^{(0)} = (S, E_P, C^{(0)})$ be the colored graph after initialization, then for each node x_i, $\mathbf{c}_{x_i} = I_{x_i} + n_s$. Since all nodes have unique identifications, $\forall x_i, x_j \in S, \mathbf{c}_{x_i} \neq \mathbf{c}_{x_j}$, $G_c^{(0)}$ is valid.

Assume $G_c^{(k-1)}$ is valid. Let the graph after the kth iteration be $G_c^{(k)}$. Since, in each iteration, only the node that has the largest color in its neighborhood can update its color, we assume, without loss of generality, that node x_p updates its color from $\mathbf{c}_{x_p}^{(k-1)}$ to $\mathbf{c}_{x_p}^{(k)}$. According to Color Updating Condition 2, $\mathbf{c}_{x_p}^{(k)} \neq \mathbf{c}_{x_q}^{(k)}$ for all x_q that are its neighbors. Thus, $G_c^{(k)}$ is valid. ■

When Algorithm 6.1 produces a valid but infeasible graph—for example, some sensor nodes have a color that is out of the feasible range $[1,\ldots,n_s]$—the sensor nodes with infeasible colors will randomly choose a feasible color regardless of their neighbors' colors.

6.4.2.2 Algorithm challenges

Several practical challenges are associated with this distributed coloring algorithm.

Loose Synchronization: The correctness of the distributed coloring algorithm holds only if at most one node in its neighborhood updates its color in each iteration. Such a condition can be guaranteed only if every node decides whether it should update its color after all color announcements are delivered. Thus, it is important to let every node have a loosely synchronized clock and to let the color announcements reach its neighbors. For synchronization, one can use the timing-sync protocol for sensor networks (TPSN) [7], a lightweight synchronization protocol. To avoid severe flooding, the coloring announcement uses time to live (TTL) to control the flooding range. The neighbors are not communication neighbors with regard to the SPG. Thus, the coloring announcement has to be broadcast beyond a one-hop neighborhood. In cases where the communication range r_c equals the sensing range r_s, the privacy pair can be located up to $2r_s$ apart and, therefore, TTL $= 2r_s/r_c = 2$.

Reducing energy through on-demand, incremental coloring: Energy efficiency is one of the main concerns when designing algorithms for sensor networks. The SPG-based coloring algorithm is energy efficient, in the sense that each node always chooses a valid color of the storage node closest to it, and it converges in at most $|S|$ steps. Additionally, the algorithm adopts the following rules to further reduce energy consumption: (1) Construct the SPG on demand. In a tracking sensor network, a few nodes will detect the target; those nodes are called *hot nodes* (S_{hot}). Instead of constructing an SPG across the whole network, only hot nodes will participate in constructing the SPG by broadcasting control messages locally. (2) Incremental coloring: To incrementally update the SPG as the target moves continuously.

The incremental coloring algorithm works in the following manner. When the target moves to location L_1 initially, all hot nodes $S_{hot}(L_1)$ will color themselves using Algorithm 6.1. In the next time window, the target moves to another location L_2, and the $S_{hot}(L_2)$ will intersect with $S_{hot}(L_1)$. The nodes that belong to the intersection $S_{hot}(L_2) \cap S_{hot}(L_1)$ keep their color unchanged, and the nodes that are part of the set $S_{hot}(L_2) - S_{hot}(L_1)$ select their colors. As such, the colors of $S_{hot}(L_2) \cap S_{hot}(L_1)$ can be treated as prior knowledge, and only nodes in the set $S_{hot}(L_2) - S_{hot}(L_1)$ need to announce and update their colors iteratively. This incremental coloring is especially beneficial in reducing energy costs when the target moves at a low speed.

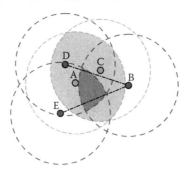

Figure 6.6: SPG and information redundancy. Nodes *B*, *D*, and *E* form privacy pairs, and their intersected sensing area is contained by the intersection of *A*'s and *C*'s sensing regions.

6.4.3 *Enhancing availability via message replication*

In a nonfailure scenario, the mobile sinks can derive $\eta^*(t)$ by acquiring data from every storage node. However, the data stored at storage nodes may be unavailable due to hardware failure or jamming attacks. The goal of maintaining high data availability is to ensure that the intersection of the information state of available storage nodes, $\bigcap_{i \in S_s} \eta_i(t)$, is close to $\eta^*(t)$. A natural way to improve high availability involves replication; for example, let a sensor node deliver a copy of the data to another storage node. However, naive duplication will increase energy costs. To replicate efficiently, the coloring algorithm must solve the following three issues: (1) Who should duplicate its messages, (2) how, and (3) where should the duplicated messages go?

Who? Only privacy pairs shall duplicate their messages. This heuristic can be illustrated by the example in Figure 6.6, which consists of two privacy pairs, (B,D) and (B,E), and isolated nodes A and C. The nodes that do not form privacy pairs with any hot nodes are usually located in between hot nodes. Their intersection (denoted by the light gray shading) is typically larger than the interaction of privacy pairs, and thus is less valuable to increasing availability. Letting privacy pairs duplicate messages allows us to spend energy on the most valuable messages.

How? Availability and privacy are conflicting objectives. Thus, the duplication probability p is used to keep a balance between two goals. Each node that is part of a privacy pair will replicate messages with probability p. In particular, in each data reporting period, a node generates a random number in the range $[0,1]$. Only if the random number is smaller than p will it send a replicated message to a second storage node. Setting $p = 0$ gives privacy higher priority, while assigning $p = 1$ favors availability.

Where? To avoid the situation that the duplicated messages from the same region are always delivered to the same storage node, the privacy pairs will randomly choose a second storage node to which to deliver their duplicated messages.

6.5 Experiment Validation

6.5.1 Simulation methodology

We implemented the SPG-based data dissemination algorithm using C++. We simulated a sensor network deployed in a 2000 m × 2000 m region with $r_s = r_c = 250$ m, and a target moved randomly throughout the network region at a speed of 25 m/s. We studied all three data dissemination strategies: shortest path, random coloring, and our SPG-based algorithm. For the SPG-based algorithm, we set the privacy factor a to 15m and measured the energy costs both for constructing the SPG and delivering data. To capture the statistical characteristics, we evaluated P, A, and E by running our experiments in ten rounds, where each round lasted for 1000 s with a 1 s sensing interval.

6.5.2 Experiment results

We performed two sets of experiments to study the impact of p and the number of storage nodes n_s, respectively.

6.5.2.1 Impact of p

We first compared the performance of the three algorithms in the scenario of 200 sensor nodes and three storage nodes when varying p from 0 to 1. The results are depicted in Figure 6.7, from which we observed that the availability of all three algorithms improves with an increasing value of p but at the cost of less privacy and higher energy costs. Compared with the other two algorithms, the energy costs of the SPG-based algorithm rise more slowly. Interestingly, when p is larger than 0.1, the energy costs of the SPG-based algorithm become smaller than those when using the shortest path scheme. This is because our SPG-based algorithm only allows privacy pairs to duplicate messages, instead of all hot nodes.

Figure 6.7b shows P and E for all three algorithms. Note that the point at $(0, 1)$ represents the (unachievable) ideal of perfect privacy with no energy costs. Figure 6.7b shows that the SPG-based algorithm accomplishes higher privacy than the shortest path scheme, which can only achieve a maximum privacy value of 0.2. Compared with the random coloring scheme, the SPG-based algorithm can achieve the same level of privacy with lower energy costs.

Finally, Figure 6.7(c) shows that the SPG-based algorithm is superior to both the shortest path and random coloring schemes with regard to A and E. That is, for the same energy costs, the SPG-based algorithm provides the highest availability.

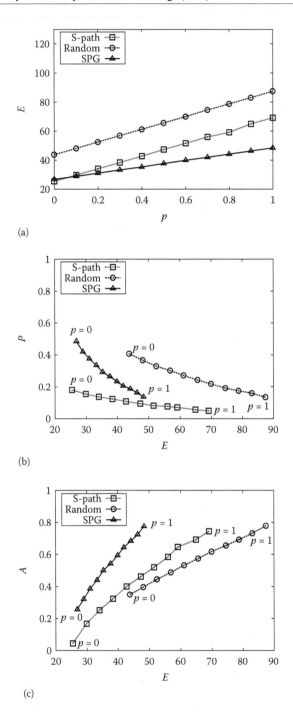

Figure 6.7: Comparison between shortest path coloring, random coloring, and the SPG-based algorithm with p changing from 0 to 1, n_n=200, $n_s = 3$: (a) p vs. energy; (b) privacy vs. energy ($0 \leq p \leq 1$); (c) availability vs. energy ($0 \leq p \leq 1$).

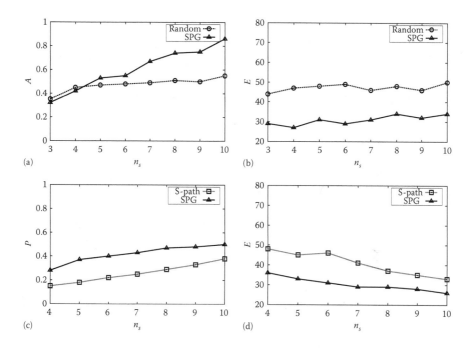

Figure 6.8: Influence of the number of storage nodes when $n_n = 200$; (a) and (b): given the requirements of $P \geq 0.4$ and $E \leq 50$, the maximum achievable A and the corresponding E; (c) and (d): given the requirements of $A \geq 0.6$ and $E \leq 50$, the maximum achievable P and the corresponding E.

6.5.2.2   Impact of n_s

Besides tuning p to balance between A and P, it is interesting to know what the maximum achievable value of A is, given the energy budget and the minimum required value of P. Figure 6.8a and b show such cases with requirements of $E \leq 50$ and $P \geq 0.4$. As n_s becomes larger than 4, the SPG-based algorithm outperforms the random coloring schemes, and uses a smaller amount of energy. Moreover, we observe that in Figure 6.8a, with an increase in the number of storage nodes, the value of availability in the SPG-based algorithm increases much faster than the value of availability in the random coloring algorithm. This confirms our analysis: Distributing messages evenly is insufficient, and the content of messages is more important than the number of messages in terms of data uncertainty. We note that the shortest path algorithm cannot achieve the requirements and does not show up in the diagrams. Similarly, as shown in Figure 6.8c and d, given the requirements of $A \geq 0.6$ and $E \leq 50$, the SPG-based algorithm achieves higher maximum privacy than the shortest path scheme and uses less energy. We note the random coloring scheme cannot find any feasible solution to meet the requirements and does not appear in the diagrams.

In summary, our SPG-based data dissemination protocol combines the advantages of two baseline dissemination schemes and can achieve better data privacy and a higher level of data availability while consuming less energy.

6.6 Related Work

Much attention has been devoted to addressing privacy issues in the context of data mining and databases [1, 12, 16]. A common technique is to perturb the data and to reconstruct distributions at an aggregate level. This type of approach is centralized and cannot be applied to resource-constrained sensor networks.

The problem of providing contextual location privacy in WSNs has been well studied. The primary concern of location privacy in WSNs is to protect the source location [10, 15, 23] and sink location information [4]. To protect the source location against a local adversary, phantom routing [10] uses a random walk before commencing with regular flooding/single-path routing. Later, Mehta et al. [15] and Yang at al. [23] studied the problem of source location privacy in the presence of a global adversary who can observe all traffic in the network. Mehta et al. proposed the use of hop-by-hop encryption to hide the message flows, and Yang et al. suggested the injection of fake messages. Deng et al. [4] proposed randomized routing algorithms and fake message injection to prevent an adversary from locating the network sink based on observed traffic patterns.

A common design goal of data dissemination protocols [2] in wireless sensor networks is to achieve energy efficiency. Ugur et al. [2] let data travel down an event dissemination tree based on a schedule to save energy. To address data privacy issues, Shao et al. [21] designed a data dissemination scheme called pDCS that can provide different levels of data privacy based on different cryptographic keys.

In the area of constructing storage systems, Gregory et al. [8] and Safe-Store [11] have addressed issues of ensuring system availability and integrity policies in the presence of component failures and malicious attacks.

Unlike prior work, the data dissemination scheme introduced in this study addresses the problem of data privacy and data availability at the same time using a noncryptographic method.

6.7 Conclusion

Preserving data privacy and data availability in WSNs cannot be achieved purely by cryptographic strategies. In this work, an SPG-based data dissemination protocol is proposed. It is complimentary to traditional cryptographic techniques and can enhance data privacy and data availability in sensor networks deployed for target tracking. We argued that data uncertainty is important to quantify data privacy and data availability, and message content is more important than the number of messages with regard to data uncertainty. As such, we provided a content-based definition of data privacy and data availability, utilizing information states. To strike a balance between two conflicting objectives, we introduced a graph called the SPG that identifies node pairs whose combined sensed data provide high certainty of the target location, and showed that the task of disseminating data to storage nodes is equivalent to the problem of coloring the SPG.

The SPG-based data dissemination protocol consists of the following steps: (1) constructing the SPG among hot nodes (nodes that detect the target) on demand; (2) coloring the SPG using our energy-efficient distributed coloring algorithm; (3) letting those nodes that provide "valuable" information replicate messages with a probability p. The experiment results have shown that the SPG-based data dissemination scheme combines the advantages of two baseline dissemination schemes: the shortest path routing and random coloring protocols. It can achieve better data privacy and a higher level of data availability while consuming lower energy than either baseline data dissemination scheme.

Bibliography

[1] R. Agrawal and R. Srikant. Privacy-preserving data mining. In *Proc. of the ACM SIGMOD Conference on Management of Data*, 439–450. ACM Press, May 2000.

[2] U. Cetintemel, A. Flinders, and Y. Sun. Power-efficient data dissemination in wireless sensor networks. In *Proceedings of Workshop on Data Engineering for Wireless and Mobile Access (MobiDe)*, 1–8, 2003.

[3] H. Chan and A. Perrig. Security and privacy in sensor networks. *IEEE Computer*, 36(10):103–105, October 2003.

[4] J. Deng, R. Han, and S. Mishra. Intrusion tolerance and anti-traffic analysis strategies for wireless sensor networks. In *Proceedings of Conference on Dependable Systems and Networks (DSN)*, 637, 2004.

[5] C. Díaz, S. Seys, J. Claessens, and B. Preneel. Towards measuring anonymity. In *Proceedings of the 2nd International Conference on Privacy Enhancing Technologies*, 54–68, 2003.

[6] M. Erdmann. Randomization for robot tasks: Using dynamic programming in the space of knowledge states. *Algorithmica*, 10:248–291, October 1993.

[7] S. Ganeriwal, Ram Kumar, and M. B. Srivastava. Timing-sync protocol for sensor networks. In *Proceedings of Conference on Embedded Networked Sensor Systems (SenSys)*, 138–149, 2003.

[8] G. Ganger, P. Khosla, M. Bakkaloglu, M. Bigrigg, G. Goodson, S. Oguz, V. Pandurangan, C. Soules, J. Strunk, and J. Wylie. Survivable storage systems. *DARPA Information Survivability Conference and Exposition*, 2: 184–195, 2001.

[9] C. Intanagonwiwat, R. Govindan, and D. Estrin. Directed diffusion: A scalable and robust communication paradigm for sensor networks. In *Proceedings of Conference on Mobile Computing and Networks (MobiCOM)*, 2000.

[10] P. Kamat, Y. Zhang, W. Trappe, and C. Ozturk. Enhancing source-location privacy in sensor network routing. In *Proceedings of the 25th IEEE International Conference on Distributed Computing Systems (ICDCS)*, 2005.

[11] R. Kotla, L. Alvisi, and M. Dahlin. Safestore: A durable and practical storage system. In *USENIX Annual Technical Conference*, 07–20, 2007.

[12] C. K. Liew, U. J. Choi, and C. J. Liew. A data distortion by probability distribution. *ACM Trans. Database Syst.*, 10(3):395–411, 1985.

[13] N. Linial. Locality in distributed graph algorithms. *SIAM J. Computing*, 21(1):193–201, 1992.

[14] S. Madden, M. Franklin, J. Hellerstein, and W. Hong. TAG: A tiny aggregation service for ad-hoc sensor networks. In *Proceedings of the Usenix Symposium on Operating Systems Design and Implementation*, 2002.

[15] K. Mehta, D. Liu, and M. Wright. Location privacy in sensor networks against a global eavesdropper. In *Proceedings of Conference on Network Protocols (ICNP)*, 314–323, 2007.

[16] N. Minsky. Intentional resolution of privacy protection in database systems. *Commun. ACM*, 19(3):148–159, 1976.

[17] J. M. O'Kane and W. Xu. Energy-efficient target tracking with a sensorless robot and a network of unreliable one-bit proximity sensors. In *Proc. IEEE International Conference on Robotics and Automation*, 2009.

[18] A. Perrig, R. Szewczyk, D. Tygar, V. Wen, and D. Culler. SPINS: Security protocols for sensor networks. *Wireless Networks*, 8(5):521–534, 2002.

[19] A. Savvides, C. Han, and M. B. Strivastava. Dynamic fine-grained localization in Ad-Hoc networks of sensors. In *International Conference on Mobile Computing and Networks (MobiCOM)*, 166–179, 2001.

[20] A. Serjantov and G. Danezis. Towards an information theoretic metric for anonymity. In *Proceedings of the 2nd International Conference on Privacy Enhancing Technologies*, 41–53, 2003.

[21] M. Shao, S. Zhu, W. Zhang, G. Cao, and Y. Yang. pDCS: Security and privacy support for data-centric sensor networks. *IEEE Trans. Mob. Comput.*, 8(8):1023–1038, 2009.

[22] W. Trappe and L. Washington. *Introduction to Cryptography with Coding Theory*. Prentice Hall, 2002.

[23] Y. Yang, M. Shao, S. Zhu, B. Urgaonkar, and G. Cao. Towards event source unobservability with minimum network traffic in sensor networks. In *Proceedings of Conference on Wireless Network Security (WiSec)*, 77–88, 2008.

Chapter 7

Privacy Preservation for IoT Used in Smart Buildings

Nico Saputro

Ali Ihsan Yurekli

Kemal Akkaya

Arif Selcuk Uluagac

CONTENTS

7.1   Introduction

The proliferation of various Internet of Things (IoT) devices has led to several innovative applications including the development of smart home and buildings. While the use of IoT devices can bring a lot of advantages in terms of efficiency, convenience, and cost, their extensive use raises several privacy concerns regarding the users and their activities inside these smart buildings. For instance, through analyzing the smart meter data, one can infer avocations, finances, occupation, credit, health, or other similar personal information about the customer or the household. In commercial buildings, the privacy concerns are mostly on user tracking and pattern detection of behavior when employees utilize their smart devices connected to Wi-Fi access points. In the same manner, the use of IoT devices in the workplace may leak information about the social fabric of that organization, which is largely hidden from direct observation. It is the interpersonal connectivity in a group that is largely created and maintained by physical interactions in the space, which can be monitored in part by analyzing the IoT traffic within the building. The details of these interactions are very sensitive from personal and organizational privacy standpoints, and thus it is important to treat them with great caution.

In this chapter, we will first provide an overview of the smart building concept and the IoT devices commonly used in smart buildings in Section 7.2. In Section 7.3, the privacy issues regarding the use of IoT devices are discussed. Then, in Section 7.4, a survey of the existing efforts to address these challenges are presented. Finally, we will conclude with future research challenges in this emerging area.

7.2   Overview of Smart Building Concept

The *intelligent building* concept has existed for more than three decades and its definition has evolved over time with new developments in technology [10]. As the definitions expanded, the term *smart building* arose and is used interchangeably with the term intelligent building. However, with the increasing use of this new term in industrial reports and academic literature in recent years, the term smart building is more popular and used instead of intelligent building. This allows smart building to have a broader scope than intelligent building and incorporates the latest trends such as smart grid. Similar to the intelligent building definition, smart building also has various definitions, which are introduced by various parties including academic institutions, companies, and organizations. Interested readers are referred to [10] and [55] for a more comprehensive discussion about various intelligent/smart building definitions.

In this section, the definition from the Institute for Building Efficiency [20] is presented to give the reader a high-level overview of smart building. Smart building is defined as "buildings that provide lowest cost and environment friendly building services that make occupants productive through the use of information technology in the building operations." The information technology interconnects various independent subsystems inside the building and enables information sharing between those subsystems. It also interacts with and empowers the building operators and occupants with actionable information. Smart buildings are usually assumed to have their own renewable power generation systems and use smart meter as the gateway to the smart grid as depicted in Figure 7.1.

Besides the interchangeable use of the term intelligent and smart, the presence of other building concepts such as *green building* [59] and *net-zero energy building* [49] concepts may add further confusion to the existing definitions. Even though an internationally agreed definition for each concept is still lacking, all these similar concepts can basically be differentiated from their goals. The green building concept focuses on environmentally friendly aspects and covers the whole building life cycle, including design, construction, operation, maintenance, renovation, and demolition. The net-zero energy building concept, on the other hand, is driven by the availability of distributed renewable energy generation and conservation efforts in the building to provide self-fulfillment of energy. Finally, the *smart*/intelligent building concept focuses on intelligence and communications capability for energy-efficient buildings. This may, as well, involve some parts of the building life cycle from design to maintenance. It is worth noting that the latter two are fundamental concepts for successful smart grid implementations. Figure 7.2 shows the distinction between these concepts.

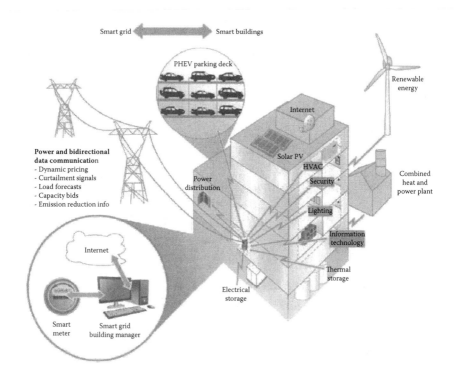

Figure 7.1: Smart buildings and smart grid. (From Institute for Building Efficiency. http://www.institutebe.com/smart-grid-smart-building/What-is-a-Smart-Building.aspx.)

7.2.1 Smart building subsystems

Smart building subsystems have evolved over time following the progress in information and communications technology and the development of new concepts such as smart grid. Current major subsystems consists of three interrelated fundamental subsystems [47], as depicted in Figure 7.3:

1. *Building automation system (BAS).* This had a long-standing evolution since the early 1940s, from a centralized control and monitoring panel to the open BAS that is compatible with the Internet or intranets [53]. BAS has adopted various commonly used Internet/intranet communications and software technologies for monitoring and controlling various building subsystems such as lighting, heating, ventilation, and air-conditioning (HVAC), security and access, fire and safety, and many more.

2. *Building energy management and grid interaction system (BEMGS).* This has emerged from building energy management systems in recent years

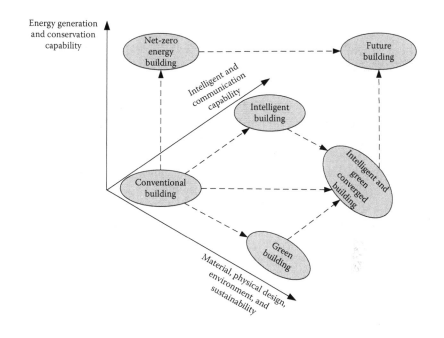

Figure 7.2: Building concepts classification. (Redrawn from J. Pan et al. *Communications Surveys Tutorials, IEEE,* **16(3), 2014.)**

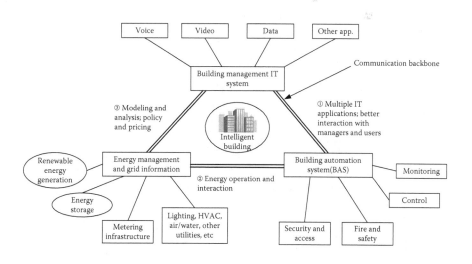

Figure 7.3: Intelligent buildings and related systems. (Redrawn from J. Pan et al. *Communications Surveys Tutorials, IEEE,* **16(3), 2014.)**

following the transformation of the legacy power grid into a *smart grid*. It is responsible for internal energy-related operations and external interaction with the smart grid.

3. *Building management information technology (IT) system (BMITS).* This enables better building functionalities and performance through two-way communications with the other two subsystems in order to achieve various goals. It provides better presentation of the current building status through video or voice applications, which in turn increases the awareness and involvement of the building manager and occupants in controlling the performance of BAS. BMITS also interacts with BEMGS by collecting power consumption data for further modeling and analysis. The results can be used for in-building energy policies or interaction with the smart grid. These policies are implemented by BAS and building energy management.

7.2.2   IoT devices used in smart buildings

The IoT devices used in a smart building environment can be classified into three types: (1) building devices, which are used in the smart building for the purpose of monitoring and controlling the buildings; (2) mobile wireless devices, which are typically used personally by the occupants, such as smartphones, personal digital assistants, personal notebooks, body sensors, digital cameras, portable game consoles, wearable devices, and so on; and (3) smart home appliances, which are typically stationary and mostly found in the residential building, such as televisions, washing machines, refrigerators, and so on. The major IoT building devices that are used in smart buildings include the following:

1. *Smart metering* is basically an advanced electronic recording device that is used to record energy consumption in the building over a certain interval (in hours or minutes) and reports these data to the utility company at certain time intervals through various types of communications technology (e.g., fiber optics, power line communication [PLC], cellular networks, wireless mesh networks, etc.). Even though the term smart metering can also be used for recording water or natural gas consumption, it is often referred to as the electric meter for the recording of electrical energy usage. The smart meter replaces the traditional electric meter and offers two-way communications between the utility company and the consumer.

2. *Wireless local area networks (LANs)* are commonly used to provide wireless access for people within a smart building. The system consists of a number of wireless access points (AP) distributed throughout the building.

3. *Radio Frequency Identification (RFID)* is a wireless short-range low-energy device that has been widely used for years. RFID is considered to be one of the enabling technologies for the IoT since it can provide a unique identity for anything (e.g., consumer goods, apparels, cars, animals, human beings, etc.). A typical RFID system consists of two components, a *reader* and a *tag*, that operate at a certain frequency. The former is an active device sending queries and the latter is an active or passive device responding to these queries. RFID readers in smart buildings are typically installed for access control, for example, for automatic door entrance. The RFID tag, which may be embedded in the employee's ID card, is used for identification before providing physical or logical access. The RFID tag can store data and transmit the data to the reader. The communications between the tag and the reader does not need to be in the line of sight and may be contactless.

4. *Video Surveillance* has commonly been used for security and access control for years. These IoT devices provide high spatial resolution for still images and video and produce a wide range of information, such as shape, color, size, texture, and so on from the captured objects. The objects must be in the direct line of sight of the camera.

5. *Various Sensors*: carbon dioxide (CO_2) sensor, passive infrared (PIR) sensor, ultrasound sensor, magnetic door sensor, and so on. The CO_2 sensor measures the carbon dioxide concentration in the air and is typically used for monitoring indoor air quality. However, the CO_2 sensor can also be used to collect some indirect occupancy information in certain areas based on the CO_2 concentration in that area. The PIR sensor measures infrared (IR) light radiating from objects in its direct line of sight. Typically, a human emits heat energy invisible to the human eyes, but can be detected by the PIR sensor. However, the direct line of sight and continuous motion requirements are the limitations of the PIR sensor, and therefore, it will not be able to detect stationary occupants. The ultrasound sensor, on the other hand, does not require these. The Ultrasound sensor is an active sensor that transmits and receives ultrasonic rays reflected from objects and obstacles.

Figure 7.4 illustrates various IoT devices used in *smart buildings*.

7.2.3 Intelligence in smart buildings

A wide variety of research has been conducted in Intelligent/Smart Buildings for more than 30 years from the independent building subsystems to the system integration of those subsystems. Among the building subsystems, the research on

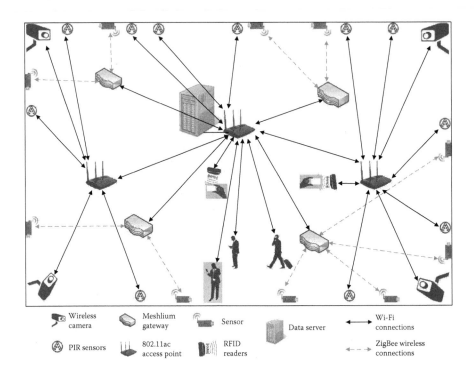

Figure 7.4: Examples of IoT devices used in smart buildings.

heating, ventilation and air-conditioning (HVAC) subsystems and lighting subsystems attract a lot of attention, since they contribute to the largest portion of total energy consumption in buildings. It has been shown from previous research that up to 40% energy saving can be achieved by adopting occupancy-based controls for HVAC subsystems and a combination of control strategies for lighting subsystems such as daylight harvesting (i.e., exploiting external light sources), occupancy sensing, scheduling, and load shedding [43].

Real-time occupancy-based control for HVAC and lighting systems have been the main research focus in Intelligent/Smart Buildings for decades. Various IoT devices have been used to collect occupancy information. These devices can be used in the form of wireless sensor networks (WSNs) which use either a single sensor type or sensor fusion (i.e., multiple sensor types). A single type of sensor may be adequate to collect the desired occupancy information; however, for most cases employing sensor fusion will give a more accurate result. For instance, binary information generated from PIR and ultrasound sensors are adequate to provide presence/absence information. Nevertheless, a more accurate occupancy information can be provided by sensor fusion through the use of PIR and magnetic door sensors [2], or PIR and image sensors [18]. Moreover, a sensor fusion that uses simple binary sensors can also provide more informative occupancy information, such as the occupant's activities, by employing PIR, chair pressure sensors, and acoustic sensors [42].

A wide range of occupancy information is available to support real-time occupancy-based control, ranging from simple binary information about the presence or absence of a person in an observed area, to more significant occupancy information [31], such as where they are (i.e., location), how many people are present (i.e., counting the number of occupants), what they are doing (i.e., activity), who they are (i.e., identity), and where they were before (i.e., tracking). Typically, each IoT device can collect a certain level of occupancy information. Additionally, an IoT device can also provide several pieces of occupancy information at once. For instance, RFID is employed in [34] to provide the estimation of the occupants' activities in real time. Additionally, occupants' identities, the number of occupants, their location, and presence or absence information can be provided. In recent years, it has also become possible to have occupancy information through implicit occupancy sensing. For instance, occupancy information from the existing IT infrastructure such as Wi-Fi [5, 14].

Real-time occupancy-based control can further be classified into two groups: (1) individualized approaches, and (2) nonindividualized approaches [34]. Individualized approaches reveal the occupants' identities and are able to track individual occupants, while nonindividualized approaches are only able to provide nonpersonal occupancy information such as presence/absence and number of occupants. Typically, nonindividualized approaches are nonintrusive, scalable, and easy to deploy, but do not work well in virtual environments (i.e., require physical environments).

Besides real-time occupancy-based control, two new research directions for smart buildings; namely, real-time occupancy-based control with the occupant's individual preferences and control based on predicted occupant behavior, have emerged in recent years [43]. In the first research direction, instead of providing uniform indoor climate or lighting at certain locations for all occupants and operating according to fixed schedules and maximum occupancy assumptions, control with the occupant's individual preferences strives to create a microclimate zone in a relatively small space around the occupant based on the occupant's personal comfort. For instance, in [12], an RFID is used as the occupant's identifier and, when the presence of this occupant is detected in a certain location, the climate and lighting condition in that location are adjusted based on his/her preferences. Interested readers can refer to [52] for a more comprehensive review. The latter research direction is driven due to the fact that climate control has a long response time, unlike lighting control. Hence, it needs to be set in advance in order to meet the occupant's comfort needs on time. The research in this area is very challenging, since an accurate and powerful predictor is needed to predict occupant behavior, which may involve identifying the occupant's activities. For instance, a smart thermostat that uses occupancy sensors is introduced in [36] to automatically turn off the HVAC when the occupant is sleeping or the home is unoccupied. A fusion sensor that consists of wireless motion and door sensors are used to infer occupant activities (e.g., sleeping, left home unoccupied, or active). The interested reader may refer to [43] for more detailed information.

7.3 Privacy Threats in Smart Buildings

Smart buildings are basically designed to enhance user comfort, to provide better access control, and security and to deliver efficient building management. As part of numerous processes taking place within a smart building, the information about the presence of the occupants and their behavior should be gathered and processed in order to provide desirable services. However, the collected information may pose some privacy issues. By using the information collected by several sensors throughout the building or by using the information obtained from personal devices, the physical location of the user can easily be detected. Furthermore, the tracking of an individual's activities can be performed by collecting the physical location information of that individual over a period of time. This would help unauthorized users and attackers to determine the behavior of users and their usage patterns.

Compared to other IoT devices used in smart buildings, the smart meter has some specific features and challenges. While all other IoT devices collect occupancy information from the building, report, and use them for internal purposes, a smart meter acts as the gateway of the building to the smart grid infrastructure and reports the collected data to the utility company or a third party for external use. Moreover, in contrast to traditional meter reading, which is mainly for billing purposes, with data collection frequency once per billing cycle, the smart meter can collect fine-grained power consumption data and report them to the utility company or a third party at a much higher frequency (e.g., per day/hour/minute) through a communications infrastructure. Such data can be used for various purposes by the utility company, such as for real-time dynamic pricing, demand forecasting, and power grid operations. Hence, fine-grained power usage data is available at different locations: at the smart meter, in transit through the communications network on its way to the utility company or the third party, and at the utility company or the third party. This situation may have a higher risk of privacy threats due to the various parties involved. For this reason, privacy issues related to the smart meter have been gaining a lot of attention from academic communities in recent years, as well as its vital role for the successful operation of the smart grid. When the real-time fine-grained power usage information is aggregated over time, it can be used to infer the number of occupants, their habits, and the rhythm of their movements. These issues are usually considered in the scope of user behavior privacy.

7.3.1 Privacy of user behavior

This type of privacy issue stems from the fact that occupants' identity can be learned and their activities can be collected, tracked, or deduced from the information generated by IoT devices.

User behavior privacy becomes an issue, in particular, when a smart meter is used in a residential building. The fine-grained energy consumption data generated from the smart meter can be disaggregated into appliance-level information. The goal of disaggregating power consumption is to provide information on the breakdown of energy consumption and to profile high-energy-usage appliances. The appliance-level information gives some benefits to many parties [3]: The consumer can get direct feedback related to his/her electric consumption and receive automated personalized recommendations, which in turn enables his/her active participation in order to reduce or alter his/her electricity demand. The utility company can obtain fine-grained data to improve economic modeling and policy recommendations. Finally, R&D institutions and manufacturers can use the fine-grained data to support redesign of energy-efficient appliances, to support energy-efficient marketing, and to improve building simulation models. However, disaggregation of data also creates privacy issues, since the process is not intrusive.

Nonintrusive load monitoring (NILM) or nonintrusive appliance load monitoring (NIALM) is a technique for analyzing and extracting appliance-level information from power consumption in a nonintrusive fashion. There have been various NILM approaches proposed ever since it was first introduced in [26]. Figure 7.5 shows an example of activities deduced using an NILM approach. Interested readers may refer to [56] and [3] for more detailed information.

7.3.2 Location privacy

Location privacy is defined as "the ability to prevent unauthorized parties from learning someone's current or past location" [35]. Sources of location information can either be various technologies used in smart buildings, such as sensors, RFID readers, video cameras, Wi-Fi access points, PIR sensors, and so on, or personal electronic devices used by the occupants themselves, such as smartphones, notebooks, tablets, body sensors, or wearables. It may not be considered an issue for a relatively small environment, like inside of a house, where a user is already known to be located and does not have a lot of internal space to move around. However, in closed public environments, such as airports or shopping centers, or in big office buildings, location privacy becomes a problem.

7.3.2.1 Privacy issues with wireless LANs

Due to the broadcast nature of the wireless LAN technology, it is much easier to obtain private information about the users. The following user data can

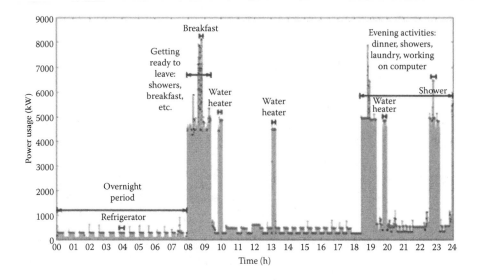

Figure 7.5: An example of activities deduced from NILM approach. (From A. Molina-Markham et al. *Proceedings of the 2nd ACM Workshop on Embedded Sensing Systems for Energy-Efficiency in Building, BuildSys '10, ACM,* **New York, 2010.)**

be disclosed to unauthorized parties during wireless communication: content of the communication, who is sending or receiving data (user identity), when the communication takes place (time) and where the communication takes place (location). While the content can be protected using encryption at application level, the rest of the information may be available to external entities, as explained below:

1. User identity can be determined from the node information (i.e., MAC and IP addresses).

2. Time information can be related to the time of the transmitted or received packet.

3. Location can be inferred from: (i) the single access point (AP) that receives the transmission, providing a rough estimation; (ii) the transmitted signal strength information from multiple APs which receive the transmission, providing more accurate location information, for instance, by the triangulation method or by fingerprint-based localization [4, 57].

   When all this information is combined together, the where, when, and who of a wireless communication event can be used for tracking and inferring user behavior.

7.3.2.2   *RFID privacy issues*

The privacy issue comes from the fact that an RFID tag and reader do not have to be in line of sight. An unauthorized RFID reader at a distance or beyond the wall(s) may try to get access to the tag information and the tag owner may not be aware that his/her tag is being read.

7.3.3   Visual privacy

Visual privacy refers to the private information in the form of image or video. Today, streets of modern cities and almost all closed public places are equipped with surveillance cameras in order to track suspicious activity and identify criminals. We expect that, in the near future, the number of cameras will increase even further with the introduction of smart cameras and vision-based intelligent surveillance systems. Surveillance cameras may also be used as part of ambient-assisted living systems in support of autonomy and well-being of older or disabled people. In any case, videos or images of a person carry the richest privacy information about a person and his/her environment. Not only the face of a person, but also the clothes, posture, gait, time, and environment can reveal sensitive information.

7.4   Privacy-Preserving Approaches in Smart Buildings

7.4.1   Wireless LAN privacy-preserving approaches

The evident solution to the privacy problem is to break the link between the user identity and the time and location information. The best way to achieve this goal is to anonymize the user or node information with frequent disposal of short-lived identifiers or pseudonyms.

Factors affecting successful use of frequent disposable identifiers for location privacy in wireless LAN are: (i) the type of environment, (ii) location resolution, and (iii) prior knowledge of the system or user by the attacker. First, if it is an open environment with a high fluctuation of users, such as an office building with several employees or in public areas such as an airport or shopping center, it is difficult to detect the changes in identifiers. However, if the user is located in a closed environment, such as a company network where all authorized clients' interface identifiers are registered, changes in identifiers are easier to detect. The second factor to consider is the location resolution, which is the accuracy of locating a user. A single access point (AP) connected to the user will provide a rough estimate of the user location. On the other hand, multiple APs may be installed in the area, providing more accurate location information detection (i.e., enabling cooperation between APs through the triangulation approach to determine the user location). The solution to this problem is to control the transmitted signal strength from the device. This will reduce the number of APs which are able to receive the transmission [28]. Finally, if the attacker has prior knowledge about the environment (e.g., building layout, office assignment, working schedule of the employees, etc.), he/she can use this information to better identify the user [25].

The goals of applying anonymization are threefold. First, the identifier should be unlinkable, that is, the new and old identifiers from the same client node should be dissociated. Secondly, anonymization should cause minimum network disruption. In order to achieve this goal, proper timing is needed. The address switching may close network connections in real-time applications such as voice over IP (VoIP) or long communication sessions like streaming media. Finally, the solution should be readily applicable to the current IEEE 802.11 standard [4]. The key challenges in anonymization are

1. *Address selection.* The addresses (any including fake ones to disguise the real ones) must still be valid and follow the standard, which requires 48-bit MAC addresses, consisting of 24 bits for the Organization Unique Identifier (OUI) and another 24 bits as assigned by the NIC vendor so that it will not be rejected or ignored due to incompatibility reasons.

2. *Address uniqueness.* All nodes or users sharing a network source should have a unique address. Thus, we need a detection and prevention mechanism for duplicate addresses. If it is a large network with many users, address collision becomes a problem, especially if each user independently generates its own fake MAC address. One solution to this problem is to configure the AP to provide a pool of MAC addresses and to assign a MAC address to the node or user that joins it. In this case, the user or client needs to request a MAC address when joining the AP. The problem here is that the request must be attributable, which means it must contain the real MAC address of the user, in which case the user identity will again be revealed. To solve this problem, Jiang et al. [28] proposed using a joint address (i.e., group address) within the request for concealment purposes and a 128-bit nonce (one-time code) to provide uniqueness.

3. *Integration with port authentication.* Other identifiers besides MAC addresses (in protocols such as EAP-TLS, CHAP, RADIUS) should also be taken into account so that eavesdroppers will not use them to track the user.

An important issue to consider is how to unlink different MAC addresses of the same user when frequent address changes are employed, that is, how to reduce the correlation of two addresses of the same user and increase the entropy in address selection.

One solution is to use a silent period after performing address changes [27]. In this approach, the users intentionally do not transmit within a certain period of time after the address change has occurred. The goal is to obscure the address change event by the presence of incoming users or clients. This is, of course, practical when user density is high enough to mask the address change event. Since forced silent periods without user intervention can disrupt communications, the concept of an opportunistic silent period is introduced [27], where address changes are performed during the idle time between users' communications, thus minimizing the negative effect on established communications, and hence enhancing the quality of service.

Another solution is employing mix-zone areas [7, 21] which can be described as the spatial version of the silent period approach so that clients are not allowed to transmit in predefined areas. This involves middleware installed on mobile devices to preset the physical location so that all users in this area are indiscernible. All clients may change their pseudonyms (e.g., MAC addresses) in the mix-zone but they are not allowed to transmit there. A mix-zone for a group of users is defined as a connected spatial region of maximum size where none of these users register for an application. In contrast, an application zone is an area where a user can register for an application callback. When a client that has just changed its pseudonym moves out from the mix-zone and starts to transmit again, an adversary or location-based service (LBS) application will not be able to relate the new pseudonym, to a specific old pseudonym, since this new detected pseudonym may come from any client that has just entered the mix-zone. This approach works well when many clients enter or exit the mix-zone at the same time. In order to increase anonymity, the application may be configured not to transmit or not to send any location update if the mix-zone has fewer than *k* users.

7.4.2 *RFID privacy-preserving approaches*

There are various proposed solutions to privacy problems caused by RFID devices, including (1) hiding and blocking and (2) rewriting and encryption [32]. In hiding and blocking, the tag is silenced through jamming the radio channel used for RFID communication and providing the reply only to readers with proper credentials. In rewriting and encryption, the access to the tag is controlled securely by using techniques such as anonymization through hash-based approaches. Using a hash-lock scheme [54], unauthorized reader access to the tag is prevented, since the tag is, by default, locked and only opened when the correct key is introduced to it. To open the tag, the reader requests the metaID (hashed ID) and tries to find the key and the ID in the back-end server. The back end sends information (key, ID) to the reader and the reader sends the key to the tag. Then, the tag hashes the key and compares it to the metaID. If there is a match, the tag is unlocked.

While preserving privacy at a certain level and having a short search time because the database is implemented by a hash table, tracking is still possible in

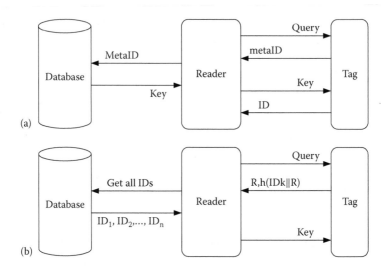

(a)

(b)

Figure 7.6: (a) Hash locking: a reader unlocking a hash-locked tag; (b) randomized hash locking: a reader unlocks a tag whose ID is *k* in the randomized hash-lock scheme. (Redrawn from S.A. Weis et al. in D. Hutter et al. [eds]. *Security in Pervasive Computing*, Springer, Berlin, 2004.)

the hash lock scheme since a fixed metaID is used (i.e., a single pseudonym). To overcome this problem, a randomized hash-lock scheme is proposed. Here, the tag output changes each time it is accessed, since each time a reader accesses the tag, the tag replies with a random string plus the hash of the concatenated tag ID, which means that the pseudonym will change in each access each time the tag is accessed and will prevent unauthorized readers tracking the user. Tags in this randomized scheme ensure full privacy. However, it is not scalable for a large number of tags, since a huge number of hash operations must be performed at the back-end database. Furthermore, this protocol does not guarantee forward privacy, since the stored information in a compromised tag reveals much data about the previous communications of that tag [11]. Figure 7.6 shows how these two approaches work.

To overcome the forward security issue, a hash-chain scheme is proposed [44], where the basic idea is to refresh the tag identifier each time the tag is queried by a reader. The scheme can be achieved via a low-cost hash-chain mechanism. However, this scheme is also not scalable because of the exhaustive search process that must be performed by the back-end server.

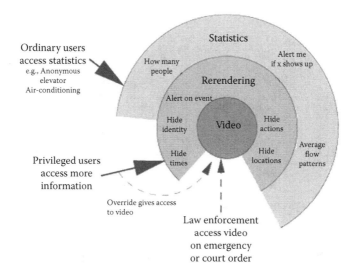

Figure 7.7: Layered approach for accessing video surveillance information. (Redrawn from A. Senior et al. *Security Privacy, IEEE,* 3(3), 2005).

7.4.3   *Video surveillance privacy-preserving approaches*

Since video surveillance and associated intelligent monitoring systems provide the richest privacy information about subjects, the solutions for preserving visual privacy should be defined accordingly, preferably starting at the design phase, such as whether to choose a high- or low-resolution camera, whether or not to use encryption, and so on.

An important issue is the definition of access control for different types of users having access to video surveillance data. As depicted in Figure 7.7, a layered approach is proposed by Senior et al. [50], providing capability to determine who can view what data under what circumstances. In this model, three different types of users have access at three different levels: Ordinary users can only access statistical information about the video; privileged users can access to rerendered and limited information; and finally, law enforcement agencies may have full access, including raw video and related individual identity information. Such a system should comprise video analysis, encoding/decoding, storage facilities, and basic security functions such as authentication, accounting, and encryption.

Considering the temporal aspect, visual privacy preservation mechanisms can be applied either in real time during the acquisition of the image or video, or after its acquisition. A real-time example proposed by Zhang et al. [58] uses two cameras, IR and RGB, to capture video simultaneously. The thermal IR camera is used to discriminate the face region and other parts of the human body based

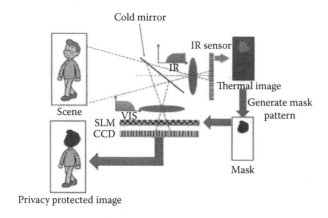

Figure 7.8: Concept of the anonymous camera system. (From Y. Zhang et al. *Pattern Recognition (ICPR), 2014 22nd International Conference on*, 2014).

on the fact that human skin radiates shorter wavelengths ($\sim$10 μm). Thermal imaging generates a mask pattern corresponding to the position of the face of the subject. A spatial light modulator (SLM) (e.g., LCD) is inserted in front of the CCD/CMOS image sensor of the RGB camera, which applies the thermal imaging mask and prevents the face of the subject being recorded (see Figure 7.8). Since this implementation only protects the subject's face or open extremities, valuable privacy information can still be obtained from the clothing of the subject or the environment if prior information is available.

To preserve privacy, applicable methods can be considered in five different categories [45]: intervention, blind vision, secure processing, redaction, and data hiding.

1. *Intervention* methods involve prevention of visual data being captured from the environment by physically interfering with the camera devices, for instance, by creating excessive illumination.

2. *Blind vision* implementation consists of image or video processing in an anonymous way using cryptographic techniques, such as secure multiparty computation (SMC), where a contributing party is using the algorithm of the another party and does not know the details of it.

3. *Secure processing* methods involve video processing techniques other than SMC to preserve privacy.

4. *Redaction* methods, with many subcategories, such as image filtering, encryption, *k*-same family, object/people removal, and visual abstraction, are the most common preservation methods, of which we will provide some examples in the following paragraphs.

5. *Data hiding* methods are based on hiding the original image data inside a cover message which can be used for retrieval if needed in the future.

In image filtering, a Gaussian blur or Gaussian smoothing filter is applied to modify each pixel in the image by using neighboring pixels. As an example, an image is divided into 8×8 pixel blocks and the average color of the pixels in that block is calculated. The result is then used as the new color for all the pixels in that block.

Encryption of video and images uses either traditional encryption, like DES, AES, and RSA, which is generally slow for real time, or lightweight encryption, which is faster but less secure. Encryption techniques help to scramble the region of interest by pseudorandomly flipping bits. They can be used for the compressed video/image (code-stream) domain, the spatial domain, and the frequency domain [9, 15].

In face deidentification techniques, the goal is to alter the face region so that face recognition systems will be unable to recognize it. One of most robust methods, the k-same family algorithm, which is an implementation of the k-anonymity concept, computes the average of k images in a set and replaces the cluster with the average image obtained (see Figure 7.9) [41]. On the other hand, object/people removal is performed by removing a private object or people from the original image. The issue here is how to refill the void area after removal, and the solution relies on using inpainting methods to restore the damaged portion. While still image inpainting is easier, since it should take care of spatial consistencies only, video inpainting has to deal with both spatial and temporal consistencies [24]. Finally, the goal of visual abstraction/object replacement is to protect privacy while maintaining the object activity, including position, pose, and orientation. For this purpose, image filtering and deidentification techniques can be used [13].

7.5 Smart Meter Privacy-Preserving Approaches

Efforts to preserve privacy for smart meter are based on the following facts: billing requires an association between the meter reading and the consumer identity, but it does not really need fine-grained meter reading. Fine-grained meter readings are necessary for grid operations, and exact consumer identity is barely needed in these cases. When the consumer identity and the fine-grained power consumption are exposed to unauthorized parties, some privacy threats may arise.

Efforts to preserve privacy may be classified into three categories: (1) approaches that attempt to disassociate consumer identity from meter-reading data (i.e., working on user identity) through anonymization; (2) approaches that endeavor to prevent NILM from obtaining appliance-level information (i.e.,

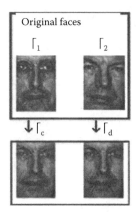

Figure 7.9: *K*-anonymity by averaging *k* distinct faces. (From E.M. Newton et al. *Knowledge and Data Engineering, IEEE Transactions on*, 17(2), 2005.)

working on meter-reading data) through the modification of the meter reading; and (3) encryption-based approaches that employ encryption and data aggregation to provide privacy protection while the data is in transit within the smart grid communications network. In addition, the third party may also be involved in efforts to preserve privacy as the data gateway which can send individual or aggregated meter readings (acting as the data aggregator also) or as an identity generator which can create pseudonym identities for smart meters.

7.5.1 Anonymization approaches

The anonymity of the consumer can be achieved by replacing the consumer identity with pseudonym(s) (i.e., identity pseudonymization), employing a trusted data gateway, or using a trusted third party (TTP) as the data collector.

7.5.1.1 Identity pseudonymization

Pseudonym(s) can be generated through TTP [16], without TTP involvement, by employing the public key infrastructure (PKI) [19] or using group anonymity [51].

In [16], TTP generates two distinct pseudonyms for every consumer, *anonymous identity* and *attributable identity*. An anonymous identity is used to send the nonbilling meter reading to the utility company or third party that requires the aggregated meter-reading data, while the attributable identity is used to send the billing meter reading to the utility company. Figure 7.10 illustrates the use of the pseudonyms. These pseudonyms are hard-coded within the smart meter and only the TTP possesses the association information. The utility company only

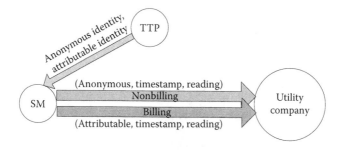

Figure 7.10: Identity pseudonymization through TTP.

knows the attributable identity. To avoid an unauthorized party discovering the association between the pseudonyms, the delivery of the pseudonyms is performed separately over a long random time schedule.

In [19], instead of using TTP, the smart meter generates one RSA key pair of a public and private key (SM_{PUB},SM_{PRV}) while the grid operator generates two RSA key pairs of public and private keys. The grid operator uses the first public and private key to create and check the blind signature (GS_{PUB},GS_{PRV}), while the second key pair is used to encrypt and decrypt the meter reading (GE_{PUB},GE_{PRV}). A blinded factor r is used to create a blinded pseudonym from the smart meter public key. This blinded pseudonym is sent to the grid operator through a secure channel. The grid operator signs the blinded pseudonym with its private key GS_{PRV}, and sends this signature to the smart meter. When the smart meter sends its meter reading, the meter reading is encrypted with the grid operator public key GE_{PUB} and signed with smart meter private key S_{PRV}. The smart meter then sends a data tuple that consists of the encrypted meter reading, its signature, the smart meter public key, and the smart meter public key signature to the utility company. To avoid the association of the pseudonym and the network address of the smart meter when sending the meter reading directly to the utility company, a peer-to-peer (p2p) overlay network [38] is employed to hide the association. In the p2p overlay network, each meter reading generated from a smart meter will pass through several other smart meters before it reaches the utility company. In this way, the utility company will never know from which smart meter the received meter reading originated. Another effort to create anonymity is by using group anonymity [51]. In this approach, a group pseudonym is used by a group of k smart meters (i.e., k-anonymity).

7.5.1.2 Anonymity through trusted neighborhood gateways

Anonymity from the utility company can also be provided by avoiding transmission of the fine-grained meter reading directly to the utility company. A trusted neighborhood gateway [40] is used as the data collector. Every smart meter sends

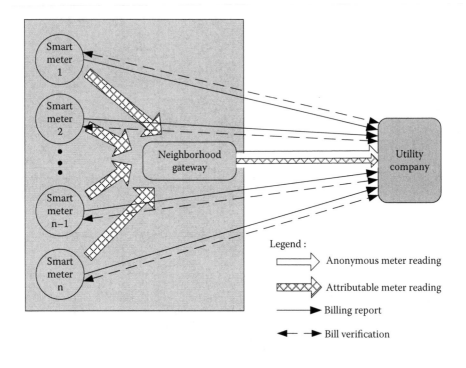

Figure 7.11: Anonymity through trusted gateways.

its attributable fine-grained power consumption to the gateway (e.g., it sends [user identity, timestamp, usage]). The gateway then relays it to the utility company in the form of anonymous power consumption (i.e., without any originator identity (e.g., [timestamp, usage]). All communications between smart meters, the gateway, and the utility are assumed to be over a secure channel that provides authenticity, confidentiality, and integrity. Since the utility company only receives anonymous power consumption, the smart meter performs the billing calculation and sends it directly to the utility company. In order to verify the correctness of the billing report, a zero-knowledge protocol [23] is employed. In each billing cycle, the smart meter must perform the registration by cryptographically designating N pseudorandom tags and a set of m keys. N is the number of meter readings needed for billing, and m is the number of verification rounds in each billing cycle. The utility company will carry out m series of challenge-response mechanisms with the smart meter for interactive billing verification. In addition, the gateway can leak a small amount of attributable power consumption to the utility company for sporadic random spot checks. The goal is to prevent the smart meter from manipulating data. Figure 7.11 illustrates the gateway operations.

Relying on smart meters for billing calculations poses an issue of software updates when there is a change in billing regulations. In such a case, millions of smart meters may need to update their software, which may not be feasible. To overcome this problem, another approach, in which a TTP replaces the gateway, is pursued [8]. In this approach, instead of sending anonymous individual meter readings to the utility company, the TTP aggregates the meter readings and sends the neighborhood-level power consumption to the utility company. At the end of each billing cycle, the TTP aggregates the individual consumption amounts from each smart meter and sends the attributable aggregated power consumption amounts from each smart meter to the utility company for billing processing.

7.5.2   Power consumption modification approaches

Power consumption modification approaches endeavor to hide the real energy consumption of the consumer in attributable energy consumption reporting. In these approaches, the identity of the consumer is not anonymous, but the consumption data is modified. In this way, an adversary cannot really achieve correct conclusions from the data, since an accurate appliance-level information cannot be deduced. The modification can be done before the power consumption data are collected by smart meter, or after they have been collected by smart meter and are sent to the utility company. Typically, internal energy sources are needed in order to modify the energy consumption data before they are collected by smart meter. Modification before they are collected collection by smart meter is known as *load signature moderation* and modification after smart meter collection is known as *power usage data masking*.

7.5.2.1   Load signature moderation

There are two types of approaches in load signature moderation: (1) battery-based load hiding (BLH) and (2) load-based load hiding (LLH). As the name implies, a rechargeable battery is employed in the first approach. This battery is used as an internal energy source to supply the power demand of the building. The battery is discharged and charged at strategic times to alter the external load recorded by the smart meter. These charging and discharging events are chosen in a such a way that the appliance load signature can be reshaped. There are several ways to achieve this. The load signature can be hidden, smoothed, changed, or obfuscated. Figure 7.12 shows three examples of load shaping strategies, namely, hiding, smoothing, and obfuscation. The load signature is completely hidden from the smart meter when the battery fully supplies the appliance's power demand. The smart meter will only record a constant power usage when the battery is slowly recharged, as depicted in Figure 7.12a. A mix of battery and external power sources that supply the appliance power

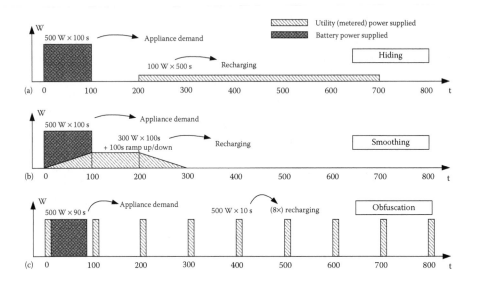

Figure 7.12: Examples of load shaping strategies. (From G. Kalogridis et al. Privacy for smart meters: towards undetectable appliance load signatures. In *Smart Grid Communications (SmartGridComm), 2010 First IEEE International Conference on.*)

demand can produce different load reshaping results, as depicted in Figures 7.12b and c, when different energy supplier compositions and recharging time are used.

Several algorithms have been proposed for load signature moderation. The best effort algorithm [29] is a deterministic algorithm that endeavors to keep the external load constant whenever possible. When the energy demand is higher than the previous reported load, the battery is discharged to provide a partial power supply to the appliance, and when the energy demand is lower than the previous reported load, the battery will be charged. This approach, however, is limited by the battery's capacity. There will be a change in the external load when the battery is empty and the energy demand is higher. In this case, the battery cannot be discharged and it may even need to be charged, which in turn increases the external load. A similar situation happens when the energy demand is lower and the battery is fully charged. In contrast to [29], the nonintrusive load leveling (NILL) algorithm [39] strives to maintain an adaptable target load profile by taking into account all battery states.

In contrast to BLH, which attempts to hide the load signature in such a way that NILM cannot detect appliance-level information from the aggregated power consumption, load-based load hiding uses a controllable energy-intensive appliance that has a daily non-user-driven power demand, such as an electric water boiler, to add some random noise to the power consumption data [17].

By randomly turning on and off the electric water boiler with the constraint of its given daily power consumption LLH obfuscates the power consumption collected by the smart meter. The advantage of LLH is that it increases appliance-level privacy, which makes NILM approaches unable to detect appliance-level information.

7.5.2.2 Power usage data masking

In this approach, a mask value is added to the power consumption data before it is sent to the utility company. In this way the, an adversary will not know the real power consumption value. The mask value is generated in such a way that, when all mask values are added together, the result is a certain known value, such as zero. The utility company can obtain the real aggregated power consumption data by adding all received obfuscated meter readings and subtracting this known value.

There are several methods for the mask value generation, such as the secret sharing method [30], using a random value from a known distribution with known variance and expectation [8], or using a distributed Laplacian perturbation algorithm [1]. The overhead and scalability of the approach should be carefully considered, given that the generation of secret values is based on certain computations.

7.5.3 Encryption-based approaches

Encryption-based approaches endeavor to provide privacy protection through end-to-end encryption while the meter reading is in transit within the smart grid communications network. The main goal is to provide confidentiality for the data. The smart meter encrypts the meter-reading data (including the consumer identity) and sends this encrypted data to the utility company. The utility company can obtain the real power consumption and consumer identity by decrypting the encrypted meter reading. Either symmetric- or asymmetric-key cryptography may be used for encryption. Even though symmetric-key cryptography is faster, storing the key in the smart meter increases the risk of the key being stolen; thus, asymmetric-key cryptography is preferred.

Privacy-preserving data aggregation is another approach that has been proposed for use with smart meters. The goal of data aggregation is to reduce the total bandwidth used, given that a huge number of smart meters send their meter reading to the utility company. In [6], hop-by-hop concatenation is used on the encrypted meter reading. Two different symmetric key pairs are used. The first key pair is used by the smart meter and the utility company to provide end-to-end encryption for meter-reading data. The second key pair is used by the aggregator and its one-hop parent node for hop-by-hop authentication. In this approach, the utility company will obtain individual meter readings. The bandwidth saving from this approach is shown on the header count.

An encryption mechanism that has a homomorphic property can also be used for privacy-preserving data aggregation. This property enables a set of operations to be carried out on the ciphertext without exposing the plaintext. Among many homomorphic encryption mechanisms, Paillier [46] homomorphic encryption is widely proposed for data aggregation in smart grids such as in-network secure data aggregation [33], fraud/leakage detection [22], and multidimensional meter-reading aggregation [37]. Paillier is the preferred choice due to its addition property, small message expansion factor, and strong security features [48].

7.6 Concluding Remarks and Future Research

In this chapter, we presented the privacy issues caused by various IoT devices used in smart buildings; in particular, to support energy-efficient and environmentally friendly building services for occupants' comfort. We identified three privacy issues that have to be solved to realize privacy-aware smart buildings: user behavior privacy, which can be inferred from fine-grained meter readings or by tracking mobile IoT devices, location privacy, and visual privacy. We surveyed several privacy-preserving approaches for identified IoT devices causing privacy concern, categorized the approaches, and provided an overview of them. We also provided references to other useful resources for interested readers.

We also discussed that the privacy issue with the smart meter arises from the disaggregation of the power consumption to obtain appliance-level information by using NILM approaches. This is not the only way to obtain this information. As a matter of fact, a remote monitoring service has been offered for several years for online monitoring of and possible online services for HVAC systems. This remote service enables a third party to access the device and collect some operational information from it for accurate fault detection and suitable proposed corrective actions. Collecting data remotely may also reveal some occupancy information, such as when the occupant is in the building or not. When this remote monitoring service is widely be adopted for various smart appliances in the near future, the similar issue will arise. The third party will have access to usage reports of each smart appliance for diagnostics and repairs. Therefore, privacy-aware remote monitoring services may become one of the future research directions.

Another possible research direction is to involve interdisciplinary research and incorporate the user perspective into privacy research. Most of the approaches are based on fixed assumptions about the user's privacy perspectives. However, each user may have a different sensitivity to privacy, which needs to be reflected in the various approaches. This requires ethnographic approaches by social scientists to understand the needs of the users. Once those needs are identified, differential privacy can be offered via novel approaches.

Bibliography

[1] G. Ács and C. Castelluccia. I have a dream!: differentially private smart metering. In *Proceedings of the 13th international conference on Information hiding, IH'11*, pages 118–132, Berlin, 2011. Springer-Verlag.

[2] Y. Agarwal, B. Balaji, R. Gupta, J. Lyles, M. Wei, and T. Weng. Occupancy-driven energy management for smart building automation. In *Proceedings of the 2nd ACM Workshop on Embedded Sensing Systems for Energy-Efficiency in Building, BuildSys '10*, pages 1–6, New York, 2010. ACM.

[3] K.C. Armel, A. Gupta, G. Shrimali, and A. Albert. Is disaggregation the holy grail of energy efficiency? the case of electricity. *Energy Policy*, 52:213–234, 2013.

[4] P. Bahl and V.N. Padmanabhan. Radar: an in-building RF-based user location and tracking system. In *INFOCOM 2000. Nineteenth Annual Joint Conference of the IEEE Computer and Communications Societies. Proceedings. IEEE*, volume 2, pages 775–784, vol. 2, 2000.

[5] B. Balaji, J. Xu, A. Nwokafor, R. Gupta, and Y. Agarwal. Sentinel: occupancy based HVAC actuation using existing WIFI infrastructure within commercial buildings. In *Proceedings of the 11th ACM Conference on Embedded Networked Sensor Systems, SenSys '13*, pages 17:1–17:14, New York, 2013. ACM.

[6] A. Bartoli, J. Hernandez-Serrano, M. Soriano, M. Dohler, A. Kountouris, and D. Barthel. Secure lossless aggregation for smart grid M2M networks. In *Smart Grid Communications (SmartGridComm), 2010 First IEEE International Conference on*, pages 333–338, Oct. 2010.

[7] A.R. Beresford and F. Stajano. Mix zones: user privacy in location-aware services. In *Pervasive Computing and Communications Workshops, 2004. Proceedings of the Second IEEE Annual Conference on*, pages 127–131, Mar. 2004.

[8] J.-M. Bohli, C. Sorge, and O. Ugus. A privacy model for smart metering. In *Communications Workshops (ICC), 2010 IEEE International Conference on*, pages 1–5, May 2010.

[9] T.E. Boult. Pico: privacy through invertible cryptographic obscuration. In *Computer Vision for Interactive and Intelligent Environment, 2005*, pages 27–38, Nov. 2005.

[10] A.H. Buckman, M. Mayfield, and S.B.M. Beck. What is a smart building? *Smart and Sustainable Built Environment*, 3(2):92–109, 2014.

[11] J.-C. Chang and H.-L. Wu. A hybrid rfid protocol against tracking attacks. In *Intelligent Information Hiding and Multimedia Signal Processing, 2009. IIH-MSP '09. Fifth International Conference on*, pages 865–868, Sep. 2009.

[12] H. Chen, P. Chou, S. Duri, H. Lei, and J. Reason. The design and implementation of a smart building control system. In *e-Business Engineering, 2009. ICEBE '09. IEEE International Conference on*, pages 255–262, Oct. 2009.

[13] K. Chinomi, N. Nitta, Y. Ito, and N. Babaguchi. Prisurv: privacy protected video surveillance system using adaptive visual abstraction. In *Proceedings of the 14th International Conference on Advances in Multimedia Modeling, MMM'08*, pages 144–154, Berlin, 2008. Springer-Verlag.

[14] K. Christensen, R. Melfi, B. Nordman, B. Rosenblum, and R. Viera. Using existing network infrastructure to estimate building occupancy and control plugged-in devices in user workspaces. *Int. J. Commun. Netw. Distrib. Syst.*, 12(1):4–29, Nov. 2014.

[15] F. Dufaux and T. Ebrahimi. Scrambling for privacy protection in video surveillance systems. *Circuits and Systems for Video Technology, IEEE Transactions on*, 18(8):1168–1174, Aug. 2008.

[16] C. Efthymiou and G. Kalogridis. Smart grid privacy via anonymization of smart metering data. In *Smart Grid Communications (SmartGridComm), 2010 First IEEE International Conference on*, pages 238–243, Oct. 2010.

[17] D. Egarter, C. Prokop, and W. Elmenreich. Load hiding of households power demand. In *Smart Grid Communications (SmartGridComm), 2014 IEEE International Conference on*, pages 854–859, Nov. 2014.

[18] V.L. Erickson, S. Achleitner, and A.E. Cerpa. Poem: power-efficient occupancy-based energy management system. In *Information Processing in Sensor Networks (IPSN), 2013 ACM/IEEE International Conference on*, pages 203–216. IEEE, 2013.

[19] S. Finster and I. Baumgart. Pseudonymous smart metering without a trusted third party. In *Trust, Security and Privacy in Computing and Communications (TrustCom), 2013 12th IEEE International Conference on*, pages 1723–1728, July 2013.

[20] Institute for Building Efficiency. What is a smart building. http://www. institutebe.com/smart-grid-smart-building/What-is-a-Smart-Building.aspx. Accessed: 2015-04-27.

[21] J. Freudiger, R. Shokri, and J.-P. Hubaux. On the optimal placement of mix zones. In *Proceedings of the 9th International Symposium on Privacy Enhancing Technologies, PETS '09*, pages 216–234, Berlin, 2009. Springer-Verlag.

[22] F.D. Garcia and B. Jacobs. Privacy-friendly energy-metering via homomorphic encryption. In *Proceedings of the 6th international conference on Security and trust management, STM'10*, pages 226–238, Berlin, 2011. Springer-Verlag.

[23] S. Goldwasser, S. Micali, and C. Rackoff. The knowledge complexity of interactive proof-systems. In *Proceedings of the Seventeenth Annual ACM Symposium on Theory of Computing, STOC '85*, pages 291–304, New York, 1985. ACM.

[24] M. Granados, J. Tompkin, K. Kim, O. Grau, J. Kautz, and C. Theobalt. How not to be seen & object removal from videos of crowded scenes. *Comp. Graph. Forum*, 31(2pt1):219–228, May 2012.

[25] M. Gruteser and D. Grunwald. Enhancing location privacy in wireless lan through disposable interface identifiers: a quantitative analysis. In *Proceedings of the 1st ACM International Workshop on Wireless Mobile Applications and Services on WLAN Hotspots, WMASH '03*, pages 46–55, New York, 2003. ACM.

[26] G.W. Hart. Nonintrusive appliance load monitoring. *Proceedings of the IEEE*, 80(12):1870–1891, Dec. 1992.

[27] L. Huang, K. Matsuura, H. Yamane, and K. Sezaki. Enhancing wireless location privacy using silent period. In *Wireless Communications and Networking Conference, 2005 IEEE*, volume 2, pages 1187–1192, Mar. 2005.

[28] T. Jiang, H.J. Wang, and Y.-C. Hu. Preserving location privacy in wireless lans. In *Proceedings of the 5th International Conference on Mobile Systems, Applications and Services, MobiSys '07*, pages 246–257, New York, 2007. ACM.

[29] G. Kalogridis, C. Efthymiou, S.Z. Denic, T.A. Lewis, and R. Cepeda. Privacy for smart meters: towards undetectable appliance load signatures. In *Smart Grid Communications (SmartGridComm), 2010 First IEEE International Conference on*, pages 232–237, Oct. 2010.

[30] K. Kursawe, G. Danezis, and M. Kohlweiss. Privacy-friendly aggregation for the smart-grid. In *Proceedings of the 11th international conference on Privacy enhancing technologies, PETS'11*, pages 175–191, Berlin, 2011. Springer-Verlag.

[31] T. Labeodan, W. Zeiler, G. Boxem, and Y. Zhao. Occupancy measurement in commercial office buildings for demand-driven control applicationsa survey and detection system evaluation. *Energy and Buildings*, 93(0):303–314, 2015.

[32] M. Langheinrich. A survey of rfid privacy approaches. *Personal and Ubiquitous Computing*, 13(6):413–421, 2009.

[33] F. Li, B. Luo, and P. Liu. Secure information aggregation for smart grids using homomorphic encryption. In *Smart Grid Communications (SmartGridComm), 2010 First IEEE International Conference on*, pages 327–332, Oct. 2010.

[34] N. Li, G. Calis, and B. Becerik-Gerber. Measuring and monitoring occupancy with an {RFID} based system for demand-driven {HVAC} operations. *Automation in Construction*, 24(0):89–99, 2012.

[35] L. Liu. From data privacy to location privacy: models and algorithms. In *Proceedings of the 33rd International Conference on Very Large Data Bases, VLDB '07*, pages 1429–1430. VLDB Endowment, 2007.

[36] J. Lu, T. Sookoor, V. Srinivasan, G. Gao, B. Holben, J. Stankovic, E. Field, and K. Whitehouse. The smart thermostat: using occupancy sensors to save energy in homes. In *Proceedings of the 8th ACM Conference on Embedded Networked Sensor Systems, SenSys '10*, pages 211–224, New York, 2010. ACM.

[37] R. Lu, X. Liang, X. Li, X. Lin, and X. Shen. Eppa: an efficient and privacy-preserving aggregation scheme for secure smart grid communications. *Parallel and Distributed Systems, IEEE Transactions on*, 23(9):1621–1631, Sep. 2012.

[38] E.K. Lua, J. Crowcroft, M. Pias, R. Sharma, and S. Lim. A survey and comparison of peer-to-peer overlay network schemes. *Communications Surveys Tutorials, IEEE*, 7(2):72–93, 2005.

[39] S. McLaughlin, P. McDaniel, and W. Aiello. Protecting consumer privacy from electric load monitoring. In *Proceedings of the 18th ACM conference on Computer and communications security, CCS '11*, pages 87–98, New York, 2011. ACM.

[40] A. Molina-Markham, P. Shenoy, K. Fu, E. Cecchet, and D. Irwin. Private memoirs of a smart meter. In *Proceedings of the 2nd ACM Workshop on Embedded Sensing Systems for Energy-Efficiency in Building, BuildSys '10*, pages 61–66, New York, 2010. ACM.

[41] E.M. Newton, L. Sweeney, and B. Malin. Preserving privacy by de-identifying face images. *Knowledge and Data Engineering, IEEE Transactions on*, 17(2):232–243, Feb. 2005.

[42] T.A. Nguyen and M. Aiello. Beyond indoor presence monitoring with simple sensors. In *PECCS*, pages 5–14, 2012.

[43] T.A. Nguyen and M. Aiello. Energy intelligent buildings based on user activity: a survey. *Energy and Buildings*, 56(0):244–257, 2013.

[44] M. Ohkubo, K. Suzuki, S. Kinoshita, et al. Cryptographic approach to privacy-friendly tags. In *RFID privacy workshop*, volume 82. Cambridge, USA, 2003.

[45] J.R. Padilla-López, A.A. Chaaraoui, and F. Flórez-Revuelta. Visual privacy protection methods: a survey. *Expert Systems with Applications*, 42(9):4177–4195, 2015.

[46] P. Paillier. Public-key cryptosystems based on composite degree residuosity classes. In *Proceedings of the 17th international conference on Theory and application of cryptographic techniques, EUROCRYPT'99*, pages 223–238, Berlin, 1999. Springer-Verlag.

[47] J. Pan, R. Jain, and S. Paul. A survey of energy efficiency in buildings and microgrids using networking technologies. *Communications Surveys Tutorials, IEEE*, 16(3):1709–1731, 2014.

[48] N. Saputro and K. Akkaya. On preserving user privacy in smart grid advanced metering infrastructure applications. *Security and Communication Networks*, 7(1):206–220, 2014.

[49] I. Sartori, A. Napolitano, and K. Voss. Net zero energy buildings: a consistent definition framework. *Energy and Buildings*, 48(0):220–232, 2012.

[50] A. Senior, S. Pankanti, A. Hampapur, L. Brown, Y.-L. Tian, A. Ekin, J. Connell, C.F. Shu, and M. Lu. Enabling video privacy through computer vision. *Security Privacy, IEEE*, 3(3):50–57, May 2005.

[51] M. Stegelmann and D. Kesdogan. Gridpriv: a smart metering architecture offering k-anonymity. In *Trust, Security and Privacy in Computing and Communications (TrustCom), 2012 IEEE 11th International Conference on*, pages 419–426, June 2012.

[52] M. Vesel and W. Zeiler. Personalized conditioning and its impact on thermal comfort and energy performance: a review. *Renewable and Sustainable Energy Reviews*, 34(0):401–408, 2014.

[53] S. Wang. *Intelligent building and building automation*. Routledge, 2009.

[54] S.A. Weis, S.E. Sarma, R.L. Rivest, and D.W. Engels. Security and privacy aspects of low-cost radio frequency identification systems. In D. Hutter, G. Müller, W. Stephan, and M. Ullmann, editors, *Security in Pervasive Computing*, volume 2802 of *Lecture Notes in Computer Science*, pages 201–212, Springer Berlin 2004.

[55] J.K.W. Wong, H. Li, and S.W. Wang. Intelligent building research: a review. *Automation in Construction*, 14(1):143–159, 2005.

[56] M. Zeifman and K. Roth. Nonintrusive appliance load monitoring: review and outlook. *Consumer Electronics, IEEE Transactions on*, 57(1):76–84, Feb. 2011.

[57] D. Zhang, F. Xia, Z. Yang, L. Yao, and W. Zhao. Localization technologies for indoor human tracking. In *Future Information Technology (FutureTech), 2010 5th International Conference on*, pages 1–6, May 2010.

[58] Y. Zhang, Y. Lu, H. Nagahara, and R.-I. Taniguchi. Anonymous camera for privacy protection. In *Pattern Recognition (ICPR), 2014 22nd International Conference on*, pages 4170–4175, Aug. 2014.

[59] J. Zuo and Z.-Y. Zhao. Green building research current status and future agenda: a review. *Renewable and Sustainable Energy Reviews*, 30(0): 271–281, 2014.

Chapter 8

Exploiting Mobility Social Features for Location Privacy Enhancement in Internet of Vehicles

Xumin Huang

Jiawen Kang

Rong Yu

Xiang Chen

CONTENTS

Abstract

As one of the promising branches of the Internet of Things (IoT), the Internet of Vehicles (IoV) is envisioned to serve as an essential data sensing, exchanging, and processing platform for future intelligent transportation systems (ITS). In this chapter, we aim to address the location privacy issue in the IoV by leveraging the mobility social features of vehicles. In traditional pseudonym-based solutions, the privacy-preserving strength is mainly dependent on the number of vehicles meeting at the same occasion. We notice that an individual vehicle actually has many chances to meet several other vehicles. In most meeting occasions, there are only a few vehicles appearing concurrently. Motivated by these observations, we propose a new privacy-preserving scheme, called *MixGroup*, which is capable of efficiently exploiting the sparse meeting opportunities for pseudonym changing. By integrating the group signature mechanism,

MixGroup constructs extended pseudonym-changing regions, in which vehicles are allowed to successively exchange their pseudonyms. As a consequence, for the tracking adversary, the uncertainty of the pseudonym mixture is cumulatively enlarged, and therefore location privacy preservation is considerably improved. We carry out simulations to verify the performance of MixGroup. Results indicate that MixGroup significantly outperforms the existing schemes. In addition, MixGroup is able to achieve a favorable performance even in low traffic conditions.

Keywords: Location privacy, Internet of Vehicles, vehicular social network, pseudonym, group signature

8.1 Introduction

With the rapid development of wireless technologies, especially dedicated short-range communications (DSRC) technology, the Internet of Vehicles (IoV) has become a dispensable data transmission platform. Note that the IoV has significantly facilitated the realization of the intelligent transport system (ITS) [1–3]. In the IoV, there are vehicles of advanced sensing and communication capability and smart roadside infrastructures of compact computation and storage capability. With the assistance of vehicular onboard units (OBUs) and roadside units (RSUs), communication in the IoV is resiliently extended to include vehicle-to-vehicle (V2V) and vehicle-to-infrastructure (V2I) data exchanges [4, 5]. This scenario has been conventionally depicted as a vehicular ad hoc networks (VANETs) [6]. Due to the high potential for a large variety of applications, the VANETs have received considerable attention from both academic and industrial fields.

Prospectively, VANETs are envisioned to integrate advanced computing intelligence (e.g., cloud computing) and social networking perspective, to efficiently support vehicle-, road-, and traffic-related data sensing, transmitting and processing for ITS applications, and eventually evolving toward the new paradigm of vehicular social networks (VSNs) [7].

Although VSN is expected to have a wide-range of applications in future ITS services, there are considerable challenging technical issues. As a crucial data-transmitting and processing platform for the ITS, VSNs should inherently protect the security and privacy of cyber-physical systems for ITS users [8–10].

However, for the sake of safety, vehicles are required to periodically broadcast their current position, speed, and acceleration in authenticated safety messages to surrounding neighbors. These messages increase the awareness of vehicles about their neighbors' whereabouts and warn drivers of dangerous situations, which poses that pose potential threats to the location privacy of vehicles. To address the problem, efficient schemes such as Mix-zone [11–13] and group signature [14] have been proposed for location privacy preservation. The central idea behind these schemes is to create opportunities for vehicles to obscure the eavesdropping of the adversary. However, Mix-zone is limited by the number of vehicles appearing at the pseudonyms changing occasions. Mix-zone may not perform very well in the places with few vehicles or low traffic. The group signature approach is restricted by the group size. A large-scale group has low efficiency in managing the signatures while a small group is weak in preserving privacy.

By observing the vehicular traces and exploiting the social features of mobility, we find that an individual vehicle actually has many chances to meet a lot of other vehicles. However, in most meeting occasions, only a few vehicles appear concurrently. This fact implies that, if the vehicle could cumulatively aggregate these meeting occasions, it has indeed sufficient opportunities for pseudonym mixture. Otherwise, if the vehicle performs pseudonym changing merely at places of crowded neighbors, a large number of opportunities will be wasted. In this chapter, we are motivated to propose a new privacy-preserving scheme that is capable of efficiently exploiting the potential opportunities for pseudonym mixture. By creating a local group, we construct an extended region with multiple road intersections, in which pseudonym exchanges are allowed to successively take place. Consequently, for the tracking adversary, the uncertainty of a pseudonym mixture is cumulatively enlarged, and hence location privacy preservation is substantially improved.

8.1.1   Related work

For driving safety, vehicles have to broadcast periodical messages, which consist of four-tuple information $\{Time, Location, Velocity, Content\}$. If the real identities of vehicles are used in the safety messages, their location privacy will be easily eavesdropped. For this reason, vehicles should use pseudonyms instead of their real identities. Moreover, the vehicles should randomly change their pseudonyms when driving, since the irrelevance of these pseudonyms can guarantee the location privacy of vehicles [12]. However, under consecutive adversary tracking, the pseudonym schemes are still vulnerable if vehicles keep using identical pseudonyms for a long time or change their pseudonyms at an improper occasion.

As shown in Figure 8.1, three vehicles run on a straight road. If only one vehicle changes its pseudonym from P_3 to A_1 during Δt, an adversary can easily link A_1 with P_3 since P_1 and P_2 are unchanged. Even if all three vehicles simultaneously change their pseudonyms, the location and velocity information embedded in safety messages could still provide a clue for adversaries to link the pseudonyms. Then, the pseudonyms may fail to protect location privacy. To address this privacy protection problem, previous work has proposed three major types of schemes: (1) Mix-zone, (2) group signature, and (3) silent period [15].

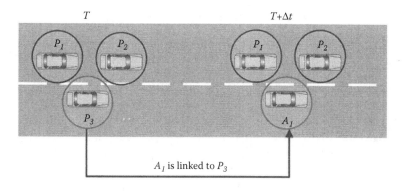

Figure 8.1: The pseudonyms are linkable.

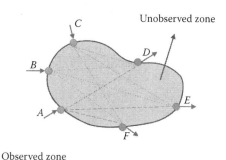

Figure 8.2: The illustration of Mix-zone scheme.

The nature of all these schemes is to obscure the mapping relationship between vehicles' real identities and their factitious identities.

The concept of Mix-zone is firstly presented in the context of location privacy in [16], and its variants are discussed in [11, 17, 18]. The vehicle uses different pseudonyms to guarantee location privacy by the unlinkability of pseudonyms. However, if a vehicle changes its pseudonym at an improper occasion, the scheme will fail to protect location privacy. The adversary could still link a new pseudonym with the old one by continuously overhearing the surrounding vehicles and inferring the pseudonym changing. In [11], the authors divide the road network into an observed zone and an unobserved zone. The unobserved zone (the gray zone as shown in Figure 8.2) works as a Mix-zone region. In this region, it is difficult for the adversaries to track vehicles because the vehicles change and mix their pseudonyms in this zone. Therefore, the Mix-zone constructs an appropriate time and location for vehicles to change their pseudonyms. Typically, at an intersection of multiple entries, the vehicles are allowed to change their pseudonyms and separately depart from different exits, which achieves the unlinkability of pseudonyms.

More specifically, there are three entrances (i.e., A, B, C) and three exits (i.e., D, E, F) in Figure 8.2. A vehicle enters the Mix-zone coverage through A and broadcasts its safety messages with the help of RSUs. The vehicle changes its pseudonyms in the coverage, and then the vehicle departs from any one exit, which ensures the unlinkability of pseudonyms. The road intersections or parking lots can naturally be assigned as Mix-zones [19]. The limitation of the Mix-zone scheme is the concurrent appearance of vehicles in the same intersection. On roads with minimal traffic, the scheme may not perform well.

For the group signature scheme, a vehicle joins a group and signs for messages using the group identity, thereby protecting its location privacy. Using a group signature scheme, the members of a group can sign a message with their respective secret keys. The resulting signature can be verified by anyone who knows the common public key, but the signature does not reveal any information about the signer except that he or she is a member of the group. Essential to a group signature scheme is a group leader, who is the trusted entity. The group leader knows the true identity of vehicles, and has the right to track down any of the group members if necessary. However, if the size of a group is too large [20], it is challenging to manage all the group members efficiently.

For the silent period scheme, a target vehicle enters a region of interest, where it initially broadcasts safety messages, then keeps silent and updates its pseudonym from P_1 to P_2 for a random silent period during moving from locations L_1 to L_2 (Figure 8.3). The vehicle finally broadcasts safety messages using P_2 in L_2. At the same time, if one of its neighboring vehicles happens to update its

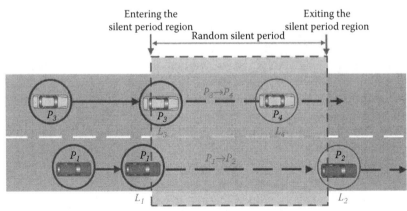

Vehicle with pseudonym P_1 broadcasts
saftey messages using updated P_2

Figure 8.3: The illustration of silent period scheme.

pseudonym (from P_3 to P_4) from proximity locations L_3 to L_4, then the adversary will be misled to treat the neighboring vehicle as the target. The random silent period scheme is efficient in resisting the adversary tracking. However, the maximum silent period is limited by the safety message broadcast period [21]. With the maximum silent period constrained to the order of hundredths of milliseconds, it is still possible to track vehicles by inferring the temporal and spatial relationship of the vehicles.

8.1.2 Contributions and organization of the paper

In this chapter, we aim to address the problem of location privacy preservation in VSNs. The main contributions of our work are presented as follow.

- First, we provide observations on vehicle traces: although social spots crowded with vehicles exist, each vehicle tends to meet others sporadically and mostly outside the social spots. Following the observations, we propose a new scheme, MixGroup, to cumulatively exploit the meeting opportunities for pseudonym changing and improve the location privacy preservation.

- Second, by leveraging group signature, we construct an extended pseudonym-changing region, namely, group-region, in which vehicles are allowed to use the group identity instead of pseudonyms, meanwhile cumulatively exchanging their pseudonyms with each other. The usage of group identity efficiently covers the procedure of pseudonym exchange.

- Third, to facilitate the operation of pseudonym exchange among vehicles, we devise an entropy-optimal negotiation procedure. In the procedure, each vehicle will evaluate its benefit and risk in taking part in pseudonym exchange. The benefit and risk during pseudonym exchange are quantitatively measured by the predefined pseudonym entropy.

The rest of this chapter is organized as follows. In Section 8.2, we introduce the network model, the threat model, and the location privacy requirements. In Section 8.3, the proposed location privacy-preserving scheme, called *MixGroup*, is presented. Firstly, two observations from vehicle traces are described. Then, we provide a brief overview of MixGroup. After that, the detailed operations and protocols of MixGroup are elaborated. In Section 8.4, the performance analysis and optimization are discussed. A performance evaluation is provided in Section 8.5. Finally, we conclude our work in Section 8.6.

8.2 System Model

8.2.1 Network model

As shown in Figure 8.4, we consider a vehicular social network deployed in an urban area. The VSN consists of a number of vehicles, roadside infrastructures,

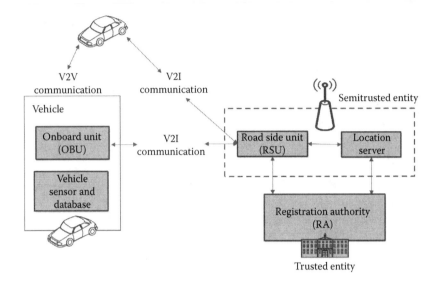

Figure 8.4: Architecture of a vehicle social network.

and an intelligent transportation system (ITS) data center. These components are explained as follows.

■ *Vehicle.* A large number of vehicles run on the roads in the urban area of interest. Each vehicle is equipped with an OBU, which allows the vehicles to communicate with each other or with the roadside infrastructures for data exchange. Each vehicle will periodically broadcast its location information for the purposes of driving safety. To protect its location privacy, each vehicle should identify itself by a predefined pseudonym instead of its real identity when broadcasting location-related safety messages.

Moreover, two hardware modules are needed for security in OBU, that is, a tamperproof device (TPD) and an event data recorder (EDR). The TPD possesses cryptographic processing capabilities and the EDR provides storage for the TPD. The EDR records the critical data of a vehicle during emergency events, such as its speed, location, time, etc. The EDR is similar to the "black box" in airplane. The EDR can be extended to record safety message broadcasts during driving. As the electronic devices are easily accessible by a driver and some mechanics, the cryptographic keys of a vehicle should be protected in the TPD. The TPD is a safe hardware to store all cryptographic material and perform cryptographic operations. The TDP stores a set of cryptographic keys with the identity binding of a given vehicle. These keys in the

TDP guarantee the accountability property. The TDP includes its own clock and has a rechargeable battery that is periodically recharged by the vehicle [22].

■ *Roadside Infrastructure.* To collect ITS-related data (e.g., the condition of the traffic, vehicles, and roads) from vehicles, roadside infrastructures are deployed along the roads of the urban area of interest. A roadside infrastructure has two main components: an RSU as a wireless communication interface and a front-computing unit (FCU) for local data processing. A roadside infrastructure can extend the communication of VANETs by redistributing or sending the information to other roadside infrastructures. The roadside infrastructure also provides Internet connectivity to OBUs, and runs safety applications, for example, accident warning or blacklist broadcasting [23]. For economic reasons roadside infrastructures are placed sparsely along the road. As a consequence, there is only intermittent coverage on the road for vehicles to access. All roadside infrastructures are connected to the ITS data center by wired backhauls.

■ *Data Center.* All ITS-related data are aggregated to the data center. The trusted registration authority, the location server, and the pseudonym database are located in the data center. The registration authority is a trusted third party operated by governmental organizations. It is responsible for the VSN, and manages the identity and credentials of all vehicles registered with it. The data center is responsible for global decision-making, such as pseudonyms generation and revocation.

Regarding their moving traces, the vehicles in a VSN exhibit inherent social features, which may be exploited for designing the privacy protection scheme. To describe the social features of the spatial distribution of vehicles, we propose the concepts of social hot spot and individual hot spot in the following.

■ *Global Social Spot.* From the perspective of a VSN, a global social spot is the place where a number of vehicles meet at a certain time. For example, a road intersection of a busy street in a Central Business District (CBD) is a typical global social spot, where many vehicles wait at red lights. It is noteworthy that global social spots are usually selected as Mix-zones in many existing works e.g., [11, 18, 19, 24].

■ *Individual Social Spot.* From the perspective of a specific vehicle, an individual social spot refers to the place where the vehicle frequently visits. For example, a road intersection near the vehicle owner's workplace and a supermarket parking lot near the vehicle owner's home are usually potential individual social spots. Actually, vehicles may share common individual social spots. For example, for people working in the same company, their vehicles have the same parking lot as a common individual social spot. In this sense, if a place is a common individual social spot of many vehicles, it is indeed a global social spot. Note that, for a specific vehicle, its individual social spots are candidate places for pseudonym changing, if it happens to meet enough vehicles there.

8.2.2   Threat model

To broadcast safety-related messages periodically, the radio of the OBU cannot be switched off when a vehicle is running on the road. As a result, an eavesdropper may track a specific vehicle and monitor its location information by leveraging these periodical safety messages [19, 25]. Location privacy protection is therefore necessary to deal with potential adversaries. In our threat model, we consider both external and internal adversaries. More specifically, two types of external adversaries, namely, a global passive adversary (GPA) and a restricted passive adversary (RPA), and two types of internal adversaries, namely, an internal betrayal adversary (IBA) and an internal tricking adversary are considered.

■ *Global Passive Adversary* (GPA). The GPA (e.g., "Big Brother" surveillance [21]) can locate and track any vehicle in a region of interest by eavesdropping its broadcasts.

■ *Restricted Passive Adversary* (RPA). The RPA (e.g., a compromised service provider) is limited in its location tracking capability in a region of interest, since it can only exploit the deployed infrastructure RSUs for eavesdropping and estimating the locations of vehicle broadcasts. Hence, the region over which the RPA can track vehicles is dependent on the vehicle transmission range and the distance between any two successive deployed RSUs [26].

■ *Internal Betrayal Adversary* (IBA). For the group signature based scheme, an internal adversary is a compromised group member who becomes an adversary after being a group member. The IBA will collude with a GPA or RPA to track a target vehicle. After exchanging privacy-related information (e.g., the pseudonyms) with the target vehicle, an IBA will leak the information to the GPA and RPA, resulting in the reconstruction of the target vehicle's trace if the target vehicle only exchanges once in the MixGroup.

For example, a vehicle V_i has some pseudonyms, denoted as PID_i. The vehicle exchanges its pseudonyms with an adversary (e.g., a compromised group member), who owns a set of pseudonyms PID_j. Finally,

V_i gets PID_j, and the adversary obtains PID_i. The adversary leaks out the pseudonym's information to a GPA or RPA. Then, the adversaries can restructure the historical trajectory of V_i by analyzing the eavesdropped record of safety messages signed by PID_i. If V_i no longer exchanges PID_j with others after departing the MixGroup zone, V_i will use PID_j to broadcast safety messages. By monitoring the safety messages signed by PID_j, the adversaries can infer the real trace of the target vehicle and continue to track the target vehicle.

■ *Internal Tricking Adversary* (ITA). Unlike the IBA, the ITA will tautologically use the pseudonyms, which had been exchanged with others more than once. The victim obtains useless pseudonyms and may exchange with others without knowing. The number of victims depends on the number of vehicles that exchange information with the ITA.

There are other methods for an eavesdropper to track a target vehicle. For example, a video-based approach using traffic-monitoring cameras is able to visually identify the target, using color, size, or license plate number. Another physical-layer approach may use specialized hardware to capture and process electromagnetic signatures, such as signal strength, or commercial-off-the-shelf hardware to passively track multiple vehicles. However, these approaches require significant efforts like expensive cameras with sufficiently high resolution to track even a single target vehicle. The adversary has to undertake the overwhelming cost of the entire system. In this chapter, we consider the adversary using the aforementioned radio-based eavesdropping, which involves only a moderate system expense.

8.2.3 Location privacy requirements

To preserve the location privacy of vehicles in vehicular social networks against the four types of adversaries mentioned previously, the requirements should be satisfied as follows [19].

■ *Identity privacy*: Identity privacy is a prerequisite for the success of location privacy. Each vehicle should use pseudonyms instead of a real identity to broadcast safety messages for the preservation of identity privacy.

■ *pseudonyms*: Each vehicle should periodically change its pseudonyms to weaken the relationship between the former and the latter locations of a vehicle. The vehicles should choose appropriate times and locations to periodically change the pseudonyms to avoid continuous adversary tracking.

■ *Conditional tracking*: Location privacy should be conditional in this chapter. The pseudonyms of vehicles should be trackable to the trusted register authority (RA). The RA is capable of disclosing the real identity as well as the location of any vehicle in the VSN. The adversaries should be held accountable for illegal activity by the RA.

In the following section, a location privacy-preserving scheme, which achieves the above requirements, is proposed and discussed for VSNs.

8.3 Proposed Location Privacy Preservation Scheme: MixGroup

In this section, we present the design of MixGroup for preserving the location privacy of vehicles in VSNs. Our discussion begins with the characteristics of vehicular social networks and two interesting and intuitive observations from real vehicle traces. The notations used in this paper are listed in Table 8.1.

Table 8.1   Standard definition of symbols used in this chapter

Notation	Description
v_i	The ith vehicle in the VSN.
$PID_{i,k}$	The kth pseudonym of vehicle i. Each vehicle has w pseudonyms, $\{PID_{i,k}\}_{k=1}^{w} = \{PID_i\}$.
G_j	The jth group of vehicles in the VSN.
GL_j	A group leader of the jth group in the VSN.
GID_j	The identity of jth group.
$SK_{G_{j,i}}, Cert_{G_{j,i}}$	Group private key of group ID and corresponding certificate for vehicle i.
$\{x\}$	A set with element x.
$L_s^{v_i}$	The sth location of vehicle v_i.
$C_k^{v_i}$	The kth exchange location of vehicle v_i.
$i \rightarrow j$	Vehicle v_i sends a message to v_j.
$x \| y$	Element x concatenates to y.
RSU_k	The kth RSU in the VSN.
$PK_i, SK_i, Cert_i$	Public and private key pair of vehicle v_i and corresponding certificate.
$PK_i', SK_i', Cert_i'$	Public and private key pair of vehicle v_i's temporary identity and the corresponding certificate.
$PK_{e,i}, SK_{e,i}, Cert_{e,i}$	Public and private key pair of vehicle v_i for pseudonym exchange and the corresponding certificate.
$E_{PK_x}(m)$	Encryption of message m with public key of entry x.
$E_{SK_x}(m)$	Encryption of message m with private key of entry x.
$Sign_{SK_x}(m)$	Digital signature on message m with private key of entry x.
$dual\text{-}signature_{i \rightarrow j}$	Dual signature from vehicle v_i and vehicle v_j.
$TimeRecord$	Time record of pseudonym exchanging event.

8.3.1 Characteristics of vehicular social networks

In a densely populated region, many people spend one or more hours every day driving between home, workplaces, and commercial districts. Since the mobilities of vehicles are restricted by the road networks, the trajectories of vehicles are predictable and regular. Day after day, the same people travel along the same roadways almost at the same time. Therefore, there are opportunities to form periodic virtual mobile communities. These virtual communities are called vehicular social networks (VSNs) [27].

A VSN is one kind of VANET, which also include traditional V2V communications and V2I communications. Compared with other VANETs, VSNs take human factors into consideration. Vehicles are driven by humans in the road networks, so that the mobility of vehicles directly reflects humans' intentions. Humans' intentions are shown by some social characteristics. The social characteristics of VSNs are as follows [7]:

- Shortest path-based movement: A vehicle randomly chooses a start point and a destination on the road networks. The vehicle uses Dijkstra's algorithm to calculate the shortest path to the destination.

- Social hot spots based model: In a VSN, there are several spots that have high social attractivity in a road network. The social attractivity is decided by the number of vehicles that are currently stopping in the spot, for example, a supermarket downtown.

- Spatiotemporal mobility model: The vehicles driven by people travel to different spots at different times every day, but almost in a periodic manner. For example, people go to the office in the morning, to the restaurant at noon, and home in the evening. Day after day, the mobility of vehicles shows some spatiotemporal laws.

8.3.2 Two observations from real vehicle traces

Through trace-based experiment and analysis [28], we have the following two observations:

8.3.2.1 Observation one

Only a few vehicles meet in global social spots, while most vehicles meet sporadically. The mobility of vehicles is spatially restricted by the shape and distribution of the roads. Usually, vehicles gather in parking lots or road intersections when the traffic lights are red. In this paper, we choose 40 major road intersections as *social hot spots* in San Francisco and observe the number of vehicles that pass by the observed intersections from 8:00 a.m. to 12:10 p.m. every 10 min. As shown in Figure 8.5a, during the 250 min of interest, about

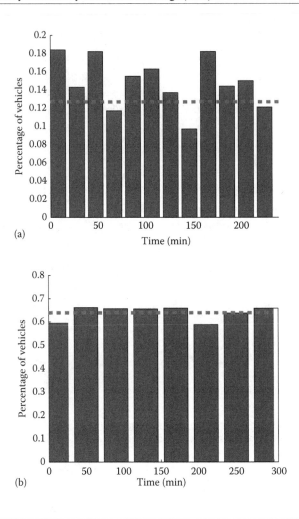

Figure 8.5: Statistics of vehicles in social spots: (a) observation one; (b) observation two.

13% of observed vehicles collectively pass by the social hot spots as a crowd every 10 min (an aggregation of more than 10 vehicles is considered a crowd). Moreover, the vehicles in geographical proximity tend to meet frequently. The other 87% vehicles navigate sparsely. Each of these vehicles meets other vehicles sporadically in different road intersections, but not necessarily in the social hot spots.

8.3.2.2   Observation two

Most vehicles always visit their individual social spots, where they meet most other vehicles that they may meet in one day. Most vehicles move with highly regular patterns every day. Each vehicle usually passes by several fixed places, marked as *individual social spots*. Furthermore, the time when they arrive at each of these places is fairly similar every day. This is because people's social behavior patterns usually remain stable within a relatively long interval [29]. We focus on the meetings of the vehicles and find that each vehicle tends to meet 64% of other vehicles that it may meet in one day in its individual social spots (see Figure 8.5b), but only 13% of other vehicles in global social spots. The above two observations jointly reveal the fact that vehicles have individual social features as well as common social features. The individual social features have a major impact on vehicles' movement patterns.

The two observations on the vehicles' mobility features could be traced back to the Pareto principle (also known as the 80–20 rule). More specifically, roughly 80% of the vehicles meet others in 20% of the social spots (i.e., the hot spots). Our observations match with the Pareto principle and go a step further to reveal the fact that the hot spots can be divided into global hot spots and individual hot spots. In roughly 80% (it is actually 77%) of the vehicles, 64% meet others in the individual hot spots, while only 13% meet others in the global social spots.

In designing location privacy preservation schemes, it is important to exploit both the common and individual social features of vehicles' movement patterns.

8.3.3   MixGroup: Brief overview

As we have pointed out, the main concern in designing location privacy protection in VSNs is to increase the number of meeting vehicles and hence maximize the uncertainty of pseudonym mixture. In traditional schemes, pseudonym changing happens only at global social spots. Consequently, a lot of mixture opportunities are *wasted*, as we know from the aforementioned two observations. In this chapter, we are motivated to propose a new location privacy preserving scheme, namely MixGroup, which aims to efficiently aggregate the potential opportunities for changing pseudonyms along vehicles' moving paths. To be more concrete, let us consider the scenario in Figure 8.6. There are global and individual social spots along the path of vehicle v_i. In the traditional scheme, v_i is allowed to change its pseudonyms in the global social spot S_3 where there are eight other vehicles at the intersection. Actually, there are still three, three and four other vehicles at the intersections of the individual social spots S_1, S_2, and S_4, respectively. To efficiently leverage these potential opportunities, the proposed scheme strategically combines the spots S_1 to S_4 to constitute an extended social region R_1. Vehicle v_i is then allowed to accumulatively exchange pseudonyms with vehicles that it meets in R_1. For instance, it may exchange pseudonyms with

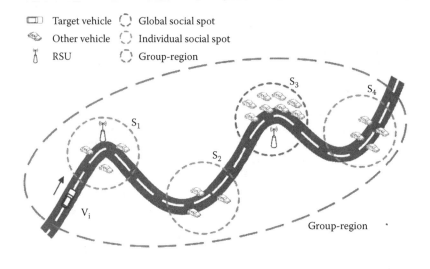

Figure 8.6: Illustration of group-region.

vehicle v_b in S_1 and then with v_c in S_3 subsequently. Theoretically, since v_i will meet a total of 18 other vehicles, the opportunities for pseudonym mixture are considerably enlarged from 8 to 18. As a consequence, the privacy preservation is much increased.

To implement the proposed scheme, four key mechanisms are devised: (1) the pseudonym mechanism, (2) the group signature, (3) temporary in-group identity, and (4) the encryption and authentication mechanisms, as explained below.

- *Pseudonym Mechanism.* In MixGroup, the usage of pseudonyms is the fundamental mechanism to protect the location privacy of vehicles. For vehicle v_i, it will be allocated with w pseudonyms. For example, $PID_{i,k}$ $(k = 1, \cdots, w)$ represents the kth pseudonym of v_i. The pseudonym is used outside the group-region for safety message broadcasting. In a group-region, vehicles will use group identities instead of pseudonyms. Pseudonyms are changed among vehicles in a group-region.

- *Group Signature.* By leveraging the mechanism of group signature, MixGroup constructs extended pseudonym-mixing regions (i.e., group-regions), in which vehicles are allowed to accumulatively change their pseudonyms. Each group has a group identity GID_j and a group leader GL_j. When vehicle v_i enters a group-region, the group leader GL_j will deliver the group identity GID_j and the corresponding group private key $SK_{G_{j,i}}$ and certificate $Cert_{G_{j,i}}$ to the vehicle after authentication. Vehicle v_i will use GID_j, $SK_{G_{j,i}}$, and $Cert_{G_{j,i}}$ for broadcasting safety messages and subsequently changing pseudonyms.

■ *Temporal In-Group Identity.* During the procedure of pseudonym exchange, each vehicle needs a dedicated identity to indicate itself and exchange pseudonyms with others. To avoid associating the real identity with the identity of pseudonym exchange and adversary tracking, neither the real identity nor the current pseudonym can be set as the dedicated identity. For this reason, we define a new ID called temporary in-group identity (TID) for each vehicle. When a vehicle v_i enters the group-region, the group leader will allocate a set of TIDs, $Pk'_{i,l}$, $SK'_{i,l}$ ($l = 1, \ldots, L$), to it. After that, TIDs will be used for sending requests and responses in the pseudonym exchanging procedures. Usually, each TID is expected to be used only once for pseudonym exchange. As a result, the adversary cannot establish the mapping relationship between a vehicle's real identity and pseudonym exchanging identity.

■ *Encryption and Authentication.* To protect wireless communication security and exclude illegal vehicles, MixGroup uses restrict encryption and authentication mechanisms. For each vehicle v_i, there are three sets of public and private keys and certificates, respectively, for real identity, TID, and pseudonym exchange. Specifically, $\{PK_i, SK_i, Cert_i\}$ are used in V2I communications through which the RA can authenticate the vehicle's real identity; $\{PK'_i, SK'_i, Cert'_i\}$ are used according to TID for sending requests and responses before pseudonym exchange; and $\{PK_{e,i}, SK_{e,i}, Cert_{e,i}\}$ are used to authenticate the validity of the two sides during pseudonym exchange.

In the system, vehicles broadcast exchanging requests to each other without location relating to their requests. Two vehicles exchange pseudonyms and relevant data by encryption with the exchanging key, for example, V_i uses $PK_{e,i}, Cet_{e,i}$ as its exchange key. A *dual-signature* is produced during the exchange process, which is used to authenticate the validity of exchange data by two sides. An Event Record device is used to record the event of exchange between two vehicles and ensure accountability. Vehicles have the right to use exchanged pseudonyms only after authorization via the RA.

8.3.4   MixGroup: Detailed operations

MixGroup mainly consists of six operations: system initialization, key generation, group join, pseudonym exchange, group leaving, and revocation. Figure 8.7 shows a state diagram of vehicles in MixGroup to explain how vehicles transit from one state to another.

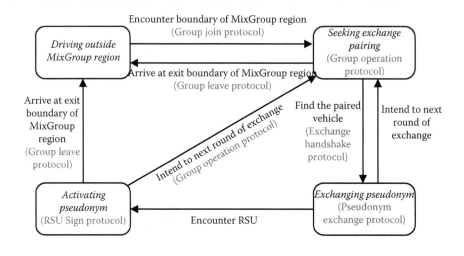

Figure 8.7: State diagram of vehicles.

8.3.4.1 System initialization and key generation

In MixGroup, we employ the efficient Boneh–Boyen short signature scheme in [20, 30] for system initialization and key generation. In the scheme, vehicle v_i with identity ID_i joins the system and gets its public/privacy key and certificate, denoted as PK_i, SK_i, and $Cert_i$, respectively. The RA stores (ID_i, PK_i) in the tracking list. The vehicle is provided with a set of w pseudonyms $\{PID_{i,k}\}_{k=1}^{w}$ by RA and accordingly public/private key pair $(PK_{PID_{i,k}}, SK_{PID_{i,k}})$ and certificates $Cert_{PK_{PID_{i,k}}}$ for each pseudonym $PID_{i,k}$. The group public key of group G_j and group private key for vehicle v_i are denoted as $\{GID_j, SK_{G_{j,i}}, Cert_{G_{j,i}}\}$, respectively. In this chapter, TIDs are generated by the RSA algorithm. After that, TIDs are delivered through RSUs (which are located at the boundary of the MixGroup region) to the vehicles when they enter the MixGroup region. It is noteworthy that TIDs are only used for sending requests and responses during the procedure of pseudonym exchange.

8.3.4.2 Group join

Before entering a group-region and joining a group, each vehicle v_i periodically broadcasts safety messages with its own pseudonyms $\{PID_i\}$ given by the RA. Upon hearing the broadcast messages from the nearby RSU, say, RSU_k, v_i will propose to the group leader GL_j through RSU_k, requesting membership of group G_j. The group leader, who is responsible for distributing and managing group identity (GID) and the associated keys and certificates, is elected by the RSUs of group G_j. The group leader GL_j verifies the legality of v_i (identity parameters of v_i included in the request) with the help of the RA. Then, GL_j provides v_i with parameters of group identity (GID) and the associated private key and certificate and also the parameters of a temporary in-group identity (TID) used during pseudonym exchange with others. After that, v_i becomes a group member and will broadcast safety message using GID_j instead of $\{PID_i\}$ to prevent the possible continuous tracking of pseudonyms from potential adversaries. In order to ensure liability of the message originator and the safety of the message receiver, each vehicle signs its safety message with a timestamp to ensure message freshness and includes the group private key and certificate to enable verification. The pseudocode of the group join protocol is presented here: Group Join Protocol (GROUP_JOIN)

---

1. v_i: `listen to the messages from neighboring` RSU_k, $RSU_k \in G_j$;
2. v_i: verify the legitimate identity of RSU_k, and change its pseudonyms from $PID_{i,k-1}$ to $PID_{i,k}, PID_{i,k} \in$
    $\{PID_i\}$;
3. $v_i \rightarrow GL_j$:
        $request = RSU_k||E_{GL_j}(join\_request||PID_{i,k}$
            $||Cet_{PID_{i,k}})||TimeStamp,$
    where $join\_request = PK_{PID_{i,k}}||location_{v_i}||velocity_{v_i}$
              $||acceleration_{v_i}||TimeStamp;$
4. **if** (**verified** $PID_{i,k}$) and ($location_{v_i}$ **is within range of** RSU_k, $RSU_k \in G_j$)
    $GL_j \rightarrow i:$
    $reply = E_{PK_{PID_{i,k}}}(Group\_key||TID\_key||Cet_{GL_j})$
        $||TimeStamp,$
    where $Group\_key = GID_j||SK_{G_{j,i}}||Cert_{G_{j,i}},$
        $TID\_key = PK_i'||SK_i'||Cert_i';$
    else
        GL_j: do not reply;
    endif
5. **if** (received reply within T_{max})
    v_i: broadcast by GID_j instead of $PID_{i,k}$,
        $broadcast = GID_j||navigation\_data_i||Sign_{SK_{G_{j,i}}}$
              $(navigation\_data_i)||Cet_{G_{j,i}},$
    where $navigation\_data_i = location_{G_j}||velocity_{G_j}$
        $||acceleration_{G_j}||TimeStamp,$
    v_i: go to GROUP_OPERATION when meeting other vehicles;
    else
        v_i: go to step 3;
    endif

8.3.4.3 Pseudonym exchange

When vehicle v_i navigates as a group member of G_j, it will periodically broadcast safety messages with the identity GID_j. Once vehicle v_i meets other group members of G_j, there is an opportunity to exchange their pseudonyms. At this moment, the vehicle will broadcast a pseudonym exchange request. In traditional Mix-zone [11], vehicles change their pseudonyms at road intersections under the assistance of RSUs. The operations of pseudonym changing in MixGroup are different from that in Mix-zone. Two vehicles of a same group are allowed to directly exchange their pseudonyms without the involvement of RSUs. This means that pseudonym changing can be performed outside the coverage area of RSUs. Furthermore, the newly exchanged pseudonyms would not be used immediately but after the vehicles leave the group-region. Instead, the group identity is still used for broadcasting safety messages. The usage of group identity is beneficial to "cover" the procedure of pseudonym exchange. By leveraging the group signature mechanism, pseudonym changing in MixGroup may take place anywhere as a vehicle meets with other vehicles.

The procedure of pseudonym exchange has several steps. First, if vehicle v_i finds out that there are other vehicles in the proximity (by hearing safety messages) and it attempts to exchange pseudonyms, v_i will broadcast a pseudonym exchanging request message associated with its public key of TID PK_i'. After receiving the request messages from other vehicles, vehicle v_i will compute its own exchange benefit and decide whether to exchange at this time or not. In this paper, the exchange benefit is quantitatively evaluated by the *pseudonym entropy*. The procedure of negotiation on the participation of pseudonym exchange will be elaborated in Section 8.4. If vehicle v_i decides to exchange with others, it will randomly select a neighbor vehicle, say, v_j (actually indicated by the TID), and send a pseudonym exchange proposal to v_j, which is encrypted with public key of v_j's TID (i.e., PK_j') in its broadcast request. With the agreement from v_j, v_i will receive and verify the response of v_j including exchanging public key $PK_{e,j}$ and the associated certificate. The pseudocode for the operations of pseudonym exchange is presented here, including three protocols: GROUP_OPERATION, EXCHANGE_HANDSHAKE, and PSEUDONYM_EXCHANGE.

Group Operation Protocol (GROUP_OPERATION)

```
1. vᵢ: receive and verify broadcast messages from neighbors;
2. vⱼ: receive and verify broadcast messages from neighbors;
3. if (vᵢ wants to exchange {PIDᵢ} with neighbor vⱼ)
      vᵢ: broadcast request = (exchange_request||PKᵢ'
                ||Certᵢ'||Cert_{Gⱼ,ᵢ}||TimeStamp)
      and go to EXCHANGE_HANDSHAKE protocol;
   else
      vᵢ: go to step 1;
   endif
```

Exchange Handshake Protocol (EXCHANGE_HANDSHAKE)

```
1. vᵢ: receive pseudonym exchanging request from neighbors;
2. vᵢ: verify and evaluate the benefit to decide whether to exchange right now;
3. vᵢ: if (exchange)
      3.1 vᵢ: randomly choose a vehicle vⱼ with PKⱼ' ;
      3.2 vᵢ → vⱼ:
          proposal = PKⱼ'||E_{PKⱼ'}(exchange_proposal
                 ||Certᵢ'||Sign_{SKᵢ'}(Certᵢ'))||TimeStamp);
      3.3 vⱼ → vᵢ:
      if (vⱼ agrees to exchange)
          response = PKᵢ'||E_{PKᵢ'}(reponse_confirm||M
                 ||Sign_{SKⱼ}(M)||Certⱼ'),
      where M = PK_{e,j}||Cet_{e,j}||Sign_{SK_{e,j}'}(PK_{e,j}||Cet_{e,j})
                 ||TimeStamp;
```

3.4 $v_i \rightarrow v_j$:
$$reply = E_{PK_{e,j}}(PK_{e,i}||Cert_{e,i}||SIG||TimeStamp),$$
where $SIG = Sign_{SK_{e,i}}(PK_{e,i}||Cert_{e,i}||TimeStamp);$
3.5 go to PSEUDONYM_EXCHANGE protocol;
else
    $v_j \rightarrow v_i$:
$$response = PK'_i||E_{PK'_i}(disagree||Cert'_j$$
$$||Sign_{SK'_j}(Cert'_j)||TimeStamp);$$
3.6 v_i: go to step 3.1;
endif
else
    3.7 v_i: go to step 1;
endif

Pseudonym Exchange Protocol (PSEUDONYM_EXCHANGE)

1. $v_i \rightarrow v_j$:
$$Pseudonyms_{i \rightarrow j} = E_{PK_{e,j}}(data\_1|Sig\_1||Cet_{G_j}$$
$$||TimeStamp),$$
where $data\_1 = PID_{i,k}||Cert_{PID_{i,k}}||Sign_{SK_{PID_{i,k}}}$
$$(PID_{i,k}||Cert_{PID_{i,k}})$$
$$Sig\_1 = Sign_{SK_{e,i}}(data\_1);$$
2. v_j: verify and store data from v_i;
3. $v_j \rightarrow v_i$:
$$Pseudonyms_{j \rightarrow i} = E_{PK_{e,i}}(data\_2||Sig\_2||Sig\_1$$
$$||Dual\text{-}signature_{j \rightarrow i}||TimeStamp),$$
where $data\_2 = PID_{j,k}||Cert_{PID_{j,k}}||Sign_{SK_{PID_{j,k}}}$
$$(PID_{j,k}||Cert_{PID_{j,k}}),$$
$$Sig\_2 = Sign_{SK_{e,j}}(data\_2),$$
$$Dual\text{-}signature_{j \rightarrow i} = E_{SK_{e,j}}(Sig\_1||TimeStamp);$$
4. v_i: verify and store from v_j;
$v_i \rightarrow v_j$:
$$data_i = E_{PK_{e,j}}[Dual\text{-}signature_{i \rightarrow j}||Sig\_2$$
$$||TimeRecord||TimeStamp],$$
where $Dual\text{-}signature_{i \rightarrow j} = Sign_{SK_{e,i}}(Sig\_2$
$$||TimeStamp);$$
5. v_j: verify and store data from v_i ;
6. v_i: $Record\_1 = E_{PK_{RA}}(Cert_{e,i}||Cert_{e,j}$
$$||\{PID_i\}\ ||\{PID_j\}||Add\_data),$$
v_j: $Record\_2 = E_{PK_{RA}}(Cert_{e,i}||Cert_{e,j}||\{PID_i\}$
$$||\{PID_j\}||Add\_data),$$
where $Add\_data = TimeRecord;$
7. v_i: send $Record\_1$ to v_j;
v_j: send $Record\_2$ to v_i;
8. v_i: compare received $Record\_2$ with $Record\_1$,
if ($Record\_2$ and $Record\_1$ are identical)
    $i \rightarrow j$:
$$R_2 = E_{PK_{e,j}}(Record\_2||SigR_{i \rightarrow j}||TimeRecord),$$
where $SigR_{i \rightarrow j} = Sign_{SK_{e,i}}(Record\_2||Timestamp);$
9. v_j: verify and store data from v_i;
10. v_j: compare received $Record\_1$ with $Record\_2$,
if ($Record\_2$ and $Record\_1$ are identical)
    $v_j \rightarrow v_i$:
$$R_1 = E_{PK_{e,i}}(Record\_1||SigR_{j \rightarrow i}||TimeRecord),$$
where $SigR_{j \rightarrow i} = Sign_{SK_{e,j}}(Record\_1||Timestamp);$
11. v_i: verify and store data from v_j.

8.3.4.4 RSU signing protocol

As mentioned above, a vehicle may meet and exchange its pseudonyms and the associated certificates with other vehicles. However, before having permission to use the exchanged pseudonyms, the vehicle should firstly activate the pseudonyms by the RA through the RSUs. After pseudonym exchange with the last vehicle, say, v_j, v_i will listen to the broadcast messages of RSUs nearby. When connected to an RSU, say RSU_m, v_i will send a signing request to it with *Exchange_data* and *Personal_data* that are encrypted by the public key of the RA. The Exchange_data includes exchanged pseudonyms and a dual-signature signed by v_i and v_j to prevent forgery. The RA validates the Personal_data to verify the legal identify of v_i and distributes a new exchanging key pair for the next exchange and renewed certificates of $\{PID_j\}$ to v_i. The RA will keep a record of these data, while v_i will also verify and store them. If the Exchange_data is invalid, the RA will redistribute valid pseudonyms and certificates in its backup list to v_i. The pseudocode of the RUS_SIGN protocol is illustrated here:

RSU Sign Protocol (RSU_SIGN)

```
1. v_i: receive and verify broadcast from RSU_m and decide to activate the new
pseudonyms (RSU_m ∈ G_j);
2. v_i → RSU_m   (RA):
request_sign = RSU_m||E_PK_RA (Exchange_data
            ||Personal_data)||TimeStamp,
where Personal_data = PK_e,i||Cert_e,i,
Exchange_data = (PID_j||Cert_e,i
            ||Dual − signature_j→i||Sig_j→i),
SigR_j→i = Sign_SK_e,j (Record_1||Timestamp);
3. RA: if (validate Personal_data and v_i)
         go to REVOCATION
     else
       3.1 if (Exchange_data valid)
           send new exchanging keys and certificates,
           RA → v_i:
update = E_PK_i (new_key||new_certification
            ||new_pseudonyms||Cert_RA)||TimeStamp
       where new_key = Hash(PID_j||SK_i
                   ||Cert_i||TimeStamp)
         new_certification = Hash(PID_j||Cert_i
                   ||TimeStamp)
         new_pseudonyms = PID_j
       3.2 v_i: validate and store renewed data;
         else
       3.3 RA: redistribute pseudonyms for v_i
           go to REVOCATION;
           endif
       3.4 go to GROUP_LEAVE;
       endif
```

8.3.4.5 Group leaving

After moving out of the group-region, a vehicle will broadcast safety messages using the newly changed pseudonyms. The procedure of group leaving is described in the following. When v_i receives the message from the RSU located at the region boundary, it will prepare for the group leaving by sending the newly changed pseudonyms to the RA as the RSU signing protocol. As long as vehicle v_i passes by the boundary RSU and cannot receive the signal, it will replace the group identity GID_j with $PID_{j,k}$ for safety message broadcast. For the RSU and RA, when GL_j does not receive any safety messages from v_i with certificate $Cert_{G,i}$ for a maximum time T_{max}, GL_j believes that v_i has left the group. As a result, it will delete the entry v_i from the group member list. When leaving the group, v_i will determine by itself whether it is necessary to find a new group (in the next group-region) or remain using the pseudonyms for a while. The pseudocode of GROUP_LEAVING is presented below.

Group Leave Protocol (GROUP_LEAVE)

```
1. vi: compute distance from zone boundary of Gj;
2. vi: if (before going out of Gj at leave_time, t)
3. vi: randomly choose t to use PIDj instead of GIDj
   go to GROUP_JOIN
4. GLj: if (no broadcast received from vi during Dmax )
GLj: delete entry vi from current group member list
      endif
    else
5. go to GROUP_OPERATION
   endif
```

8.3.4.6 Revocation protocol

In MixGroup, any violation of vehicles will be monitored and accused by neighboring vehicles or RSUs. For example, if a compromised vehicle v_k in the group is detected by v_i, v_i will record the violation actions of v_k and report to the group leader GL_j. There is vital evidence in the report. If v_i stays in the group-region, the report will include information such as the type of violation of v_k, the group certificates of v_i and v_k, and the messages signed by v_i. If v_i has left the group, it will integrate the pseudonym $PID_{i,n}$, the public key $PK_{PID_{i,n}}$ and the certificate $Cert_{PID_{i,n}}$ into the report. After receiving the report, the group leader GL_j will check the validity of the report as well as the identity of v_i and then forward them to the RA. The RA will validate the report and repeal the true identity of v_k by the tracking list. If the violation is confirmed, the RA will add v_k to its blacklist and broadcast a new blacklist to all RSUs and vehicles in the VSN.

The pseudocode of REVOCATION protocol is shown here:

Revocation Protocol (REVOCATION)

```
1. v_i: if (being in G_j);
2. v_i → GL_j: accuse vehicle v_k to GL_j;
report_1 = E_{GL_j}(VIO{type||Mess_1}||Cert_{G_{j,i}}
           ||TimeStamp)
where
Mess1 = (GID_j||message||Sig_{G_j}(message)
         ||Cert_{G_{j,k}}||TimeStamp)
3. GL_j → RA: validate report_1 and send report to RA
report = E_{PK_{RA}}(Mess||GL_j||Cert_{GL_j}||TimeStamp)
where Mess = Mess_1||Cert_{G_{j,i}}
4. RA: validate report, repeal v_k and add v_k into blacklist;
5. RA → GL_j and all vehicles: broadcast newest blacklist;
   else
6. v_i → RA: accuse vehicle v_k to RA
report = E_{PK_{RA}}[VIO{type||Mess}||PK_{PID_{i,n}}
         ||Cert_{PID_{i,n}}||TimeStamp]
where Mess = (PID_k||message||Sig_{SK_{PID_k}}(message)
             ||TimeStamp);
7. RA: validate report, repeal v_k and add v_k into blacklist;
8. RA → GL_j and all vehicles: broadcast newest blacklist;
   endif
```

8.3.4.7 Conditional tracking

When a vehicle is in a group G_j, its periodical broadcasting message includes the safety-related data and the group certificate $Cert_{G_{j,i}}$. Although group members of G_j only can verify the validity of the safety message, the RA can link all the messages with certificates to the true identities of vehicles by checking the tracking list. When a vehicle is out of any group and uses its own pseudonyms for communication, its safety message also includes the certificate, which can be identified by RA. In other words, the true identity of each vehicle is totally unconcealed for the trust RA, but conditionally private for the group leader and unknown for the other common vehicles.

8.3.4.8 Discussions

In the proposed scheme, there are two separate procedures related to the pseudonym changing: the pseudonym exchange procedure and the pseudonym activation procedure. These two procedures are efficiently integrated to allow distributed pseudonym changing. The pseudonym exchange procedure only involves vehicles. Vehicles of the same group are allowed to exchange their pseudonyms directly and out of the coverage area of RSUs. In addition, a vehicle is allowed to accumulatively exchange its pseudonym with others without the involvement of RSUs. During the pseudonym activation procedure, vehicles have to activate their pseudonyms through RSUs. After exchanging its pseudonym with others, a vehicle will activate the new pseudonym whenever it meets the RSUs. In this sense, it is unnecessary to have continuous RSU radio coverage. Eventually, when a vehicle encounters the RSUs at the boundary of the MixGroup region, it will have a final check to ensure the pseudonym activation procedure is carried out.

8.4 Security Analysis

In this section, we discuss the possible attacks and the corresponding defense measures in MixGroup. In addition, we present the optimization of pseudonym exchange to improve the pseudonym entropy against location privacy tracking.

8.4.1 *Attack and defense analysis*

In principle, the strength of location privacy preservation in pseudonym-based schemes depends on the uncertainty (i.e., entropy) in mapping pseudonyms to real vehicle identities from the perspective of an adversary. Accordingly, the central idea of MixGroup is to combine the successively located individual hot spots of the target vehicle into an extended pseudonym-changing region. Since the area of the region is considerably enlarged, and vehicles are allowed to accumulatively change their pseudonyms, the uncertainty of pseudonym mixture is significantly improved, and, the privacy preservation is consequently enhanced.

MixGroup has favorable defense ability against many security and privacy attacks. For example, due to the encryption and authentication mechanisms, the adversary is computationally bounded and unable to launch brute-force cryptanalytic attacks on the encrypted messages. Furthermore, since all messages are authenticated, it is difficult for the adversary to emulate the legal vehicles. The replay attacks would not be successful due to the usage of timestamps. Meanwhile, the adversary cannot simulate an RSU or forge the RSU messages and therefore cannot create a fictitious MixGroup with valid keys it controls.

In the following, we will further discuss several essential attacks and the defense measures of MixGroup.

8.4.1.1 *GPA and RPA*

For GPA and RPA, the adversary passively eavesdrops vehicles' safety messages and observes the times and the locations of the entering and exiting vehicles in order to derive a probability distribution over the possible mappings. If there are few vehicles in the pseudonym-changing place, the adversary will still have a high probability of following the target vehicle. However, MixGroup is not limited to one pseudonym changing in one place. When a vehicle enters a group-zone, it meets many vehicles during navigation. By using a uniform group signature, the vehicle is allowed to exchange pseudonyms with any vehicle passing by. In this case, for GPA and RPA, it is hard to track a target if it is "mixed" with a sufficiently large number of vehicles. All these vehicles look identical under the protection of group signature. As a consequence, the GPA and RPA will be lost in tracking a target.

8.4.1.2   Incorrect data attack

The internal adversary can perform an attack on vehicle safety by misbehaving and broadcasting incorrect data to attack neighboring vehicles. However, in MixGroup, since each vehicle signs the safety messages (see Step 5 of GROUP_JOIN protocol), the adversary will be held liable for providing incorrect data. In order to detect such attacks, each vehicle must be able to detect the incorrect safety messages. In [31], an efficient scheme is proposed to detect incorrect data, by enabling each vehicle to maintain its own observations of the neighborhood (such as estimated locations of neighboring vehicles) and checking data received from neighbors for any inconsistencies.

8.4.1.3   Liability attack

The adversary may perform an attack on the vehicle liability. In order to evade liability, the adversary can counterfeit a random pseudonym in the VSN. Actually, such an attack is prevented in MixGroup. It is mentioned that safety messages from each vehicle must contain valid certificates and, furthermore, be signed by a legal group signature if inside the group or by an authenticated pseudonym if outside the group. The vehicles can authenticate the validity of the safety messages. The adversary can also attempt to impersonate the target vehicle using one of its overheard pseudonyms and the associated certificate [21]. Such impersonation attacks are avoided in our model by making each vehicle sign on the safety message and include a valid certificate from the RA according to the pseudonym in usage.

8.4.1.4   IBA and ITA

In this chapter, there are two kinds of special internal adversaries, namely internal betrayal adversary (IBA) and internal tricking adversary (ITA). For the IBA, it exchanges its pseudonyms, which had been used, with the target vehicle. The IBA finds out the pseudonyms that the target had used, and the target vehicle may be tracked if it no longer exchanges the pseudonyms from the IBA when it is out of side the group-regions. The IBA may share this information with the GPA: which pseudonyms the target will use when it is outside the group-regions and which pseudonyms the target used before entering the group-regions. The GPA can link the locations of vehicle by eavesdropping safety messages signed by these pseudonyms. However, it can be easy to resist if the target vehicle exchanges pseudonyms with one or more vehicles. The adversary cannot precisely link the pseudonyms to the target. As we know, the more vehicles it exchanges with, the lower the risk, but the higher the overhead. Additionally, with the help of neighboring group members and the group leader, the compromised group members would soon be accused. By then, the adversary would be expelled from the system.

In internal tricking adversary (ITA), the adversary will tautologically use its pseudonyms that have been exchanged with others and repeatedly perform the PSEUDONYMS_EXCHANGE protocol. The victim vehicles receive overused pseudonyms and exchange with others. The number of victims depends on the vehicles' number of exchanges with the ITA. For this adversary, with the help of *dual-signature* and signed record *SigR*, the RA can detect these adversaries through *SigR*, which is unchanged by the adversary because of encryption with the RA public key (as shown in the PSEUDONYMS_EXCHANGE protocol in Section 8.3.4.3). When the vehicles detect these adversaries, they report to the RA. The adversaries are put on a blacklist and charged with responsibility later.

8.4.2 *Entropy-optimal pseudonym exchange*

The meetings of vehicles are underlying opportunities for vehicles to enhance their location privacy. However, there are potential threats from internal IBA and ITA attacks, by which the pseudonym information of a legitimate vehicle may be copied and leaked out. Therefore, it is not always beneficial for a vehicle to exchange its pseudonyms with others. In this chapter, we define *pseudonym entropy* to measure the strength of location privacy protection for vehicles. Consider a road intersection where a collection of vehicles, denoted by $V = \{v_1, v_2, \ldots, v_K\}$, will exchange pseudonyms with each other. Let p_i represent the successful tracking probability of vehicle v_i after pseudonym exchanges. The pseudonym entropy for v_i is presented by

$$H_{v_i} = -\log_2 p_i. \tag{8.1}$$

The pseudonym entropy for the collection V is given by

$$H_V = -\sum_{i=1}^{K} p_i \log_2 p_i. \tag{8.2}$$

Clearly, the successful tracking probability p_i depends on the number of internal adversaries (IBA or ITA) inside the collection V. Suppose there are a total of N vehicles in the VSN of interest, and B of them are internal adversaries, denoted by collection V_{IA}. The probability that v_i happens to select v_j, which is an internal adversary for pseudonym exchange, is derived by

$$\Pr\{v_j \in V_{IA}\} = \sum_{i=1}^{B} \frac{\binom{B}{i}\binom{N-1}{N-1-i}}{\binom{N-1}{K-1}} \frac{i}{K-1} = \frac{B}{N-1}. \tag{8.3}$$

By pseudonym exchange, the increase of v_i's pseudonym entropy is given by

$$\Delta h = \sum_{i=1}^{B} \frac{\binom{B}{i}\binom{N-1}{N-1-i}}{\binom{N-1}{K-1}} \log_2(K-i). \tag{8.4}$$

After the kth pseudonym exchange, the pseudonym entropy of v_i is represented by

$$H_{v_i}(k) = \begin{cases} 0, & v_j \in V_{IA}, \\ H_{v_i}(k-1) + \Delta h, & v_j \notin V_{IA}. \end{cases} \tag{8.5}$$

Following Equation 8.5, each vehicle will evaluate the benefit and the risk in pseudonym exchange. For vehicles that already have high pseudonym entropy, they tend to skip the pseudonym exchange; while for vehicles of low pseudonym entropy, they expect to take the opportunity to enhance their location privacy. More concretely, a vehicle is willing to exchange pseudonyms if the possible increase of its pseudonym entropy is sufficiently large, that is,

$$\Delta h > \frac{\Pr\{v_j \in V_{IA}\}H_{v_i}(k-1)}{1 - \Pr\{v_j \in V_{IA}\}}. \tag{8.6}$$

To facilitate the decision-making of pseudonym exchange among vehicles, we elaborately devise the following negotiation procedure.

■ *Sending Pseudonym Exchange Request.* Vehicles will broadcast pseudonym exchange requests periodically and, meanwhile, listen to other vehicles' requests. Given a number of vehicles at a road intersection, the negotiation takes several rounds.

■ *Evaluating Pseudonym Exchanging Benefit.* In each round of negotiation, the vehicle will first observe the number of candidate vehicles for exchange and then evaluate the benefit using Equation 8.6. If the condition of (8.6) is satisfied, it will send out a pseudonym exchange confirmation message; otherwise, it will broadcast a pseudonym exchange ending message to indicate it will skip the opportunity.

■ *Observing Pseudonym Exchanging Candidates.* A vehicle observes the pseudonym exchanging candidates by listening to the pseudonym exchange requests and confirmation/ending messages of its neighboring vehicles. Initially, all vehicles are treated as candidates.

■ *Selecting Pseudonym Exchanging Candidates.* After receiving the confirmation messages of all candidates, each vehicle will randomly select one of the candidates for exchange. If a vehicle is selected by multiple vehicles, it has the right to choose one from them. Then, the two vehicles send their exchanging public keys and associated certificates to each other. During the procedure of pseudonym exchange, vehicles are paired to exchange pseudonyms.

When there are an odd number of vehicles, the unpaired vehicle may randomly select a paired vehicle for pseudonym exchange. In this case, the selected vehicle will sequentially exchange pseudonyms twice. Alternatively, the unpaired vehicle may skip the current exchange procedure until meeting other vehicles. As observed in Section 8.3.2, each vehicle in the MixGroup region has enough chances to meet and exchange pseudonyms with others. Even if a vehicle leaves a MixGroup region without exchanging pseudonyms with others, the adversary cannot identify whether the vehicle has exchanged with others. Therefore, the vehicle can also protect its privacy in this case.

8.5   Performance Evaluation

In this section, we study the performance of a proposed MixGroup scheme using a self-developed network simulator based on NS-3 [32] and SUMO [33]. We use synthetic vehicle traces and road maps to simulate different traffic conditions and group-region coverage ratios. Specifically, we consider a city region of 20 km². We investigate different traffic conditions: 500 vehicles for low traffic load, 1000 vehicles for medium traffic load, and 1500 vehicles for high traffic load. RSUs are placed at road intersections evenly with different density: 0.5/km² for sparse deployment, 1/km² for medium deployment, and 2/km² for dense deployment. The radio coverage radius of RSUs and OBUs is set to be 500 m, which is a typical range of the IEEE 802.11p WAVE protocol. It is noteworthy that, by integrating group signature and pseudonym changing, the proposed scheme is operated in a distributed way. This means that, even if the city road map has a larger size, the proposed scheme still works efficiently when we deploy more MixGroup regions. Table 8.2 shows the simulation parameters, most of which are common settings in existing work [24].

8.5.1   Global pseudonym entropy of VSN

Figure 8.8 shows the global pseudonym entropy of the VSN. For comparison, we set the launch time of the MixGroup at time 0, while the global pseudonym entropy is reset to 0. We know from Figure 8.8a that the global pseudonym entropy increases rapidly as vehicles start to exchange their pseudonyms within

Table 8.2   Parameter setting in the simulation

Parameter	Setting
Safety distance	10 m
Node density	[10, 160] vehicles/street
Node speed	[25, 70] km/h
Meeting frequency	[10, 30] times/h
Region	10×10 uniform street grid, 0.5 km street separation, 40 intersections, two-lane one-way street or two-lane two-way street, 3 m lane separation

the group-regions. More importantly, we find that the traffic conditions have a significant impact on the increasing rate of the global pseudonym entropy. This fact is easy to understand. The more vehicles on the roads, the more opportunities for pseudonym exchange. In the case of medium traffic conditions of a total of 1000 vehicles, the global pseudonym entropy is 35% larger than that in low traffic conditions of 500 vehicles. In the case of high traffic conditions of 1500 vehicles, the global pseudonym entropy is only 10% larger than that in medium traffic. This is due to the increasing traffic load; the traffic congestion will slow down the frequency of pseudonym exchanges among vehicles.

In Figure 8.8b, different tracking attack strengths are considered. In the simulation, we suppose that all adversary vehicles are of the four types of attacks: GPA, RPA, ITA, and IBA. Regarding the high traffic conditions, the cases of 10, 30, and 50 adversary vehicles are investigated as weak, medium, and strong attacks, respectively. We can see that the global pseudonym entropy under different attack strengths initially have the same increasing rate, but finally converge to different values. The global pseudonym entropy under weak attack is more than twice that under strong attack.

The global pseudonym entropy under different group-region coverages is reported in Figure 8.8c. The ratio of group-region coverage is set according to RSU density. For example, 50%, 30%, and 15% coverage of group-region are set to dense, medium, and sparse RSU deployment, respectively. We can observe from the figure that, in a large group-region coverage, vehicles tend to meet each other more frequently and, therefore, the resulting global pseudonym entropy is clearly larger than that in a small group-region coverage.

8.5.2   Pseudonym entropy of target vehicle

The second simulation is carried out to evaluate the pseudonym entropy of a specific target vehicle. Both the expected and actual pseudonym entropy are investigated. We select a vehicle of active social activity and track the variation of

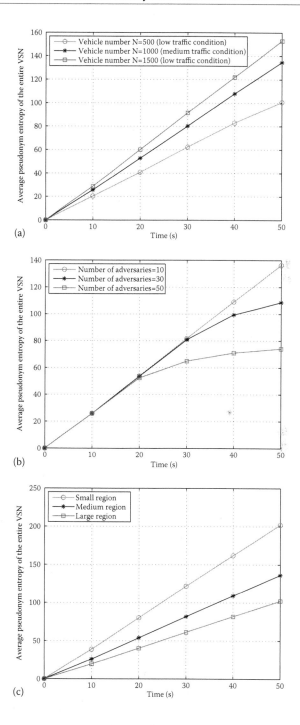

Figure 8.8: Global pseudonym entropy of the entire VSN under (a) different traffic conditions, (b) different attack strengths, and (c) different city region sizes.

its pseudonym entropy. The pseudonym entropy is reset to 0 at the beginning of the simulation. After that, the vehicle enters group-regions for pseudonym exchange. Figure 8.9a shows that the vehicle will meet more vehicles and its pseudonym entropy is improved faster in heavy traffic conditions. There is a gap between the actual pseudonym entropy and the expected pseudonym entropy. The reason is that the presence of an adversary poses a potential risk to the vehicle. Especially for the ITA and IBA, if the target vehicle happens to choose an ITA or IBA for pseudonym exchange, its location privacy will be violated as the pseudonym entropy is reset to 0. We also know from Figure 8.9b that both the expected and actual pseudonym entropy of the target vehicle decrease rapidly with the increase in attack strength. In addition, the denser the group-region coverage, the larger the pseudonym entropy of the vehicle, as shown in Figure 8.9c.

8.5.3 Comparison with existing schemes

We also compare our proposed MixGroup with two existing location privacy protection schemes: Mix-zone and PCSS. Mix-zone [11] is a well-known scheme for preserving vehicle location privacy. PCSS [19], referring to pseudonym changing at social spots, is an efficient scheme that exploits the social feature of vehicles and performs pseudonym changing at social spots (actually mentioned as *global social spots* in this chapter). In the simulation, two types of RSU coverage density are considered, which accordingly have dense and sparse coverage of Mix-zone/group-region.

In Figure 8.10, the global pseudonym entropies of the VSN in the three schemes are compared. We observe that, in dense coverage, the global pseudonym entropy in MixGroup is about 56% and five times higher than that in PCSS and Mix-zone, respectively. While in sparse coverage, the global pseudonym entropy in MixGroup is approximately 28% and four times higher than that in PCSS and Mix-zone, respectively. In Figure 8.11, the actual pseudonym entropy of a target vehicle is investigated. As the figure has shown, in dense coverage, the actual pseudonym entropy in MixGroup is 47% and 96% higher than that in PCSS and Mix-zone, respectively. In sparse coverage, the actual pseudonym entropy in MixGroup is 29% and 3.8 times higher than that in PCSS and Mix-zone, respectively.

From the above results, we know that MixGroup significantly outperforms the other two schemes. The advantage of MixGroup over the other two schemes remains remarkable in the case of sparse coverage. In low traffic situations, few vehicles appear at road intersections concurrently. However, MixGroup has the natural ability to accumulatively exploit the vehicle meeting opportunities. The number of aggregated meeting vehicles stays at a moderate level, even in low traffic conditions. As a consequence, MixGroup still has satisfying performance in low traffic conditions.

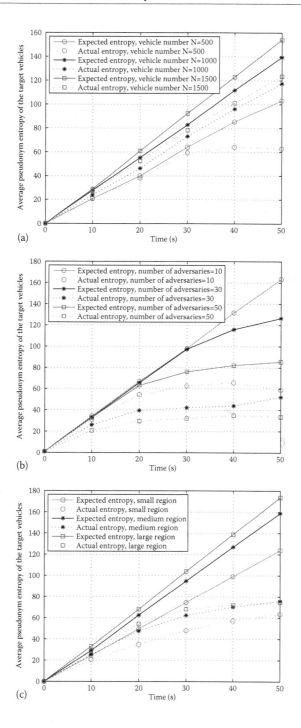

Figure 8.9: Expected and actual pseudonym entropy of a target vehicle under (a) different traffic conditions, (b) different attack strengths, and (c) different group-region coverage.

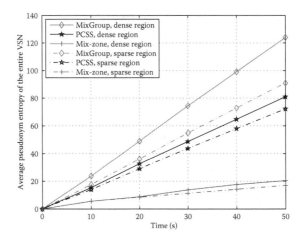

Figure 8.10: Performance comparison of global pseudonym entropy of the VSN.

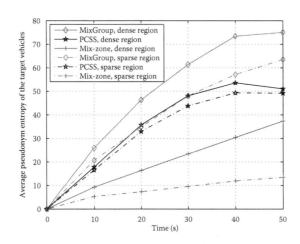

Figure 8.11: Performance comparison of actual pseudonym entropy of a target vehicle.

8.6   Conclusion

In this paper, we exploit the mobility social features of vehicular traces and then propose a new location privacy protection scheme call MixGroup in the IoV. MixGroup integrates the mechanism of group signature and constructs an extended pseudonym-changing region. The pseudonym entropy of vehicles is consecutively increased by accumulatively exchanging pseudonyms in the MixGroup region. As a consequence, the location privacy is substantially enhanced. Moreover, we propose the entropy-optimal negotiation procedure to facilitate local pseudonym exchange among vehicles. Simulation results indicate that MixGroup works very well even under low traffic conditions. Meanwhile, through comparison, MixGroup is shown to significantly outperform existing schemes.

Acknowledgment

The work is supported in part by programs of the National Nature Science Foundation of China (NSFC) under grant nos. 61422201, 61370159, and U1201253; Guangdong Province Natural Science Foundation under grant no. S2011030002886; High Education Excellent Young Teacher Program of Guangdong Province under grant no. YQ2013057; and Science and Technology Program of Guangzhou under grant no. 2014J2200097 (Zhujiang New Star Program).

Bibliography

[1] L. B. Othmane, H. Weffers, M. M. Mohamad, and M. Wolf, A survey of security and privacy in connected vehicles, in D. Benhaddou and A. Al-Fuqaha (eds) *Wireless Sensor and Mobile Ad-Hoc Networks*, pp. 217–247, Springer, New York, 2015.

[2] R. G. Engoulou, M. Bellaïche, S. Pierre, and A. Quintero, Vanet security surveys, *Computer Communications*, vol. 44, pp. 1–13, 2014.

[3] R. Hochnadel and M. Gaeta, A look ahead network (lanet) model for vehicle-to-vehicle communications using dsrc, in *Proc. of the ITS World Congress*, vol. 8, 2003.

[4] M. H. Eiza and Q. Ni, An evolving graph-based reliable routing scheme for vanets, *IEEE Transactions on Vehicular Technology*, vol. 62, no. 4, pp. 1493–1504, 2013.

[5] M. H. Eiza, Q. Ni, T. Owens, and G. Min, Investigation of routing reliability of vehicular ad hoc networks, *EURASIP Journal on Wireless Communications and Networking*, vol. 2013, no. 1, pp. 1–15, 2013.

[6] H. Hartenstein and K. P. Laberteaux, A tutorial survey on vehicular ad hoc networks, *IEEE Communications Magazine*, vol. 46, no. 6, pp. 164–171, 2008.

[7] R. Lu, Security and privacy preservation in vehicular social networks. PhD thesis, University of Waterloo, 2012.

[8] S. H. Dau, W. Song, and C. Yuen, On block security of regenerating codes at the MBR point for distributed storage systems, in *IEEE International Symposium on Information Theory (ISIT)*, pp. 1967–1971, June 2014.

[9] S. H. Dau, W. Song, and C. Yuen, On the existence of MDS codes over small fields with constrained generator matrices, in *IEEE International Symposium on Information Theory (ISIT)*, pp. 1787–1791, June 2014.

[10] S. H. Dau, W. Song, Z. Dong, and C. Yuen, Balanced sparsest generator matrices for mds codes, in *IEEE International Symposium on Information Theory (ISIT)*, pp. 1889–1893, July 2013.

[11] J. Freudiger, M. Raya, M. Félegyházi, P. Papadimitratos, et al., Mix-zones for location privacy in vehicular networks, in *Proceedings of the First International Workshop on Wireless Networking for Intelligent Transportation Systems (Win-ITS)*, 2007.

[12] L. Buttyán, T. Holczer, and I. Vajda, On the effectiveness of changing pseudonyms to provide location privacy in VANETS, in L. Buttyan, V. Gligor, and D. Westhoff (eds) *Security and Privacy in Ad-Hoc and Sensor Networks*, pp. 129–141, Springer, 2007.

[13] C. Zhang, X. Lin, R. Lu, P.-H. Ho, and X. Shen, An efficient message authentication scheme for vehicular communications, *IEEE Transactions on Vehicular Technology*, vol. 57, no. 6, pp. 3357–3368, 2008.

[14] R. Lu, X. Lin, X. Liang, and X. Shen, A dynamic privacy-preserving key management scheme for location-based services in VANETS, *IEEE Transactions on, Intelligent Transportation Systems,* vol. 13, no. 1, pp. 127–139, 2012.

[15] L. Buttyán, T. Holczer, A. Weimerskirch, and W. Whyte, Slow: A practical pseudonym changing scheme for location privacy in VANETS, in *IEEE Vehicular Networking Conference (VNC),* pp. 1–8, IEEE, 2009.

[16] A. R. Beresford and F. Stajano, Location privacy in pervasive computing, *IEEE Pervasive Computing*, vol. 2, no. 1, pp. 46–55, 2003.

[17] B. Palanisamy and L. Liu, Mobimix: Protecting location privacy with mix-zones over road networks, in *IEEE 27th International Conference on Data Engineering (ICDE)*, pp. 494–505, IEEE, 2011.

[18] A. R. Beresford and F. Stajano, Mix zones: User privacy in location-aware services, in *IEEE International Conference on Pervasive Computing and Communications Workshops*, pp. 127–127, IEEE Computer Society, 2004.

[19] R. Lu, X. Li, T. H. Luan, X. Liang, and X. Shen, Pseudonym changing at social spots: An effective strategy for location privacy in VANETS, *IEEE Transactions on Vehicular Technology*, vol. 61, no. 1, pp. 86–96, 2012.

[20] J. Guo, J. P. Baugh, and S. Wang, A group signature based secure and privacy-preserving vehicular communication framework, *Mobile Networking for Vehicular Environments*, vol. 2007, pp. 103–108, 2007.

[21] M. Raya and J.-P. Hubaux, Securing vehicular ad hoc networks, *Journal of Computer Security*, vol. 15, no. 1, pp. 39–68, 2007.

[22] M. Raya, P. Papadimitratos, and J.-P. Hubaux, Securing vehicular communications, *IEEE Wireless Communications Magazine, Special Issue on Inter-Vehicular Communications*, vol. 13, no. LCA-ARTICLE-2006-015, pp. 8–15, 2006.

[23] S. Al-Sultan, M. M. Al-Doori, A. H. Al-Bayatti, and H. Zedan, A comprehensive survey on vehicular ad hoc network, *Journal of Network and Computer Applications*, vol. 37, pp. 380–392, 2014.

[24] K. Sampigethaya, M. Li, L. Huang, and R. Poovendran, Amoeba: Robust location privacy scheme for vanet, *IEEE Journal on Selected Areas in Communications*, vol. 25, no. 8, pp. 1569–1589, 2007.

[25] N. Lyamin, A. Vinel, M. Jonsson, and J. Loo, Real-time detection of denial-of-service attacks in IEEE 802.11p vehicular networks, *IEEE Communications Letters*, vol. 18, no. 1, pp. 110–113, 2014.

[26] M. Gruteser and D. Grunwald, Anonymous usage of location-based services through spatial and temporal cloaking, in *Proceedings of the 1st International Conference on Mobile Systems, Applications and Services*, pp. 31–42, ACM, 2003.

[27] S. Smaldone, L. Han, P. Shankar, and L. Iftode, Roadspeak: enabling voice chat on roadways using vehicular social networks, in *Proceedings of the 1st Workshop on Social Network Systems*, pp. 43–48, ACM, 2008.

[28] C. Projects. http://cabspotting.org/projects/intransit/.

[29] J. Fan, J. Chen, Y. Du, W. Gao, J. Wu, and Y. Sun, Geocommunity-based broadcasting for data dissemination in mobile social networks, *IEEE Transactions on Parallel and Distributed Systems*, vol. 24, no. 4, pp. 734–743, 2013.

[30] D. Boneh and X. Boyen, Short signatures without random oracles and the sdh assumption in bilinear groups, *Journal of Cryptology*, vol. 21, no. 2, pp. 149–177, 2008.

[31] P. Golle, D. Greene, and J. Staddon, Detecting and correcting malicious data in VANETS, in *Proceedings of the 1st ACM International Workshop on Vehicular ad Hoc Networks*, pp. 29–37, ACM, 2004.

[32] T. R. Henderson, M. Lacage, G. F. Riley, C. Dowell, and J. Kopena, Network simulations with the ns-3 simulator, SIGCOMM demonstration, 2008.

[33] D. Krajzewicz, G. Hertkorn, C. Rössel, and P. Wagner, SUMO (simulation of urban mobility), in *Proceedings of the 4th Middle East Symposium on Simulation and Modelling*, pp. 183–187, 2002.

Chapter 9

Lightweight and Robust Schemes for Privacy Protection in Key Personal IoT Applications: Mobile WBSN and Participatory Sensing

Wei Ren

Liangli Ma

Yi Ren

CONTENTS

Abstract. With the development and deployment of the Internet of Things (IoT), some key personal applications attract more and more attention, for example, wireless body sensor networks and participatory sensing. In those personal IoT applications, privacy issues have been envisioned as being of paramount importance. Currently, although many schemes have been proposed for guaranteeing personal privacy, the overall performance and robustness of the schemes may not be thoroughly tackled. In this chapter, we explore current privacy problems in key personal IoT applications such as wireless body sensor networks and participatory sensing. We especially propose some feasible schemes that are lightweight and robust.

Keywords: lightweight; robust; privacy; WBSN; participatory sensing

9.1 Introduction

As IoT technologies develop, applications of the IoT start to attract more and more attention, not only from industry but also from the world of personal computing. With the development and popularity of smartphones, smartphone-oriented IoT applications become realities for each person. For example, wearable devices such as smart watches and smart wristbands that are equipped with multiple sensors can connect and upload sensing data to a smartphone. The smartphone itself is equipped with sensors that can also generate sensing data, and upload the data to a central server.

In this chapter, we concentrate on two typical key applications in the personal IoT—the mobile wireless body sensor network (WBSN) and participatory sensing. Mobile WBSN comprises multiple sensor nodes that are implanted (or attached) into (or on) a human body to monitor health or EEG physiological indicators, such as electrocardiogram (ECG electroencephalography), glucose, toxins, blood pressure, and so on [1–7]. Those data usually need to be uploaded to a central database (e.g., a cloud computing server) instantly so that doctors or nurses can remotely access them for real-time diagnosis and emergency response. As most persons possess a smartphone and customized applications can be installed on it, it is convenient and economical to use a smartphone as a gateway between WBSN and cloud servers. Anytime, anywhere uploading of health data can thus be achieved simply by the use of smartphones. Figure 9.1 illustrates the typical and basic scenario in WBSN-Cloud IoT applications.

As health data are highly critical for personal privacy and relevant regulations such as HIPAA [8] must be conformed to, the uploaded data need to be encrypted. The privacy of the communication link between smartphones and cloud servers can be protected by underlying media access control (MAC)

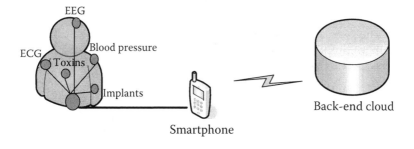

Figure 9.1: Smartphone performances as a gateway in the WBSN-cloud paradigm.

protocols. For example, IEEE 802.11 in WLAN, IEEE 802.15.4 [9] in WPAN, or WCDMA in 3G, so adversaries who seek out the communication links are defended against. However, as cloud servers are always assumed untrustworthy, extra data encryption for defending against malicious cloud servers is required. A straightforward method is to encrypt uploaded data with off-the-shelf methods such as block ciphers; for example, AES or KASUMI [10]. However, this may not be suitable and applicable in smartphones, because smartphones usually have energy constraints. Moreover, smartphones may be misused, lost, stolen, or hacked by attackers; thus privacy protection itself should be robust. Therefore, it is a critical challenge to design a lightweight and robust method to protect privacy.

Together with mobile WBSN, participatory sensing also invokes a large number of new personal IoT applications such as environmental monitoring, transportation management, and personal entertainment. For example, participants report real-time surrounding traffic to help others avoid jams in transportation systems; volunteers report parking vacancies to help others shorten parking search time. In participatory sensing, participants (usually those that have volunteered to gather information) report their sensory data on their surroundings via their smartphones. Those reported data are uploaded into central servers (e.g., cloud servers), and central servers share the data with users after data processing.

To obtain sufficient uploaded sensory data and accumulate more shared data, random volunteers may be encouraged to attend or enroll in participatory sensing as data contributors. Thus, the data are very likely uploaded by random attenders who may be potential attackers or malicious contributors. In this situation, participatory sensing poses several key security problems: (1) The trustworthiness of uploaded data should be evaluated. As participants are usually selected from a random set of volunteers, the data contributed by them may be incorrect due to mistakes, or due to malicious intent. (2) The privacy of data contributors should be protected. The data uploaded from participants should not divulge their personal private data, such as location information, user trajectory, and location dynamics over time. (3) The robustness of the overall defense system to solve the above security problems should be guaranteed. As participants could be random volunteers who may be malicious or not, the security scheme should thus defend against those internal attackers.

In this chapter, we explore privacy protection problems in two typical key applications in the personal IoT—mobile WBSN and participatory sensing. Specific research problems are pointed out and expressed in formal terms. Some lightweight and robust schemes are proposed and evaluated.

The rest of the chapter is organized as follows. Section 9.2 discusses lightweight and robust schemes for privacy protection in WBSN. In Section 9.3, we discuss a lightweight and robust scheme for privacy protection in participatory sensing. Finally, Section 9.4 concludes the chapter.

9.2 Lightweight and Robust Schemes for Protecting Privacy in Mobile WBSN

9.2.1 Related work

A large number of research studies have been conducted on wireless sensor networks but not on lightweight or energy-efficient network architecture in the eHealth domain [13]. Several studies concentrated on key management in WBSN [11]. Venkatasubramanian et al. [19] proposed a physiological signal-based key management scheme in WBSN. Law et al. [10] evaluated lightweight ciphers for wireless sensor networks. For other security problems such as security architecture, privacy, and emergency response, identity-based cryptography for WBSN was proposed [12, 14, 16–18]. Lin et al. [15] proposed a strong privacy-preserving scheme called SAGE against global eavesdropping for eHealth systems, but it relies extensively on bilinear pairing that confronts difficulties to be applied in energy constraint devices. The application of smartphones in eHealth has started to attract more and more attention [20]. Kotz et al. [21] proposed a privacy framework for ubiquitous eHealth. They pointed out several privacy policies required for the building of a privacy framework.

9.2.2 Problem formulation

9.2.2.1 Network model

The following related entities exist in typical mobile WBSN scenarios:

(1) Mobile Gateway (denoted as $\mathcal{MG}$). This is usually a mobile smartphone with Internet connection. It uploads monitoring data that are collected from WBSN to cloud servers. Although it can conveniently upload body-sensing data instantly, it imposes energy constraints.

(2) Cloud servers (denoted as $\mathcal{BC}$). This is a back-end storage server with a very large capacity via virtualization of storage resources.

(3) WBSN. This consists of body sensors that may be implanted, attached, or wearable. The sink node in WBSN periodically uploads data into $\mathcal{MG}$ via a secure channel.

(4) Accessor (denoted as $\mathcal{MA}$). This could be the mobile devices held by doctors, nurses, or guardians. They can usually access the data at $\mathcal{BC}$ in a pervasive manner.

9.2.2.2 Trust model and security requirement

$\mathcal{MG}$ is assumed to be trustworthy. Indeed, it is a minimal trust assumption and the defending scheme conducted at $\mathcal{MG}$.

The communication between $\mathcal{MG}$ and $\mathcal{BC}$ is untrustworthy. As the link privacy between them is already provided by protocols at the MAC layer or the link layer, such as IEEE 802.11 or 3GPP, the adversaries on the link can thus be ignored. Similarly, the link privacy between WBSN and $\mathcal{MG}$ can also be protected by a MAC layer protocol such as IEEE 802.15.4.

$\mathcal{BC}$ is untrustworthy. It has an interest in user privacy, but it performs properly according to certain protocols, such as the service-level agreement to store uploaded data. Thus, the major concern in this chapter is the privacy protection of data uploaded to the untrustworthy $\mathcal{BC}$.

The security requirement is that the data transferred from $\mathcal{MG}$ to $\mathcal{BC}$ should not be recovered by adversaries at $\mathcal{BC}$.

9.2.3   Proposed schemes

In this section, we investigate two schemes called OTM and OTP.

We list all major notations used in the remainder of the chapter in Table 9.1.

9.2.3.1   One-time mask (OTM) scheme

Intuitively, a straightforward method is to use an encryption algorithm, for example AES. However, this method induces a remarkable computation overhead invoked by the encryption algorithm, as energy consumption each time is large and the number of encryptions is also large. The frequency (or interval) of sensing data uploads is determined by medical requirements; thus, the only factor remaining to trade off is to reduce the power consumption for of a single encryption operation.

The most lightweight operation in encryption algorithms is exclusive-or (XOR), but naively using XOR encryption is not acceptable. Moreover, the encrypted key cannot be used more than once. We thus propose a one-time XOR-based encryption. Before describing the scheme, we firstly analyze the characteristics of $\mathcal{MG}$ and the properties of uploaded data that can be tackled. We observe the following characteristics of $\mathcal{MG}$, to facilitate energy efficiency:

Table 9.1   Major notations

n	Number of sensors in WBSN
N_i	Sensor i, where $i = 1, \ldots, n$
D_i	Sensing data from sensor i
M_i	Median value of data D_i from sensor i
O_i	Offset value (related to M_i) of data D_i from sensor i
R_i	The upper bound of offset $|O_i|$ in terms of absolute value
K_i	Mask value of data D_i for O_i from sensor i
O_i'	Masked offset value (related to M_i) of data D_i from sensor i
f	Frequency of upload from $\mathcal{MG}$ to $\mathcal{BC}$

(1) OB_1: $\mathcal{MG}$ has energy constraints. Its energy consumption comes from three sources communication, computation, and storage. Of these, the largest proportion is usually used by communication. Thus, the length of communication messages should be as short as possible. Usually, after symmetric encryption, the ciphertext has the same length as the plaintext. Thus, the message length is at least as long as the original data. XOR-based encryption consumes less energy than symmetric methods of encryption such as AES.

In addition, we observe the following properties in uploaded data:

(2) OB_2: The total number of sensing nodes (denoted as n) for a single user (namely, one $\mathcal{MG}$) is usually not very many, for example, $n < 16$. The reason is that the number of required monitoring signals is limited, for example, electrocardiogram, electroencephalography, glucose, protein, toxins, and blood pressure, to name a few. Thus, the number of pieces of source data in each upload interval is usually not very high.

(3) OB_3: The uploaded data always fall within a short range, because the sensing data on the physiology of a person rarely varies with an extremely abnormal deviation. That is, the range of data D_i ($i \in [1, n]$) is $[M_i - R_i, M_i + R_i]$, where M_i is the median (normal or average) value of sensing results and R_i is the maximal absolute offset, namely, $R_i = max(|D_i - M_i|)$.

Here we assume $D_1, D_2, \ldots, D_n$ are positive integers. If D_i is negative, it can be made positive by attaching a sign mark; for example, $s_i = 0, 1$ for a negative and a positive sign, respectively. If D_i is a noninteger it can be changed into an integer by multiplying by 10^{p_i}, where p_i is the distance of the decimal point from the rightmost position. For example, $D_i = 34.4$ can be denoted as $344, 1, 1$, and $D_i = -34.4$ is $344, 1, 0$.

(4) OB_4: The data may repeat on most occasions, or at least the median value may persist or recur most times. The reason is similar to that in OB_3.

Due to the above observations, we propose a One-Time Mask scheme (OTM), which is much securer than a naive XOR scheme (we will prove its security later), and consumes much less energy in communications. It only relies on the XOR operation for energy efficiency of encryption. The OTM scheme includes functions as follows:

9.2.3.1.1 Basic settings

According to the aforementioned $OB2$, suppose there exist n sensors in WBSN, denoted by $N_i (i = 1, \ldots, n)$. The value of n usually is smaller than 16. Each sensor uploads data to $\mathcal{MG}$ at each interval. Data from node N_i is denoted as D_i. According to $OB3$, D_i is always in the range $[M_i - R_i, M_i + R_i]$, where M_i is the median (or expectation) value of the sensing data, and R_i is the maximal absolute offset. That is, $O_i \Leftarrow D_i - M_i$ ($i = 1, \ldots, n$), where O_i is the offset value, $R_i = max(|O_i|)$. The interval for each upload time is I s. The number of upload times is thus $t = 60/I$ in 1 m.

MVT

sn	M_1	M_2	---	M_n
0001	23	8	---	120
0019	20	6	---	122

RVT

sn	R_1	R_2	---	R_n
0001	6	3	---	24
0019	5	4	---	23

KVT

sn	K_1	K_2	---	K_n
0001	3	2	---	22
0002	2	1	---	18

Figure 9.2: *MVT, RVT, KVT* at $\mathcal{MG}$. Only the first two tuples are illustrated.

9.2.3.1.2 Basic data structure

According to $OB3$, $\mathcal{MG}$ creates a median value table for all sensing data called MVT. $MVT = \langle sn, M_1, M_2, \ldots, M_n \rangle$, where sn is a unique sequence number. $\mathcal{MG}$ also creates a range value table for all sensing data called RVT. $RVT = \langle sn, R_1, R_2, \ldots, R_n \rangle$, where sn is a unique sequence number.

To facilitate the encryption, $\mathcal{MG}$ creates a mask value table for sensing data called KVT. $KVT = \langle sn, K_1, K_2, \ldots, K_n \rangle$, where sn is a unique sequence number, and K_i $(i = 1, \ldots, n)$ is a mask value for sensing data D_i. Note that a tuple may only be appended to MVT and RVT upon adjustment of M_i and R_i, but a tuple is appended to KVT for each piece of uploaded data. Figure 9.2 illustrates the major data structures MVT, RVT, and KVT.

9.2.3.1.3 Data encryption and data upload

(3.1) Upon receipt of a piece of sensing data, $\mathcal{MG}$ generates a random number in $\{0,1\}^{L_K = \sum_{i=1}^{n} \lceil \log_2 R_i \rceil}$, denoted as $\{K_1 \| K_2 \| \ldots \| K_n\}$.

(3.2) D_i is encrypted with K_i as follows: $O'_i = |O_i| \oplus K_i (i = 1, \ldots, n)$, where $| \cdot |$ is an operator for returning corresponding absolute values.

(3.3) $\mathcal{MG}$ stores $\langle sn, M_1, \ldots, M_n \rangle$ to MVT and stores $\langle sn, K_1, \ldots, K_n \rangle$ to KVT.

(3.4) Encryption is computed as follows:

$$D' \Leftarrow \{S(O_1) \| O'_1 \| S(O_2) \| O'_2 \| \ldots \| S(O_n) \| O'_n\},$$

where $S(O_i) = 1$ if $O_i \geqslant 0$, or $S(O_i) = 0$ if $O_i < 0$.

(3.5) $\mathcal{MG}$ uploads the encryption result to BC. $\mathcal{MG} \rightarrow BC : \{D'\}$.

9.2.3.1.4 $\mathcal{MA}$ access

(4.1) If An $\mathcal{MA}$ wishes to access the uploaded data, it will be securely provided segments of MVT, RVT, and KVT by $\mathcal{MG}$ that cover the data of interest.

(4.2) $M_i, K_i (i = 1, \cdot, n)$ is retrieved from MVT and KVT via sn, respectively.

(4.3) D' is decrypted via

$$\{S(O_1)\|O_1' \oplus K_i\|S(O_2)\|O_2' \oplus K_i\| \dots \|S(O_n)\|O_n' \oplus K_i\}.$$

(4.4) The data is recovered via $D_i \Leftarrow M_i + O_i$ $(i = 1, \dots, n)$.

9.2.3.1.5 Security and performance analysis for OTM

We present a formal analysis in the following propositions.

Definition 9.1 Computational data privacy ($\mathcal{CDP}$). We say that scheme S protects $\mathcal{CDP}$ if any polynomial time Turing machine (PTTM) at $\mathcal{BC}$ can reveal D from D' with only a negligible probability $negl(n)$ (n is a security parameter). That is,

$$\mathcal{CDP}^S = I(D; D') = H(D) - H(D|D') < \mathrm{negl}(n),$$

where $I(\cdot; \cdot)$ is an mutual information; $H(\cdot)$ is an entropy function; $\mathrm{negl}(n)$ is a negligible function.

Proposition 9.1
If MVT, RVT, and KVT are securely possessed, the OTM scheme can guarantee the privacy of any uploaded data. Stated formally, $\mathcal{CDP}^{\mathrm{OTM}} < negl(\sum_{i=1}^{n} \log_2 M_i + L_K)$.

Proof.

$$\mathcal{CDP}^{\mathrm{OTM}} = I(D; D') = H(D) - H(D|D')$$

$$= \sum_{i=1}^{n} (H(D_i) - H(D_i|D_i'))$$

$$= \sum_{i=1}^{n} (H(D_i) - H(D_i|M_i + (K_i \oplus O_i') + S(O_i)))$$

$$< \mathrm{negl}(\sum_{i=1}^{n} (\log_2 M_i + \log_2 K_i))$$

$$< \mathrm{negl}(\sum_{i=1}^{n} \log_2 M_i + L_K)$$

Indeed, D' is a piece of unstructured data without knowledge of RVT, which further decreases the probability of a correct guess.

Definition 9.2   Energy efficiency of communication (EEC) in scheme S (denoted as $\mathcal{EEC}^S$): $\mathcal{EEC}^S = 1 - Ratio_1$, where $Ratio_1$ is measured by defined as the communication length in scheme S divided by the communication length in naive XOR-based scheme.

Proposition 9.2
$$\mathcal{EEC}^{OTM} = 1 - \frac{n + \sum_{i=1}^{n} \lceil \log_2 R_i \rceil}{\sum_{i=1}^{n} \lceil \log_2 (M_i + R_i) \rceil}$$

Proof. As the energy constraints are only present in $\mathcal{MG}$, we thus concentrate only on the energy consumption in the sending operation at $\mathcal{MG}$. Suppose that the energy consumption of communications is proportional to the length of the message. Hence, message length is critical in analysis.

The length of the original sensing data is $\sum_{i=1}^{n} \lceil \log_2 D_i \rceil) = \sum_{i=1}^{n} \lceil \log_2 (M_i + R_i) \rceil$. The length of the uploaded data in the OTM scheme is

$$\sum_{i=1}^{n} (1 + \lceil \log_2 R_i \rceil) = n + \sum_{i=1}^{n} \lceil \log_2 R_i \rceil,$$

where 1 is a bit for the sign mark. Thus,

$$\mathcal{EEC}^{OMS} = 1 - \frac{n + \sum_{i=1}^{n} \lceil \log_2 R_i \rceil}{\sum_{i=1}^{n} \lceil \log_2 (M_i + R_i) \rceil}.$$

We further analyze the approximate value of $\mathcal{EEC}^{OTM}$. Suppose $\gamma = R_i/M_i$, which is a value depicting the data division and for the convenience of the approximation.

Proposition 9.3
If $R_i = \gamma M_i$ and $0.05 \leq \gamma \leq 0.3$, $\mathcal{EEC}^{OTM} \approx 2/(\frac{1}{n} \sum_{i=1}^{n} M_i)$

Proof. By simple mathematical transformation,

$$\mathcal{EEC}^{OTM} = 1 - \frac{n + \sum_{i=1}^{n} \lceil \log_2 R_i \rceil}{\sum_{i=1}^{n} \lceil \log_2 (M_i + R_i) \rceil}$$

$$\approx 1 - \frac{n + \sum_{i=1}^{n} \log_2 \gamma M_i}{\sum_{i=1}^{n} \log_2 (1 + \gamma) M_i}$$

$$= 1 - \frac{n + n \log_2 \gamma + \sum_{i=1}^{n} M_i}{n \log_2 (1 + \gamma) + \sum_{i=1}^{n} M_i}$$

Let $AVG_M = \sum_{i=1}^{n} M_i/n$. The above equation equals $1 - \frac{1 + \log_2 \gamma + AVG_M}{\log_2 (1 + \gamma) + AVG_M} = \frac{\log_2 \frac{1+\gamma}{2\gamma}}{\log_2 (1 + \gamma) + AVG_M}$. It approximates to $(\log_2 (1/\gamma + 1) - 1)/AVG_M$. For example suppose $0.05 \leq \gamma \leq 0.3$, thus $\log_2 (1 + \gamma)$ is constantly increasing at

$[0.07, 0.38]$ and $\log_2 \frac{1+g}{2g}$ is constantly decreasing at $[3.39, 1.12]$. Supposing that $AVG_M >> 2$, the final result is roughly $2/(\frac{1}{n}\sum_{i=1}^{n} M_i)$, as desired.

Definition 9.3   Energy efficiency of computation (EEP) in scheme S (denoted as $\mathcal{EEP}^S$): $\mathcal{EEP}^S = 1 - Ratio_2$, where $Ratio_2$ is defined as the computation cost in security scheme S divided by the communication length in a naive scheme.

Proposition 9.4

$$\mathcal{EEP}^{\text{OTM}} = 1 - \frac{n + \sum_{i=1}^{n} \lceil \log_2 R_i \rceil}{\sum_{i=1}^{n} \lceil \log_2 (M_i + R_i) \rceil}.$$

Proof. As the OTM and the naive scheme both rely on the XOR operation, the energy consumption of computation is related to the length of the plaintext. Similarly to Proposition 2, energy consumption of the computations is proportional to the length of the plaintext. It is, indeed, the same as the length of the message being sent. The length of the original sensing data is

$$\sum_{i=1}^{n} \lceil \log_2 D_i \rceil) = \sum_{i=1}^{n} \lceil \log_2 (M_i + R_i) \rceil$$

The length of uploaded data in the OTM scheme is

$$\sum_{i=1}^{n} (1 + \lceil \log_2 R_i \rceil) = n + \sum_{i=1}^{n} \lceil \log_2 R_i \rceil.$$

Thus, $\mathcal{EEP}^{\text{OTM}} = 1 - \frac{n + \sum_{i=1}^{n} \lceil \log_2 R_i \rceil}{\sum_{i=1}^{n} \lceil \log_2 (M_i + R_i) \rceil}$

Definition 9.4   Extra storage induced by scheme S (denoted as $\mathcal{ES}^S$). This is the additional storage induced by scheme S compared to the naive scheme.

Proposition 9.5
$\mathcal{ES}^{\text{OTM}}$ is trivial.

9.2.3.2   One-time permutation (OTP) scheme

In the OTM scheme, the energy consumption for communication and computation is much less than that of the naive XOR-based scheme (and, of course, also much less than a straightforward scheme such as encryption by AES). To further decrease energy consumption, we propose the use of permutation to replace XOR encryption, called One-Time Permutation (OTP). This can significantly decrease energy consumption due to the avoidance of XOR computation, but still maintain security (which will be justified later and related to $OB3$).

   The intuition of OTP is that the encryption secrecy relies on permutation styles instead of keys. The permutation style determines the arrangement of O_i in the uploaded data. The detailed design is as follows:

(1) Basic settings: These are the same as OTM.

(2) Basic data structure: MVT and RVT are required, but KVT is replaced by PVT. That is, to facilitate the permutation, $\mathcal{MG}$ stores a permutation value table for sensing data, called PVT. $PVT = \langle sn, P_1, P_2, \ldots, P_n \rangle$, where sn is a unique sequence number, and $P_i, (i = 1, \ldots, n, P_i \in [1, \ldots, n])$ is the position of sensing data D_i in n positions. Figure 9.3 illustrates major data structures MVT, RVT, PVT.

(3) Data encryption and data upload:

   (3.1) Upon receipt of sensing data, $\mathcal{MG}$ generates a random permutation in n, denoted as $\{P_1 \| P_2 \| \ldots \| P_n\}, P_i \in [1, n], i \in Z, i \in [1, n], \forall i, j \in [1, n], P_i \neq P_j$.

   (3.2) $\mathcal{MG}$ stores $\langle sn, M_1, \ldots, M_n \rangle$ to MVT. $\mathcal{MG}$ stores $\langle sn, P_1, \ldots, P_n \rangle$ to PVT.

   (3.3) Upload results are computed as follows:

   $$D' \Leftarrow \{S(O_{P_1}) \| O_{P_1} \| S(O_{P_2}) \| O_{P_2} \| \ldots \| S(O_{P_n}) \| O_{P_n}\},$$

   where $S(O_i) = 1$ if $O_i \geqslant 0$, or $S(O_i) = 0$ if $O_i < 0$.

   (3.4) $\mathcal{MG}$ uploads the encryption results to BC. $\mathcal{MG} \rightarrow BC : \{D'\}$.

(4) $\mathcal{MA}$ access:

   (4.1) If $\mathcal{MA}$ wishes to access the uploaded data, it will be securely provided segments of MVT, RVT, and PVT by $\mathcal{MG}$ that cover the data of interest.

   (4.2) $M_i, P_i (i = 1, \cdot, n)$ is retrieved from MVT and PVT via sn, respectively.

MVT

sn	M_1	M_2	---	M_n
0001	23	8	---	120
0019	20	6	---	122

RVT

sn	R_1	R_2	---	R_n
0001	6	3	---	24
0019	5	4	---	23

PVT

sn	P_1	P_2	---	P_n
0001	4	2	---	6
0002	2	5	---	9

Figure 9.3: *MVT,RVT,PVT* at $\mathcal{MG}$. Only the first two tuples are illustrated.

(4.3) D' is rearranged to

$$\{S(O_1)\|O_1\|S(O_2)\|O_2\|\ldots\|S(O_n)\|O_n\}.$$

(4.4) Data is recovered via $D_i \Leftarrow M_i + O_i \ (i = 1,\ldots,n)$.

9.2.3.2.1 Security and performance analysis for OTP

Proposition 9.6
If MVT, RVT, and PVT are securely possessed, the OTP scheme can guarantee the privacy of uploaded data. Stated formally, $\mathcal{CDP}^{OTP} < negl(\sum_{i=1}^{n}(\log_2 M_i) + \log_2(n!))$.

Proof. As MVT, PVT are securely possessed, adversaries have to correctly guess P_i for O_i and M_i $(i = 1,\ldots,n)$ to reveal a piece of uploaded data. As M_i is securely possessed, the probability of correctly guessing $M_i, (i = 1,\ldots,n)$ is negligible in $\sum_{i=1}^{n} \log_2 M_i$. Next, consider the possibility of a correct guess for P_i. The probability of recovering D_i in one interval is $1/n$. The probability of recovering all D_is in one interval is $1/n!$. Therefore, $\mathcal{CDP}^{OTP} < negl(\sum_{i=1}^{n}(\log_2 M_i) + \log_2(n!))$. Similarly, D' is a piece of unstructured data without knowledge of RVT, which further decreases the probability a correct guess.

Proposition 9.7
$$\mathcal{EEC}^{OTP} = 1 - \frac{n + \sum_{i=1}^{n} \lceil \log_2 R_i \rceil}{\sum_{i=1}^{n} \lceil \log_2 (M_i + R_i) \rceil}.$$

Proof. This is the same as $\mathcal{EEC}^{OTM}$.

Proposition 9.8
$\mathcal{EEP}^{OTP} \ll \mathcal{EEP}^{OTM}$.

Proof. As OTP avoids the XOR operation, the energy consumption of the computation only occurs during the one-time permutation generation. It generates n numbers in $[1,n]$ for each upload tuple. The energy consumption of the computation in OTM has two sources: One stems from XOR encryption, which is proportional to the length of the plaintext. The other stems from one-time key generation. Thus, the conclusion is justified.

Even only comparing the performance of random number generation, in OTM n numbers in $[1, R_i]$ are generated for each upload tuple. In OTP, n numbers in $[1,n]$ are generated for each upload tuple. Based on $OB2$ and $OB3$, we have $n < 16$. Thus, $n * n$ is much likely less than $\sum_{i}^{n} R_i$. This again justifies the conclusion.

Table 9.2 Performance comparisons between OTM and OTP

Overhead	OTM		OTP
Communication	$\mathcal{EEC}^{\text{OTM}} = 1 - \dfrac{n + \sum_{i=1}^{n} \lceil \log_2 R_i \rceil}{\sum_{i=1}^{n} \lceil \log_2 (M_i + R_i) \rceil}$		$= OTM$
Computation	$\mathcal{EEP}^{\text{OTM}} = 1 - \dfrac{n + \sum_{i=1}^{n} \lceil \log_2 R_i \rceil}{\sum_{i=1}^{n} \lceil \log_2 (M_i + R_i) \rceil}$		$\ll OTM$
Storage	$\mathcal{ES}^{\text{OTM}}$ is trivial		$< OTM$

Proposition 9.9
$\mathcal{ES}^{\text{OTP}} < \mathcal{ES}^{\text{OTM}}$.

Proof. As storage costs for MVT and RVT are the same, thus our concentration falls on the comparison between PVT in OTP and KVT in OTM. The length of one tuple (row) in table PVT is $Len(sn) + \sum_{i=1}^{n} \log_2 P_i$. Based on the observation $OB2$ ($n < 16$), $Len(sn) + \sum_{i=1}^{n} \log_2 P_i < Len(sn) + n * \log_2 16 = Len(sn) + n * 4$. The length of one tuple in table KVT is $Len(sn) + \sum_{i=1}^{n} \log_2 K_i = Len(sn) + \sum_{i=1}^{n} \log_2 R_i > Len(sn) + n * 4$, as usually on average $\frac{1}{n} \sum_{i=1}^{n} \log_2 R_i$ is greater than 4.

More specifically, as the total number of tuples in PVT is $60t/I$, the total volume of storage for PVT is less than $((60t)/I) * (Len(sn) + 16 * 4)$, $(letn = 16)$. If $t = 60 * 24 * 30 * 12 = 518400$, (namely, 1 year), and $I = 5s$, we have $60t/I < 7 * 10^6$, which is the total number of tuples for 1 year. Hence, $Len(sn) < 23$. Finally, the total volume of storage for PVT is less than $7 * 10^6 * (23 + 64)/8bit < 0.08GB$. In other words, the total volume of PVT for 1 year's data is less then $0.08GB$, which is trivial in $\mathcal{MG}$.

9.2.3.3 Comparison and numerical results

Due to the above analysis, we list the comparisons between OTM and OTP in Table 9.2 for better understanding of our design logic.

Next, we illustrate the performance of OTM in Figure 9.4. It depicts the energy efficiency of communications of OTM. It justifies the approximation in Proposition 3. As OTP costs much less than OTM in terms of communication, computation and storage, the graph also justifies the lightweight property of OTP.

9.3 A Lightweight and Robust Scheme for Privacy Protection in Participatory Sensing

9.3.1 Related work

The security in participatory sensing is attracting more and more attention [22, 23, 26–28, 30]. Boutsis et al. [24] proposed a scheme for preserving privacy

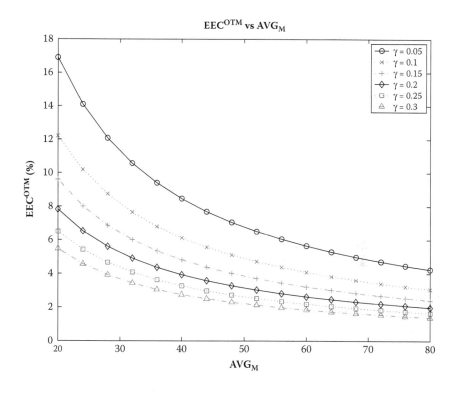

Figure 9.4: $\mathcal{EEC}^{\mathrm{OTM}}$ **as a function of** *AVG$_M$* **and** γ **(see Proposition 3).**

with a low overhead. Their scheme assumes that user data are generated and stored locally on individual smartphone devices, instead of maintained in a centralized database. Groat et al. [29] proposed a privacy protection scheme for multidimensional data that uses negative surveys. Kazemi et al. [32] proposed a privacy-aware framework called PiRi, which enables participation by users without compromising their privacy. Wang et al. [33] proposed an anonymous sensory data collection approach designed particularly for mobile environments. They think most previously proposed methods are not designed for mobile environments and thus resource constraint has not been focused on in those solutions. Huang et al. [31] proposed a reputation scheme that prevents the inadvertent leakage of with data because of the inherent relationship private reputation. They consider there exists a dilemma: privacy is often achieved by removing the links between successive user contributions but, at the same time, such links are essential for establishing trust. Christin et al. [25] proposed a framework called IncogniSense to utilize periodic pseudonyms generated using a blind signature and, which relies on reputation transfer between these pseudonyms. The current work on this topic cannot solve security goals such as data trustworthiness, reputation evaluation, privacy protection, and robustness in one solution, and especially, in a lightweight manner.

9.3.2 Problem formulation

9.3.2.1 Network model

There exist three major entities in participatory sensing: contributors, central servers, and consumers. Contributors upload sensing data to central servers; central servers manage the uploaded data, and prepare it for presentation to consumers; consumers retrieve the data presented from central servers.

Contributors may be volunteers who are willing to install application softwares in their smartphones for participatory sensing. Thus, contributors should not be constrained by attending admission control processes in advance; for example, registration. The real identities of contributors should be shielded for the protection of personal privacy such as locations, location dynamics over time, trajectory, and so on.

Central servers store the uploaded data from contributors. The data may be cleaned, refined, reorganized, and finally provided to consumers as presenting data.

9.3.2.2 Attack model and design goals

We concentrate on adversaries targeting peers instead of channels, as channels between contributors and central servers are protected by other inherent security mechanisms (e.g., encryption and integrity protection) at link layers such as IEEE802.11i, GPRS, or CDMA. As there exist three entities in the model and, among them, consumers are not our concern; we focus on contributors and central servers. The adversaries targeting contributors consist of two major types: (1) Contributors who upload forged data to misinform central servers. Thus, this kind of contributor should be detected and the forged data should be removed, which is carried out at the central servers. (2) Contributors who may intentionally bypass or breach the proposed defense scheme. In other words, the proposed scheme should defend against internal malicious contributors.

We assume the central servers may leak contributor privacy data such as location, trajectory, behaviors, and habits. Thus, the actual identification of contributors should be hidden to central servers. The trajectory and other dynamics over time should be concealed.

The design goals have three facets, as follows: confirming the trustworthiness of uploaded data in the presence of possible malicious contributors; protecting contributor privacy without admission control; maintaining the robustness of the proposed defense system to impede those malicious contributors who intend to subvert it. In next section, we propose a scheme called LibTip (lightweight and robust for trustworthiness and privacy) for those design goals.

9.3.3   Proposed scheme

9.3.3.1   Data trustworthiness

Definition 9.5   Uploaded data. These are the data sent from contributors to central servers to report on surroundings.

Definition 9.6   Actual data on surroundings. These are the actual data correctly reporting on surroundings.

Definition 9.7   Trusted contributors. These are contributors whose uploaded data are accurate data on surroundings.

Definition 9.8   Bad-mouth contributors. These are contributors whose uploaded data are inaccurate data on surroundings.

Definition 9.9   Bad-mouth attacks. Such attacks are launched by bad-mouth contributors, whose uploaded data are inaccurate.

As the participatory sensing system may be "open", anyone who installs the application (e.g., APP) on a smartphone can upload data on to central servers. The open system has no admission control, to promote more data uploads; it cannot distinguish trusted contributors and bad-mouth contributors from any prior information. Therefore, the task of distinguishing the data has to rely on the observation of contributors at central servers after uploading by subsequent information.

Definition 9.10   Central servers' observations. These are a series of uploaded data received by central servers and sent from contributors.

To distinguish between trusted contributors and bad-mouth contributors, a reputation system has to be established at central servers. The central servers evaluate contributors' reputation according to their observations.

Definition 9.11   Contributor reputation. This is a value to evaluate the likelihood of a contributor being a trusted contributor or a bad-mouth contributor. The value is stored in a reputation system at central servers and calculated after central servers' observations.

Definition 9.12   Reputation system. This is a series of calculating and managing methods to establish and evaluate the contributor reputation of each contributor to distinguish trusted and bad-mouth contributors.

We can state a more general principle to clarify our motivations or the necessary condition of the proposed scheme.

Proposition 9.10
An "open" system in which there exists no prior information (e.g., admission control information) must rely on a reputation system to distinguish between trusted and others.

Proof. Roughly speaking, as an open system has no prior information, trusted contributors and others cannot be distinguished at admission stage. Distinguishing them thus has to rely on the observations of their behaviors after admission. To distinguish between trusted contributors and others, a distinguishing system has to record and evaluate the observation, and make a judgment on the contribution, which in the end forms a reputation system to make judgments.

To build a reputation system, "good behavior" and "bad behavior" should be judged on each observation. The reputation system can thus evaluate the dynamics of behaviors, usually metrics for reputation evaluation. Before "good behavior" and "bad behavior" are defined, the criteria for judgment should be identified first. The judgment may be based on intuition and inferring information.

Definition 9.13 Inferred actual surrounding data. These are an approximation of actual surrounding data estimated by central servers, from the uploaded data by other contributors at similar locations and timestamps.

Example Contributors named A, B, and C upload the data D_a, D_b, and D_c, at similar locations (i.e., $|L_b - L_a| < \delta_1$, $|L_c - L_a| < \delta_1$, where L_a, L_b, L_c are the locations of A, B, and C, respectively; δ_1 is a threshold value for distance) and similar timestamps (i.e., $|T_b - T_a| < \delta_2$, $|T_c - T_a| < \delta_2$, where T_a, T_b, T_c are the upload timestamps of A, B, and C, respectively; δ_2 is a threshold value for time). The central servers will try to estimate contributor A's actual surrounding data. The inferred actual surrounding data is the function of D_b and D_c. That is to say, $D \Leftarrow \mathsf{Inf}(D_b, D_c)$, where D are inferred actual surrounding data; Inf is an inference function, taking D_b, D_c as input and outputting D.

Next, we propose detailed methods to deduce inferred actual surrounding data.

Suppose the uploaded data at similar locations (within δ_1) and similar timestamps (within δ_2) are $< D_i, L_i, T_i, C_i >$, where D_i are uploaded data; L_i is a location id; T_i is a timestamp; C_i is a contributor id; $i = 1, ..., n$. $D \Leftarrow \mathsf{Inf}(D_i)$, where D are inferred actual surrounding data, and $\mathsf{Inf}()$ is an inference function taking as input D_i and output D.

As policies are highly related to the types of uploaded data, we leave it as an open context-aware component and propose five typical inference policies, as follows:

9.3.3.1.1 (Inf-policy-I) average

$D = \mathsf{Avg}(D_i) = \sum_{i=1}^{n} D_i$, where $\mathsf{Avg}()$ is a standard function computing the average value of input parameters D_i. This policy may be used for all types of uploaded data.

9.3.3.1.2 (Inf-policy-II) median

$D = \mathsf{Med}(D_i)$, where $\mathsf{Med}(D_i)$ is a standard function returning the median value of input parameters D_i. This policy may be used for all types of uploading data.

9.3.3.1.3 (Inf-policy-III) distance average

(i) Suppose the inferred location of the inferred actual surrounding data is L. The space Euclidean distance between L_i and L is computed, denoting them as $SD_i, i = 1, ..., n$.

(ii) SD_i is sorted from the largest to the smallest value; the two end values are denoted as SD_{max} and SD_{min}, respectively. The corresponding uploaded data at these two locations (distances) are denoted as D_{min} and D_{max}, respectively.

(iii) The summation of total distance is computed as $SD_{sum} = \sum_{i=1}^{n} SD_i$.

(iv) The summation of total uploaded data is computed $D_{sum} = \sum_{i=1}^{n} D_i$.

(v) The value of $(\frac{D_{max} - D_{min}}{SD_{max} - SD_{min}} * SD_{sum} + D_{sum})/n$ is computed.

This policy is suitable for uploaded data that degrade with distance, for example, temperature or noise.

Proposition 9.11
Inf-Policy-III is sound.

Proof. Let k be the degradation rate over distance. Suppose x is the inferred value at L. We have

$$x - D_{min} = k * SD_{max}, x - D_{max} = k * SD_{min}$$

Thus, $k = (D_{max} - D_{min})/(SD_{max} - SD_{min})$. Also, $SD_{sum} = \sum_{i=1}^{n} SD_i, D_{sum} = \sum_{i=1}^{n} D_i$. We have

$$nx - D_{sum} = k * SD_{sum}.$$

That is, $x = (k * SD_{sum} + D_{sum})/n$. Further,

$$x = (\frac{D_{max} - D_{min}}{SD_{max} - SD_{min}} * SD_{sum} + D_{sum})/n.$$

9.3.3.1.4 (Inf-policy-IV) time average

Similarly, the procedures are as follows:

(i) Suppose the inferred time of inferred actual surrounding data is T. The time spans between T_i and T are computed, and denoted as $TS_i, i = 1, ..., n$.

(ii) TS_i is sorted from the largest to the smallest vale; the, denoted two end values are denoted as TS_{max} and TS_{min}, respectively. The corresponding uploaded data at these two timestamps are denoted as D_{min} and D_{max}, respectively.

(iii) The summation of the total distance is computed as $TS_{sum} = \sum_{i=1}^{n} TS_i$.

(iv) The summation of the total uploaded data is computed as $D_{sum} = \sum_{i=1}^{n} D_i$.

(v) The value of $\left(\frac{D_{max} - D_{min}}{TS_{max} - TS_{min}} * TS_{sum} + D_{sum} \right)/n$ is computed.

This policy is suitable for uploaded data that degrade with time, for example volumes of traffic or crowds.

Proposition 9.12
Inf-Policy-IV is sound.

Proof. The proof is similar to Proposition 2.

(Inf-Policy-V) Reputation weighted average.

(i) Suppose $R_i, i = 1, ..., n$ are the reputations of contributors who upload $D_i, i = 1, ..., n$. The summation of all reputation values is computed as $R_{sum} = \sum_{i=1}^{n} R_i$.

(ii) The weighting of each value is computed as $w_i \Leftarrow R_i / R_{sum}$.

(iii) The value of $D = \sum_{i=1}^{n} D_i * w_i$ is computed.

This policy may be used for all types of uploaded data. Also, $R = \sum_{i=1}^{n} R_i * w_i$ may be computed, which is the reputation of inferred actual surrounding data. (We will state how to create the reputation system later.)

Next, we define "good upload" and "bad upload" behaviors.

Definition 9.14 Good (bad) uploading. The reputation system within the central servers judges whether the uploaded data is good if and only if the distinction between the uploaded data and the inferred actual surrounding data is within a threshold value. The reputation system calls this upload from a contributor a "good upload." Otherwise, the reputation system calls upload a "bad upload."

We next propose a typical judgment policy for "good upload" and "bad upload" as follows:

Definition 9.15 Threshold judgment. Suppose the uploaded data is U, and the inferred actual surrounding data is D. This upload is a "good uploading", if and only if $|U - D|/D > Th$, where Th is a threshold value in system parameters. Otherwise, this upload is a "bad uploading".

Definition 9.16 Data trustworthiness of uploaded data. This is a value to evaluate the bias between uploaded data (denoted as U) sent from contributors and actual surrounding data (denoted as A). It is defined as $|U - A|/A$. Actual surrounding data is approximated by inferred actual surrounding data at the central servers, namely, $A \Leftarrow D$.

9.3.3.2 Reputation evaluation

Suppose the current contributor reputation is R. To evaluate contributor reputation dynamics, we propose following the evaluation policies:

9.3.3.2.1 (Eva-policy-I) threshold bias linear adjustment

A threshold judgment is used. If the bad upload occurs, $R \Leftarrow R - 1$. Otherwise, $R \Leftarrow R + 1$.

9.3.3.2.2 (Eva-policy-II) exponential bias linear adjustment

Suppose the uploaded data is U, and the inferred actual surrounding data is D. Compute $Bia = |U - D|/D$ is computed. Suppose the threshold value is Th.

If $Bia > Th$ and $|Bia - Th|/Th \in [A_i, A_{i+1})$, let $R \Leftarrow R - i$, where $A_i, i = 1, ..., n$ are system parameters. $A_i < A_{i+1}, i = 1, ..., n - 1$. For example, $A_i = 0.1 * a^{i-1}, a = 2$. If $Bia < Th$ and $|Bia - Th|/Th \in [A_i, A_{i+1})$, let $R \Leftarrow R + i$.

9.3.3.2.3 (Eva-policy-III) exponential bias exponential adjustment

Suppose the uploaded data is U, and the inferred actual surrounding data is D. The value of $Bia = |U - D|/D$ is computed. Suppose the threshold value is Th.

If $Bia > Th$ and $|Bia - Th|/Th \in [A_i, A_{i+1})$, $R \Leftarrow R - a^{i-1}, a = 2$ where $A_i, i = 1, ..., n$ are system parameters. $A_i < A_{i+1}, i = 1, ..., n - 1$. For example, $A_i = 0.1 * a^{i-1}, a = 2$. If $Bia < Th$ and $|Bia - Th|/Th \in [A_i, A_{i+1})$, we have $R \Leftarrow R + a^{i-1}, a = 2$.

Definition 9.17 Presented Data (PD). These are the data presented to consumers at the central servers.

Definition 9.18 Data trustworthiness of presented data. This is a value to evaluate the bias between the presenting data (denoted as P) for consumers and actual surrounding data (denoted as A). It can be defined as $|P - A|/A$. Actual surrounding data is estimated by inferred actual surrounding data at the central servers, namely, $A \Leftarrow D$.

The reputation system is not only used to deduce inferred actual surrounding data so as to compute the data trustworthiness of uploaded data, but is also used to create presenting data and computing its data trustworthiness. It poses two situations:

9.3.3.2.4 (Situation-I) inferred surrounding data are available

Suppose the uploaded data of the contributor is U, and the reputation of the contributor is r. Suppose D is the inferred actual surrounding data computed from contributors with similar locations and timestamps, and R is the reputation of D.

The presenting data are $P \Leftarrow \mathsf{FunP}(U, D, r, R)$, where $\mathsf{FunP}()$ is a function taking U, D, r, R as input and outputting presented data, denoted as P. The trustworthiness of this data is $T \Leftarrow \mathsf{FunT}(r, R)$, where $\mathsf{FunT}()$ is a function taking r, R as input and outputting the data trustworthiness of the presented data, denoted as T.

Example $\mathsf{FunP}(U, D, r, R) = U * r/(r + R) + D * R(r + R)$. $\mathsf{FunT}(r, R) = r/(r + R)$.

9.3.3.2.5 (Situation-II) inferred surrounding data are unavailable

The presenting used data have to be U, as inferred surrounding data are unavailable. The data trustworthiness provided to customers is calculated by r/R_{max}, where R_{max} is the current maximal reputation value in the reputation system. Or, the data trustworthiness is Λ, to denote that inferred surrounding data are unavailable.

9.3.3.3 Contributor privacy protection

Definition 9.19 Contributor actual identification. This is the essential identification of a contributor for uniquely distinguishing him/her, for example, student ID driver's license ID, social security number, and so on.

Definition 9.20 Contributor privacy ($\mathcal{CNP}$). This is the probability that central servers correctly identify the contributor actual identification after observing the uploaded data of the contributor. In shorthand,

$$\mathcal{CNP} = \Pr\{Id \Leftarrow CS | CS \leftarrow d\},$$

where $\Pr\{A|B\}$ denotes the probability that event A happens after event B happens; $A \Leftarrow B$ means "A is derived by B"; $A \leftarrow B$ means "A receives B"; ID is contributor actual identification; CS is central servers; d is uploaded data of contributors.

Definition 9.21 Contributor Perfect Privacy. This is guaranteed if and only if $\mathcal{CNP} = 0$.

We propose to use contributor anonymous identity instead of contributor actual identity in participatory sensing to protect contributor privacy.

Definition 9.22   Contributor Anonymous Identity. This is a unique identity to distinguish each contributor in the reputation system.

The procedures for contributor privacy protection consist of the following steps:

(PP-Step1) Initial key preparation.

When a contributor sends uploaded data for the first time, its contributor reputation is set as an initial value of r_0. It belongs to an initial group with a group identity of $gid = gid_0$, and has an initial group authentication key of $gak = gak_0$. Both gid and gak have been deployed previously by application software on smartphones.

(PP-Step2) Contributors generate their contributor anonymous identity.

The contributor anonymous identity is randomly generated with a fixed length, when each contributor sends uploaded data to the central servers.

(PP-Step3) Contributors upload data to central servers.

The uploaded data from a contributor to the central servers has six tuples

$$< cai, l, t, d, h(gak\|cai), gid >$$

where cai is the contributor anonymous identity; l is the location identity of the uploaded data; t is the time stamp of uploaded data; d is the data of the surroundings; $h()$ is a one-way and collision-free function.

(PP-Step4) Central servers verify the validity of contributors.

Central servers search gak by gid, and verify whether $h(gak\|cai)$ is correct or not. If it is correct, central servers deem that the contributor possesses the group gid, and thus have the corresponding reputation value of that group.

(PP-Step5) Central servers update reputation.

The reputation system in the central servers stores the contributor reputation of each contributor, and updates reputation values for contributors via the aforementioned reputation evaluation policies. That is to say, each contributor has a corresponding contributor reputation value that is computed and maintained by the reputation system.

(PP-Step6) The central servers update gak and gid.

Reputation The reputation system maintains an update period. It is a period determined by the central servers for updating to update all group authentication keys and group identities. For example, suppose the update period is 24 hours. The update time of the group authentication keys and group identities is at 12:00PM each day.

Suppose at the end of the updating period, each contributor has a reputation value, denoted by r. All current contributors are grouped by their reputation value. The group authentication key and group identity are both randomly generated by central servers. Central servers store $< gid, gak, r >$, and send new *gid* and *gak* values to the corresponding contributors confidentially.

(PP-Step7) Contributors update *gak* and *gid*.

The contributors in the same group receive the same group authentication key (*gak*) and group identity (*gid*). The contributor replaces the old values of *gid* and *gak* with the new ones.

9.3.3.4 Robustness enhancement

First, we analyze the potential attacks on our proposed scheme for data trustworthiness and contributor privacy protection. In the previous section, we pointed that out the adversaries among contributor peers are known as bad-mouth contributors; next we point out another possible malicious type of contributor in the current context.

Definition 9.23 Traitor contributors. These are the contributors who leak the group authentication key to other contributors, so that other contributors can obtain advantages; for example, easily obtain a higher reputation value.

Definition 9.24 Key leakage attack. Traitor contributors leak the group authentication key to other contributors, so that other contributors can obtain a corresponding reputation directly, avoiding to avoid any reputation evaluation procedure.

To further enhance the robustness of the scheme, we propose the following two methods.

(ROB-M1) Counting group members.

At the end of each updating period, the central servers record the total number of group members. In the next period, when a member with a different contributor anonymous identity joins the group, the central servers will decrease the count. Once the count reaches zero, newcomers who ask to join the group are not permitted to do so.

This method can limit the influence of the leaking of the group authentication key and detect the key-leaking attack.

(ROB-M2) Traitor tracing.

It is appropriate that central servers can trace the traitor who exposes the authentication group key to other contributors. The naive method is to change the group authentication key. For example, this can be achieved by making the group authentication key consist of two parts: one is the group authentication key generated by the central servers; the other is the private key generated by the contributors. The traitor can be traced through the distinct group authentication key.

The components of the scheme LibTip are listed in Table 9.3.

Table 9.3   Components of LibTip

Data Trustworthiness	Inf-Policy I-V
	Average
	Median
	Distance average
	Time average
	Reputation weighted average
Reputation Evaluation	Eva-Policy I–III adjustment
	Threshold bias linear
	Exponential bias linear
	Exponential bias exponential
	Situation-I–Situation-II
Privacy Protection	PP-Step1–PP-Step7
robustness	ROB-M1–ROB-M2

9.3.3.5   Analysis

Proposition 9.13

The contributor anonymous identity is necessary for the solution.

Proof. Contributor anonymous identities have to be generated to conceal contributor actual identities. In addition, contributor anonymous identities have to be identical to those in the reputation system to enable reputation evaluation. That is to say, the reputation computation is dedicated to a representative identity (i.e., the contributor anonymous identity) during an updating period. Thus, the contributor anonymous identity is necessary for the design goals solution.

Proposition 9.14

The group authentication key is necessary for the solution.

Proof. The group authentication key has to be used to anonymously authenticate contributors for their current reputation value; thus, the reputation value of a particular contributor can be continually evaluated and updated in the reputation system.

Proposition 9.15

The group identity is necessary for the solution.

Proof. The group identity has to be used to sort the group authentication key at the central servers. As the group identity is randomly generated and periodically updated, adversaries at channels cannot trace certain groups or their group members after link layer encryption.

Proposition 9.16
LibTip is lightweight.

Proof. In the LibTip scheme, the extra inducing items in $< cai, l, t, d, h(gak\|cai), gid >$ are cai, gak, gid. As cai, gak, gid are both necessary, LibTip only induces extra items that are necessary. Thus, LibTip is a lightweight solution.

Proposition 9.17
Contributor perfect privacy is guaranteed (namely, $\mathcal{CNP} = 0$).

Proof. The central servers can only view contributor anonymous identities; thus, contributor actual identities are unknown to them. In any case contributor anonymous identities are generated randomly. Thus, the linkage between contributor actual identities and contributor anonymous identities, and that between individual contributor anonymous identities, are both broken. That is to say, $\mathcal{CNP} = \Pr\{Id \Leftarrow CS | CS \leftarrow d\} = \Pr\{Id | cai\} = 0$.

Proposition 9.18
The risk of the exposure of a contributor's trajectory within an updating period is $f(\min(|G|), \max(e))$, where $f(\cdot, \cdot)$ is a function; $|G|$ is the group size; e is the number of uploads by this contributor within the group in this period.

Proof. Within one period, the contributor anonymous identities are is unchanged; thus, the trajectory of one contributor can be traced. The risk of the exposure of a contributor's trajectory is related to two elements as follows: If the number of uploading times is larger, the trajectory contains more information such as locations and timestamps. If the number of group members is smaller, the risks of trajectory exposure of a contributor's anonymous identity is larger. Thus, the risk is a function of $\min(|G|)$ and $\max(e)$. However, as the contributor actual identity is unknown, the trajectory cannot be linked to any actual identity.

9.4   Conclusions

In this chapter, we reviewed the importance of lightweight and robust security in the IoTs. We also proposed lightweight and robust security schemes in key IoT applications such as WBSN and participatory sensing. Extensive analysis on performance in terms of communication, computation, and storage verified that the OTM and OTP schemes are lightweight for WBSN. We also proposed a further lightweight scheme, LibTip to guarantee data trustworthiness, reputation evaluation, contributor privacy protection, and robustness against internal attackers in participatory sensing. LibTip provides an integral solution package consisting of a set of methods, policies, and procedures.

Some results in the chapter are published in [34, 35].

Acknowledgment

Wei Ren's research was financially supported by the National Natural Science Foundation of China (61170217).

Bibliography

[1] M. Seyedi, B. Kibret, D. Lai, and M. Faulkner, "A Survey on intrabody communications for body area network applications," *IEEE Trans. Biomed. Eng.*, vol. 60, no. 8, pp. 2067–2079, 2013.

[2] J. Liu, Z. Zhang, X. Chen, and K. Kwak, "Certificateless remote anonymous authentication schemes for wireless body area networks," *IEEE Trans. Parallel Distrib. Syst.*, vol. 25, no. 2, pp. 332–342, 2014.

[3] T. Ma, P. L. Shrestha, M. Hempel, D. Peng, H. Sharif, and H. Chen, "Assurance of energy efficiency and data security for ECG transmission in BASNs," *IEEE Trans. Biomed. Eng.*, vol. 59, no. 4, pp. 1041–1048, 2012.

[4] Z. Zhang, H. Wang, A.V. Vasilakos, and H. Fang, "ECG-cryptography and authentication in body area networks," *IEEE Trans. Inf. Technol. Biomed.*, vol. 16, no. 6, pp. 1070–1078, 2012.

[5] A. Banerjee, K. Venkatasubramanian, T. Mukherjee, and S. Gupta, "Ensuring safety, security, and sustainability of mission-critical cyber–physical systems," *Proceedings of the IEEE*, vol. 100, no. 1, pp. 283–299, 2012.

[6] M. Patel and J. Wang, "Applications, challenges, and prospective in emerging body area networking technologies," *Wireless Commun.*, vol. 17, no. 1, pp. 80–88, 2010.

[7] J. Biswas, J. Maniyeri, K. Gopalakrishnan, L. Shue, J. Phua, H. Palit, Y. Foo, L. Lau, and X. Li, "Processing of wearable sensor data on the cloud – a step towards scaling of continuous monitoring of health and well-being," in *Proc. 2010 Annual Int'l Conf. of the IEEE Engineering in Medicine and Biology Society (EMBC)*, September 2010, pp. 3860–3863.

[8] U.S. Deptarment of Health & Human Services, "The health insurance portability and accountability act of 1996 (HIPAA)," 1996.

[9] H. Tseng, S. Sheu, and Y. Shih, "Rotational listening strategy (rls) for ieee 802.15.4 wireless body networks," *IEEE Sensors J.*, vol. 11, no. 9, pp. 1841–1855, 2011.

[10] Y. W. Law, J. Doumen, and P. Hartel, "Survey and benchmark of block ciphers for wireless sensor networks," *ACM Trans. Sens. Netw.*, vol. 2, pp. 65–93, 2006.

[11] S. Keoh, E. Lupu, and M. Sloman, "Securing body sensor networks: Sensor association and key management," in *Proc. IEEE Int'l Conf. on Pervasive Computing and Communications (PerCom '09)*, March 2009, pp. 1–6.

[12] J. Sun, Y. Fang, and X. Zhu, "Privacy and emergency response in e-healthcare leveraging wireless body sensor networks," *Wireless Commun.*, vol. 17, no. 1, pp. 66–73, 2010.

[13] H. Wang, D. Peng, W. Wang, H. Sharif, H. hwa Chen, and A. Khoynezhad, "Resource-aware secure ecg healthcare monitoring through body sensor networks," *Wireless Commun.*, vol. 17, no. 1, pp. 12–19, 2010.

[14] C. Tan, H. Wang, S. Zhong, and Q. Li, "Ibe-lite: A lightweight identity-based cryptography for body sensor networks," *IEEE Trans. Inf. Technol. Biomed.*, vol. 13, no. 6, pp. 926–932, 2009.

[15] X. Lin, R. Lu, X. Shen, Y. Nemoto, and N. Kato, "Sage: a strong privacy-preserving scheme against global eavesdropping for ehealth systems," *IEEE J. Sel. Areas Commun.*, vol. 27, no. 4, pp. 365–378, 2009.

[16] Y. Zhu, S. L. Keoh, M. Sloman, and E. Lupu, "A lightweight policy system for body sensor networks," *IEEE Trans. Netw. Service Manag.*, vol. 6, no. 3, pp. 137–148, 2009.

[17] S. Nabar, J. Walling, and R. Poovendran, "Minimizing energy consumption in body sensor networks via convex optimization," in *Proc. 2010 Int'l Conf. on Body Sensor Networks (BSN '10)*, June 2010, pp. 62–67.

[18] M. Quwaider, J. Rao, and S. Biswas, "Body-posture-based dynamic link power control in wearable sensor networks," *IEEE Commun. Mag.*, vol. 48, no. 7, pp. 134–142, 2010.

[19] K. Venkatasubramanian, A. Banerjee, and S. Gupta, "Pska: Usable and secure key agreement scheme for body area networks," *IEEE Trans. Inf. Technol. Biomed.*, vol. 14, no. 1, pp. 60–68, 2010.

[20] F. X. Lin, A. Rahmati, and L. Zhong, "Dandelion: a framework for transparently programming phone-centered wireless body sensor applications for health," in *Proc. Int'l Conf. on Wireless Health (WH '10)*, 2010, pp. 74–83.

[21] D. Kotz, S. Avancha, and A. Baxi, "A privacy framework for mobile health and home-care systems," in *Proc. of the First ACM Workshop on Security and Privacy in Medical and Home-Care Systems*, 2009, pp. 1–12.

[22] T. Dorflinger, A. Voth, J. Kramer, and R. Fromm, "My smartphone is a safe! the user's point of view regarding novel authentication methods and gradual security levels on smartphones," in *Proc. of 2010 Int'l Conf. on Security and Cryptography (SECRYPT '10)*, July 2010, pp. 1–10.

[23] K. Vu, R. Zheng, and J. Gao. Efficient algorithms for k-anonymous location privacy in participatory sensing. In *Proc. of IEEE INFOCOM 12*, 2012, pp. 2399–2407.

[24] I. Boutsis and V. Kalogeraki. Privacy preservation for participatory sensing data. In *Proc. of 2013 IEEE International Conference on Pervasive Computing and Communications (PerCom '13)*, pp. 103–113, 2013.

[25] D. Christin, C. Rosskopf, M. Hollick, L.A. Martucci, and S.S. Kanhere. Incognisense: An anonymity-preserving reputation framework for participatory sensing applications. In *Proc. of 2012 IEEE International Conference on Pervasive Computing and Communications (PerCom '12)*, pp. 135–143, 2012.

[26] C. Costa, C. Laoudias, D. Zeinalipour-Yazti, and D. Gunopulos. Smarttrace: Finding similar trajectories in smartphone networks without disclosing the traces. In *Proc. of 2011 IEEE 27th International Conference on Data Engineering (ICDE '11)*, pp. 1288–1291, 2011.

[27] E. Cristofaro and C. Soriente. Participatory privacy: Enabling privacy in participatory sensing. *IEEE Network*, vol. 27, no. 1, pp. 32–36, 2013.

[28] S. Gao, J. Ma, W. Shi, G. Zhan, and C. Sun. Trpf: A trajectory privacy-preserving framework for participatory sensing. *IEEE Transactions on Information Forensics and Security*, vol. 8, no. 6, pp. 874–887, 2013.

[29] M. Groat, B. Edwards, J. Horey, W. He, and S. Forrest. Enhancing privacy in participatory sensing applications with multidimensional data. In *Proc. of 2012 IEEE International Conference on Pervasive Computing and Communications (PerCom '12)*, pp. 144–152, 2012.

[30] S. Hachem, A. Pathak, and V. Issarny. Probabilistic registration for large-scale mobile participatory sensing. In *Proc. of 2013 IEEE International Conference on Pervasive Computing and Communications (PerCom '13)*, pp. 132–140, 2013.

[31] K. Huang, S. Kanhere, and W. Hu. A privacy-preserving reputation system for participatory sensing. In *Proc. of 2012 IEEE 37th Conference on Local Computer Networks (LCN '12)*, pp. 10–18, 2012.

[32] L. Kazemi and C. Shahabi. Towards preserving privacy in participatory sensing. In *Proc. of 2011 IEEE International Conference on Pervasive Computing and Communications Workshops (PerComW '11)*, pp. 328–331, 2011.

[33] C. Wang and W. Ku. Anonymous sensory data collection approach for mobile participatory sensing. In *Proc. of 2012 IEEE 28th International Conference on Data Engineering Workshops (ICDEW '12)*, pp. 220–227, 2012.

[34] W. Ren, J. Lin, Y. Ren, LibTiP: A lightweight and robust scheme for data trustworthiness and privacy protection in participatory sensing, International Journal of Embedded Systems, 2015, preprint.

[35] W. Ren, uLeepp: An ultra-lightweight energy-efficient and privacy-protected scheme for pervasive and mobile WBSN-cloud communications, Ad Hoc & Sensor Wireless Networks, vol. 27, no. 3-4, pp. 173–195, 2015.

TRUST AND AUTHENTICA-TION

Chapter 10

Trust and Trust Models for the IoT

Michael Schukat

Pablo Cortijo Castilla

Hugh Melvin

CONTENTS

10.1 Introduction

The aim of this chapter is to investigate frameworks that ensure trust as well as communication security between nodes in an IoT deployment. The former captures trust in the identity and in the credentials or privileges of a communication peer, therefore dealing with authentication and authorization, while the latter provides some guarantee about the privacy and integrity of the data exchanged between peers. Trust does not refer to the trust in the validity of the data itself; for example, the question of whether a sensor or other data source provides correct readings or not. This relates to the problem of sensor and device reputation, as, for example, discussed by Ganeriwal et al. [1].

Trust and security are based on tokens or credentials, provided by a trust management infrastructure, which are embedded in and potentially shared between devices (note that this chapter will use the terms peers, devices, and (end) entities to describe IoT nodes). The integrity and robustness of these tokens (which can, for example, be symmetric keys or digital certificates) are the cornerstone of trust and security. They are useful in deflecting external attacks initiated by entities that are not in possession of credentials, but fail to deflect internal attacks, where credentials or nodes that own credentials have been compromised.

10.1.1   Trust and security from a device perspective

IoT devices are vulnerable in many aspects, so providing and maintaining trust and security (e.g., providing token integrity over time) is a difficult endeavor. Once token integrity is compromised, for example, by recovering a secret network key from a device and using it to fabricate malicious nodes, the entire network is vulnerable to internal attacks.

On the physical level, device enclosures are often not tamperproof; devices can be opened and their hardware can be accessed via probes and pin headers. Device central processing units (CPUs) are low-cost components that often have no sophisticated means to protect their code, data, and tokens from external access, that is, via its Joint Test Action Group (JTAG). This allows an attacker to clone entire devices or manipulate software and data; for example, to manipulate a glucometer so that it will provide incorrect readings. If the device is deployed in an unsupervised environment, it may be accessed and manipulated by a malicious third party without notice.

Furthermore, IoT devices are often based on low-power hardware and may only be able to process tokens with a low complexity. This can have an implication on the robustness of a token, as it can be reengineered or recovered via a brute-force attack.

As a result of this, any trust management system for IoT deployments must have the ability to dynamically withdraw trust of individual devices. Likewise, individual devices must be dynamically able to validate the trustworthiness of other nodes they engage with.

When trust and security credentials are distributed at the time of manufacturing or deployment, a device is seen as initially trustworthy. This trustworthiness, which might degrade over time, is based on many assumptions and prerequisites, including

- The device's hardware, as well as all stages of its manufacturing/ integration, is trustworthy and sound. For example, it must not have JTAG pin headers that allow the extraction of program code and data.

- Likewise, the firmware and its development process (from specification to test) is trustworthy and follows best practices. For example, devices must not have undocumented software back doors that have been deliberately left by developers.

- The generation, management, and deployment of tokens is trustworthy and sound. For example, pseudorandom number generators must have sufficient entropy to avoid the generation of weak and predictable keys.

A general problem in the context of trustworthy firmware is that many embedded processors (even if they operate under a modern multitasking operating system) do not provide process encapsulation via memory virtualization. As a result, malicious code in a firmware image can access and manipulate credentials used by other system processes to initiate an internal attack. Therefore, it is not sufficient to determine the trustworthiness of firmware components individually, but the firmware image as a whole must be validated.

Devices with a static ("factory-flashed") firmware image can maintain a higher degree of trustworthiness over time than devices that can be updated dynamically in the field (i.e., via firmware download), if the upload mechanism itself poses a potential back door for attacks. If such a mechanism allows the upgrade of individual firmware components, the number of image variations can increase exponentially, which makes the validation of all firmware images variations a very cumbersome task. Nonetheless, a secure device firmware updating or patching mechanism, as, for example, found in embedded Linux systems, is an integral component to maintain security, as otherwise a single vulnerability can compromise a number of systems. A network-wide update mechanism will preferably incorporate a smooth and effective patching process, which includes robust integrity and authenticity checks, minimizes service outages, and allows for a version rollback if needed.

10.1.2 Secure key storage

Secure storage facilities (also known as keystores) increase the robustness of trust tokens used both within an IoT system and its trust management infrastructure (like a certificate authority or a trust center). Passive keystores provide a means to securely save and retrieve credentials; cryptographic operations are executed outside these stores by the device's CPU. Active keystores in contrast allow the internal execution of cryptographic operations via an application program interface (API), so the credentials are never exposed. The following sections will describe various types of key stores.

10.1.2.1 Hardware stores

The high-end representatives of this category are hardware security modules (HSM). HSMs have a place in trust management infrastructures with extensive cryptographic requirements.

General-purpose HSMs provide a thoroughly secure, generally configurable administration; a security level that can be somewhat adjusted to needs; and tools that cover the whole life cycle of the HSM (such as secure key backup).

Their main disadvantage is their lack of flexibility if uncommon token formats or algorithms are used.

Cryptographic smart cards (embedded or otherwise) and cryptographic universal serial bus (USB) dongles are low-cost HSMs. They are particularly adequate for resource-constrained nodes or low-cost trust management infrastructures. On the downside, smart cards and dongles may not have as high-security certifications as HSMs; their default administrative options are generally limited and their security-level settings are not as flexible.

10.1.2.2 Trusted platform modules

Trusted platform modules (TPMs) are dedicated processors that offer both an interesting complement and an alternative to the options presented above. They are meant to protect hardware (by authenticating devices, or possibly attesting a certain hardware is present), booting processes, and so on, and can also be used in a more general way to store and retrieve credentials after booting has taken place. Their interfaces, however, are different from the ones found in the above HSMs and, in the case of devices that comply with TPM 1.1b, are vendor specific.

10.1.2.3 Software stores

The natural place for software stores is in devices with low security requirements or low-cost embedded systems that have no provisions to physically connect hardware modules.

There is a plethora of both active and passive software stores that can be used in IoT systems. PKCS#12 stores are based on the homonymous public key cryptography standard (PKCS), initially defined by RSA Security (now part of EMC Corporation) and later expanded and corrected by several request-for-comments documents (RFCs), such as RFC 7292 [2]. The standard defines a data structure syntax that can contain cryptographic objects (keys, certificates, etc.) and, optionally, arbitrary data, encrypted and signed. In principle, PKCS#12 defines two types of integrity/privacy modes, the asymmetric cryptography and the password-based modes.

Privacy-enhanced electronic mail (PEM) stores are files that contain Base 64 versions of ASN.1 formatted certificates and (encrypted) keys, enclosed by human-readable headers for convenience.

Java stores are part of a much larger programming framework, the Java cryptography architecture/Java cryptography extension (JCA/JCE). This framework defines a provider-based, pluggable architecture that includes, among many other things, keystore implementations. One such implementation is provided by the Sun provider and included in all distributions since the early versions of Java. It implements keystores as proprietary password-protected Java KeyStore (JKS) files.

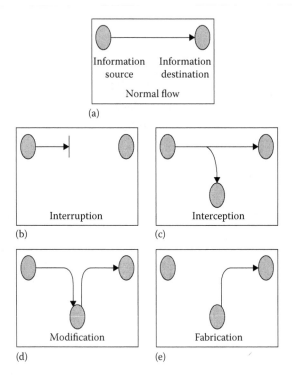

Figure 10.1: Principal attack vectors in IoT device communication.

10.1.3 Trust and security from a network perspective

During the operation of a network, devices set up static or dynamic (e.g., short-lived) communication links with other peers. These links can be either point-to-point or incorporate a group of nodes. From a device's perspective, the challenge is to validate the authenticity and authority of the other peer(s) and to set up a secure communication link to avoid attack scenarios, as shown in Figure 10.1. For this purpose, trust tokens are exchanged and validated, or new session tokens are created (i.e., session keys derived from a master key).

Overall the following requirements must be fulfilled:

■ Data: confidentiality

■ Data: integrity

■ Peer: authenticity

■ Peer: Proof of authorization

- Communication: service and system availability

- Communication: nonrepudiation

The assurance of data integrity, optionally in combination with data confidentiality via encryption, provides trustworthiness in the data a node sends or receives. For example, in a body area network, a wireless glucometer sends glucose readings to an integrated insulin pump. This information must be protected from accidental or deliberate tampering, while patient privacy considerations require the data to be encrypted. Data integrity and confidentiality provide a foundation (complementary to additional protocol-specific features like sequence numbers or timestamps) to deal with the principal attack vectors of interception, interruption, and modification.

Data confidentiality is usually provided via symmetric encryption (with the Advanced Encryption Standard [AES] algorithm as a de facto industry standard) often implemented directly in hardware, while data integrity is provided via message authentication codes or cryptographic hashes that are attached to the data payload.

Peer authenticity relates to the problem of how a peer can validate another peer's identity before a communication link is established; that is, an insulin pump must be able to validate that it actually connects to a trusted glucometer (and subsequently receives data from it) and not from a malicious device.

Peer authenticity can go hand in hand with system availability. For example, denial-of-service (DoS)-style attacks are typically external attacks (e.g., they are launched by external nodes outside the jurisdiction of an IoT deployment), so the ability to qualify and if necessary to discard data or connection requests (i.e., SYN flood attack for transmission control protocol [TCP] connections) at an early stage can help to alleviate such attacks.

Proof of authorization provides assurance that a peer has the authority to (a) communicate with another peer and (b) conduct a certain action; for example,

- A glucometer will only accept data requests from an insulin pump (and not from the blood pressure monitor). Furthermore, both glucometer and pump must be from the same manufacturer.

- A reset command sent to the glucometer sensor by the insulin pump (after a sensor reconfiguration) should only be executed if the insulin pump has the required authorization level.

Therefore, proof of authorization is a viable mechanism to protect against fabrication.

Nonrepudiation—for example, the ability to ensure that communicating peers cannot deny the authenticity of their action—is linked back to peer tokens.

Data integrity and data confidentiality are based on credentials only known to the communicating peers (i.e., shared secret keys). If these credentials are created dynamically on the fly, they must be mutually authenticatable during the key generation phase (to avoid man-in-the-middle [MitM] style attacks, as possible in Diffie–Hellman key exchanges).

Likewise, peer authentication and authenticity are provided via additional device descriptors that are mutually available and can be mutually validated.

10.2 Trust Model Concepts

The following section will present three trust models. They provide the conceptual basis for a trust management infrastructure.

10.2.1 Direct trust model

In a direct trust model, a peer obtains credentials of other peers in such a way that it is immediately convincing to them. A common approach is the predistribution of peer credentials before the network is deployed. Two approaches will be described here, if only briefly: one based on symmetric keys and another one that makes use of static whitelists.

The first option uses as credentials pairwise shared symmetric keys (installed during manufacturing or system integration), which provide data confidentiality and integrity as well as implicit peer authenticity and proof of authorization—the latter can be expanded via additional peer descriptor tables in each device that associate further attributes with each peer.

A direct trust model, which is based on pairwise shared symmetric keys for n nodes, requires a total of $n*(n-1)/2$ keys, with $(n-1)$ keys stored in every node, making it unsuitable for large-scale deployments. Also, the revocation or renewal of tokens is very tedious, as every node has to be notified.

The second option, discussed in [3], uses asymmetric keys and whitelists containing references to certificates (as further discussed below). Here, each device is equipped with its own certificate (entailing its identity, a public key, and further device attributes) signed by an authority, and a complementary whitelist that contains unforgeable certificate identifiers of all peers with which it is allowed to communicate. An identifier can be a certificate's hash value, its public key, or its serial number.

While this solution substantially reduces the number of credentials distributed in a network (each device would have exactly one certificate containing one key pair), the management of the whitelists is impractical for large or nonstatic developments. Likewise, the revocation or renewal of tokens is very tedious.

Overall, a direct trust model approach is only feasible in small and static networks because of its management constraints and memory requirements.

10.2.2   Web-of-trust model

In a web-of-trust model, a peer accepts the credentials of another peer if these credentials have been validated (e.g., signed) by another, already trusted peer [4]; that is, in a body area network, a glucometer will accept the credentials of an external programming device (and subsequently establish a connection to it) if these credentials were signed by a trusted insulin pump. The web-of-trust model is implemented in Pretty Good Privacy (PGP), where individual users maintain a list of credentials (e.g., public keys) in a key ring. When a key from another peer is inserted, the user assigns the key legitimacy which can hold the value as complete (e.g., complete confidence that the credential is owned by the other peer), marginally, or not trusted.

However, in an IoT environment, a web-of-trust model is not feasible for nonstatic networks, as it does not enable new previously unknown nodes to join a network. Unmanaged key servers, as used in PGP, for example, are vulnerable to identity spoofing and do not solve this problem.

Furthermore, the withdrawal of trust is tedious, as it has to be propagated across the network to reach all nodes.

Also, IoT devices may operate in very regimented environments (i.e., medical or critical infrastructure), where a web-of-trust model is simply not acceptable and a tight, centralized trust management, as discussed in the next section, is required.

10.2.3   Hierarchical trust model

Here, trust is managed by one or more trust anchors, whereby multiple anchors form a hierarchical infrastructure.

10.2.3.1   Trust center infrastructures

In a trust center infrastructure (TCI) one or more dedicated trust anchors manage on-the-fly connection requests between network nodes. The network authentication system Kerberos [5] is based on this approach, whereby individual clients receive tickets (with a certain life span) that allow them to authenticate and authorize themselves to other nodes.

Figure 10.2 shows an example for a TCI based on a single dedicated trust anchor (TA). The TA has a unique shared token (a symmetric key $K[x]$) with each node x in the network. It also optionally maintains a descriptor table $D[x]$ for each device.

Whenever two nodes NI and ND set up a network connection, the initiating peer NI refers first to the TA (Step 1) (i) to validate NDs identity ID[ND], (ii) to obtain ND's descriptor table D[ND] (to resolve authentication and authorization issues), and (iii) to obtain a randomly generated session key S[NI-ND] to be shared later with ND. The same information about NI is provided for ND, but encoded using the key K[ND] shared between TA and ND. Both components form a response which is encoded using K[NI] before transmission (Step 2).

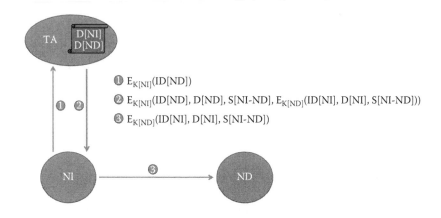

Figure 10.2: Node authentication in a trust center infrastructure.

NI receives and decodes the response, validates ND's identity and authorization level, and sends the second (still encoded) component to ND (Step 3), which in turn will validate NI's identity and authorization level. Finally, both peers use S[NI-ND] to set up a secure communication link.

The revocation of nodes (in the sense of marking them as untrusted) is straightforward when a TA is used. Also, each node only requires a single key K[x] and the TA's identity (i.e., its media access control [MAC] or Internet protocol [IP] address), making it a very resource-efficient approach from a device's perspective.

The downside is that the TA poses a single point of failure; for example, a compromised trust anchor (which, for example, is a victim of a DoS attack) will compromise the integrity and availability of the entire network. The inability to revoke the TA complicates things further.

Since the TA holds information about all devices managed by the TCI, this approach is only suitable for static networks or environments where an operator can add and remove device details on the fly.

10.2.3.2 Public key infrastructures

Public key infrastructures (PKIs) are another implementation of the hierarchical trust concept. PKIs are less susceptible to attacks on availability, as they provide network nodes with verifiable credentials (also known as public key certificates) prior to deployment, which can be validated without accessing a TA. In contrast to TCIs they also demonstrate better scalability and manageability.

According to the Internet Engineering Task Force (IETF) PKIX working group, a PKI is "the set of hardware, software, people, policies and procedures needed to create, manage, store, distribute, and revoke Public Key Certificates." [6].

A PKI, therefore, does not limit its scope to strictly technical elements such as hardware, software, networking infrastructures, protocols, or algorithms. It permeates into organizations that manage and use them by requiring the participation of other agents and resources: people, policies, and procedures.

Public key certificates (also called digital certificates or identity certificates), the public key cryptography procedures and technologies that substantiate them, and trust relations form the basis of a PKI. Trust relations based on the issuance of certificates, which in turn rely on public key cryptography, are what differentiates PKIs over other forms of security constructs, and what determines their properties.

The life cycle of certificates is at the core of PKIs. Typically, an entity called the certification authority (CA) issues identity certificates by digitally signing a set of (identity-related and other) attributes including a public key (of a public–private key pair in the context of public key cryptography). The act of issuing such a certificate constitutes a proof of the linkage between the attributes and possession of the public key. By signing the certificate with his own private key of a public–private-key pair, a CA states that the attributes are tied to the entity that owns the public key pair.

Issuing certificates that, in turn, issue other certificates is common practice, in what are called multilevel PKI hierarchies. PKIs also allow for the issuance and management of other types of certificates, such as attribute certificates.

The issuance of certificates is the germinal event that constitutes a PKI, and is often present in successive steps throughout its existence. Other events give shape to the life cycle of a PKI as well.

Overall, the life cycle of a certificate includes the following events:

- End-entity registration

- Issuance

- Publication

- Revocation of certificates

- Generation of revocation state data

- Archival and recovery of certificates and key material

Trust plays a crucial role in PKIs. The key concept in this case is transitivity of trust. Due to the mathematical properties of the algorithms that underpin PKI technologies, a well-formed certificate directly or indirectly (through intermediate certificates, that form an unbroken chain of issuances) issued by a trusted certificate can in turn be trusted. In the simplest scenario, a single initial trusted certificate is all that has to be interchanged through fully verifiable channels. These channels can vary, but have to meet these two conditions:

- The authenticity of the data must be assured by already established and particularly trustworthy methods, in proportion with the importance of the hierarchy that is to be established.

- The mechanism must not rely on the trust of any component of the hierarchy it is meant to establish.

As a result of all the above, previously unknown entities can communicate securely. In fact, there is no theoretical limit to the number of previously unknown entities that could securely communicate, hence the excellent scalability properties of a PKI. But, of course, as trust can be transmitted down trust chains with relative ease, the initial source of trust, or root trust anchor, has to be protected at all costs.

10.3   PKI Architecture Components

The following components are usually found in PKIs:

- Certification authorities

- Registration authorities

- Validation authorities

- Central directories

Optionally, PKIs can also incorporate timestamping authorities and certificate revocation authorities.

In addition, but on an entirely different level, a PKI comprises a series of policies (of which certificate/certification policies are the most salient example), procedures, and personnel.

10.3.1   Certification authorities

CAs form the backbone and the trust anchors of a PKI. They issue certificates and, in many cases, revocation status data (for instance, certificate revocation lists, CRLs) regarding the certificates they issue, and publish both types of products. Certification authorities are typically structured in levels, thus forming a hierarchical PKI.

10.3.2   Registration authorities

Registration authorities (RAs) act as the front end of certification authorities, in that they are responsible for identifying and authenticating entities that request certificates, and then dispatching certificate requests to CAs and routing back the certificate(s) to the requesting entity. In some cases, RAs are just a specialized component of CAs.

10.3.3   Validation authorities

Validation authorities (VAs) allow for the validation of certificates. Validating a certificate actually comprises several steps (verifying signatures, possibly obtaining certificates, checking revocation status). It is generally assumed that VAs only provide services in relation to checking revocation status, typically via online certificate status protocol (OCSP) services. VAs are, therefore, usually OCSP servers.

10.3.4   Central directories

Central directories make certificates available to other entities. Since other data, such as policies, or CRLs need to be published as well, central directories store and make all these data available. They are often implemented as lightweight directory access protocol (LDAP) servers.

10.3.5   Timestamping authorities

Timestamping authorities (TSAs) are characterized by their ability to issue PKI-based trusted timestamps. Trusted timestamps can prove that data existed prior to the issuance of the timestamp; as a result, "time-aware" validation mechanisms (i.e., those that take into account the moment timestamped signatures were generated) can be used. That property makes it possible for timestamps to form an important part of certain advanced signature mechanisms, such as CAdES-T/CAdES-X or XAdES-T/XAdES-X [7].

10.3.6   Certificate revocation authorities

Certificate revocation authorities (CRAs) are specific authorities that allow for the revocation of certificates. Normally, revocation duties are carried out by a dedicated service that belongs to each CA. However, whenever either the number of issued certificates is high, or the complexity of revocation procedures increases, or so does the number and variety of CAs, specialized authorities, CRAs, come to play, whereby a single, centralized CRA can substitute equivalent revocation services on multiple CAs.

CRAs provide a number of benefits:

■ CRAs decouple certificate issuance from revocation, thus easing the use of resources. They can contribute to overall system availability, as they provide revocation information about already issued certificates, even if certification authorities are not available. Providing revocation information within a certain time frame is critical for the correct functioning of many PKIs. What is more, policies generally set specific time frames for the renewal of that information, so the consequences of a failure to provide can go beyond system unavailability.

- CRAs make the end of life of CAs easier, by continuing to revoke certificates after the CA has been effectively been decommissioned. A revocation authority can even revoke every single certificate a CA has issued, if necessary, thus emulating the revocation of the certificate issuance CA certificate.

- PKIs which contain multiple active CAs based on different products or technologies can find the use of CRAs especially beneficial. A CRA can provide services for all certificates, without the burden of having to revoke them using a specific CA.

- Separating CRA services from CA services improves system security, by preventing unnecessary exposure of CA services to entities that only require revocation services or data. Protecting CAs is particularly critical as compared to protecting CRAs, as attacks to the former could lead to unwanted issuance of certificates, whereas attacks to a CRA would, at the most, mean an unwanted retrieval of certificates.

10.4   Public Key Certificate Formats

10.4.1   X.509 certificates

X.509 identity certificates are specified in the X.509 recommendation of the telecommunication standardization sector of the International Telecommunication Union (ITU-T) [8]. They are self-descriptive entities that use Abstract Syntax Notation One (ASN.1) as the specification language. Its high-level structural representation (in ASN.1) is as follows:

```
Certificate ::= SEQUENCE {
tbsCertificate TBSCertificate,
signatureAlgorithm AlgorithmIdentifier,
signatureValue BIT STRING }
```

TBSCertificate further defines (in ASN.1 notation) the data fields represented in a certificate (Figure 10.3), while *Algorithm Identifier* is described via an object identifier (OID); that is, for example, "1.2.840.113549.1.1.4" for "MD5withRSA," a combination of the MD5 hash algorithm encoded using the CA's private RSA key. The signature (a signed hash value) is itself stored as a bit string.

X.509 certificates are encoded using Distinguished Encoding Rules (DER) [9] (see Figure 10.4) and stored as an ASCII string.

The X.509 standard distinguishes between three identity certificate versions (see Figure 10.5), with version 3 certificates being the most common and versatile ones.

In relation to IoT devices, a digital certificate must contain the following information at a minimum:

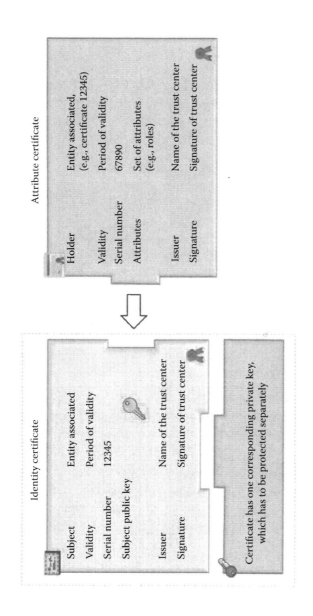

Figure 10.3: Identity certificates and attribute certificates.

```
0000 : 30 82 03 05                    ; SEQUENCE (305 Bytes)
0004 :    30 82 01 f1                    ; SEQUENCE (1f1 Bytes)
0008 :    | a0 03                    ; OPTIONAL [0] (3 Bytes)
000a :    | | 02 01                    ; INTEGER (1 Bytes)
000c :    | |    02
000d :    | 02 10                    ; INTEGER (10 Bytes)
000f :    | | 6e 92 35 46 0e db b5 94   4d 59 f9 f1 a8 f1 cf e6
001f :    | 30 09                    ; SEQUENCE (9 Bytes)
0021 :    | | 06 05                    ; OBJECT_ID (5 Bytes)
0023 :    | | | 2b 0e 03 02 1d
         |    |  | |    ; 1.3.14.3.2.29 sha1RSA (shaRSA)
0028 :    | | 05 00                    ; NULL (0 Bytes)
002a :    | 30 1a                    ; SEQUENCE (1a Bytes)
002c :    | | 31 18                    ; SET (18 Bytes)
002e :    | |    30 16                    ; SEQUENCE (16 Bytes)
0030 :    | |    06 03                    ; OBJECT_ID (3 Bytes)
0032 :    | |    | 55 04 03
         |    |  |    ; 2.5.4.3 Common Name (CN)
0035 :    | |    13 0f                    ; PRINTABLE_STRING (f Bytes)
0037 :    | |    4d 6f 72 67 61 6e 20 53   69 6d 6f 6e 73 65 6e
         |    |  ; "Morgan Simonsen" (3 Bytes)
0046 :    | 30 1e                    ; SEQUENCE (1e Bytes)
0048 :    | | 17 0d                    ; UTC_TIME (d Bytes)
004a :    | | | 31 33 30 34 31 36 30 38 35 37 31 37 5a
         |    | | | ; 16.04.2013 10:57
0057 :    | | 17 0d                    ; UTC_TIME (d Bytes)
```

Figure 10.4: ASN.1 DER encoded certificate [10]. (From: *Morgan Simonsen's Blog.* https://morgansimonsen.wordpress.com/2013/04/16/understanding-x-509-digital-certificate-thumbprints/.)

■ The name of the subject—for example, the identity of the device—to which the public key in the certificate is bound. Note that there is a conceptual difference between a typical server-side certificate (in which the entity is identified by its domain name system [DNS] name), and a device certificate, which can also be identified by a uniform resource identifier (URI), a MAC address, or an IP address.

■ Its public key and the cryptographic algorithm it relates to.

■ The certificate's serial number (for revocation purposes) and validity.

■ The name of the issuer (e.g., the CA).

■ The purpose and restrictions of the public key in the certificate.

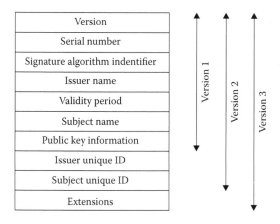

Figure 10.5: X.509 Identity certificate versions 1, 2, and 3.

X.509 certificates are rather large (∼2 Kbyte), have a complex structure (as shown in Figure 10.4), and require a complex parser; resource-constrained IoT devices may have difficulty handling them, in terms of both memory and computational requirements. Therefore, alternative formats are discussed in the following two sections.

10.4.2 Self-descriptive card verifiable certificates

Self-descriptive card verifiable certificates (CVCs) are very compact public key certificates suitable for resource-constrained devices like smart cards. While CVCs are still DER encoded (and are therefore self-descriptive), they only contain a subset of the fields of an identity certificate [11]:

```
cvcBody ::= SEQUENCE {
 profileId UNSIGNED INTEGER,
 issuer   CHARACTER STRING,
 pubKey   CHARACTER STRING,
 subject  CHARACTER STRING,
 notBefore DATE,
 notAfter  DATE }
```

10.4.3 Non-self-descriptive card verifiable certificates

These certificates are not DER encoded and therefore do not have type tags. Information about their internal structure is provided via header fields that are kept separate from the certificate itself. Non-self-descriptive CVCs can be represented in their simplest form by a static abstract data type; for example, a structure in the programming language C:

```
typedef struct simpleNonSelfDescriptiveCVC {
char version;
char owner[20];
char issuer[20];
char alg; // Note that OIDs are omitted here
...
} tSimpleNonSelfDescriptiveCVC;
```

A memory-efficient format that incorporates dynamic length fields, but requires a simple parser for processing (as it cannot be mapped onto a fixed-length data structure any more), appears as follows:

```
typedef struct dynamicNonSelfDescriptiveCVC {
char version;
char ownerLength;
char owner[ownerLength];
char issuerLength;
char issuer[issuerLength];
char alg; // Note that OIDs are omitted here
...
} tDynamicNonSelfDescriptiveCVC;
```

10.4.4 *Attribute certificates*

An attribute certificate only assigns—in contrast to a conventional public key or identity certificate—privileges to end entities. X.509 attribute certificates are specified in [12]. They have the following ASN.1 structure:

```
AttributeCertificateInfo ::= SEQUENCE {
version                   AttCertVersion -- version is v2,
holder                    Holder,
issuer                    AttCertIssuer,
signature                 AlgorithmIdentifier,
serialNumber              CertificateSerialNumber,
attrCertValidityPeriod    AttCertValidityPeriod,
attributes                SEQUENCE OF Attribute,
issuerUniqueID            UniqueIdentifier OPTIONAL,
extensions                Extensions OPTIONAL }
```

An attribute certificate is issued and signed by an attribute authority; it has a certain life span and binds an authorization (of whatever nature) to the end entity.

It has also optional extension fields. However, as can be seen in the listing, it does not contain a public key. Instead, it is tied to an identity certificate, as shown in Figure 10.3.

This separation allows identity and attribute certificates to have different life spans, which is, for example, extensively used in digital rights management. Here, a consumer acquires the right to access certain digital content over a potentially limited period of time via an attribute certificate. The consumer himself is identified via his identity certificate.

10.5   Design Considerations for Digital Certificates

10.5.1   *Device identifiers*

In a digital identity certificate, both its owner and the CA that signed the certificate must be uniquely identified. While there will be a relatively small number of CAs (with each CA being able to potentially manage millions of certificates), there is a need for a scalable naming scheme suitable for billions of nodes.

Device identifier construction schemes can be based on various methods. These methods incorporate either (i) random data, (ii) a hierarchy identifier, (iii) the encoding of additional information (e.g., the manufacturer), or (iv) the use of cryptographic operations (e.g., hash of public key) [16]. One scheme can apply several methods at the same time, as shown in Table 10.1 [15].

In today's Internet, the URI is the de facto naming scheme to identify the name of a web resource. At the network level, a device is identified via its static or dynamic (v4 or v6) IP address. The DNS translates the hostname into a URI into an IP address.

However, while this approach is suitable for a hierarchy of certificate authorities, it does not necessarily scale for IoT networks, as (i) such networks can be isolated without having access to a DNS service and (ii) the anticipated number of IoT devices makes a classical URI approach unworkable. Furthermore, machine-to-machine (M2M) communication does not necessarily require human-readable URIs.

An alternative solution is the use of a device's IPv6 address as its unique device identifier. Such an address consists of 16 octets; the overall address space is in the order of 10^{38} possible addresses [13].

The underlying IPv6 network communication protocol is already widely embraced by standard (IEEE 802.3 Ethernet and IEEE 802.11 Wi-Fi based) networks and has found its way into the IoT via the low-power wireless personal area networks (6LoWPAN) communication standard, so it is a potential candidate for an IoT naming scheme.

Table 10.1 Device identifier construction schemes and their underlying methods

Project or Architecture	Naming Scheme	Method Applied
IPv6	URI	Hierarchy identifier; encoding additional information
IPv6	IPv6	Hierarchy identifier
Glowbal IP Protocol	AAID	Encoding additional information
GS1	GS1 identification keys	Random data; encoding additional information
SWE	Sensor UID	Encoding additional information
IoT@Work	Name of a node within a namespace	Hierarchy identifier; encoding additional information
NDN	Name of the data	Hierarchy identifier; encoding additional information; cryptographic operations
Mobility First	GUID	Encoding additional information; cryptographic operations
RFID	RFID	Random data
802.15.4	MAC address	Random data

An IPv6 address is ideally broken into two 64-bit segments, with the first segment being the network's subnet address. In most IoT networks, this address will be assigned during deployment, so it cannot be anticipated when a certificate is generated during manufacturing. Furthermore, the address can potentially change over time. The second 64-bit segment, however, is the device's MAC address, which is in fact unique and available for certificate generation during manufacturing.

Table 10.1 lists other naming schemes that could be potentially considered. These include:

■ The radiofrequency identification (RFID) naming scheme, which is based on a unique 64–92-bit identifier [13]

■ GS1 identification keys [14]

■ Sensor web enablement and sensor UID [15]

■ IoT@Work naming scheme [16]

■ Mobility First [17]

10.5.2   Certificate validity

X.509 certificates have a limited life span which is encoded in the validity field. The field contains the two dates *notBefore* and *notAfter*, both containing a timestamp in the UTCTime encoding format.

Checking the validity of a certificate requires access to accurate time, and since low-cost oscillators found in embedded systems have a significant drift in the order of up to several seconds per day [18], the use of time synchronization protocols like Network Time Protocol (NTP) or Precision Time Protocol (PTP) should be considered.

If an end-entity certificate has expired and the device is not decommissioned, the certificate needs to be renewed. This causes significant logistical and technical challenges, as the PKI must have the ability to provide and manage dynamically certificates to a potentially large number of devices, while the devices themselves need a secure download and storage mechanism. Furthermore, an underlying trust mechanism must ensure that only authentic certificates are accepted and reflashed on a device.

Similarly, CA certificates can expire as well, which has implications for the validity of a signature provided by a device; for example, the initial handshake in a peer-to-peer authentication protocol during operation. Overall there are three different validity models in place:

- The shell model as outlined in RFC 5280 [19] prescribes that a signature provided by an end entity is only deemed valid, if all certificates of the entire CA chain (up to the root CA) are valid at the time when the signature is validated.

- The chain model only requires the end-entity certificate to be valid at the time of signature creation. The certificates of the CA chain only need to be valid at the time of creation of the end-entity certificate itself.

- The modified shell or hybrid model as outlined in RFC 5126 [20] dictates that an end-entity signature is valid if it is valid in the shell model at the time of creation; for example, at the time of signature creation the entire CA chain is valid.

The drawback of the latter two models is that an end-entity signature is still valid even when the underlying trust chain is compromised; for example, after one of the certificates of the CA chain has been revoked because of (for instance) a compromised key.

10.5.3   Public key cryptosystems

public key cryptosystems provide pairs of keys, whereby the public encryption key differs from the secret decryption key. Such cryptosystems are at the core of PKI, as they (a) provide a means to digitally sign (e.g., encrypt) the hash value of a digital certificate using a CA's private key; (b) provide a means to validate the integrity of a digital certificate, via decoding the previously encoded hash value using a CA's public key and comparing it with the hash value calculated over the presented certificate (therefore providing proof of authorization); and (c) allow a device to digitally sign or decrypt messages (therefore providing message confidentiality, message integrity, and peer authentication).

The two popular public key cryptosystems are RSA and Elliptic Curve Cryptography (ECC). RSA is based on the practical difficulty of factoring the product of two large prime numbers, while ECC is a relatively new approach to public key cryptography based on the algebraic structure of elliptic curves over finite fields.

The RSA algorithm has been widely used in PKIs for many years and has a significantly lower algorithmic complexity than ECC-based algorithms, but requires a longer key size to provide equivalent security. As a result, ECC is deemed to be faster than RSA and has become the public key cryptosystem of choice for resource-constrained embedded systems. For example, a 3072-bit RSA key has a similar cryptographic strength as a 256-bit ECC key or a 128-bit symmetric AES key.

ECC has been adopted by Suite B Cryptography, a set of cryptographic algorithms promulgated by the U.S. National Security Agency (NSA). There are a range of standardized ECC curves and parameters defined with effective key length between 160 and 512 bits [21]. However, there are some unresolved patent and licensing issues around some ECC algorithms.

10.5.4   Hash functions

Cryptographic hash functions are one-way functions that convert a bit string of variable length into a fixed-length hash value. They are used to digitally sign a certificate. Hash functions have four important mathematical and algorithmic properties: (i) they should have a low computational complexity, while (ii) being irreversible "one-way" functions. Furthermore, it must be (iii) infeasible to modify an input without changing the hash and (iv) it must be infeasible to find two different inputs with the same hash. The latter two requirements are also called strong collision resistance and weak collision resistance.

There are a number of different future-proof hash algorithms in use, most notably SHA-2 and SHA-3 with customizable hash lengths of between 224 and 512 bits. Legacy hash functions such as MD5 and SHA-1 are being phased out, as they are not deemed to be sufficiently secure (e.g., collision resistant) any more [22].

10.6 A Public Key Reference Infrastructure for the IoT

10.6.1 Certificate format

Self-descriptive or non-self-descriptive CVCs have the advantage of lower resource requirements in comparison to DER-encoded certificates. However, most off-the-shelf and open-source authentication protocol implementations (i.e., OpenSSL) solely support the latter, and there are only very few implementations that support non-DER-encoded certificates [23].

Therefore, many IoT implementations, particularly if they are based on the TCP/IP protocol stack, or if they require Internet interoperability, will have to use standard DER-encoded X.509 certificates.

Alternatively, in situations where storage space for certificates is scarce, but DER-encoded certificates are a necessity, the following certificate translation process could be considered:

- A CA signs a DER-encoded certificate using its private key and returns it to the RA.

- The RA executes a certificate parser, which extracts device-specific and nongeneric certificate fields, including the signed hash, and copies it into the appropriate fields of a non-self-descriptive CVC. The assumption is that certain fields are static and identical or predictable for all devices in a certain deployment (i.e., the version field or the reference to the root CA).

- The RA forwards this CVC to the device, where it is stored.

- Whenever a device has to present its original DER certificate, it will parse the CVC and rebuild it.

10.6.2 Certificate life cycle and number of device certificates

Multiple certificates embedded in a single device can be

- Valid over different time periods (therefore reducing the risk of compromising a certificate during its lifetime)

- Remotely activated/disabled on demand, if a certificate has been identified as being compromised, or

- Used for different purposes (i.e., network control vs. device control operations)

Apart from the additional storage requirements, the above scenarios also imply access to secure time or a management interface to enable/disable certificates, therefore opening other potential back doors for cyberattacks.

It is therefore envisaged that, unless the above issues are addressed, an IoT device contains only a single universal certificate, whose lifetime is the anticipated operational life span of the device. Compromised devices (e.g., devices with compromised certificates) must either be discarded or reflashed in the factory.

Similarly, the validity of the root CA certificate must extend over the operational lifetime of all networks under its control, while the validity period of an intermediate CA (iCA) certificate can be more restrictive and, for example, limited by the anticipated lifetime of the network(s) it serves.

All these certificates operate under the modified shell validity model, which is complemented by an online certificate validation mechanism, as discussed in Section 10.6.7.

10.6.3   Combined identity and attribute certificates

IoT device authentication and authorization require additional and customizable device credentials that go beyond the capability and purpose of standard fields in X.509v3 identity certificates. Attribute certificates, on the other hand, are not a viable solution to fill this gap either, as

■ IoT devices do not change over time; therefore, device attributes and identity certificates can have an identical life span; therefore, there is conceptually no need to support two different certificate types.

■ Multiple certificates require additional resources for both certificate storage and parsing.

■ Multiple certificate types increase the management overheads for a PKI.

However, the storage of additional device attributes in an identity certificate can be achieved via extension fields of X.509v3 certificates. An extension field consists of the following components:

■ An OID that identifies the type of extension. For example, device certificates issued by the Irish cybersecurity group OSNA use certificate extensions with the OSNA OID prefix 1.3.6.1.4.1.44409.

■ A flag that indicates whether the extension is critical, that is, if the extension holds vital information. A relying party shall consider a certificate invalid if it does not recognize a critical extension, that is, it has no support for the extension. If an extension is labeled noncritical, it can be ignored if not understood.

■ The actual extension field.

In the absence of a standardized framework to encode device attributes entailing authorization credentials in a certificate, customized domain-specific OID extensions must be defined. For example, in the context of device attributes for wearable medical sensors in a body area network, the following extensions (under the above OSNA OID) and their possible values could be considered (Table 10.2).

With every device having a customised combined identity and attribute certificate with the above descriptors and an appropriate authentication/ authorisation protocol implementation, it could be assured that

- A network controller will only connect to sensors of type "1", "2" or "3."

- Sensors can only have peer-to-peer connections to a network controller with device type "4." A glucometer (device type "1"), however, will also accept connection requests from an insulin pump.

- The insulin pump has the peer communication privilege level "2" and can retrieve readings from the glucometer, which in turn has (like the electrocardiogram [ECG] monitor) the privilege level "1."

- The network controller has privilege level "3" and has full control over the entire network.

Similarly to the device attributes, an X.509v3 certificate also allows the specification of the purpose and scope of its public key via two additional extension fields, *KeyUsage* and *ExtendedKeyUsage*. In relation to the above example, these fields indicate if the network controller is allowed to sign and distribute firmware updates or to generate and sign revocation lists for the other network devices.

10.6.4 Peer authentication protocols for the IoT

In an IoT deployment, a device can have multiple simultaneous peer-to-peer connections with one or more nodes. Each connection will be based on a specific

Table 10.2 Medical device descriptors and their OIDs

OID Suffix	Name	Value: Meaning
1	Device type	"1": Glucometer
		"2": Single-channel ECG monitor
		"3": Insulin pump
		"4": Network controller
2	Peer communication privilege	"1": No data access to peer device
		"2": Read-only access to sensor data
		"3": Full control over peer device

authorization level of the end points; for example, to execute privileged commands or to have access to classified sensor data. Here, secure device authentication and communication must—from a protocol stack perspective—reach from data end point to data end point; for example, from process to process. Therefore, an application layer authentication and communication protocol should be chosen over a data-link or network protocol.

In an IP-based environment, the application layer protocol of choice is the Transport Layer Security (TLS) protocol. TLS goes hand in hand with X.509 identity certificates, as during its initial handshake phase the certificates of two peers are exchanged, validated, and, using the public keys encoded in the certificates, a session key is negotiated.

TLS went through various iterations, but for an IoT deployment it is recommended to use its most recent version (v1.2) as specified in RFC 5246 [24]. In particular, it must accommodate the recommendations of RFC 6176 [25]; for example, TLS sessions will not negotiate the use of Secure Sockets Layer (SSL) v2.0, which has known security flaws, including a cryptographically weak hash function (e.g., MD5), unprotected handshake messages (which allows a MitM to trick the client into picking a weaker cipher suite than it would normally choose), and the termination of sessions via MitM TCP FIN insertions.

Also, in the context of IoT communication, TLS must be configured to provide a client-authenticated handshake, whereby both peers exchange and validate the other peer's certificate. This is in contrast to the typical use of TLS (i.e., in secure Internet browsing), where only the server certificate is validated by the client.

The recent discovery of the Heartbleed bug in the OpenSSL implementation of RFC 6520 [26], as well as recent revelations about the widespread capture and storage of encrypted network communication (for later cryptanalysis) by some government agencies, emphasizes the need for perfect forward secrecy (PFS). PFS is a property of cryptographic systems which ensures that a session key derived from a set of public and private keys will not be compromised if one of the private keys is compromised in the future. TLS can PFS by enforcing the use of ephemeral Diffie–Hellman key exchange to establish session keys.

Also, current TLS implementations do not provide functionality to validate authorization levels encoded in certificates, as these are customized extensions. However, various TLS implementations support optional callback functions during the handshake phase that allow the integration of such functionality.

10.6.5 CA hierarchy

Trust in today's Internet is provided by more than 600 publicly operating certification authorities, which in turn are interconnected via CA hierarchies. This tree-like hierarchy consists of root CAs with self-signed certificates at the top and a set of intermediate CAs, whose certificates are signed either by another intermediate CA or a root CA. To facilitate the process of verifying a chain of trust, every certificate includes the fields *IssuedTo* and *IssuedBy*. This network of trust allows, in principle, every device certificate issued by any CA to be validated as part of the TLS handshake. However, recent high-level breaches in CA organizations (like Comodo in 2011) [27] and the theft or generation of counterfeit certificates has shown that this distributed network of trust has some significant flaws.

An IoT deployment may only be operational within its own perimeter, for example, there might no need to issue certificates that can be globally validated. Also an IoT-PKI must tightly control the generation of device certificates, while giving suppliers and system integrators the ability to issue certificates on the fly.

Therefore it is suggested to use a two-tier CA hierarchy consisting of a root CA and a set of intermediate CAs (iCAs) as shown in Figure 10.6. Each iCA is used by a defined set of stakeholders (e.g., device manufacturers). Depending on its scope, such a PKI would issue devices for a single deployment (e.g., a private home-automation network), a single application type (e.g., wireless medical device networks), or a single client (e.g., the smart meter/smart grid infrastructure of a single utility company).

From a device's perspective, the main advantage of such a two-tier organization is that each node only requires two CA certificates (i.e., the root certificate and the certificate of its iCA that signed its certificate) to authenticate every single certificate issued by the PKI, even if it was signed by a different iCA.

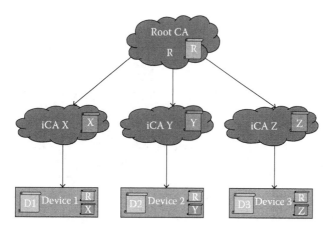

Figure 10.6: Certificate distribution in a two-tier PKI with CA certificates R, X, Y, and Z and device certificates D1, D2, and D3.

10.6.6 Certificate generation

Embedded systems are typically not suited to generate their own public–private key pairs, as they do not have sufficient entropy to provide sufficiently random numbers. Therefore, the registration authority—optionally in combination with cryptographic hardware support—as the interface between the device and the CA has to step in and provide such data. The RA has to dispose of any generated key material after the certificate has been issued to the device, as otherwise the integrity of the entire deployment could be affected if the RA is compromised.

Likewise, each RA requires a suitable interface to transmit key materials securely back to the device (e.g., via a JTAG interface) as well as a sound mechanism to acquire device-specific attributes (e.g., MAC address, device capabilities, etc.) which are inserted into a device certificate [28].

The issuing and storage of device certificates must take place in a controlled, auditable, and secure environment; for example, a manufacturing floor, and not in the field during network deployment or integration, where this process can be more easily compromised.

10.6.7 Certificate validation

IoT deployments must be resilient from external cyberattacks and are, therefore, isolated from the Internet as far as possible, as well as being self-contained. This is accommodated by the above two-tier CA hierarchy, which allows for a straightforward certificate validation during the TLS handshake:

- Each device has—as already mentioned—a copy of the (self-signed) root CA certificate, as well as its device certificate and a copy of the certificate of the iCA that signed the device certificate.

- During the initial handshake, two devices exchange their device certificates. If both have been signed by the same iCA, the public key embedded in the iCA's certificate is used for their validation.

- In situations where devices have received their certificates from different iCAs, these iCA certificates are exchanged and validated using the root CA's public key.

A certificate within the trust chain can be invalidated (e.g., revoked) by a PKI before its expiration time; for example, if the private key that corresponds to an iCA's certificate's public key has been compromised. It is best practice to perform a certificate status check before completing the handshake, as highlighted in a recently discovered configuration flaw in the Java runtime environment, which resulted in malicious Java code (signed by a compromised and revoked certificate) to be executed on client computers [29].

As outlined before X.509 provides two principal mechanisms to validate the status of a certificate:

- A certificate revocation list (CRL) is a list of revoked certificates signed by a CA or CRA. CRLs are regularly updated with newly revoked certificates being inserted into the CRL or compiled into a delta CRL. Devices that wish to obtain revocation information need to download the CRL from some repository and process it locally. It is therefore not a suitable solution for resource and bandwidth-limited IoT nodes.

- OCSP, as defined in RFC 2560 [30], allows clients to query the status of an individual certificate in real time via an OCSP server. Revoked certificates can be added on the fly, making OCSP far more responsive than CRL and more suitable for real-time certificate status validation in IoT deployments. However, devices should only retain a connection to another client if the OCSP server returns an "OK" message for its certificate. This makes an OCSP server a single point of failure, as a DoS attack on the server will prevent it from responding to requests.

RFC 6066 [31], also called OCSP stapling, provides a solution to the DoS attack vulnerability of OCSP. Here, a client requests a validation of its own certificate by an OCSP server on a regular basis and keeps the response (which is timestamped and digitally signed by the server) locally in storage. Whenever a new TLS handshake is initiated, the device sends its own certificate as well as the OCSP response to the other peer. By doing so, a temporarily unavailable OCSP server can be compensated for.

However, OCSP stapling only operates during the handshake of a TLS connection; for example, it does not support the validation of a certificate once a connection is established. Since it is not always economic for an IoT device to terminate a connection and redo the handshake with a given peer, an extension of OCSP stapling should be considered that allows certificate validation for established connections. This feature could be implemented similar to the TLS Heartbeat extension in RFC 6520, so that OCSP update queries and responses between two peers are implemented on the record layer of TLS.

10.7 Summary

Introducing and managing trust will be a major challenge for the IoT, as in the absence of robust, versatile, and verifiable trust credentials provided by a trust management infrastructure, the fundamental requirements of data confidentiality and integrity in interdevice communication, as well as peer authentication and authorization, will not be met sufficiently.

This chapter provided an overview of perceived problems and potential solutions with regard to the provision of trust and proposed a scalable and robust trust management solution suitable for the IoT. This solution is based on a tightly curtailed public key infrastructure in combination with combined identity/attribute certificates based on X.509v3 and customized extension fields.

Bibliography

[1] S. Ganeriwal, L. K. Balzano, M.B. Srivastava, Reputation-based framework for high integrity sensor networks. *ACM Transactions on Sensor Networks (TOSN)*, 4(3), 15, 2008.

[2] K. Moriarty, S. Parkinson, A. Rusch, M. Scott, PKCS #12: Personal Information Exchange Syntax v1.1, RFC 7292, July 2014.

[3] R. Falk, S. Fries, Managed certificate whitelisting—A basis for internet of things security in industrial automation applications. *SECURWARE 2014, The Eighth International Conference on Emerging Security Information, Systems and Technologies*, 2014, Munich, Germany.

[4] G. Guo, J. Zhang, Improving PGP web of trust through the expansion of trusted neighborhood. *IEEE/WIC/ACM International Conference on Web Intelligence and Intelligent Agent Technology (WI-IAT)*, 2011, University of Saskatchewan, Canada.

[5] C. Neuman, T. Yu, S. Hartman, K. Raeburn, The Kerberos Network Authentication Service (V5), IETF RFC 4120, July 2005.

[6] A. Arsenault, S. Turner, Internet X.509 public key infrastructure PKIX roadmap, IETF Roadmap, September 8, 1998.

[7] Technical Specification. XML Advanced electronic signatures (XAdES), ETSI TS 101 903, June 2009.

[8] Recommendations X.509 ITU-T, Information Technology—Open Systems Interconnections—The Directory: Public Key and Attribute Certificate Frameworks, August 2005.

[9] Recommendations X.690 ITU-T, Information Technology—ASN.1 Encoding Rules: Specification of Basic Encoding Rules (BER), Canonical Encoding Rules (CER) and Distinguished Encoding Rules (DER), July 2002.

[10] Morgan Simonsen's Blog. https://morgansimonsen.wordpress.com/2013/04/16/understanding-x-509-digital-certificate-thumbprints/

[11] J. A. Buchmann, E. Karatsiolis, A. Wiesmaier, *Introduction to Public Key Infrastructures*. New York: Springer Verlag, 2013.

[12] S. Farrell, R. Housely, An internet attribute certificate profile for authorisation, IETF RFC 3281, April 2002.

[13] L. Atzori, A. Iera, G. Morabito, The internet of things: a survey. *Computer Networks*, 54(15), 2787–2805, 2010.

[14] M. Bourlakis, I. P. Vlachos, V. Zeimpekis (editors), *Intelligent Agrifood Chains and Networks*, Wiley-Blackwell, 2011.

[15] Y. Li, Naming in the Internet of Things, Washington University in St. Louis, 2013.

[16] M. Bauer, P. Chartier, K. Moessner, Catalogue of IoT Naming, Addressing and discovery schemes in IERC projects V1.7, IERC-AC2-D1, 2013.

[17] J. Li, Y. Zhang, K. Nagaraja, D. Raychaudhuri, Supporting efficient machine-to-machine communications in the future mobile internet, *Wireless Communications and Networking Conference Workshop (WCNCW)*, 2012, IEEE, New York.

[18] J. Shannon, H. Melvin, A. G. Ruzzelli, Dynamic flooding time synchronisation protocol for WSNs, IEEE GLOBECOM, 2012.

[19] D. Cooper, S. Santesson, S. Farrell, S. Boeyen, R. Housley, W. Polk, Internet X.509 Public key infrastructure certificate and certificate revocation list (CRL) profile, IETF RFC 5280, May 2008.

[20] D. Pinkas, N. Pope, J. Ross, CMS advanced electronic signatures (CAdES), IETF RFC 5126, February 2008.

[21] E. Barker, L. Chen, A. Roginsky, Recommendation for pair-wise key establishment schemes using discrete logarithm cryptography, NIST Special Publication 800-56A Revision 2, May 2013.

[22] W. Yu, C. Jianhua, H. Debiao, A new collision attack on MD5, *International Conference on Networks Security, Wireless Communications and Trusted Computing (NSWCTC '09)*, 2009.

[23] M. Schukat, Securing critical infrastructure, *The 10th International Conference on Digital Technologies*, 2014, Zilina, Slovakia.

[24] T. Dierks, E. Rescorla, The transport layer security (TLS) protocol version 1.2, RFC 5246, August 2008.

[25] S. Turner, T. Polk, Prohibiting secure sockets layer (SSL) version 2.0, RFC 6176, March 2011.

[26] R. Seggelmann, M. Tuexen, M. Williams, Transport layer security (TLS) and datagram transport layer security (DTLS) Heartbeat extension, RFC 6520, February 2012.

[27] P. Hallam-Bake, Comodo SSL affiliate; the recent RA compromise. https://blogs.comodo.com/uncategorized/the-recent-ra-compromise/.

[28] A. R. Metke, R. L. Ekl, Security technology for smart grid networks. *IEEE Transactions on Smart Grid*, 1(1), 99–107, 2010.

[29] E. Romang, When a signed Java JAR file is not proof of trust. http://eromang.zataz.com/2013/03/05/when-a-signed-java-jar-file-is-not-proof-of-trust/.

[30] M. Myers, R. Ankney, A. Malpani, S. Galperin, C. Adams, X.509 Internet Public Key Infrastructure Online Certificate Status Protocol—OCSP, RFC 2560, June 1999.

[31] D. Eastlake, Transport layer security (TLS) extensions: Extension definitions, RFC 6066, January 2011.

Chapter 11

Trustable Fellowships of Self-Organizing "Things" and Their Software Representatives: An Emerging Architecture Model for IoT Security and Privacy

Antonio Marcos Alberti

Edielson Prevato Frigieri

Rodrigo da Rosa Righi

CONTENTS

11.1   Introduction

Many Internet of Things (IoT) challenges will require more than incremental solutions or the application of models already established. New architectural models will be demanded, since the current cloud and networking technologies and their protocols are inherently limited for several expected scenarios [25]. Some examples we will address in this chapter are entities naming, identification, mobility, decoupling of devices' identifiers from locators, scalability, control and management, data integrity, provenance, and joint physical and virtual resources orchestration, among others.

Current Internet naming is very limited and does not favor security [6, 14, 36]. Unique identification of "things" is nonexistent. Name resolution is limited to resolve domain names to Internet protocol (IP) addresses. There is no support for service names, among many other relevant names for IoT scenarios. Naming has also an important role in data integrity and provenance, as we will further discuss.

Host mobility causes variations to services session states [5], leading to unstable application behavior. IP addresses have two simultaneous purposes: host identification and location. When a node moves from one network to another, its location should change to enable datagram delivery to new position; however, the IP address change affects upper-layer sockets, which employ IP addresses as host identifiers. Identifier/locator splitting is an approach that enables nodes to move without changing their identifiers, maintaining session state invariance [5]. The scalability of addressing and routing in the current Internet are also concerns [5, 6, 14].

Joint orchestration of physical-world resources, services, and contents is also a requirement far ahead of current Internet support. The current model for device control and management was designed in an epoch where the number of devices was orders of magnitude smaller. Control and management in the IoT scenario will require self-driven approaches, reducing operational costs and meeting the upcoming scales on device numbers, interactivity, and traffic.

Emerging convergent information paradigms will be required to face these issues in the next decades. Incremental solutions could be intrinsically limited, since they were not projected with any thought for the amazing interaction between physical and virtual worlds we are going to experience in a few years. Many science-fiction scenarios are becoming real at an impressive speed. Biometric sensors, implants, monitoring devices, wearable electronics, smart clothes, residences that welcome us and make our lives easier are among the technologies that are arriving. Augmented reality, tactile systems, haptic interfaces, virtual reality, cyborgs, robotics, self-assembly machines, ubiquitous computing are examples of technologies on the border between the physical world and computing systems.

In this chapter we discuss the path to a futuristic scenario, where these emerging technologies converge synergistically, gracefully, quietly. We discuss security issues from an architectural perspective, instead of specific points (as is usually done in the literature). We start with a survey of some important current architecture limitations that could limit IoT potential (Section 11.2). We also present contemporary paradigms that are emerging in the literature to address these identified shortcomings. Then, we present a new IoT architecture model that integrates these paradigms toward a future IoT architecture (Section 11.3).

In this model, "swarms of things" are represented and controlled by trustable "swarms of services," which self-organize to establish the required security, privacy, and trust levels. Since the number of devices expected is going to be extremely high, more autonomic behavior is expected [28], reducing the degree of human intervention on the control and management of IoT devices. The role of naming and name resolution will be revisited and related to source authentication, as well as data integrity and provenance. The support for distributed storage of name bindings enables the representation of real-world relationships among things, services, and contents. What is required is the integration of the life cycling physical resources, contents, and services, life cycling using emerging trust-based security and privacy approaches.

This emerging architecture model is being developed in the context of an information and communications technologies (ICT) architecture called NovaGenesis (NG) [1]. NG started in 2008 and aims at integrating many future Internet (FI) ingredients toward a convergent information architecture (CIA). A CIA integrates in only one design information processing, storage, and exchange [2]. It is broader than an Internet, which was designed to put computer networks

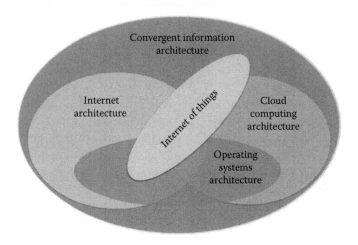

Figure 11.1: Operating systems and cloud computing architectures are focused on computer nodes, that is, intranode information processing, storage, and exchange. Internet architecture emerged to interconnect computer networks, that is, global internode information exchange. Convergent information architecture integrates all previous information processing, storage, and exchange architectures with global scope.

together, enabling end-to-end computer programs (processes) to communicate. CIAs address IoT challenges more deeply than the current Internet, since IoT requirements spread not only through networking technologies, but also through cloud, distributed, and mobile computing. Figure 11.1 illustrates this idea.

NG is a CIA that integrates information-centric networking (ICN) [38], service-centric networking (SCN) [8], service-oriented architecture (SOA) [26], software-defined networking (SDN) [1, 20], among other ICT hot topics. In NG, context-aware services establish contract-based coordination toward fulfilling network operator objectives, rules, and regulations. Energy awareness and disruptive/delay tolerant communication is enabled. IoT devices, services, and contents are named and synergistically integrated to address the most challenging IoT prerequisites. NG proposes a new control and management model where physical devices are represented by named services called proxy gateway controllers (PGCs) [1]. These PGCs expose node capabilities, negotiate and establish contracts, encapsulate NG messages, and configure devices according to software-implemented controllers.

NG has an experimental proof-of-concept implemented. Several aspects of the proposed model have already been tested. In this context, this chapter finishes with Section 11.4 with an example scenario, where an illustration of the proposed architecture is provided based on current proof-of-concept design.

11.2   Current Technologies Limitations and Emerging Solutions for IoT

Nowadays, technologies for the IoT are already available, among them [24]:

■ IEEE 802.15.4: This is a wireless communication standard for low-power, low-data rate, and short-distance radio coverage sensor and actuator networks [15]. It was developed within the IEEE 802.15 personal area network (PAN) group. Its typical data rate is 250 kb/s with maximum packet size of 127 bytes, which limits the available payload to about 86 and up to 116 bytes. It defines a physical layer (16 channels with direct sequence spread spectrum) and media access control (MAC) layer. It can be applied to multihop networks, but requires the radio to be on all the time. It employs single-channel operation, which suffers with multipath fading and shadowing.

■ IEEE 802.15.4e: This employs a time-synchronized channel hopping (TSCH) technique to avoid interference, shadowing, and multipath fading [24]. The IEEE redesigned MAC protocol supports centralized or distributed scheduling of time slots for communication between neighboring nodes. A time-frequency structure is used to create virtual links among neighboring stations using specific time slots/frequency channels. The standard does not define how the schedule of the time slot/frequency pairs for a certain virtual link is carried out.

■ Message queue telemetry transport (MQTT) [17]: This is a "lightweight" messaging protocol to run over the transmission control protocol/Internet protocol (TCP/IP). It follows a publish/subscribe hub-and-spoke paradigm where a broker server asynchronously forwards messages to one or more interested nodes. Named topics are used to share information. All messages addressed to a named topic (e.g., "myhome/groundfloor/livingroom/temperature") by publishers will be delivered to a broker. Subscribers to a certain topic get published information from the brokers. It provides an agnostic binary payload. Nodes must connect to brokers.

■ Advanced message queuing protocol (AMQP) [35]: AMQP is an open standard message middleware specification, based on a topic-oriented message queue paradigm, where products written for different platforms and in different languages can exchange messages. Despite being a standardized protocol, not all implementations are fully compliant with the standard. The complete AMQP standard is composed by publishers, subscribers, and brokers that have internal routing capabilities. The AMQP specification (version 1.0) defines a wire-level protocol for publishers'/subscribers' communication with their message brokers. The broker can modify incoming messages and, based on a set of rules or criteria,

IETF CoAP	←→	IETF CoAP
IETF TCP/UDP	←→	IETF UDP/TCP
IETF IPv6	←→	IETF IPv6
IETF 6LoWPAN	←→	IETF 6LoWPAN
IETF 6top	←→	IETF 6top
IEEE 802.15.4e MAC	←→	IEEE 802.15.4e MAC
IEEE 802.15.4 PHY	←→	IEEE 802.15.4 PHY

Figure 11.2: Possible IoT stack combining IETF and IEEE standards.

decide to which queues the messages need to be forwarded to arrive at one or more subscribers.

■ Data distribution service (DDS) [27]: DDS has a global data space (GDS) to where nodes can publish/subscribe to (pub/sub) data using topics and keys. Data objects are manipulated using natural language. Local object caches can be fed by available global data. There is also support for quality of service (QoS) contracts. DDS provides automatic discovery of publishers and subscribers using a protocol called simple discovery protocol (SDP).

■ Internet Protocol version 6 (IPv6) over low-power personal area network (6LoWPAN): IPv6 packets are too big for IEEE 802.15.4. 6LoWPAN provides an adaptation layer to segment and reassemble IPv6 datagrams. It provides IPv6 header compression and belongs to the Internet engineering task force (IETF) stack for IoT [34], as illustrated in Figure 11.2. 6LoWPAN protocol data units (PDUs) can be encapsulated directly over IEEE 802.15.4. However, for 802.15.4e the 6top adaptation protocol was created.

■ 6TiSCH operation sublayer (6top): IPv6 over the TSCH mode of IEEE 802.15.4e (6TiSCH) provides the mechanism to admit or revoke a node from a TSCH network, including the scheduling of virtual channels to this node using available time slots/frequency channels with neighboring nodes. It also makes the adjustments to support the IETF routing protocol for low-power and lossy networks (LLNs) (RPL) over 802.15.4e nodes.

■ Constrained application protocol (CoAP): Provides a specialized web transfer protocol for LLNs that conforms to the REST style. There is a uniform resource identifier (URI) for every device. Contrary to the hypertext transfer protocol (HTTP), it employs the user datagram protocol

(UDP) instead of TCP. It enables asynchronous message exchange with low complexity parsing. HTTP-CoAP mapping is standardized.

■ GS1 EPCglobal: EPCglobal is a GS1 initiative to develop industry-driven standards for the electronic product code (EPC) to support the use of radiofrequency identification (RFID) in today's fast-moving, information-rich, trading networks. In particular, EPC information services (EPCIS) is an EPCglobal standard designed to enable EPC-related data sharing within and across enterprises. In this way, at least, data cryptography is pertinent when transferring data between different companies, since common Internet links are used in this case.

MQTT is typically used for machine-to-server (M2S) scenarios, while AMQP is typically employed on server-to-server (S2S). DDS is focused on machine-to-machine (M2M) communications. While MQTT does not support real-time operation, DDS is focused on supporting timely data distribution among devices. Concerning the relation with the current Internet, MQTT and AMPQ relay on TCP/IP sockets, while DDS specifies a reliable real-time pub/-sub wire protocol DDS interoperability wire protocol specification (DDS-RTPS) aimed at a UDP/IP stack or other data encapsulations (TCP/IP or direct shared memory inside a node). Besides the traditional TCP/IP stack, CoAP can also run over 6LoWPAN.

MQTT also enables the bypassing of TCP/IP by using a modified standard called MQTT for sensor networks (MQTT-SN). This standard supports direct transport of MQTT-SN messages over 6LoWPAN or even ZigBee. AMQP connections are always established by publishers/subscribers to the message broker. First, they must open a TCP socket and the initial exchanged messages define the capabilities and limitations of each side. For constrained networks, it could be expensive to exchange this information on each connection, so AMPQ has a mechanism to omit some negotiation messages on consecutive connections. Although the dependence on the current Internet stack brings a lot of benefits, it also has a number of limitations, as we shall see in the following subsections.

A common aspect of MQTT, AMQP, and DDS is the adoption of the pub/-sub communication paradigm. In this paradigm, information owners (principals) publish measured data to authorized peers. The interoperability between publishers and subscribers can be a problem in MQTT, since the message payload format needs to be agreed among peers before any data transfer. DDS is an example of pub/sub being applied to mission-critical applications, where constrained delay is required. A challenge here is to deal with a large number of pub/sub nodes. Kyoungho An et al. [4] explore the scalability of the DDS pub/sub discovery protocol. Pub/sub has a direct relation with architecture security and privacy. The IETF stack for IoT does not adopt the pub/sub model. Rather, it is based on the classical request/reply model.

Considering particularly the logistics and supply-chain field, we can highlight the adoption of the EPCglobal Class 1 Gen 2 standard [18], which provides the notion of the EPC to uniquely identify a physical object stored in an RFID tag. Briefly, the main objective of EPCglobal is to provide an architecture to collect vast amounts of raw data from a heterogeneous RFID environment, filter them, compile them into usable data structures, and send them to computational systems. To accomplish this, EPCglobal defines the following components [30]: (i) RFID readers (also denoted as RFID sensors); (ii) application level events (ALE) for filtering and collecting EPC data; (iii) EPC information services (EPCIS) to store EPC data, as well as to exchange this data along the EPCglobal network; (iv) EPC capturing applications, as a box-in-the-middle between ALE and EPCIS, regulating how the former sends data to the latter. Each company has its own set of components, so the idea is to generate value by providing a standard way of capturing data from objects between the partners involved in a particular application field (including, for example, suppliers, enterprises, resellers, clients, buildings, and users).

Given the potential size of the data generated by sensors and related devices, a trade-off will need to be found between in-network processing and aggregation techniques versus streaming data to the external support system. This trade-off is not an easy one. It depends on the capabilities of the distributed sensor network, the communication channel between sensor network and support system, and the support system itself. In some cases, a networking delay or the intermittent connectivity could hinder external support, requiring more computing power at the IoT nodes. This balance may affect the amount of energy spent on nodes, limiting the energy they could spend on security issues.

In the following subsections, we will analyze how these technologies relate to emerging paradigms for future ICT architectures. The idea is that the IoT faces the same challenges that the current Internet does. We will highlight some of the aforementioned technologies' limitations from the perspective of these state-of-the-art paradigms. We contend that more synergistic approaches are required to maximize IoT potentials.

11.2.1 Naming and name resolution

Names inhabit the human mind. People like to denote things by names. *Names* are symbols used to denote one or more individual existences. In this case, *to denote* means to represent something by signals. By definition, names denote meaning and sense. However, there are names that are almost randomly generated, having "weak semantics." One can call a car "xyzwertyu"; however, this name makes sense only for the owner or other closely related people that have been introduced to it. Another example: One can denote a car by a sequence of symbols that typically carry "weak semantics," that is, the numbers and letters on its license plate (e.g., 1ABC234 in Figure 11.3). Or yet, one can call a car

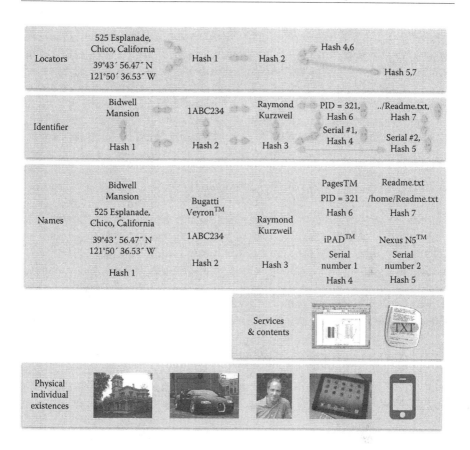

Figure 11.3: People attribute "weak semantics" and meaningful names to physical (e.g., a car or a house) and virtual existences (a computer program or a file). If they are unique in some scope, they can be used as identifiers and locators. Therefore, bindings among names (or name bindings) can capture all sort of relationships between virtual and physical existences. They can represent semantic relationships like "contains," "is contained," or "close to." In this example scenario, the car is "close to" the house, and "contains" the tablet and smartphone. Also, the person "is contained" in the car.

by its brand, for instance "Bugatti Veyron," which has more "strong semantics." A last example is the binary word obtained at the output of a hashing algorithm. This binary word (also called hash code) can be used as a name—a self-verifying name (SVN). In this case, the binary input of the hash function can be the physical existence of an item itself (e.g., computer program executable, source code, or information files) or other binary input related to the entity being named (e.g., entities' immutable attributes). In the first case, the name is said to be self-certifiable, because at any time the existence's binary words can be hashed again to get exactly the same name. In the second case, the perennial physical existence attributes can be digitalized again to certify the name. Figure 11.3 illustrates some hash names calculated for physical and virtual existences. "Hash 1," for example, can be obtained from the perennial attributes of "Bidwell Mansion" in the USA, such as its physical proportions. "Hash 2" can be obtained from car attributes, like the chassis number or serial numbers. "Hash 3" could be based on the biometrics of human body. "Hash 4" may be obtained from device serial numbers or processors' unique IDs. "Hash 6" may be generated from entire executable binaries.

Emerging paradigms, like information-centric networking (ICN) [12, 14, 38] and service-centric networking (SCN) [3, 8, 37], put naming at the core of architecture design. According to these approaches, the host-centric Internet is no longer appropriate for modern requirements, like content distribution, in-network caching, name-based routing, name resolution, and named-services chaining. By name-based routing we mean a routing approach that uses content or service names instead of IP network addresses in packet headers. Name resolution means to locate an entity by its name, as with in the current Internet where domain names are resolved to IP addresses. Name-based service chaining means to create a chain of services using their names instead of their locations. These authors contend that contents and services should be named directly, independently of host naming. Sockets and uniform resource identifiers (URIs) are too limited for this. The only way to implement these new ideas on the current Internet is to use the World Wide Web as an overlay. What these paradigms defend is to replace the current TCP/IP stack "narrow waist" by names. The current Internet naming solution, either v4 or v6, will have a great impact on the IoT. The IoT also requires the aforementioned improvements, for instance, information objects naming, in-network caching or named-services chaining, to improve its efficacy, security, provenance, mobility supports.

Some of the current IoT technologies propose incremental solutions to overcome Internet naming limitations. MQTT employs a unicode transformation format (UTF-8) string to create hierarchically named topics that facilitate publishers' and subscribers' meeting. The MQTT brokers filter messages according to the topics. A topic example is: "Brazil/Minas Gerais/Santa Rita do Sapucai/Inatel/Room II-17/temperature." However, MQTT topics fall into the human-readable names category [14] and therefore suffer from: (i) weak binding to the real-world entity that produced the information; (ii) security dependence on name trustworthiness; and (iii) vulnerability to phishing attacks where malicious names are created similar to real ones to confuse people.

Sastry and Wagner analyzed security issues for IEEE 802.15.4 [32]. Addresses are IEEE-defined 64-bit extended unique identifiers (EUI-64). These addresses contain numbers that identify organizations and companies behind nodes. Since IEEE is directly involved in generating these IDs, improved name binding to real-world authority is provided. The same is true for IEEE 802.15.4e.

IPv6 addressing is considered by many people as the main claim to change between IPv4 and IPv6. This is due to the depletion of globally valid IPv4 addresses. IPv6 is used for naming hosts. Due to its large size (128 bits), 6LoW-PAN emerged. The IETF 6top standard allocates 16-bit identifiers for nodes that join a network. It is not clear how these IDs are generated.

CoAP URIs are defined as: *coap[s]: <host>: <port>/<path> <query>*. Observe that a URI has the same dependence on host names. CoAP follows the classical request/response model of HTTP. IETF standardized HTTP-CoAP (HC) mapping using proxies. TCP connections need to be mapped to UDP segments in the HTTP to the CoAP direction. URI mapping is required to map URIs between two different protocols—a complex task—an example that adopting two stacks can create more complexity.

The DDS global name space (GNS) provides data-centric communication among nodes. Data objects (content) are addressed by topic name and key. Communication among publishers and subscribers only happens if there is a topic match. DDS topics can have different syntaxes and are specified using script programming languages, such as interface definition language (IDL), extensible markup language (XML), or unified modeling language (UML), among others. Topic names are typically natural language strings, like "TempSensorTopic" for a temperature-related topic. Topic names are bound to domains, which have 32-bit integer identifiers. The key is also an integer used to identify records of the same topic.

These examples illustrate the diversity of naming in IoT technologies. Naming has an important role in the security and privacy of ICT architectures, and the IoT is no exception. Ghodsi et al. [14] contend that self-verifying names (SVNs) have better security properties than natural language names (NLNs), since they allow straight verification of the binding between name and entity. The convergence of SVNs and the IoT are two topics which are largely underexplored nowadays. None of the aforementioned technologies use SVNs right now currently and many cannot use them even if desired. This is a huge gap to be overcome for the IoT.

11.2.2 Identifier/locator splitting

Names can be used as identifiers if they are unique in some scope. The scope can be a domain, a city, or a country, and so on. Therefore, to be used as an identifier in some scope, a name must be unique in that scope. For example, in a certain small city, the name "John Smith" can be used as an identifier, while in other major cities more than one person could have this name. Thus, we can define an identifier as symbols used to unambiguously identify some individual existence from others to some extent. The name "Raymond Kurzweil" identifies the famous entrepreneur and inventor worldwide.

A locator denotes the current position at which an individual existence inhabits or is or attached to in some space. A space is the set of all possible positions which some individual existence can inhabit or be attached to. Therefore, from a certain space definition, one can determine how close or far apart two existences are. Interestingly, a name can be a locator if it is possible to derive notions of distance from its interpretation. For instance, a geographic coordinate systems is composed of three names: latitude, longitude, and elevation. In Figure 11.3, consider the famous "Bidwell Mansion" in the USA. The name "$39°43'$ $56.47''N121°50'$ $36.53''W$" can be used as a locator for this physical existence, as well as the address "525 Esplanade, Chico, California." Even the identifier "Bidwell Mansion" may be used as a locator if it is unique nationwide. Interestingly, using some mapping system it is possible to derive the notion of distance between the entities named as "Bidwell Mansion," "$39°43''$ $56.47'N$," and "1ABC234."

The IP network address has a double functionality [5]. It works not only as a locator for datagram routing on an IP network (or subnetwork), but also as a host identifier for upper layers on a TCP/IP stack. The IP address is a component of Internet sockets, together with port numbers and the information about the used transport layer protocol (TCP or UDP). Services identify other target services using sockets. Thus, when a computer moves from one network to another, its IP address changes, affecting established sockets, generating instability in the session state [5]. Also, observe that this solution enmeshes service names with host locators, hindering services to communication independently of host locations. In addition, it makes identifiers opaque, since they will be restricted to autonomous systems—behind a network address translator (NAT) barrier.

With ID/LOC splitting, IDs are used by the application and transport layers to identify a node, while the locators are used by the network layer to logically locate them in the topology and route packets to/from the nodes. Mobility is supported by rebinding the name used to identify the node to the new locators. Figure 11.4 illustrates the current situation of Internet node mobility and what would be a future solution with SVNs as IDs and LOCs.

The majority of current IoT technologies do not support ID/LOC splitting for sensor and actuator nodes. The exception could be 6LoWPAN with mobile IPv6 (MIPv6). Montavont et al. [22] contend that MIPv6 can be successfully used together with 6LoWPAN. Kim et al. [19] propose an approach for mobility support on wireless sensor networks (WSNs) using ID/LOC splitting. They also contend that ID/LOC splitting can be supported by IoT nodes despite its energy fingerprint.

Regarding IoT services, CoAP does not decouple URIs from locators. DDS employs SDP for service discovery. SDP is based on special topics to provide service advertising and discovery. Therefore, it supports identifiers decoupled from underlaying network locators. For this, topic names should be mapped to DDS-RTPS locators.

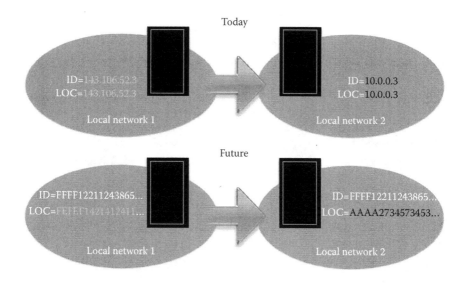

Figure 11.4: In the current Internet, when a node moves from one network to another, its IP address changes (from 143.106.52.3 to 10.0.0.3), affecting the node ID and LOC. There is no problem with changing the LOC, but changing the ID causes inconsistent states in upper-layer applications. In future architectures, the ID and LOC are decoupled. The ID never changes unless the entity itself changes. Many emerging ICT approaches use SVNs as IDs or LOCs, but this is far from reality in the IoT panorama.

One can expect that a significant portion of IoT devices will be mobile. Therefore, IoT architectures should support devices ID/LOC splitting. However, networked devices are not the only things that will move in IoT scenarios. Mobility of services is also a prerequirement. Service ID/LOC is a widely unexplored problem on in Internet-based SOA. IoT services need to have perennial IDs, making them accessible independently of their locators. In addition, to improve traceability and provenance of information, unique identifiers for services and devices are required. An IoT with NAT creates opaque IDs, which discontinue end-to-end traceability.

11.2.3 Resources, services, and content orchestration

We envision the challenge related to spontaneous interactions among devices and the support of a distributed system for that. For example, associations between devices are routinely created and destroyed by identifying and locating a device, such as a sensor. According to Presser et al. [28], the idea of "social devices" requires not only the unique identification of nodes, but also the nodes' capacity to discover peers and establish trustable relations, including service-level agreements (SLAs).

The DDS standard provides mechanisms to expose node interests, as well as the kind of information each node can provide. DDS automatically connects to subscribers to topic-related publishers. Also, the nodes (or a support system) must be capable of semantically interpreting information, allowing them to collaborate with each other toward a common objective. MQTT provides a similar topic-based coordination among services. Interestingly, MQTT-SN allows multiple broker discovery.

CoAP provides representational state transfer (REST) [7] web services adequate for LLNs. REST-style resource discovery on constrained environments was defined in IETF RFC 6690 [33]. This RFC proposes the concept of constrained RESTful environments (CoRE), which aims to perform efficient REST suitable for current IoT nodes. The aim is to discover IoT resources behind a CoAP web server, as well as their attributes and formats. The responses to queries are CoRE-specific links that identify and provide metadata about IoT resources. CoRE provides the means to create IoT-resource directories.

The IoT requires joint orchestration of physical resources (sensors and actuator nodes), services, and content/information. Current technologies implement the orchestration of these separately, duplicating systems to deal with the life cycle of each of these architectural components.

The support for natural language names (NLNs) is required for "semantic-rich" orchestration. However, NLNs need to be bound to SVNs, to provide increased security, as previously discussed. The dissemination of unencrypted NLNs or SVNs to express intent can violate users' or nodes' privacy. While disclosure of topics of interest is fundamental to the exchange of contextualized information, the public disclosure of all interests may affect people's and machine privacy.

In addition, SLA support for the orchestration of physical resources and services is missing in all approaches. It is required to tie peers together and create trust networks among IoT resources/services. The absence of agreements governing the degree of privacy and confidentiality among publishers and subscribers is noticeable in current standards. Content processing, exchange, and storage need to be governed by secured contracts (SLAs), which create the required trust networks for reliable, private, and secure operation, control, and management.

Physical-world resources need to be securely exposed by software, which represent them and negotiate SLAs with trustable peers. This process needs to be more automatic, less dependent on human interference. This is very important to meet the impressive IoT scales we are going to achieve in the next decades. The secure, integrated life-cycle management of services and contents, together with physical resources, will require innovative approaches and is required for the success of the IoT. Reputation systems, trust network formation, entities' social behavior, innovative distributed algorithms, naming, name resolution, and many other emerging approaches should be elegantly integrated. The balance between expressiveness, "semantic-rich" exposition and subscription, constrained resources, and security issues poses a fantastic challenge for designers of future IoT technologies.

11.2.4 Security, privacy, and trust

The IoT, as a distributed system, would have the same performance hurdles and security threats as distributed mobile ad hoc networks (MANETs) [9]. Some common attacks that represent a critical challenge to trust and reputation systems are described by Zhang in [39].

Sensitive information can be used by control systems that may represent a threat to safety, or malicious nodes may attempt to disseminate false or corrupted information. The system should be designed to minimize selfish and malicious behavior, as well as to support flexible security and privacy mechanisms.

Applications may require a broad range of protocols and security mechanisms. From simple end-to-end secure channels through public key infrastructures (PKIs) to distributed reputation and voting systems, ingenious protocols can be employed using symmetric and asymmetric cryptosystems, cryptographic key management, authentication, authorization and accounting (AAA) systems, threshold cryptography, and so on.

On the other hand, security mechanisms to verify authenticity, integrity, and reputation features may be required for the operation and management of the network itself. Several distributed cryptographic, trust, reputation, and currency systems can be combined to promote an integral trust solution, making them ideal to be employed in applications built by service composition.

Sophisticated trust and reputation systems have been proposed, some of them even providing distributed reputation and quality assurance for any node, message, or piece of information. Some models designed include data-centric trust establishment (DCTE) [31] frameworks and distributed emergent cooperation through adaptive evolution (DECADE) [21].

The capacity to securely exchange data and learned knowledge is also a prerequirement asserted by Presser et al. [28]. MQTT v3.1.1 [10] only provides general guidance on security. There is no standardized mechanism for MQTT security. However, it recommends brokers that implement transport layer security (TLS) via TCP Port 8883. According to this standard, MQTT is just a transport protocol and security mechanisms are out of scope. Neisse et al. [23] discuss MQTT security issues and propose policy enforcement rules at the MQTT layer.

AMQP security is based on TLS over TCP and the simple authentication and security layer (SASL) or the traditional secure socket layer (SSL) can be used. Authentication may require human intervention to provide a username and password when accessing resources' URIs. However, digital certificates can be adopted when SSL is selected together with TCP. The AMQP payload can be encrypted to increase security in communication.

DDS standardized a more general security model in 2014 [29], where (i) user requirements are specified; (ii) mechanisms to secure topics and data objects are provided; and (iii) authorization exists to perform topics and data objects access or manipulation. The aim is to secure the entire DDS global data space. According to DDS security standard 1.0 [29], DDS provides secrecy, integrity, and nonrepudiation of data objects, as well as authentication and authorization of data writers and readers. There are limited functionalities such as domain joining, definition of new topics, publishing or subscribing from a specific topic, or even writing/reading topic values identified by topic keys.

IEEE 802.15.4(e) addresses link layer security issues. Sastry et al. [32] mention that the aims are (i) authentication of devices; (ii) secrecy and integrity of messages; and (iii) protection against replay attacks. Symmetric cryptography is used to create a checksum that is transmitted on frame headers and verified at the receiver's end. Therefore, a key is secretly shared between transmitter and receiver. Confidentiality is based on the semantic security technique, which uses nonces to introduce variability into encryption process. Replay protection is based on sequence numbering.

Although many of the current IoT technologies provide security solutions, usually they do so incrementally and are focused on individual requirements, often depending on the notoriously problematic protocols we have today. Broader architectural solutions are hardly possible, because new technologies often have to live with the older ones—many of them designed in times when security and privacy concerns did not exist. We advocate for more deep rethinking of architectures to truly address IoT security, privacy, and trust challenges.

11.3   Introducing NG as an IoT Architecture

The NG[1] project started in 2008 with this question in mind: Imagine there is no Internet architecture right now; how could we design it using the best contemporary technologies? A vast survey of emerging paradigms for new Internet architecture was carried out, resulting in a selected list of foundational ingredients. Among them there was the so-called IoT. The project aims to integrate these ingredients into a cohesive design, where one ingredient favors others, catalyzing the overall potential. In this sense, the IoT was related to many other NG ingredients such as name-based content and service orchestration.

The recent development in the future Internet architecture shows that IP-based Internet architecture has limitations when it comes to interconnection of devices in the world of IoT objects and devices. Scalability and portability are two points where NG can score over other Internet architectures. NG architecture will have native support of distributed systems and the ability to evolve

---

[1]http://www.inatel.br/novagenesis/.

its functionality to accommodate new, as yet unforeseen, requests over time for exchange and distribution of data.

A simple NG service is being developed to be embedded at IoT nodes. This service aims to implement some NG novelties at sensors and actuators, enabling them to exchange name-based messages, as discussed at subsection 11.3.1. For small-capacity IoT nodes, proxy/gateway (PG) services can represent them in the NG cloud, enabling dynamic contract establishment in the name of the "things."

The PG service (PGS) model provides a distributed gateway and interoperability solution adequate for the heterogeneity of IoT platforms, protocols, and device implementation. The PGS can also be extended to change configurations at controlled IoT nodes, as well as to detect their status. This model goes in the direction of a software-defined IoT, where nodes are controlled by NG services.

The Internet architecture follows a "narrow-waist" design, which has a great impact on the success of the present Internet. It forces that applications and protocols to be made above the waist, supporting the physical media, physical layers, and access technologies below the waist. But this has a drawback, especially regarding the dual semantics of IP addresses and obsolete fields in headers. As technology evolves, we are talking about the interconnection of billions of devices. This is where the present Internet architecture faces problems. This is where IPv6 comes into play, but it also faces problems because of the same "narrow waist" and the large size of datagram headers.

The NG pub/sub "narrow waist" resembles the DDS link protocol, but with the advantage of integrating several FIA ingredients, such as the naming structure, binding resolutions, software-defined, mobility-friendly, self-organizing, service-oriented designs. In fact, NG extends DDS in many ways, including "semantic-rich" integrated orchestration of contents, services, and IoT resources. NG provides a renewed naming scheme with dynamic messaging, where identifiers are decoupled from locators, supporting mobility by rebuilding name bindings. NG protocols are implemented as services, enabling dynamic protocol orchestration, self-adaptation, and evolution. They enable the emergence of more efficient and modern protocols for the IoT, which can operate aware of several issues such as energy, delay, communication opportunities, and so on.

11.3.1 Naming and name resolution

NG uses natural language names (NLNs) and self-verifying names (SVNs) to identify entities (physical or virtual) at some scope. When a service is initialized, it publishes the bindings among several NLNs and SVNs. It may also publish descriptors exposing its features. Representative services can reveal physical-world resource capabilities and states.

Every service can be addressed by subscribing to these initial bindings recursively. As the combination of a port number and an IP number addresses a port in the current Internet uniquely, SVN tuples allow the same for NG services.

For example, consider a proxy service that inhabits some OS. This service can generate an SVN, let us say A1, and assume that this name is an address for this service inside of the local OS.

Likewise, the OS can generate an SVN, let us say B1, and assume that this name is an address for this OS inside a certain host. The host can also have an SVN, let us say C1, which can be used to address this host inside a domain (D1). The resultant tuple, A1–B1–C1–D1, enables any other service to address a message to this proxy service globally. Additionally, natural language names linked to this tuple facilitate search and discovery of service access points.

On the current Internet, when a host moves from an autonomous system to another, its IP address could change, causing a change in the identity of the host. This results in an undesirable loss of traceability, as well as possible loss of connection. In the NG approach, there is no loss of traceability, since the host remains with the same SVN after movement. Suppose the host of the aforementioned proxy service moves to a new domain, let us say D2. The SVN tuple changes to A1–B1–C1–D2, while the host continues with the SVN C1 despite the movement. Therefore, the mobility of a host in the NG approach requires the removal of the first name binding (C1–D1) from the name resolution service and the publication of a new name binding between C1 and D2. This solution is self-similar, since it could be applied for the mobility of any existences, including content, services, hosts, and so on.

NG SVNs are generated from entities' immutable patterns, as illustrated in Figure 11.5. As long as an entity maintains its immutable attributes, its SVN will be the same. Therefore, even in ephemeral ad hoc networks, entities can preserve their SVNs, while opportunistically connecting and communicating. NG services maintain contracts that are bound to entities' SVNs. Thus, an entity's reputation can be determined based on contract analysis. Additionally, the NG pub/sub service can support new techniques like data-driven trust [31].

11.3.2 Identifier/locator splitting

NG allows names to be used as identifiers and locators. NG borrowed the idea of adopting SVNs as identifiers and locators from other ICT architectures, particularly NetInf [11] and XIA [16]. As previously mentioned, a locator should provide a notion of distance between entities in some space. As one might expect, it is not possible to derive such a notion of distance from SVNs—they are flat (semantic-free names). NG provides a notion of distance by using SVN bindings.

SVNs are certified. They can be checked anytime for integrity. They are flat, since they do not depend on a network hierarchy. Only the bindings among SVNs change according to the network hierarchy. SVNs can be globally unique, avoiding the current lack of addresses on the IPv4 Internet. In the case of host mobility, the services' addresses (tuples) change, but their identifiers, that is, SVNs, remain the same (see Figure 11.4). In other words, only the name bindings change.

Natural language names to approximate machines from human language.

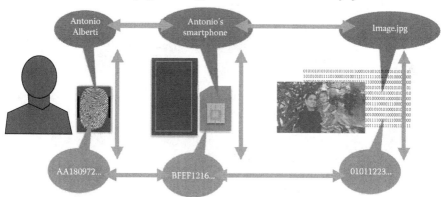

Self-verifying names generated from immutable patterns of entities to improve security.
Name bindings to create a network of relationship representations.

Figure 11.5: Naming and name binding on NovaGenesis. Natural language (meaningful) names are related each other creating an ontology. Self-verifying names (SVNs) are also bound to create a graph of meaningful free names. NLN to SVN bindings or reverse bindings create the link to accommodate not only "semantic-rich" orchestration, but also the provenance of the contents.

The SVNs remain the same while the entities do not change their binary patterns or attributes. In summary, NG generalizes ID/LOC splitting to all entities.

11.3.3 Resources, services, and content orchestration

An NG service can publish its name (NLNs and SVNs) bindings to other services. This is similar to publishing a graph of names. This publication can reveal services' relationships to devices, people, and contents. Figure 11.6 illustrates this process. Services can reveal their features, interests, and intents publicly or privately. This is much more extensive and useful to IoT developers than publishing data in a topic or forwarding topic-based messages. NG accommodates entire service using NLNs and SVNs.

After revealing their name graphs, services look for possible peers. Figure 11.7 illustrates the NG service discovery phase. Services subscribe to NLNs related to their contract interests. Of course, developers will need to provide meaning full (semantic-rich) keywords to facilitate finding good candidates. If a service discovers a good candidate (it needs to evaluate this), it publishes a contract/SLA offer.

Observe that one can develop representative services for "things." These services can reveal the physical features of sensors and actuators, negotiating contracts in the name of "things," and configuring and managing devices to reflect

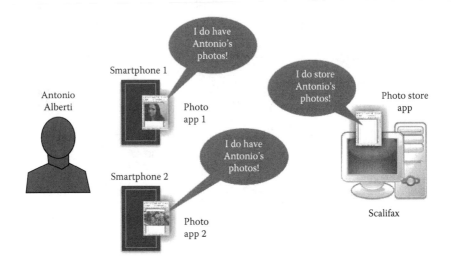

Figure 11.6: Several of Antonio's applications publish their name bindings using the NovaGenesis pub/sub service. We call this service the exposition phase. In this particular example, "Photo App 1" and "Photo App 2" announce that they have Antonio's photos, while "Photo store app" announces that it stores Antonio's photos. Obverse that natural language names are bound to each entity's self-verifiable names. Thus, semantics orchestration employs NLNs first, and SVNs thereafter, to improve security.

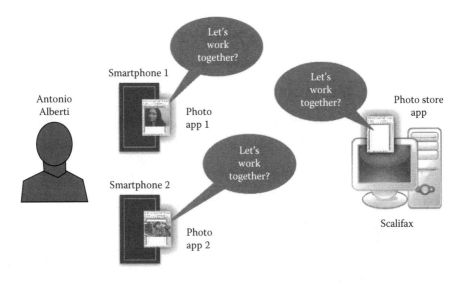

Figure 11.7: Antonio's services discover each other using meaningful keywords (NLSs) and publish contract/SLA offers to candidate peers. Pub/sub may be encrypted using asymmetric cryptography. Therefore, SLA offers can be kept secret.

the required QoS, energy restrictions, and tuning for constrained environment. This approach is adequate for the IoT, since one cannot expect that contemporary IoT devices will be able to establish contracts by themselves.

NG enables services to form trust networks, where every service has a reputation, as in online e-commerce websites that we have today; for example, eBay. Every service has a reputation and this reputation is verified before establishing service contracts (SLAs). Thus, services are evaluated regarding possible threats and risks. Secure services of good quality will prosper, while bad services, suspected of being unsafe, will have their reputation reduced, naturally forcing them to improve or disappear.

Services may hire agreements with other services to evaluate the reputation of their mutual SLAs. These reputation services (RpSs) can distributively provide reputation and quality assurance for any node, message, or piece of information. Information is secured per se and its dissemination depends on the contracting establishment and traditional secrecy and integrity mechanisms.

It is only after having established agreements that the services start the secure exchange of Named information. This is illustrated in Figure 11.8. Interestingly, to those entities authorized to view it, content provenance can be broadly verified, as illustrated in Figure 11.9. NG publishes name bindings from a distributed web of relationship representations, enabling authorized services to navigate among contents, services, and hardware relationships. Authorized services can

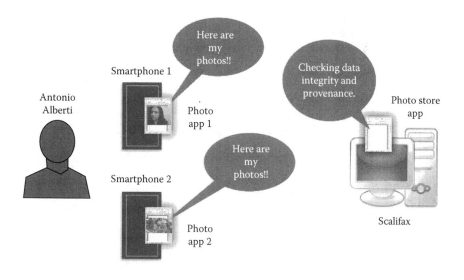

Figure 11.8: After SLA establishment, services can securely publish and subscribe data to their peers. Now, data integrity can take advantage of self-verifying names and their bindings. The two photo applications send their pictures to the "Photo store app."

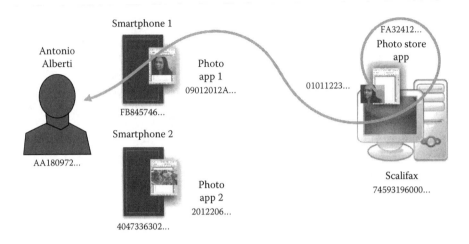

Figure 11.9: Provenance can be determined by decrypting SVNs backward from the subscriber service up to the original content publisher. For instance, the SVN "01011223..." of the female picture can be bound to the SVN "FA32412..." of the "Photo store app." By resolving this name binding, an authorized entity can determine the provenance and integrity of the exchanged data. This approach can be allied to the established SLAs among services, enabling the creation a trustable fellowship of "things," representative software in the IoT context.

derive complete graphs of relationships, clearly determining provenance, nonrepudiation, and other security properties. Name bindings can be encrypted using asymmetric cryptography, linking entities to SVNs, as recommended by Ghodsi et al. [14].

Traditional security services, like PKI, distributed reputation, and voting systems need to be mapped to NG abstractions. Novel approaches can emerge when combining the NG service framework with contemporary security techniques. However, this subject is still in its beginnings and intensive research is required.

11.3.4 Security, privacy, and trust

The NG security model is founded on the following cornerstones:

- Self-verifiable names: These have better security properties than natural language names [14]. They enable data integrity checks and can be aggregated using name bindings, providing very good scalability for message/ packet forwarding or routing.

- Pub/sub communication model: Represents a change from the traditional "receiver accepts all" paradigm to a loosely-coupled pub/sub model [13]. In the pub/sub paradigm, contents are published by services and subscribed to by others. Thus, a service publishes a content and authorizes other services to subscribe. A target communicating service needs to be authenticated and obtain authorization to have access to certain information. Pub/sub allows a secure, asynchronous rendezvous between publishers and subscribers. NG extends this model with SLAs among services. It also enables revocation of published bindings and data, as well as changes to authorizations.

- Contract-based model: Serviceable information is transferred using pub-/sub only after establishing an SLA. This enables the formation of trust networks among services, especially representative services for physical-world "things." "Things" cannot compute by themselves (maybe in future they will in the future), so they require software representatives. Representative services can establish SLAs in the name of "things," exposing its features, capabilities, constraints, status, and so on to peer orchestration services. Policy enforcement can be carried out in the context of negotiated SLAs.

- Self-organizing services: This allows the formation of a service's "social behavior," which can facilitate recognition of illegal/misbehaving services and malicious content. It enables the incorporation of immunological systems aspects in IoT scenarios. One can expect that malicious devices will try to get illegal access to or threaten IoT services, causing real damage, especially in smart home, e-Health, smart grid, and public service environments. In addition, many of the required actions to maintain safety and privacy will be intensive and on huge scales; therefore, self-organization based on user policies is a good premise [28].

- Unbiased contract, reputation, and trust evaluation: Such resources will allow a virtuous cycle of increasingly enhanced solutions for security and privacy. Autonomous decision cycles require precise evaluation of SLA results obtained. Reliability and risk can be better determined if precise and trustworthy estimates of reputation and trust are available.

- Built-in policy definition and enforcement: Service contracts can implement user/machine policies, enforcing their application in the autonomic cycle of "things," representatives.

- Distributed algorithms: Pub/sub and SLA-based orchestration creates an environment that favors distributed key generation and cryptography. Voting and coordination of entities can be established toward "social devices," security, privacy, and trust. Distributed/hierarchical certification chains are also a possibility.

- Deterministic building: Self-verifiable naming claims for deterministic building (or compilation) of source codes. Such deterministic compilation guarantees that the same SVN will be generated for a certain program every time it is compiled. Therefore, if any additional executable code is inserted into a service, its SVN will change, indicating possible back doors.

What NG can offer for IoT scenarios is the synergistic integration of these foundations toward a modern architecture that addresses IoT challenges as deeply as required. Current technologies will face limitations regarding naming, name resolution, communication models, mobility support, flexibility, elasticity, scalability, among other aspects. What this model offers are the mechanisms to create emerging trustable fellowships of social, self-organizing "things," together with their software-as-a-service representatives, creating a state-of-the-art architecture to face the security, privacy, and trust requirements one may expect on the future Internet.

11.4   Example Scenario

In the next decades, the quantity of devices on the Internet will increase exponentially. The "things" will be the majority of devices—there will be no Internet anymore without the things. Therefore, the Internet will certainly feel the pressures of an army of network-enabled devices that will require scalability, unique naming, addressing, secure and private handling of information, mobility, and so on. The benefits of the IoT will be huge, as well as the challenges behind it. The "things" will be the sensorial and actuating system of our converging human–machine technologies. Every application will have detailed physical-world information to make better decisions. Software-defined micro- and nano-"things" will ultimately lead to the emergence of so-called programmable matter.

However, to illustrate the ideas addressed in this chapter, we selected an ordinary scenario we have today, which converges a smart home environment with climate monitoring systems. Figure 11.10 illustrates this scenario, applying NG's aforementioned paradigms. In this figure, representative services named with natural language and self-verifiable names reveal "things," capabilities and status, representing them in contract establishment and pub/sub information exchange and storage.

The required actions emerge as a "social behavior" of "things," their software representatives, and smart assistants. The aim is to close a room window if there is nobody at home and a violent storm is coming. A service represents the window. Many other software applications could represent the presence sensors that determined there was nobody at home. All these representative software applications should be in the same "fellowship" as a smart home assistant. The assistant can correlate the available knowledge and decide to close the window or not in the case of some triggering event. The owners can specify policies beforehand

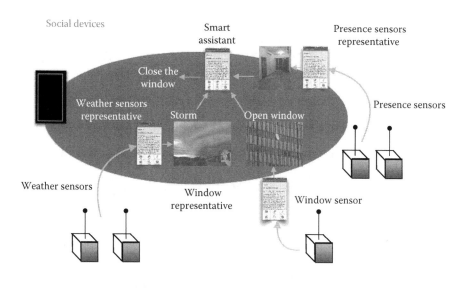

Figure 11.10: An example scenario of NovaGenesis model for the IoT.

for managing and controlling their "things," including safety procedures in the case of natural disasters.

This solution scales very well, since it is distributed and bottom up, taking advantage of pub/sub and self-verifiable naming. Imagine now you have not just one window, but thousands of "things" forming a trustable fellowship of devices at your home. Who is going to manage or control these devices? You? No, they will manage themselves according to policies you (or some operator) define. The IoT requires a secure auto-pilot—a self-driven society of devices, representatives, and assistants/controllers; a smart solution capable of dealing with billions of devices, their privacy, secrecy, and content provenance. This is illustrated in Figure 11.11.

We envision the future Internet with "swarms" of physical-world representative services to represent our physical-world resources; let us say computers, cars, roads, streets, energy systems, forests, transportation systems, farms, and so on. These representatives will form name-based, trustable fellowships, which will enrich the decision-making of assistants, controllers, and managers. Following the right policies, these "swarms" of IoT services will help our information society to address the most important problems we have today, including environmental, social, economic, safety, and so on.

In summary, a smart ICT architecture for the IoT will enable people to express their intents, preferences, and policies, which will drive intelligent applications toward better use and sharing of our fixed or mobile physical-world resources, creating a self-organizing solution, where protocols are dynamically

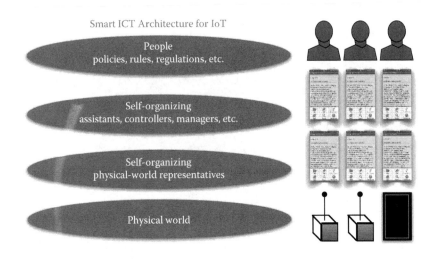

Figure 11.11: NovaGenesis approach for a secure, trustable IoT.

changed to maximize efficiency, security, and privacy, alleviating error-prone, ordinary tasks, as the Internet scales to much higher astronomical numbers. NG offers an exciting environment for developing and deploying these ideas, enabling us to integrate "things," services, and information security, privacy, and trust.

Acknowledgments

This work was partially supported by Finep/Funttel Grant No. 01.14.0231.00, under the Radiocommunication Reference Center (Centro de Referência em Radiocomunicações, CRR) project of the National Institute of Telecommunications (Instituto Nacional de Telecomunicações – Inatel), Brazil.

Bibliography

[1] A.M. Alberti, V.H. de O Fernandes, M.A.F. Casaroli, L.H. de Oliveira, F.M. Pedroso, and D. Singh. A novagenesis proxy/gateway/controller for openflow software defined networks. In *Network and Service Management (CNSM), 2014 10th International Conference on*, 394–399, November 2014.

[2] A.M. Alberti. Searching for synergies among future internet ingredients. In Geuk Lee, Daniel Howard, Dominik Slezak, and YouSik Hong (Eds.),

Convergence and Hybrid Information Technology, vol. 310 of *Communications in Computer and Information Science*, 61–68. Berlin, 2012. Springer.

[3] A.M. Alberti, A. Vaz, R. Brandão, and B. Martins. Internet of information and services (IoIS): A conceptual integrative architecture for the future internet. In *Proceedings of the 7th International Conference on Future Internet Technologies, CFI '12*, 45, New York, 2012. ACM.

[4] K. An, A. Gokhale, D. Schmidt, S. Tambe, P. Pazandak, and G. Pardo-Castellote. Content-based filtering discovery protocol (CFDP): Scalable and efficient OMG DDS discovery protocol. In *Proceedings of the 8th ACM International Conference on Distributed Event-Based Systems, DEBS '14*, 130–141, New York, 2014. ACM.

[5] R. Atkinson, S. Bhatti, and S. Hailes. Evolving the Internet architecture through naming. *Selected Areas in Communications, IEEE Journal on*, 28(8):1319–1325, October 2010.

[6] M.F. Bari, S. Chowdhury, R. Ahmed, R. Boutaba, and B. Mathieu. A survey of naming and routing in information-centric networks. *Communications Magazine, IEEE*, 50(12):44–53, December 2012.

[7] F. Belqasmi, R. Glitho, and Chunyan Fu. Restful web services for service provisioning in next-generation networks: A survey. *Communications Magazine, IEEE*, 49(12):66–73, December 2011.

[8] T. Braun, V. Hilt, M. Hofmann, I. Rimac, M. Steiner, and M. Varvello. Service-centric networking. In *Communications Workshops (ICC), 2011 IEEE International Conference on*, 1–6, June 2011.

[9] L. Buttyan and J.-P. Hubaux. *Security and Cooperation in Wireless Networks: Thwarting Malicious and Selfish Behavior in the Age of Ubiquitous Computing*. Cambridge University Press, New York, 2007.

[10] A. Banks and R. Gupta (Eds.). MQTT version 3.1.1, OASIS Standard.

[11] C. Dannewitz. NetInf: An information-centric design for the future internet. In *Proc. 3rd GI ITG KuVS Workshop on The Future Internet*, 2009.

[12] C. Dannewitz, D. Kutscher, B. Ohlman, S. Farrell, B. Ahlgren, and H. Karl. Network of information (NetInf): An information-centric networking architecture. *Comput. Commun.*, 36(7): 721–735, April 2013.

[13] N. Fotiou, G. Marias, and G. Polyzos. Publish–subscribe Internet working security aspects. In Luca Salgarelli, Giuseppe Bianchi, and Nicola Blefari-Melazzi (Eds.), *Trustworthy Internet*, 3–15. Milan, 2011. Springer.

[14] A. Ghodsi, T. Koponen, J. Rajahalme, P. Sarolahti, and S. Shenker. Naming in content-oriented architectures. In *Proceedings of the ACM SIGCOMM Workshop on Information-centric Networking, ICN '11*, 1–6, New York, 2011. ACM.

[15] J.A. Gutierrez, M. Naeve, E. Callaway, M. Bourgeois, V. Mitter, and B. Heile. IEEE 802.15.4: A developing standard for low-power low-cost wireless personal area networks. *Network, IEEE*, 15(5):12–19, September/October 2001.

[16] D. Han, A. Anand, F. Dogar, B. Li, H. Lim, M. Machado, A. Mukundan, W. Wu, A. Akella, D. G. Andersen, J. W. Byers, S. Seshan, and P. Steenkiste. XIA: Efficient support for evolvable internetworking. In *Proc. 9th USENIX NSDI*, San Jose, CA, April 2012.

[17] IBM Corporation. Message queue telemetry transport (MQTT), June 2014.

[18] M. Kang and D.-H. Kim. A real-time distributed architecture for RFID push service in large-scale EPCglobal networks. In Tai-hoon Kim, Hojjat Adeli, Hyun-seob Cho, Osvaldo Gervasi, StephenS. Yau, Byeong-Ho Kang, and JavierGarcía Villalba (Eds.), *Grid and Distributed Computing*, vol. 261 of *Communications in Computer and Information Science*, 489–495. Berlin, 2011. Springer.

[19] J. Kim, J. Lee, H. K. Kang, D. S. Lim, C. S. Hong, and S. Lee. An ID/locator separation-based mobility management architecture for WSNs. *Mobile Computing, IEEE Transactions on*, 13(10):2240–2254, October 2014.

[20] N. McKeown, T. Anderson, H. Balakrishnan, G. Parulkar, L. Peterson, J. Rexford, S. Shenker, and J. Turner. OpenFlow: Enabling innovation in campus networks. *SIGCOMM Comput. Commun. Rev.*, 38(2):69–74, March 2008.

[21] M. M. Mejia, N. M. Peña, J. L. Muñoz, O. Esparza, and M. A. Alzate. DECADE: Distributed emergent cooperation through adaptive evolution in mobile ad hoc networks. *Ad Hoc Networks*, 10(7):1379–1398, 2012.

[22] J. Montavont, D. Roth, and T. Noël. Mobile IPv6 in internet of things: Analysis, experimentations and optimizations. *Ad Hoc Netw.*, 14:15–25, March 2014.

[23] R. Neisse, G. Steri, and G. Baldini. Enforcement of security policy rules for the Internet of Things. In *Wireless and Mobile Computing, Networking and Communications (WiMob), 2014 IEEE 10th International Conference on*, 165–172, October 2014.

[24] M.R. Palattella, N. Accettura, X. Vilajosana, T. Watteyne, L.A. Grieco, G. Boggia, and M. Dohler. Standardized protocol stack for the Internet of (important) Things. *Communications Surveys Tutorials, IEEE*, 15(3): 1389–1406, 2013.

[25] J. Pan, S. Paul, and R. Jain. A survey of the research on future internet architectures. *Communications Magazine, IEEE*, 49(7):26 –36, July 2011.

[26] M.P. Papazoglou, P. Traverso, S. Dustdar, and F. Leymann. Service-oriented computing: State of the art and research challenges. *Computer*, 40(11): 38–45, November 2007.

[27] G. Pardo-Castellote. OMG data-distribution service: Architectural overview. In *Proceedings of the 2003 IEEE Conference on Military Communications - Volume I, MILCOM'03*, 242–247, Washington, DC, 2003. IEEE Computer Society.

[28] M. Presser, P. Daras, N. Baker, S. Karnouskos, A. Gluhak, S. Krco, C. Diaz, I. Verbauwhede, S. Naqvi, F. Alvarez, and A. A. Fernandez-Cuesta. Real world Internet. Technical report, *Future Internet Assembly*, 2008.

[29] OMG. DDS security 1.0. Document number: ptc/2014 06-01. June 2014.

[30] D. C. Ranasinghe, M. Harrison, and P. H. Cole. EPC network architecture. In Peter H. Cole and Damith C. Ranasinghe (Eds.), *Networked RFID Systems and Lightweight Cryptography*, 59–78. Berlin, 2008, Springer.

[31] M. Raya, P. Papadimitratos, V. D. Gligor, and J.-P. Hubaux. On data-centric trust establishment in ephemeral ad hoc networks. In *INFOCOM*, 1238–1246. IEEE, 2008.

[32] N. Sastry and D. Wagner. Security considerations for IEEE 802.15.4 networks. In *Proceedings of the 3rd ACM Workshop on Wireless Security, WiSe '04*, 32–42, New York, 2004. ACM.

[33] Z. Shelby. Constrained RESTful environments (CoRE) link format. RFC 6690 (proposed standard), August 2012.

[34] Z. Sheng, S. Yang, Y. Yu, A. Vasilakos, J. McCann, and K. Leung. A survey on the IETF protocol suite for the Internet of Things: standards, challenges, and opportunities. *Wireless Communications, IEEE*, 20(6):91–98, December 2013.

[35] S. Vinoski. Advanced message queuing protocol. *IEEE Internet Computing*, 10(6):87–89, 2006.

[36] W. Wong and P. Nikander. Secure naming in information-centric networks. In *Proceedings of the Re-Architecting the Internet Workshop, ReARCH '10*, 12:1–12:6, New York, 2010. ACM.

[37] Q. Wu, Z. Li, J. Zhou, H. Jiang, Z. Hu, Y. Liu, and G. Xie. SOFIA: Toward service-oriented information centric networking. *Network, IEEE*, 28(3): 12–18, May 2014.

[38] G. Xylomenos, C.N. Ververidis, V.A. Siris, N. Fotiou, C. Tsilopoulos, X. Vasilakos, K.V. Katsaros, and G.C. Polyzos. A survey of information-centric networking research. *Communications Surveys Tutorials, IEEE*, 16(2):1024–1049, 2014.

[39] J. Zhang. A survey on trust management for VANETs. In *Advanced Information Networking and Applications (AINA), 2011 IEEE International Conference on*, 105–112, March 2011.

Chapter 12

Preventing Unauthorized Access to Sensor Data

Liu Licai

Yin Lihua

Guo Yunchuan

Fang Bingxing

CONTENTS

Abstract

In MANETs, cooperative authentication, requiring the cooperation of neighbor nodes, is a significant authentication technique. However, when nodes participate in cooperation, their location may easily be tracked by misbehaving nodes; meanwhile, their resources will be consumed. These two factors lead to selfish nodes, reluctant to participate in cooperation, and will decrease the probability of correct authentication. To encourage nodes to take part in cooperation, we propose a bargaining-based dynamic game model for cooperative authentication to analyze dynamic behaviors of nodes and help nodes decide whether or not to participate in cooperation. Further, to analyze the dynamic decision-making of nodes, we discuss two situations: complete information and incomplete information, respectively. Under complete information, subgame perfect Nash equilibriums are obtained to guide nodes to choose their optimal strategy to maximize their utility. In reality, nodes often do not have good knowledge about others' utility (this case is often called "incomplete information"). To deal with this case, the perfect Bayesian Nash equilibrium is established to eliminate the implausible equilibriums. Based on the model, we designed two algorithms for complete information and incomplete information, respectively, and the simulation results demonstrate that, in our model, nodes participating in cooperation will maximize their location privacy and minimize their resource consumption with an increased probability of correct authentication. Both the algorithms can improve the success rate of cooperative authentication and extend the network lifetime to 160%–360.6% of the present value.

Keywords: incentive strategy; cooperative authentication; dynamic game; MANET

12.1 Introduction

The mobile ad hoc network (MANET), recognized as a ubiquitous approach for many applications such as habitat surveillance and environment monitoring, has become a focus of research in recent years [1]. Technically, MANET is a multihop wireless autonomous system without fixed infrastructures [2] and has three important features: (1) Node resources (e.g., computing and communication resources) are limited; (2) they are interconnected through wireless links, such as those formed by Bluetooth and Wi-Fi in ad hoc mode; and (3) they are often deployed within openly hostile environments [3]. Thus, MANET suffers from an increasing number of security threats (e.g., unauthorized access and injection of false data) with high risks.

In order to cope with those security threats, cooperative authentication has been proposed in recent years [4–13]. Generally, there are three kinds of nodes in cooperative authentication: source nodes, neighbor nodes, and a sink node (these are discussed in detail in Section 12.3). If a source node wants to prove the authenticity of its message to the sink node, it requests its neighbor nodes to participate in cooperation. If all neighbor nodes believe the message is true, then the sink node also believes it is true. Such an approach can effectively enhance the *probability of correct authentication*[1] (PCA). Generally, the more neighbor nodes participate in cooperation, the higher the value of the PCA. Cooperative authentication not only drastically enhances the PCA, but also mitigates the verification overheads of the sink node.

Although cooperative authentication demonstrates these advantages, selfish nodes may be unwilling to participate in cooperation due to the following reasons. (1) Leakage of location privacy: Generally, communication between nodes relies mostly on open wireless channels, and the locations of nodes can be easily exposed to a misbehaving node [14, 15]. (2) Consumption of resources: Participating in cooperation tends to consume more of the node's resources and decreases its overall lifetime. Those two factors make nodes disinclined to participate in cooperation and reduce the PCA. Thus, incentivizing an appropriate number of nodes to participate in cooperation is a key issue.

In order to solve the above problem, we propose a bargaining-based dynamic decision to balance the conflict between increasing the PCA and decreasing the loss of nodes participating in cooperation. Our core idea for this issue is to incentivize an appropriate number of neighbor nodes to participate in cooperation by using a virtual currency and to maximize their benefits at an acceptable cost via a dynamic game. In summary, our main contributions are as follows:

1. To encourage an appropriate number of neighbor nodes to participate in cooperation, we proposed a *bargaining-based dynamic game model*[2] *for*

---

[1] Authentication is correct if true or false messages are correctly recognized.

[2] We assume that neighbor nodes are rational individuals; that is, they decide locally whether or not to participate in cooperation based on their utility.

cooperative authentication to analyze the dynamic behaviors of all nodes and help nodes decide whether or not to take part in cooperation.

2. To analyze the dynamic decision-making of nodes, we discussed two situations of the dynamic game, with complete and incomplete information, respectively. We obtained subgame perfect Nash equilibriums and perfect Bayesian equilibrium under complete and incomplete information, respectively, to guide nodes to choose an optimal strategy to maximize their utility.

3. Based on our model, we designed two algorithms under complete and incomplete information, and the simulation results show that nodes participating in cooperation will maximize their location privacy and minimize their resource consumption while ensuring the value of the PCA. They can improve the success rate of authentication and extend the network lifetime to 160%–360.6% of the current value.

The remainder of this chapter is organized as follows: In Section 12.2, we present existing related work. Section 12.3 introduces the mechanism of cooperative authentication in MANET. Section 12.4 proposes a bargaining-based dynamic game model for cooperative authentication. The fifth section analyzes the dynamic game and develops two algorithms based on analysis results. Section 12.6 conducts experimental results to demonstrate the effectiveness of our model. Finally, Section 12.7 concludes the chapter and presents future work.

12.2 Related Work

The authentication problem, privacy problem, and dynamic behavior problem of MANETs have been studied by many researchers. In this section, we present the existing research studies related to the cooperative authentication mechanism and location privacy protection in mobile ad hoc networks, incentive strategy, and game theory.

12.2.1 Cooperative authentication

Due to limited resources and weak computing ability, most authentication mechanisms that work effectively for the Internet are unfit for wireless networks. To solve this problem, lots of authentication mechanisms have been proposed; cooperative authentication is an important one.

Nyang et al. [6] presented a cooperative public key authentication scheme, where a node stores a few hashed keys for other nodes and uses them to authenticate messages cooperatively. This scheme avoids cryptographic operations and can be used in the network with constrained resources. However, it is only designed for one-hop authentication, which makes it impractical and inefficient for a conventional multihop wireless network. To solve this problem, Moustafa et al. [12] employed a Kerberos authentication model, where the Kerberos server is managed by the network service provider and plays the role of a trusted third party for ad hoc nodes.

Although these schemes provide a degree of authentication, they increase the authentication burden and the nodes' computation overhead. To deal with this problem, Zhu et al. [7] used a hash message authentication code to authenticate messages cooperatively and alleviate the authentication burden by only verifying a small number of messages. Additionally, Hao et al. [5, 8, 13] proposed a cooperative message authentication protocol in vehicle networks. They aim to alleviate vehicles' computation overhead by means of sharing verification results.

Those mechanisms can alleviate the authentication burden and computation overhead of nodes. However, as they rely on the sharing of verification results between nodes, and reliable result sharing requires trust transitivity, the accuracy and reliability of authentication may be low. To solve this problem, Lu et al. [4] proposed a bandwidth-efficient cooperative authentication scheme to detect and filter injected false data with a high en-routing filtering probability. This scheme adopts a cooperative neighbor and router-based filtering mechanism. Additionally, Vijayakumar et al. [10] proposed highly secured cooperative trusted communication using an object link state routing protocol and message authentication between nodes. However, too many cooperative nodes lead to greater authentication overheads. Lin et al. [9] proposed a cooperative authentication scheme to eliminate redundant authentication of the same message by different vehicles. This scheme can reduce the authentication overheads of individual vehicles and shorten authentication delay.

Despite the fact that these mechanisms can improve the accuracy and reliability of authentication, they require the unselfishness and cooperation of nodes. These requirements often lead to location privacy leakage and resource consumption for cooperative nodes, and further reduce the willingness of nodes to cooperate. Thus, a conflict is raised between increasing willingness of cooperation and decreasing the loss of cooperative nodes. Few studies have looked at such a conflict. Now, balancing this conflict has become a critical challenge in the area of cooperative authentication.

12.2.2   Cooperation incentive

In order to encourage nodes to be cooperative, various incentive strategies have been proposed, such as price-based [16–18] and reputation-based [19]. The basic idea of a price-based incentive strategy involves providing incentives by way of virtual currency paid to nodes for offering services. Zhang et al. [16] considered bandwidth exchange as payment to encourage cooperation. Zhang et al. [17] proposed controlled coded packets as a virtual commodity currency to induce cooperative behaviors and reduce overheads. The reputation-based incentive strategy uses the historical behaviors of nodes to assess their reputation, and then distinguishes the cooperative nodes from the malicious (selfish) nodes by setting a reputation threshold. Refaei et al. [19] introduced a time-slot mechanism and proposed an adaptive reputation-based incentive mechanism to monitor the changes in node behavior quickly and accurately. Considering all factors affecting willingness to cooperate, we used fortune as a virtual currency to provide cooperation incentives.

12.2.3 Conflict balancing

Game theory is a mathematics theory which is adept at modeling conflict situations, analyzing the behavior of participants and predicting their decision. Manshaei et al. [20] overviewed existing research on security and privacy in networks using game-theoretic approaches. Freudiger et al. [15] analyzed the conflict between location privacy protection and the costs of pseudonym changes in MANET and achieved balance between maximum location privacy and minimum cost. Chen et al. [21] used coalitional game theories to evaluate cooperation in VANETs, while presenting a scheme to stimulate cooperation in message forwarding. We built a static game model for cooperative authentication to help nodes make decisions in our previous research [22] and assumed that the players chose actions simultaneously. However, this assumption did not fit the situation of dynamic decision-making, where each player performs a sequence of actions according to others' serial strategies. The reason is as follows: When a later player makes a decision, it will naturally adjust its strategy selection according to the strategies of the earlier players. So, it is important to study a dynamic game model to help nodes decide whether or not to participate in cooperation.

12.3 Preliminaries

As shown in Figure 12.1, cooperative authentication consists of a sink node n_s and a set of mobile nodes MN = $\{n_0, n_1, \ldots\}$ randomly deployed in a certain area [4], where n_s is a data collection unit with sufficient resources and any two nodes share a key pair for authentication.

In Figure 12.1, if n_0 (also called the source node) wants to send a message m to n_s via an established routing path and prove its authenticity, it first selects k neighbor nodes (denoted as k-NNs = $\{n_1, \ldots, n_k\}$) from all neighbor nodes[3] (denoted as NNs = $\{n_1, \ldots, n_N\}$, where $k \leqslant N$[4] and N is the number of NNs),

---

[3] The NNs of a node refer to the nodes within its one-step transmission range.

[4] If $k > N$, the single process of cooperative authentication fails; while if $k > N$ is always true, it means that the network has expired.

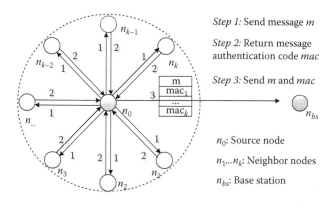

Step 1: Send message *m*

Step 2: Return message authentication code *mac*

Step 3: Send *m* and *mac*

n_0: Source node

$n_1...n_k$: Neighbor nodes

n_{bs}: Base station

Figure 12.1: Basic network model of cooperative authentication.

and then sends *m* to its *k*-NNs and requests them to cooperatively authenticate *m*. The *k*-NNs return a one-bit message authentication code (MAC) which denotes whether or not *m* is true. After receiving MACs, *n* sends *m* and *k*-bit MACs to n_s. If all *k*-NNs believe *m* is true, then n_s also believes it is true. Otherwise, it is false and is rejected. Generally, nodes may be compromised by adversaries with probability ρ. As shown in [4], any false identity/message will be recognized if the following conditions are satisfied simultaneously: (1) at least one uncompromised NN participates in cooperation; and (2) adversaries cannot completely and correctly guess all MACs generated by uncompromised NNs. The PCA is formulated as Equation 12.1:

$$\text{PCA} = 1 - \sum_{i=0}^{k} \binom{k}{i} \times \rho^i \times (1-\rho)^{k-i} \times \frac{1}{2^{k-i}} \qquad (12.1)$$

Given the PCA and ρ, we can calculate *k* from Equation 12.1; that is, the least number of NNs (denoted as *minCNN*) that should participate in cooperation. The higher *minCNN* is, the more resources are consumed and the more location privacy is exposed. So, our goal is to incentivize precisely *minCNN* nodes to participate in cooperation.

12.4   Bargaining-Based Dynamic Game Model for Cooperative Authentication

In this section, we propose a bargaining-based dynamic game model for cooperative authentication. We present, in detail, an improved bargaining mechanism [22] to incentivize an appropriate number of neighbor nodes to be cooperative and design a dynamic game to support dynamic decision-making.

12.4.1 Bargaining mechanism

In order to incentivize selfish nodes, we regarded authentication services provided by cooperative nodes as "goods" and proposed a bargaining mechanism to improve the cooperation willingness of nodes. In our mechanism, the buyer is n_0 and the sellers are NNs.

12.4.1.1 Factors affecting price

First, we discuss factors affecting the asking/bidding price.

1. The attribute of message m:

   The attributes of the authenticating message, including the message length l_m, lifetime of message TTL_m, and the importance of message Imp_m, are other important factors impacting the reservation price. The longer and more important the message, the higher the reservation price.

2. Leakage of location privacy:

   Here, similar to most approaches concerning quantifying location privacy [23], the adversary's uncertainty is used to measure the location privacy level $Priv_i$ of n_i, as in Equation 12.2a.

$$
\begin{aligned}
(a) : Priv_i &= -\sum_{d=1}^{M} p\left(loc_d | loc_i\right) \log_2 p\left(loc_d | loc_i\right) \\
(b) : Priv_{\text{max}}^i &= \log_2 M \\
(c) : DLPP_i &= \frac{Priv_i}{Priv_{\text{max}}^i}, 0 \leqslant DLPP_i \leqslant 1
\end{aligned}
\tag{12.2}
$$

   where $p(loc_d | loc_d)$ denotes the conditional probability with which the predictive location loc_d corresponds to the true location loc_i $(1 \leqslant i \leqslant N)$ and M is the number of locations.

   If the conditional probability is of a uniform distribution, then $Priv_i$ reaches the maximum $Priv_{\text{max}}^i$, as in Equation 12.2b. $DLPP_i$ in Equation 12.2c denotes the degree of location privacy preservation of n_i. For simplicity, $Priv_{\text{cons}}^i$ denotes the location privacy leakage for a cooperative process and $DLPP_{\text{min}}$ is a lower threshold to expose location privacy.

3. Node energy:

   We use three metrics to measure the energy of n_i: the initial energy $Ener_{\text{max}}^i$, the current remaining energy $Ener_{\text{remn}}^i$ and the consumed energy $Ener_{\text{cons}}^i = \alpha_m + \beta_m \times l_m$ (where α_m and β_m are weights) for a cooperation authentication process. Let $Ener_i = Ener_{\text{remn}}^i / Ener_{\text{max}}^i$ denote the fraction of remaining energy of n_i and $Ener_{\text{min}}$ is a survival threshold.

4. Bandwidth:

We assume the bandwidth of the channel is BW_{max}. For a given value m, $BW_m = l_m/TTL_m$ is the required bandwidth. The utilization of bandwidth is denoted by $BW_i = BW_m/BW_{max}$.

5. Required number of cooperative neighbor nodes:

To guarantee that the PCA reaches a given threshold value, we must ensure that a given number of neighboring nodes participate in the process of cooperative authentication. Please note that higher numbers of cooperative nodes do not necessarily imply a better service quality: the more neighboring nodes that participate in cooperative authentication, the higher the value of PCA that is reached and at the same time the more resources are consumed. Given the positive authentication probability PCA, the required number of cooperative neighboring nodes can be obtained via Equation 12.1 with the number represented by *minCNN* and requiring $minCNN \leqslant N$.

6. Fortune:

We use fortune as a virtual currency to pay for each authentication service and provide a cooperation incentive. Let FT_i denote the fortune of n_i and FL_i represent the fortune level of n_i, where FL_i is defined as in Equation 12.3.

$$FL_i = \begin{cases} FT_i/pl & \text{if } FT_i \leqslant pl \\ (2FT_i + wl - pl)/(pl + wl) & \text{if } pl < FT_i \leqslant wl \quad (12.3) \\ 2(wl - pl)/(pl + wl) + FT_i/wl & \text{if } FT_i > wl \end{cases}$$

where:

pl denotes the "poverty line"

wl denotes the "wealth line"

$pl < wl.FL_{min}$ is the payment capability threshold for nodes

12.4.1.2   Bargaining-based price

The bargaining-based price consists of the price offered by the buyer and the price asked by the seller.

1. Price offered by buyer:

If n_0 requests NNs authenticate m, n_0 first calculates cost price C_0, reservation price R_0 (R_0 is the highest price that n_0 agrees to pay for the authentication service), and loss of no authentication $LONA_0$ based on the attributes of the message and authentication request, and then offers a bidding price B_0 depending on C_0 and R_0, as in Equation 12.4.

$$C_0 = w_{C_0} \times minCNN \times l_m \times BW_m, \quad C_0 > 0$$
$$R_0 = w_{R_0} \times Imp_m \times FL_0 \times C_0, \quad R_0 \geqslant C_0$$
$$B_0 = w_{B_0} \times R_0 + (1 - w_{B_0}) \times C_0, \quad C_0 \leqslant B_0 \leqslant R_0 \text{ and } B_0 \leqslant FT_0$$
$$LONA_0 = w_{L0} \times Imp_m, \quad LONA_0 > 0$$

$$(12.4)$$

where $w_{C_0}, w_{R_0}, w_{B_0}, w_{L0}$ are weights.

2. Price asked by seller:

Before the seller n_i ($1 \leqslant i \leqslant N$) participates in cooperation, it calculates its own cost price C_i and reservation price R_i (R_i is the lowest price for which n_i agrees to provide the authentication service), and then offers an asking price A_i depending on C_i and R_i as in Equation 12.5.

$$C_i = w_{C_i} \times l_m \times BW_m \times \left(\begin{array}{c} v_{C_i} \times Priv^i_{cons} + \\ (1 - v_{C_i}) \times Ener_{cons} \end{array} \right), C_i > 0$$
$$R_i = w_{R_i} \times (1 - DLPP_i) \times (1 - Ener_i) \times BW_i \times C_i/FL_i, R_i \geqslant C_i \quad (12.5)$$
$$A_i = w_{A_i} \times R_i, \qquad\qquad C_i \leqslant R_i \leqslant A_i$$

where $w_{C_i}(w_{C_i} > 0)$, $v_{C_i}(0 \leqslant v_{C_i} \leqslant 1)w_{R_i}$ and w_{A_i} are weights. We assume that nodes care about both $Priv^i_{cons}$ and $Ener_{cons}$ equally and set v_{C_i} to 0.5.

12.4.1.3 Bargaining procedure

When n_0 requests its NNs authenticate m, the price bargaining between the buyer and potential sellers is conducted as in the following procedures.

1. The buyer offers a bidding price:

The buyer n_0 first calculates C_0, R_0 and $LOAF_0$ from Equation 12.4. It selects and offers a suitable B_0. Then, n_0 broadcasts an authentication request with the parameters of m (l_m, TTL_m, Imp_m, and $Ener_{cons}$) to NNs.

2. The potential sellers offer an asking price:

After n_i ($1 \leqslant i \leqslant N$) receives an authentication request, it calculates C_i and R_i from Equation 12.5, and then price A_i is selected and offered.

3. The buyer selects a sellers' coalition:

Let $C = \{C \in 2^{NNs} | \sum_{n_i \in C} A_i \leqslant B_0 \text{ and } |C| \geqslant minCNN\}$ denote the set of optional sellers' coalitions, where C is a coalition whose members meet the conditions that $\sum_{n_i \in C} A_i \leqslant B_0$ and the number of sellers is not less than $minCNN$. If $|C| \geqslant 1$, then the buyer chooses the coalition $SC = \arg\min_{C^j \in c} \sum_{n_i \in C^j} A_i$ as the sellers' coalition. If $|C| = 0$, then the bargain fails. In this case, to bargain successfully, n_0 can increase B_0 with the constraint that $B_0 \leqslant R_0$.

4. The buyer pays for the authentication service:

If *SC* exists, the bargain is struck at the agreed price *AP* as in Equation 12.6. The buyer pays *AP* for the authentication service.

$$AP = \alpha \times B_0 + (1 - \alpha) \times \sum_{n_i \in SC} A_i, (0 \leqslant \alpha \leqslant 1)$$

$$AS_i = A_i + (AP - \sum_{n_i \in SC} A_i)/|SC| \qquad (12.6)$$

5. The sellers authenticate:

Each cooperative seller $n_i \in SC$ receives an allocation price AS_i, as in Equation 12.6, as payment and authenticates the message of n_0. Other nodes receive nothing.

12.4.2 Dynamic game

We propose a dynamic game for cooperative authentication with consideration of location privacy leakage and resource consumption in a rational environment. Each node seeks to obtain most benefit at least cost.

Definition 12.1 *G = (P, S, U) denotes the dynamic game for cooperative authentication, where **P**, **S**, and **U** are the set of players, strategies, and utility function, respectively.*

12.4.2.1 Players

$\mathbf{P} = \{P_i\}_{i=0}^{N}$ is the set of players, where P_0 denotes n and P_i $(1 \leqslant i \leqslant N)$ represents n_i (where $n_i \in$ NNs).

12.4.2.2 Strategy

$\mathbf{S} = \{s_i\}_{i=0}^{N}$ is the strategy set of all players. s_i $(0 \leqslant i \leqslant N)$ is the strategy of P_i and the strategy set $\mathbf{s}_{-i}$ denotes strategies chosen by other players. For simplicity, $\mathbf{S}$ can be rewritten as $\mathbf{S} = (s_i, \mathbf{s}_{-i})$. In cooperative authentication: (1) when P_0 has m requiring authentication, it has two options: *cooperation* (*CP*), which represents that it requests P_i $(1 \leqslant i \leqslant N)$ to authenticate m, and *noncooperation* (*NC*), which expresses that it refuses to send m to P_i for authentication. (2) When P_i $(1 \leqslant i \leqslant N)$ receives an authentication request, it also has two choices: *cooperation* (*CP*), which indicates that it would like to authenticate m, and *noncooperation* (*NC*), which shows that it rejects the request to authenticate m. Thus, the set S_i $(0 \leqslant i \leqslant N)$ for strategies of P_i is $\{CP, NC\}$.

12.4.2.3 Utility function

$U = \{u_i\}_{i=0}^{N}$ is a set of utility functions, $u_i(s_i, \mathbf{s}_{-i})$ denotes the utility function of P_i under s_i, and $\mathbf{s}_{-i}$. $u_i(s_i, \mathbf{s}_{-i})$ $(0 \leqslant i \leqslant N)$ are defined as Equations 12.7 and 12.8:

$$u_0(s_0, s_{-0}) = \begin{cases} R_0 - AP & \text{if } s_0 = CP \text{ and } \exists SC \\ -C_0 & \text{if } s_0 = CP \text{ and } \not\exists SC \\ -LONA_0 & \text{if } s_0 = NC \end{cases} \quad (12.7)$$

The term $u_0(s_0, \mathbf{s}_{-0})$ in Equation 12.7 shows that: (1) P_0 earns the difference between R_0 and AP when the bargain succeeds; (2) if P_0 chooses CP, but the bargain fails, P_0 should pay C_0 as punishment for the failure caused by its unreasonable offer B_0. This punishment is realistic, as it can make P_0 offer a reasonable B_0 to improve the possibility of a successful bargain; (3) if P_0 chooses NC, P_0 should pay for $LONA_0$

$$u_i(s_i, s_{-i})(1 \leqslant i \leqslant N) = \begin{cases} AS_i - R_i & \text{if } s_i = CP \text{ and } P_i \in SC \\ -C_i & \text{if } s_i = CP \text{ and } P_i \notin SC \\ 0 & \text{if } s_i = NC \text{ or } s_0 = NC \end{cases} \quad (12.8)$$

The term $u_i(s_i, \mathbf{s}_{-i})$ $(1 \leqslant i \leqslant N)$ in Equation 12.8 implies that: (1) If the bargain succeeds and $P_i \in SC$, it receives the utility as the difference between AS_i and R_i; (2) if P_i chooses CP, but $P_i \notin SC$, it pays C_i as punishment for its unreasonable asking price A_i. This punishment has realistic significance in $\mathbf{G}$, as it can inhibit the behavior of too many players trying to participate in cooperation, which results in the consumption of extra resources, and make P_i offer a reasonable A_i to improve the possibility of a successful bargain; (3) if P_i refuses to be cooperative, it receives no utility.

12.5 Analysis of Dynamic Game Model for Cooperative Authentication

In this section, we discuss two situations in $\mathbf{G}$: the dynamic game with complete information (C-$\mathbf{G}$) and incomplete information (I-$\mathbf{G}$), respectively. C-$\mathbf{G}$ requires that each player can observe actions and have common knowledge about the strategy spaces and utility functions of other players. Each player in I-$\mathbf{G}$ knows all strategy types of the "nature" players[5] and the probability corresponding to each type, but it does not know which type the actions of other players belong to.

---

[5] The "nature" player refers to a player who assigns a random variable, which could take values of types for each player, to each player and associates probabilities or a probability density function with those types.

12.5.1 Dynamic game with complete information

In *C*-**G**, each rational player intends to choose the optimal strategy that maximizes its utility.

Definition 12.2   *The best response of P_i to the strategies of other players is a strategy s_i^* such that $s_i^* = \arg \max\limits_{s_i} u_i(s_i, s_{-i})$.*

Definition 12.3   *A strategy profile S_i^* is the Nash equilibrium (NE); if, for each P_i $(0 \leqslant i \leqslant N), u_i(s_i^*, s_{-i}^*) \geqslant u_i(s_i, s_{-i}^*)$.*

In *C*-**G**, each player takes sequential actions according to the serial strategies of other players and follows a *sequential rationality* premise; that is, when the buyer makes a decision, it naturally adjusts its strategy selection according to the actions of the sellers while each seller rationally expects this situation and considers the effect of its strategy selection on the buyer. Such a premise requires that any player should dynamically choose its optimal strategy according to circumstances, rather than sticking to its existing strategy. This derives the concept of the subgame[6] and leads us to the essence of the *subgame perfect Nash equilibrium* (SPNE) [24].

Definition 12.4   *A strategy profile $S^* = (S_1^*, \ldots, S_n^*)$ described by an extensive game tree (also called extensive form) is a SPNE if each subgame of the original game is NE.*

In *C*-**G**, a strategy profile, which consists of a *CP* strategy taken by players who belong to a successful bargaining for more utility and an *NC* strategy taken by other players for less loss, satisfies SPNE.

Theorem 12.1
Let C^k (where $C^k \subset (\boldsymbol{P} - P_0)$) be a set of cooperative players such that $\sum\limits_{P_i \in C^k} A_i \leqslant B_0$ and $|C^k| \geqslant minCNN$. There is a strategy profile s_i^* satisfying SPNE for $C-\boldsymbol{G}$ if there exists such C^k where

$$
s_i^* = \begin{cases} CP \text{ if } P_i = P_0 \\ CP \text{ if } P_i \in SC (where 1 \leqslant i \leqslant N) . \\ NC \text{ else} \end{cases}
$$

---

[6]The subgame of a finite game in extensive form is a part of the original game that consists of an initial node within a singleton information set and all subsequent successors.

Proof. In the condition $\sum_{P_i \in C^k} A_i \leqslant B_0$ and $C^k| \geqslant minCNN$, no player, P_0 or $P_i \in C^k$, has an incentive to unilaterally deviate from cooperation to non-cooperation; then no player who is not in $P_0 \cup C^k$ (the set of such players is denoted as D) unilaterally deviates from noncooperation to cooperation. For P_0, its utility is equal to $u_0 = R_0 - AP > 0$ when its strategy is CP and is greater than the utility $- LONA_0$ when its strategy is noncooperation (because $\sum_{P_i \in C^k} A_i \leqslant B_0$ and $0 \leqslant \alpha \leqslant 1$, thus $AP = \alpha \times B_0 + (1 - \alpha) \times \sum_{n_i \in C^k} A_i \leqslant B_0$; and as $R_0 \geqslant B_0$ and $LONA_0 > 0$, so $R_0 - AP > 0 > -LONA_0$). So the best strategy for P_0 is CP. For any player $P_i \in C^k$, its utility is $u_i = AS_i - R_i = A_i + (\alpha \times B_0 + (1 - \alpha) \times \sum_{n_i \in C^k} A_i - \sum_{n_i \in C^k} A_i) / |C^k| - R_i = A_i - R_i + (\alpha(B_0 - \sum_{n_i \in C^k} A_i)) / |C^k| > 0$ (because $\sum_{P_i \in C^k} A_i \leqslant B_0$ and $R_i \leqslant A_i$) when it chooses the cooperative strategy; while if its strategy is NC, its utility is equal to zero. So, it does not unilaterally deviate from cooperation to noncooperation. Similarly, if any player $P_i \in D$ unilaterally changes its strategy from NC to CP, then its utility $u_i = -C_i < 0$, because $C_i > 0$ is always smaller than zero, which equals the utility for its NC strategy. Hence, no player unilaterally changes its strategy to gain more utility and the strategy profile s_i^* achieves SPNE when $|C^k| > 1$.

Note: The precondition of Theorem 12.1 is that C^k exists. However, for the situation that C^k does not exist, P_i $(1 \leqslant i \leqslant N)$ always selects the NC strategy for more utility, while the decision of P_0 varies with its cost price and $LONA_0$. So, we are able to deduce two lemmas.

Lemma 12.1

Let C^k (where $C^k \subset (\mathbf{P} - P_0)$) be a set of cooperative players such that $\sum_{P_i \in C^k} A_i \leqslant B_0$ and $|C^k| \geqslant minCNN$. There is a strategy profile s_i^* satisfying SPNE for $C-\mathbf{G}$ if such C^k does not exist and $LONA_0 > C_0$, where

$$s_i^* = \begin{cases} CP & \text{if } P_i = P_0 \\ NC & \text{if } P_i(1 \leqslant i \leqslant N) \end{cases}$$

Proof. Similar to the proof of Theorem 12.1, P_0 does not unilaterally deviate from cooperation to noncooperation, as its utility $u_i = -LONA_0$ would be less than C_0 (because $LONA_0 > C_0 > 0$). Similarly, there is no player P_i $(1 \leqslant i \leqslant N)$ that has an incentive to unilaterally change its strategy from noncooperation to cooperation as its utility $u_i = -C_i < 0$ (because $C_i > 0$) is always smaller than zero, which equals the utility for its NC strategy. Hence, no player unilaterally changes its strategy to gain more utility and the strategy profile s_i^* achieves SPNE.

Lemma 12.2

Let C^k (where $C^k \subset (P - P_0)$) be a set of cooperative players such that $\sum\limits_{P_i \in C^k} A_i \leqslant B_0$

and $|C^k| \geqslant minCNN$. There is a strategy profile s_i^* satisfying SPNE for C-G if such C^k does not exist and

$$LONA_0 \leqslant C_0, \text{ where } s_i^* = \begin{cases} NC & \text{if } P_i = P_0 \\ NC & \text{if } P_i(1 \leqslant i \leqslant N) \end{cases}$$

Proof. Similarly to the proof of Lemma 12.1, P_0 does not unilaterally change its strategy from noncooperation to cooperation, as its utility $u_i = -C_0$ would be less than $-LONA_0$ (because $LONA_0 \leqslant C_0$). Similarly, for any player P_i ($1 \leqslant i \leqslant N$), its utility is $u_i = -C_i < 0$ (because $C_i > 0$) if it chooses the cooperative strategy CP; while, if its strategy is NC, then its utility is equal to zero. So, there is no player P_i ($1 \leqslant i \leqslant N$) that has an incentive to unilaterally change its strategy from noncooperation to cooperation. Hence, no player unilaterally changes its strategy to gain more utility and the strategy profile s_i^* achieves SPNE.

In **C-G**, each player performs sequential actions according to others' serial strategies, and implausible Nash equilibriums (incredible threats and promises) arising in the static game with perfect and complete information would be eliminated by using the concept of SPNE. So, a single SPNE will certainly be reached and always be selected to maximize the players' utility. Based on these analyses, we propose a dynamic game algorithm with complete information as Algorithm 12.1.

12.5.2 Dynamic game with incomplete information

Before playing the game, each player in *I*-**G** establishes its own preliminary judgment according to all the strategy types of the other players and the probability distribution corresponding to each type. When playing the game, each player can obtain practical information on what action to take by observing the actions of other players, and then correct its initial judgments and choose its optimal strategy according to such changes in judgments. This derives the concept of Bayesian inference, formulated as Equation 12.9, and leads us to the essence of the perfect Bayesian Nash equilibrium (PBNE).

$$Prob_i\left(\theta_j \mid a_j^h\right) = \frac{Prob\left(a_j^h, \theta_j\right)}{\sum\limits_{\tilde{\theta}_j \in \Theta_j} Prob\left(a_j^h \mid \tilde{\theta}_j\right) Prob\left(\tilde{\theta}_j\right)} \tag{12.9}$$

where $Prob\left(\tilde{\theta}_j\right)$ is the probability that P_j is of type θ_j and determined by "nature". P_j takes action a_j^h with the probability $Prob\left(a_j^h \mid \theta_j\right)$ when it is of type θ_j. If P_i observes an action a_j^h of P_j at information set h_i, then we can derive the beliefs that $Prob_i\left(\theta_j \mid a_j^h\right)$ (also denoted as $\tilde{Prob}_{ih}\left(\theta_j\right)$) of P_i in P_j is of type θ_j with the condition of action a_j^h at h_i, as in Equation 12.9.

Algorithm 12.1: Dynamic Game Algorithm with Complete Information for Cooperative Authentication

---

Required Parameters: Given message m with l_m, TTL_m and Imp_m requiring authentication, n_0 selects a suitable PCA and calculates $minCNN$ f Equation 12.1 and the coefficients α, pl, wl are selected; n_0 chooses suitable weights w_{C_0}, w_{R_0}, w_{B_0}, w_{L_0}; each n_i $(1 \leqslant i \leqslant N)$ selects suitable weights w_{C_i}, v_{C_i}, w_{R_i}, w_{A_i}.

1. n_0 calculates $Ener_{cons}$ and BW_m, FL_0 from Equation 12.3 and C_0, R_0, and $LONA_0$ from Equation 12.4; n_0 broadcasts an authentication request with parameters $(m, l_m, TTL_m, Imp_m, Ener_{cons})$ to NNs.

2. For each $n_i \in$ NNs, n_i collects the parameters ($Ener_i$, BW_i, $Priv^i_{cons}$, $DLPP_i$, FL_i) and calculates C_i and R_i from Equation 12.5.

3. n_0 calculates and submits B_0 by Equation 12.4. Each $n_i \in$ NNs calculates and submits A_i from Equation 12.5.

4. Let $C = \{C \in 2^{NNs} \,\|\, C \geqslant |minCNN \text{ and } \sum_{n_i \in C} A_i \leqslant B_0\}$, $SC = \underset{C^j \in c}{\arg\min} \sum_{n_i \in C^j} A_i$.

5. If $|C \geqslant| \geqslant 1$, then $S_0^* = CP$ and $S_i^* = CP$ $n_i \in SC$, $S_i^* = NC(n_i \notin SC)$; otherwise, $S_i^* = NC$ and $S_0^* = CP(LONA_0 > C_0)$, $S_0^* = NC(LONA_0 \leqslant C_0)$.

6. If $|C \geqslant| \geqslant 1$, a bargain is concluded at AP, authenticating m and allocating the utility AS_i to $n_i \in SC$ according to Equation 12.6; otherwise, the bargain fails.

Definition 12.5 *In I-G, a belief profile $\tilde{P}rob = (\tilde{P}rob_1, \ldots, \tilde{P}rob_n)$ and a type-dependent strategy profile $S^*(\theta_1, \ldots, \theta_n) = (S_1^*(\theta_1), \ldots, S_n^*(\theta_n))$ constitute a PBNE if, for each P_i $(0 \leqslant i \leqslant n)$ at information set h,*

$$S_i^*(\theta_i)|_h = \underset{S_i(\theta_i)|_h}{\arg\max} \sum_{\theta_{-i}} \tilde{P}rob_i\left(\theta_{-i}|a^h_{-i}\right) u_i\left(S_i(\theta_i)|_h, S^*_{-i}(\theta_{-i}), \theta_i, \theta_{-i}\right)$$

where:

$\tilde{P}rob$ is a set of the prior probabilities $Prob_i(\theta_{-i}|\theta_i)$, so $\tilde{P}rob_i = Prob_i(\theta_{-i}|\theta_i)$ is the profile consisting of all beliefs $\tilde{P}rob_{ih}$ of P_i at the information set h

Θ_i is the type space of P_i

$\theta_i \in \Theta_i$ is a type of P_i

$u_i = u_i(S_1^*(\theta_1), \ldots, S_n^*(\theta_n), \theta_i, \theta_{-i})$ is the type-dependent utility function of P_i

In summary, PBNE combines the strategies with the beliefs of all players in the game. One player chooses its optimal strategy according to the given beliefs of each player concerning the types of the other players. On the equilibrium

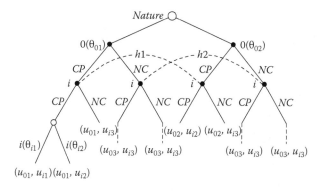

Figure 12.2: Extensive form of *I-G*. While $u_{01} = R_0 - AP, u_{02} = -C_0,$ $u_{03} = -LOAF_0, u_{i1} = AS_i - R_i,$ $u_{i2} = -Ci, u_{i3} = 0.$

path, $\tilde{Prob}_{ih}$ can be derived from observed information ($Prob_i(\theta_{-i}|\theta_i)$, a_{-i}^h and $S_{-i}^*(\theta_{-i})$) and from Equation 12.9.

The *I-G* can be represented as an extensive form as shown in Figure 12.2; both h_1 and h_2 are the information set of $P_i(1 \leqslant i \leqslant N)$. From Equation 12.7, we know that the strategy for P_0 is related to its B_0, the sum of all A_i and $|C|$. We can obtain type-dependent strategies $S_0^*(\Theta_0)$ of P as $S_0^*(\theta_{01}) = CP$, $S_0^*(\theta_{02}) = NC$ where θ_{01} is $\{\sum_{P_i \in SC} A_i \leqslant B_0 \text{ and } |SC| \geqslant minCNN\}$ and θ_{02} is $\{\sum_{P_i \in SC} A_i > B_0 \text{ or } |SC| < minCNN\}$. We can derive the belief of P_i in the type of player P_0 at a given information set and type-dependent strategies by applying Bayes' rule, as in Equation 12.10.

$$\tilde{Prob}_{ih1} = \left(\tilde{Prob}_i\left(\theta_{01}|CP_0^{h_1}\right), \tilde{Prob}_i\left(\theta_{02}|CP_0^{h_1}\right) \right)$$
$$\tilde{Prob}_{ih2} = \left(\tilde{Prob}_i\left(\theta_{01}|NC_0^{h_2}\right), \tilde{Prob}_i\left(\theta_{02}|NC_0^{h_2}\right) \right)$$

(12.10)

In the same way, the strategy for player P_i is also related to its asking price. We obtain type-dependent strategies $S_i^*(\Theta_i)$ of P_i as $S_i^*(\theta_{i1}) = CP$ $S_i^*(\theta_{i2}) = NC$, where θ_{i1} is

$$\{\sum_{P_i \in SC} A_i \leqslant B_0 \ \& \ |SC| \geqslant minCNN \ \& \ A_i \leqslant (B_0 - \sum_{P_j \in SC, j \neq i} A_j)\}$$

and θ_{i2} is

$$\{\sum_{P_i \in SC} A_i > B_0 \text{ or } |SC| < minCNN \text{ or } A_i > (B_0 - \sum_{P_j \in SC, j \neq i} A_j)\}.$$

We can derive the belief of P in the type of P_i at a given information set f_1, f_2 and type-dependent strategies as in Equation 12.11.

$$\tilde{Prob}_{0f_1} = \left(\tilde{Prob}_0 \left(\theta_{i1} | CP_i^{f_1} \right), \tilde{Prob}_0 \left(\theta_{i2} | CP_i^{f_1} \right) \right)$$
$$\tilde{Prob}_{0f_2} = \left(\tilde{Prob}_0 \left(\theta_{i1} | NC_i^{f_2} \right), \tilde{Prob}_0 \left(\theta_{i2} | NC_i^{f_2} \right) \right)$$

(12.11)

So, we are able to deduce the following theorem.

Theorem 12.2

*In I-**G**, there is a strategy profile*

$$S_i^* = \begin{cases} CP & if \ P_i = P_0 \ \& \ BP \\ CP & if \ P_i \in NNs \ \& \ BP_i \\ NC & else \end{cases}$$

that results in a PBNE
where:

$$BP_0 \equiv \left(\begin{array}{l} \tilde{Prob}_0 \left(\theta_{i1} | CP_i^{f_1} \right) > (C_0 - LONA_0) / \left(R_0 - EAP^{f_1} + C_0 \right) \ \& \\ \tilde{Prob}_0 \left(\theta_{i1} | NC_i^{f_2} \right) > (C_0 - LONA_0) / \left(R_0 - EAP^{f_1} + C_0 \right) \end{array} \right)$$

$$BP_i \equiv \left(\tilde{Prob}_i \left(\theta_{01} | CP_0^{h_1} \right) > C_i / \left(Prob_i \left(ISC \right) \left(EAS_i - R_i + C_i \right) \right) \right)$$

$$Prob_i (ISC) = Pr \left(\sum_{P_i \in SC} A_i \leqslant B_0 \ \& \ |SC| \geqslant minCNN \ \& \ A_i \leqslant \left(B_0 - \sum_{\substack{P_j \in SC, \\ j \neq i}} A_j \right) \right)$$

is the probability with which
    P_i *belongs to the sellers' coalition*
    EAS_i *is the except value of* AS_i
    EAP^{f_1} *and* EAP^{f_2} *are the except values of AP at the given* f_1 *and* f_2, *respectively*

Proof. From Equation 12.7, we know that the strategy for P_0 is related to its B_0, the sum of all A_i, and the number $|C|$ of cooperative nodes. We obtain type-dependent strategies $S_0^* (\Theta_0)$ of P_0 as $S_0^* (\theta_{01}) = CP$, $S_0^* (\theta_{02}) = NC$, where θ_{01} is

$$\left\{ \sum_{P_i \in C^k} A_i \leqslant B_0 \ and \ |C| \geqslant minCNN \right\} \ and \ \theta_{02} \ is \ \left\{ \sum_{P_i \in C^k} A_i > B_0 \ or \ |C| < minCNN \right\}.$$

Given the information set h_1, h_2 of $P_i (1 \leqslant i \leqslant N)$, we can derive the belief of P_i about the type of player P at a given information set and type-dependent strategies by applying Bayes' rule, as in Equation 12.12.

$$\tilde{Prob}_{ih1} = (\tilde{Prob}_i(\theta_{01}|CP_0^{h_1}), \tilde{Prob}_i(\theta_{02}|CP_0^{h_1}))$$
$$\tilde{Prob}_{ih2} = (\tilde{Prob}_i(\theta_{01}|NC_0^{h_2}), \tilde{Prob}_i(\theta_{02}|NC_0^{h_2}))$$
(12.12)

We assume $Prob_i(ISC)$ is the probability that P_i belongs to the sellers' coalition. For a given $S_0^*(\Theta_0)$, the expected utility of player P_i when it chooses strategy CP at information set h_1 is as in Equation 12.13:

$$u_i(CP) = \tilde{Prob}_i\left(\theta_{01}|CP_0^{h_1}\right) \left(\begin{array}{c} Prob_i(ISC)(EAS_i - R_i) + \\ (1 - Prob_i(ISC))(-C_i) \end{array} \right)$$
$$+ \tilde{Prob}_i\left(\theta_{02}|CP_0^{h_1}\right)(-C_i)$$
(12.13)

where EAS_i is the except value of AS_i at a given information set h_1. Similarly, the expected utility of P_i when it chooses strategy NC at information set h_1 is as in Equation 12.14:

$$u_i(NC) = \tilde{Prob}_i\left(\theta_{01}|CP_0^{h_1}\right)(0) + \tilde{Prob}_i\left(\theta_{02}|CP_0^{h_1}\right)(0) = 0$$
(12.14)

For a given $S_0^*(\Theta_1)$, the expected utility of player P_i when it chooses strategy CP at information set h_2 is as in Equation 12.15:

$$u_i(CP) = \tilde{Prob}_i\left(\theta_{01}|NC_0^{h_2}\right)(0) + \tilde{Prob}_i\left(\theta_{02}|NC_0^{h_2}\right)(0) = 0$$
(12.15)

Similarly, the expected utility of player P_i when it chooses strategy NC at information set h_2 is as in Equation 12.16:

$$u_i(NC) = \tilde{Prob}_i\left(\theta_{01}|NC_0^{h_2}\right)(0) + \tilde{Prob}_i\left(\theta_{02}|NC_0^{h_2}\right)(0) = 0$$
(12.16)

When $u_i(CP) \geqslant u_i(NC)$, P_i would choose CP for more utility. To make sure the player does not deviate from its strategy of CP at information set h_1, h_2 and type-dependent strategies $S_0^*(\Theta_0)$, the condition required is as in Equation 12.17:

$$\tilde{Prob}_i\left(\theta_{01}|CP_0^{h_1}\right) > C_i / (Prob_i(ISC)(EAS_i - R_i + C_i))$$
(12.17)

So, for a given information set h_1, h_2 and type-dependent strategies $S_0^*(\Theta_0)$, the strategy

$$S^* = \left[\begin{array}{c} (CP, NC), (CP, NC), (Prob(\theta_{01}), Prob(\theta_{02})), \\ \tilde{Prob}_i\left(\theta_{01}|CP_0^{h_1}\right) > C_i / (Prob_i(ISC)(EAS_i - R_i + C_i)), \\ \tilde{Prob}_i\left(\theta_{01}|CP_0^{h_1}\right) \leqslant C_i / (Prob_i(ISC)(EAS_i - R_i + C_i)) \end{array} \right]$$

is a PBNE.

In the same way, the strategy for player P_i is also related to its asking price A_i, B_0, the sum of all A_i, and the number $|C|$ of cooperative nodes. We obtain type-dependent strategies $S_i^*(\Theta_i)$ of P_i as

$$S_i^*(\theta_{i1}) = CPS_i^*(\theta_{i2}) = NC$$

where θ_{i1} is

$$\left\{ \sum_{P_i \in C^k} A_i \leqslant B_0 \ \& \ |C| \geqslant minCNN \ \& \ A_i \leqslant \left(B_0 - \sum_{P_j \in C^k, j \neq i} A_j \right) \right\} \frac{-b \pm \sqrt{b^2 - 4ac}}{2a}$$

and θ_{i2} is

$$\left\{ \sum_{P_i \in C^k} A_i > B_0 \ or \ |C| < minCNN \ or \ A_i > \left(B_0 - \sum_{P_j \in C^k, j \neq i} A_j \right) \right\}$$

We can derive the belief of P_0 about the type of player P_i at a given information set f_1, f_2 and type-dependent strategies by applying Bayes' rule, as in Equation 12.18:

$$\tilde{Prob}_{0f_1} = \left(\tilde{Prob}_0 \left(\theta_{i1} | CP_i^{f_1} \right), \tilde{Prob}_0 \left(\theta_{i2} | CP_i^{f_1} \right) \right)$$
$$\tilde{Prob}_{0f_2} = \left(\tilde{Prob}_0 \left(\theta_{i1} | NC_i^{f_2} \right), \tilde{Prob}_0 \left(\theta_{i2} | NC_i^{f_2} \right) \right) \tag{12.18}$$

The expected utility of player P_0 when it chooses strategy CP at information set f_1 is as in Equation 12.19:

$$u_0(CP) = \tilde{Prob}_0 \left(\theta_{i1} | CP_i^{f_1} \right) \left(R_0 - EAP^{f_1} \right) + \tilde{Prob}_0 \left(\theta_{i2} | CP_i^{f_1} \right) (-C_0) \tag{12.19}$$

where EAP^{f_1} is the except value of AP at the given information set f_1.

Similarly, the expected utility of player P_0 when it chooses strategy NC at information set f_1 is as in Equation 12.20:

$$u_0(NC) = \tilde{Prob}_0 \left(\theta_{i1} | CP_i^{f_1} \right) (-LONA_0) + \tilde{Prob}_0 \left(\theta_{i2} | CP_i^{f_1} \right) (-LONA_0) \tag{12.20}$$

The expected utility of player P_0 when it chooses strategy CP at information set f_2 is as in Equation 12.21:

$$u_0(CP) = \tilde{Prob}_0 \left(\theta_{i1} | NC_i^{f_2} \right) \left(R_0 - EAP^{f_2} \right) + \tilde{Prob}_0 \left(\theta_{i2} | NC_i^{f_2} \right) (-C_0) \tag{12.21}$$

where EAP^{f_2} is the except value of AP at the given information set f_2.

Similarly, the expected utility of player P_0 when it chooses strategy NC at information set f_2 is as in Equation 12.22:

$$u_0(NC) = \tilde{Prob}_0 \left(\theta_{i1} | NC_i^{f_2} \right) (-LONA_0) + \tilde{Prob}_0 \left(\theta_{i2} | NC_i^{f_2} \right) (-LONA_0) \tag{12.22}$$

When $u_0(CP) \geqslant u_0(NC)$, P_0 chooses CP for more utility. In order to make sure the player does not deviate from its strategy of CP at information set f_1, f_2 and type-dependent strategies $S_i^*(\Theta_i)$, the condition required is as in Equation 12.23.

$$\tilde{Prob}_0\left(\theta_{i1}|CP_i^{f_1}\right) > (C_0 - LONA_0)/\left(R_0 - EAP^{f_1} + C_0\right) \text{ and}$$
$$\tilde{Prob}_0\left(\theta_{i1}|NC_i^{f_2}\right) > (C_0 - LONA_0)/\left(R_0 - EAP^{f_2} + C_0\right) \tag{12.23}$$

So, for given values of f_1, f_2 and $S_i^*(\Theta_i)$, the following strategy is a PBNE:

$$S^* = \begin{bmatrix} (CP,NC),(CP,NC),(Prob\,(\theta_{i1})),(Prob\,(\theta_{i2})), \\[2mm] \tilde{Prob}_0\left(\theta_{i1}|CP_i^{f_1}\right) > (C_0 - LONA_0)/\left(R_0 - EAP^{f_1} + C_0\right) \\[2mm] \tilde{Prob}_0\left(\theta_{i1}|NC_i^{f_2}\right) > (C_0 - LONA_0)/\left(R_0 - EAP^{f_2} + C_0\right), \\[2mm] \tilde{Prob}_0\left(\theta_{i1}|CP_i^{f_1}\right) \leqslant (C_0 - LONA_0)/\left(R_0 - EAP^{f_1} + C_0\right) \| \\[2mm] \tilde{Prob}_0\left(\theta_{i1}|NC_i^{f_2}\right) \leqslant (C_0 - LONA_0)/\left(R_0 - EAP^{f_2} + C_0\right) \end{bmatrix}$$

Therefore, we are able to deduce the theorem that the strategy

$$S_i^* = \begin{cases} CP & \text{if } P_i = P_0 \ \& \ BP \\ CP & \text{if } P_i \in NNs \ \& \ BP_i \\ NC & \text{else} \end{cases}$$

results in a PBNE.

On the whole, the PBNE can help a player in *I*-**G** to decide whether or not to participate in cooperation and to maximize its utility based on its belief about the types of other players. The belief can be obtained from the given information set (such as the history record of observed actions, the probability distributions of types) by applying Bayes' rule. Based on the above analysis, a dynamic game algorithm with incomplete information for *I*-**G** is designed as shown in Algorithm 12.2.

12.6 Experimental Results

In this section, we evaluate the performance of the bargaining-based dynamic game model for cooperative authentication by using a MATLAB® simulation.

Algorithm 12.2: Dynamic Game Algorithm with Incomplete Information for Cooperative Authentication

---

Required Parameters: Suitable coefficients α, pl, wl are selected. The information set (f_1, f_2, h_1, h_2), type-dependent strategies $(S_0^*(\Theta_0), S_i^*(\Theta_i))$, and probability distributions of B_0 and A_i are given. Given message m with l_m, TTL_m and Imp_m requiring authentication, n_0 selects a suitable PCA and calculates $minCNN$ from Equation 12.1 and chooses suitable weights $w_{C_0}, w_{R_0}, w_{B_0}, w_{L_0}$; each $n_i (1 \leqslant i \leqslant N)$ selects suitable weights $w_{C_i}, v_{C_i}, w_{R_i}, w_{A_i}$.

1. n_0 calculates $Ener_{cons}$ and BW_m, FL_0 from Equation 12.3, C_0, R^0, and LNA_0 from Equation 12.4, and then predicts the beliefs $\tilde{Prob}_{0f1}$ and $\tilde{Prob}_{0f2}$; n_0 broadcasts an authentication request with parameters $(m, l_m, TTL_m, Imp_m, Ener_{cons})$ to NNs.

2. For each $n_i \in$ NNs:

   n_i collects parameters $(Ener_i, BW_i, Priv^i_{cons}, DLPP_i, FL_i)$, calculates C_i and R_i from Equation 12.5, and then predicts the beliefs $\tilde{Prob}_{ih1}$ and $\tilde{Prob}_{ih2}$.

3. n_0 selects and submits B_0 by Equation 12.4. Each $n_i \in$ NNs selects and submits A_i by Equation 12.5.

4. Let $C = \{C \in 2^{NNs} \,||\, C \geqslant |minCNN \text{ and } \sum_{n_i \in C} A_i \leqslant B_0\}$, $SC = \arg\min_{C^j \in c} \sum_{n_i \in C^j} A_i$

5. If BP_0 is true, $S_0^* = CP$;

   otherwise $S_0^* = NC$;

   if BP_i is true, $S_i^* = CP$;

   otherwise $S_i^* = NC$.

6. If $|C| \geqslant 1$, a bargain is concluded at AP, authenticating m and allocating the utility AS_i to $n_i \in SC$ according to Equation 12.6; otherwise, the bargain fails.

In our simulation study, we consider a network topology where 2000 nodes with a transmission range $R = 50$ are randomly distributed in an area of 1000 m $\times$ 1000 m. Given PCA = 99.8% and $\rho = 2\%$, we set $minCNN = 8$ using Equation 12.1, which means that at least eight NNs should be encouraged to participate in cooperation. In order to demonstrate that our scheme can effectively decrease the leakage of location privacy and resource consumption, we compare it with two schemes. One is "all nodes cooperate," where all nodes participate in cooperation and the other is "nodes randomly cooperate," where nodes choose to participate in cooperation randomly. In the simulation, we set coefficients w_{B_0} and w_{A_i}, which follow B(6,6) and N(2,1), respectively.

12.6.1 Location privacy leakage

Regarding the aspect of location privacy leakage, the simulation results, as shown in Figure 12.3, demonstrate that the average privacy decreases near-linearly with an increasing number of successful cooperative authentications in the three strategies. The indicator in our strategy decreases at far lower speed than in the other strategies.

12.6.2 Resource consumption

Regarding the aspect of resource consumption, the simulation results, as shown in Figure 12.4, indicate that the average energy decreases near-linearly with an

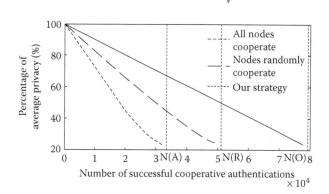

Figure 12.3: Average privacy varies with the number of successful cooperative authentications in the three strategies.

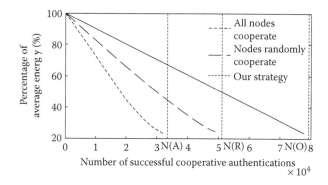

Figure 12.4: Average energy varies with the number of successful cooperative authentications in the three strategies.

increasing number of successful cooperative authentications in the three strategies. The indicator in our strategy decreases at far lower speed than in the other strategies.

From Figures 12.3 and 12.4, the average privacy and energy in our strategy drop to 20% after N(O) times of successful cooperative authentication, while the numbers in the "all nodes cooperate" scheme and "nodes randomly cooperate" scheme are N(A) and N(R), respectively. We can calculate that N(O) > N(R) > N(A). The reason is that, in our strategy, the node takes into account and measures the loss of location privacy leakage and resource consumption when it calculates utility, and then decides whether or not to participate in cooperation according to such utility. In the other two schemes, no approach is adopted to reduce the leakage of location privacy and resource consumption.

12.6.3   *Network survival*

In MANETs, survival is an important indicator due to limited resources. In our simulation, given node n_i, if $DLPP_i \geqslant DLPP_{min}$, $Ener_i \geqslant Ener_{min}$ and $FL_i \geqslant FL_{min}$, we call n_i a survival node. We evaluate the performance of the survival nodes and the network lifetime in our strategy.

The percentage of survival nodes varies with the number of initiating cooperative authentication in the three strategies and the network lifetime of the three different strategies.

As shown in Figure 12.5a, we can calculate that the percentage of survival nodes rapidly decreases with an increasing number of initiating cooperative authentications in the three strategies. The indicator in our strategy decreases at far lower speed than the other strategies. The simulation results imply that there are more survival nodes in our model after the same number of initiating cooperative authentications (N(R)) than in the other strategies. So, an initiating cooperative authentication in our model will be successful with higher probability after (N(R)) occurrences of cooperative authentication than in the other strategies.

Figure 12.5b expresses the network lifetime in the three strategies. The network lifetime in our model is more than twice as long as that in the "all nodes cooperate" scheme and 60% longer than in the "nodes randomly cooperate" scheme. The simulation results demonstrate that in contrast with the "all nodes cooperate" scheme and the "nodes randomly cooperate" scheme, our proposed strategy can increase the success rate of cooperative authentications and extend the network lifetime.

12.7   Conclusion

With the conflict between the improvement of the probability of correct authentication and the noncooperation of neighbor nodes caused by location privacy

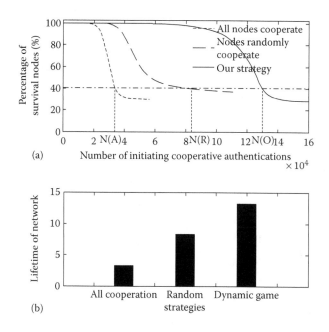

Figure 12.5: Compare the survival nodes and network lifetimes of three strategies.

leakage and resource consumption when it participates in authentication, two key issues, the incentive strategy of willingness to cooperate and the balance of conflict, need to be solved. We discussed the incentive strategic of cooperative authentication with a dynamic game and considered nodes as rational individuals that decide locally whether or not to participate in cooperation. In this chapter, we proposed a dynamic game model for bargaining-based cooperative authentication. It includes an improved bargaining-based cooperative authentication mechanism that is inherited from our previous research to encourage nodes to participate in cooperation, and a dynamic game for cooperative authentication to analyze dynamic behaviors of all nodes and help nodes decide whether or not to participate in cooperation. We discussed two variations of the dynamic game, with complete and incomplete information, respectively. The SPNEs are obtained to guide the node to choose its optimal strategy to maximize its utility under complete information. As nodes do not know others' utility in reality, then the dynamic game with incomplete information is considered and the perfect Bayesian equilibrium is established to eliminate the implausible equilibriums. We designed two algorithms based on the analysis results and the simulation results demonstrate that our algorithms realized the goal that nodes participating in cooperation will maximize their location privacy and minimize their resource consumption while increasing the probability of correct authentication. They can improve the success rate of cooperative authentication, and extend the network lifetime to 160%–360.6% of the present value.

In future, more novel strategies should be studied to improve the performance of cooperative authentication, such as predicting the players' strategy by observing and learning from opponents' historical actions rather than by the probability distribution of types.

Bibliography

[1] C. E. Perkins. *Ad Hoc Networking*. New York: Addison-Wesley Professional, 2008.

[2] N. Devi. Mobile ad-hoc networks for wireless systems. *International Journal of Computer Science & Communication*, 3(1): 245–248, 2012.

[3] N. Islam and Z. A. Shaikh. Security issues in mobile ad hoc network. In S. Khan and A.-S.K. Pathan (Eds), *Wireless Networks and Security, SCT*, Berlin: Springer-Verlag, 49–80, 2013.

[4] R. Lu, X. Lin, H. Zhu, X. Liang, and X. Shen. BECAN: A bandwidth-efficient cooperative authentication scheme for filtering injected false data in wireless sensor networks. *IEEE Transactions on Parallel and Distributed Systems*, 23(1): 32–43, 2012.

[5] Y. Hao, T. Han, and Y. Cheng. A cooperative message authentication protocol in VANETs. In *IEEE Global Communications Conference*, GlobeCom, IEEE, Anaheim, CA, 5562–5566, 2012.

[6] D. Nyang and A. Mohaisen. Cooperative public key authentication protocol in wireless sensor network. In *International Conference on Ubiquitous Intelligence and Computing, UIC*, Beijing, 864–873, 2006.

[7] X. Zhu, S. Jiang, L. Wang, and H. Li. Efficient privacy-preserving authentication for vehicular ad hoc networks. *IEEE Transactions on Vehicular Technology*, 63(2): 907–919, 2014.

[8] W. Shen, L. Liu, X. Cao, Y. Hao, and Y. Cheng. Cooperative message authentication in vehicular cyber-physical systems. *IEEE Transactions on Emerging Topics in Computing*, 1(1): 84–97, 2013.

[9] X. Lin and X. Li. Achieving efficient cooperative message authentication in vehicular ad hoc networks. *IEEE Transactions on Vehicular Technology*, 62(7): 3339–3348, 2013.

[10] A. Vijayakumar and K. Selvamani. Node cooperation and message authentication in trusted mobile ad hoc networks. *International Journal of Engineering and Technology*, 6(1): 388–397, 2014.

[11] X. Li, X. Liang, R. Lu, S. He, J. Chen, and X. Shen. Toward reliable actor services in wireless sensor and actor networks. In *IEEE International Conference on Mobile Ad-Hoc and Sensor Systems, MASS*, IEEE, Valencia, 351–360, 2011.

[12] H. Moustafa, I. Moulineaux-cedex, and G. Bourdon. Authentication and services access control in a cooperative ad hoc environment. In *International Conference on Broadband Communications, Networks and Systems, IEEE Broadnets*, Raleigh, NC, 38–40, 2008.

[13] Y. Hao, Y. Cheng, Z. Chi, and S. Wei. A distributed key management framework with cooperative message authentication in VANETs. *IEEE Journal on Selected Areas in Communications*, 29(3): 616–629, 2011.

[14] R. Shokri, G. Theodorakopoulos, C. Troncoso, J.-P. Hubaux, and J.-Y. Le Boudec. Protecting location privacy: Optimal strategy against localization attacks. In *ACM Conference on Computer and Communications Security, CCS*, Denver, CO, 617–627, 2012.

[15] J. Freudiger, M. H. Manshaei, J.-P. Hubaux, and D. C. Parkes. Non-cooperative location privacy. *IEEE Transactions on Dependable and Secure Computing*, 10(2): 84–98, 2013.

[16] D. Zhang, R. Shinkuma, and N. Mandayam. Bandwidth exchange: an energy conserving incentive mechanism for cooperation. *IEEE Transactions on Wireless Communications*, 9(6): 2055–2065, 2010.

[17] C. Zhang, X. Zhu, Y. Song, and Y. Fang. C4: A new paradigm for providing incentives in multi-hop wireless networks. In *INFOCOM, 2011 Proceedings IEEE*, University of Florida, Gainesville, FL, 918–926, 2011.

[18] M. Guizani, A. Rachedi, and C. Gueguen. Incentive scheduler algorithm for cooperation and coverage extension in wireless networks. *IEEE Transactions on Vehicular Technology (TVT)*, 62(2): 797–808, 2013.

[19] M. T. Refaei, L. A. Dasilva, M. Eltoweissy, and T. Nadeem. Adaptation of reputation management systems to dynamic network conditions in ad hoc networks. *IEEE Transactions on Computers*, 59(5): 707–719, 2010.

[20] M. H. Manshaei, Q. Zhu, T. Alpcan, T. Basar, and J.-P. Hubaux. Game theory meets network security and privacy. *ACM Computing Surveys (CSUR)*, 45(3): 1–39, 2013.

[21] T. Chen, L. Zhu, F. Wu, and S. Zhong. Stimulating cooperation in vehicular ad hoc networks: A coalitional game theoretic approach. *IEEE Transactions on Vehicular Technology*, 60(2): 566–579, 2011.

[22] G. Yunchuan, Y. Lihua, L. Licai, and F. Bingxing. Utility-based cooperative decision in cooperative authentication. In *IEEE International Conference on Computer Communications, INFOCOM*, IEEE, Toronto, ON, 1006–1014, 2014.

[23] R. Shokri, G. Theodorakopoulos, J.-Y. Le Boudec, and J.-P. Hubaux. Quantifying location privacy. In *IEEE Symposium on Security and Privacy, SP*, IEEE, Berkeley, CA, 247–262, 2011.

[24] M. J. Osborne. *Introduction to Game Theory*. New York: Oxford University Press, 2009.

Chapter 13

Authentication in IoT

Hong Liu

CONTENTS

> The raw math of shifting from IPv4 to IPv6 enables the universe of connectivity to go from the size of a golf ball to the size of the sun. Today, we have not proved our ability to manage security for a golf ball. What are we going to do when we inhabit the "sun"—when everything around us is a connection point, and thus an entry point for an attacker?
>
> **Emily Frye**
>
> *Principal Engineer with MITRE Corporation, Opening statement for the Cyber Security Panel at the 2015 IEEE International Symposium on Technologies for Homeland Security*

Authentication defends the universe of connectivity against attackers by verifying identities at entry points to manage security. This identification applies to both entities that manipulate data and information itself that data carry. Communicating entities should identify one another. Information exchanged during communication should be validated as regards its origin, time, content, and so on. Therefore, authentication is usually divided into two major classes: entity authentication and message authentication.

This chapter, after explaining the fundament of authentication, considers entity authentication and message authentication pertinent to the Internet of Things (IoT) that connects everything around us. For each class, a case study of IoT applications for industries such as transportation or healthcare is examined. The last section considers key management in a body area network (BAN), an IoT application for healthcare. Authentication is often supported with encryption techniques, which in turn require key management. Symmetric-key cryptography has to establish a shared secret key between the two parties who wish to communicate confidentially, and the knowledge of the shared secret can serve the purpose of authenticating the participants' identities. Asymmetric-key cryptography involves a trusted third party to bind the identity of an entity to its public key for other entities to communicate with it confidentially, and the binding serves as the certificate to authenticate the entity. These traditional key management methods are not suitable for a BAN due to the limitation of computation resources and power consumption. The demand for a high level of security in healthcare, for the sake of human lives, challenges the design of the BAN. However, the human body offers unique opportunities for a new authentication methodology with biometrics.

13.1   Fundament of Authentication

Authentication refers to the process to guarantee that an entity is who it claims to be or that information has not been changed by an unauthorized party. Authentication is classified by the security objective specific to a service, such as *message authentication, entity authentication, key authentication, nonrepudiation,* and *access control*. Message authentication assures the integrity and origin of the information. As synonyms of message authentication, *data integrity* preserves the information from unauthorized alteration, while *data origin authentication* assures the identity of the data originator; data origin authentication implies data integrity because the originator is no longer the source of the modified message. Entity authentication, also named *endpoint authentication* or *identification*, assures both the identity and the presence of the claimant at the time of the process. The timely verification of one's identity is either *mutual*, when both parties—sender and receiver, for example—are confirmed with each other, or *unilateral*, if only one party is assured of the other's identity. Key authentication assures the linkage of an entity and its key(s), which extends to broader aspects of *key management* from key establishment/agreement, key distribution, key usage control, and the key life cycle. Key authentication plays a vital role in the Internet age when users cannot meet face-to-face to exchange keys or know each other personally to verify the keys. Trusted third parties step in as the *certification authority (CA)* responsible for vouching for the key's authenticity, such as binding keys to distinct individuals, maintaining certificate usage, and revoking certifications [1]. Nonrepudiation prevents an entity from denying its previous action; often, a trusted third party is needed to resolve a dispute due to an entity denying that it committed a certain action or no action. Access control or *authorization*, following successful entity authentication, posts selective restrictions on an entity to use data/resources.

To clear up the confusion in terms of authentication, this book classifies authentication by *timeliness* into two categories, from which the others can be derived:

1. *Entity authentication* in real time: Alice and Bob, both active in the communication, assure each other's identity with no time delay.

2. *Message authentication* in an elastic time frame: Alice and Bob exchange messages with assurance of the integrity and the origin of the messages even at a later time.

Traditionally (before the mid-1970s), authentication was intrinsically connected with secrecy. For example, password authentication during ancient wartime was kept as a shared secret, such as a word between parties; demonstrating the knowledge of this secret by revealing the word proved the corroboration of the entity's identity and then granted the entity a pass into the territory. Fixed-password schemes, involving time-invariant passwords, are considered *weak authentication*, subject to attacks by eavesdropping and exhaustive searching. Various techniques are applied to fixed-password schemes to strengthen secrecy. Instead of a clear text password, the password is encrypted to make it unintelligible or is salted/augmented with a random string to increase the complexity of dictionary attack.

However, authentication does not require secrecy, as the discovery of *hash functions* and *digital signatures* showed. A hash function is a one-way function that maps a binary string of arbitrary length to a binary string of fixed length, called a hash value, which serves as a compact representative of the input string. Two features that make hash functions useful for authentication are

1. It is computationally infeasible to find two distinct inputs with the same hash values, that is, two colliding inputs x and y such that $h(x) = h(y)$.

2. It is computationally infeasible, given a specific hash value v, to find an input x with the hash value v, that is, given v, to preimage x such that $h(x) = v$.

Symmetric-key encryption is one-key cryptography with a shared secret key; asymmetric-key encryption is two-key cryptography with a pair of one public key and one private key; a hash function is unkeyed cryptography with no key.

Hash functions may be used for data integrity to authenticate messages without keeping the secrecy of the messages. A typical process of data integrity with a hash function works as follows:

- Alice computes the hash value corresponding to a message and then sends the message to Bob, along with its hash value.

- Bob computes the hash value corresponding to the received message and compares his computed hash value with the extracted hash value. The comparison verifies if the message has been altered or not.

If Eva altered the message en route, Bob would be able to detect the modification, thus preserving data integrity without the need to keep the message secret from Eva. Note the inability to find two inputs with the same hash value satisfies the security requirement for data integrity. Otherwise, Eva would substitute another message with the same hash value to fool Bob from detecting modification. Keyed hash functions, encrypting hash values with a shared secret key, are named *message authentication code (MAC)* algorithms whose specific purpose is message authentication (data origin authentication as well as data integrity).

Hash functions may also be used for digital signatures. A digital signature binds an entity's identity to an information with a tag called the signature. A typical process is shown here:

- Alice signs a long message by computing its hash value and then sends the message to Bob along with its hash value, usually encrypted as her signature.

- Bob receives the message, computes its hash value, and verifies that the received signature matches the hash value.

Note, again, that the noncollision property of hash functions prevents Alice from claiming later to have signed another message because the signature on one message would not be the same as that on another. In addition, it is not necessary to keep the message secret from Eva for the purpose of data signature, since the hash value, not the message itself, is encrypted to increase the strength of nonrepudiation.

The third cryptographic use of hash functions is identification or entity authentication. Using a one-way (nonreversible) function of the shared key and the challenge, a claimant proves its knowledge of the shared key by providing a verifier with the hash value rather than the key, and the verifier can check if the delivered hash value matches the computed hash value to assure the claimant's identity. The challenge is to prevent replay attacks.

Though the terms identification and entity authentication are considered synonymously, they can be distinguished as *identification* only for a claimed (stated) identity, whereas *entity authentication* (or identity verification) is used to corroborate an identity. Likewise, a digital signature is closely related to entity authentication, but it involves a variable message to be signed for nonrepudiation after the fact, while entity authentication uses a fixed message such as a claimed identity to grant/deny instant access with no lifetime.

Parties in entity authentication:

- *Claimant* (prover): An entity that declares its identity as a message, often in response to an earlier message as *challenge–response protocols*, to demonstrate that it is the genuine entity.

- *Verifier*: Another entity that corroborates that the identity of the claimant is indeed as declared by checking the correctness of the message, thereby preventing impersonation.

- *Trusted third party*: An entity that mediates between two parties to offer an identity verification service as a trusted authority.

Objectives of entity authentication:

- *Conclusive*: The outcome of entity authentication is either *completion with acceptance* of the claimant's identity as authentic or *termination as rejection*.

- *Transferability*: Identification is *not transferable* so as not to allow a verifier reuse an identification exchange with a claimant to impersonate the claimant to a third party.

- *Impersonation*: There is a negligible probability that any entity, other than the claimant, can play the role of the claimant to cause a verifier to provide completion with acceptance of the claimant's identity; that is, no entity can *impersonate* a claimant. Nonimpersonation remains true even if an adversary has participated in previous authentications with either or both the claimant and the verifier in multiple instances.

Factors of entity authentication:

- *Something known*: The claimant demonstrates the knowledge of a secret by such means as passwords, personal identification numbers (PINs), shared secret keys, or private keys.

- *Something possessed*: The claimant typically presents a physical token functioning as a passport. Examples are magnetic-stripe cards, smart/IC cards, and smartphones to provide time-variant passwords.

- *Something inherent*: The claimant provides the biometrics inherited in human physical characteristics and involuntary actions. Examples are fingerprints, retinal patterns, walking gait, and dynamic keyboarding characteristics. These techniques have now been extended beyond authentication of human individuals to device fingerprints.

Levels of entity authentication:

- *Weak authentication*: Entity authentication schemes are considered *weak* if previously unknown parties verify their identities without involving trusted third parties. Single-factor authentication may not be weak: a *one-time password*, for example, is viewed as unbreakable against eavesdropping and later impersonation. A one-time password, as the "something known" factor, ensures that each password is used only once.

- *Strong authentication*: Entity authentication techniques using at least two factors are called *strong authentication*. Challenge–response protocols are strong authentication, in which a claimant proves its identity to a verifier by demonstrating knowledge of a secret known to be associated with the claimant, without revealing the secret itself to the verifier during protocol execution. Since the claimant's response to a time-variant challenge depends on both the claimant's secret (such as its private key) and the challenge (such as a random nonrepeating number called a *nonce*), two factors are used in the protocols.

- *Zero-knowledge (ZK) authentication*: Authentication protocols based on zero knowledge do not reveal any partial information at execution. Simple password schemes reveal the whole secret since, after a claimant gives a verifier the password, the verifier can impersonate the claimant by replaying the password. Challenge–response protocols improve this aspect by demonstrating knowledge of the secret in a time-variant manner without giving away the secret itself, so that the information is not directly reusable by an adversarial verifier. However, some partial information about the claimant's secret has been revealed, making challenge–response protocols susceptible to chosen-text attacks. ZK protocols allow a claimant to demonstrate knowledge of a secret while revealing no information of use to a verifier for impersonation. Therefore, the claimant only proves the truth of an assertion, similar to an answer obtained from a trusted oracle. However, the ZK property does not guarantee that a protocol is secure unless its attack problem is computationally hard.

Properties of entity authentication that are of interest to users are

- *Reciprocity of identification*: Both parties corroborate each other as mutual authentication, or one party corroborates the other as unilateral authentication. Some unilateral authentications, such as fixed-password schemes, are susceptible to an adversary posing as a verifier to capture a claimant's password for replay attacks.

- *Computational efficiency*: Computational complexity of an authentication protocol.

- *Communicational efficiency*: Communicational overhead of a protocol.

- *Third party*: Entity authentication techniques may involve a third party between two parties wishing to communicate in a trusted manner.

- *Timeliness of involvement*: The third party may stay online to provide authentication services in real time, such as the Kerberos protocol that distributes common symmetric keys to communicating parties for entity authentication. A CA often works offline to issue or revoke public-key certificates.

- *Nature of trust*: The third party could be an untrusted directory service for distributing public-key certificates. The nature of trust required in a third party includes trusting the third party's delivery of correct outcomes.

- *Nature of security guarantees*: Examples are provable security and ZK properties.

- *Storage of secrets*: This refers to where and how to store critical keying materials; examples are local disks, smart cards, or clouds in software or hardware.

13.2   Entity Authentication: Node Eviction in VANET

Vehicular networking features high-speed mobility, short-lived connectivity, and infrastructureless networking, forming vehicular ad hoc networks (VANET). Figure 13.1 depicts a typical network architecture of VANET, where roadside units (RSUs) operate in two modes: infrastructure and ad hoc. RSUs, operating in infrastructure mode, connect to network infrastructure such as the Internet or cellular networks for services provided by external components such as travel advertisement and electronic toll collection. An RSU will communicate with vehicles' onboard units (OBUs) sporadically in ad hoc mode. OBUs also communicate among themselves in ad hoc mode. An OBU will contain OBD-II as a set of sensors to measure the vehicle's own status such as its brake, GPS to identify its location, radar to detect other vehicles nearby, and transceiver (TRX) to communicate with RSUs and other vehicles. These components feed information to the codriver, a special-purpose computer, which monitors road safety and processes travel services. Thus, VANET is an exemplary IoT, with cars as some largest things to be connected on the IoT.

Beyond faulty nodes, such as malfunctioning OBUs, hindering VANET performance with fatal consequences in safety applications, malicious nodes intentionally inject faulty messages into VANET with the potential of massive destruction [2]. It is of paramount importance to remove errant nodes from VANET immediately. Node-eviction schemes accompany authentication mechanisms in network security. Traditionally, a centralized CA, such as the

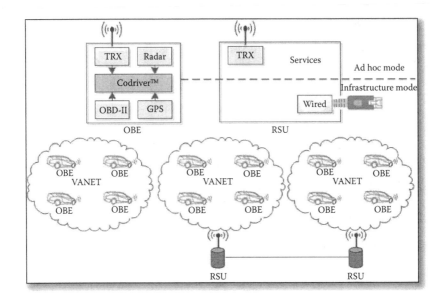

Figure 13.1: Network architecture of VANET.

Motor Vehicle Registry, revokes an errant node's certificate. However, the nature of VANET renders CA-based approaches ineffective. Current node-eviction schemes in VANET allow nodes to make decisions and take action against other errant nodes, both distributed and locally. Local node-eviction schemes can be classified into five categories.

1. Reputation: In the absence of a strong authentication infrastructure in VANET, simple node misbehavior could severely degrade VANET with catastrophic consequences. For example, a selfish node may flood fake congestion messages upstream, subverting traffic to clear its own way but possibly leading to a chain of accidents. As a security mechanism, an individual node forms/updates a reputation metric of other nodes with which it has interacted through its own direct observation and information provided by other nodes. Individuals will disengage from nodes of which they have had bad experiences. Eventually, nodes with a bad reputation will be excluded from VANET. CORE is a typical collaborative reputation mechanism that enforces nodes' proper behavior to remain in a mobile ad hoc network [3]. Reputation-based approaches are resilient from false detection but respond to incidents slowly.

2. Vote: Raya et al. proposed a local eviction of attackers by voting evaluators (LEAVE) protocol [4]. The CA collects accusations from different nodes that have witnessed a node's misbehavior and, on reaching a threshold, revokes the node being accused. LEAVE augments this

infrastructure-based revocation protocol with a misbehaving detection system, enabling individual nodes to safeguard themselves. Vote schemes equip individuals with a rapid reaction and self-protection. However, voting becomes an injustice when there exist more deceptive nodes than honest ones.

3. Suicide: To ensure the accountability of accusers, the suicide class allows a single node to unilaterally revoke another node at the cost of itself being revoked, known as *karmic suicide* [5]. The motivation comes from nature, in that a bee stings, losing its life, to respond to a perceived threat against its hive. The karmic-suicide revocation scheme offers an incentive to the nodes which have committed suicide through a periodically available trust authority (TA) rewarding a node for its justified suicide by reinstating it back into VANET. Suicide schemes inherit the speedy revocation process of vote schemes while increasing accuracy.

4. Abstinence: At the extreme of reputation schemes, the abstinence class keeps its ratings of others to itself. On experiencing a bad node's misbehavior, the node takes a passive role of staying away from the bad node but provides no reporting, expecting other nodes to eventually remove the bad node from the network. Each node can take one of the three actions in a revocation process: abstain, vote, or commit suicide. Optimal revocations in ephemeral networks (OREN) is a game-theoretic framework for local revocation, based on reputation, which dynamically adapts its cost parameters to guarantee a successful revocation in the most socially efficient manner [6].

5. Police: The police class is effective for revocation in transportation, but largely unexplored in VANET. A special vehicle, such as a police car, patrols the network of roads and revokes any misbehaving nodes immediately on detection [7]. This class is accurate, as the evidence is first hand, but its speed depends on the chance of a node being caught, though the eviction is made instantly.

Various factors affect the performance of node-eviction schemes. The topology of roads, spread of RSUs, speed of vehicles, drivers' behavior, and number of malicious nodes are just some examples.

Using an agent-based approach, we simulate the node-eviction schemes described above. The choice of agent-based simulation is due to its richness in flexibility and emergence such as being able to model behaviors and goals of individual nodes such as mobility and scheme configuration. This is useful for the modeling of systems that are very complex, such as intelligent transportation systems that involve driver behaviors, vehicle speeds, and individual goals. We used the recursive porous agent simulation toolkit (Repast) as our agent simulation toolkit because of its platform independence, seamless GIS integration, huge learning resource base, user friendliness, and programmer control [8].

The simulation scenario consists of a circular road setup in the grid, where vehicles at different speeds cycle around the road and communicate with one another or the RSU when in close proximity. The RSU relays information to the CA. The behavior of the system components is dependent on the scheme used.

The node-eviction scheme and frequency of contact was implicit in our model. The frequency of contact refers to how often the nodes come into contact with each other and exchange messages. This has a significant impact on the performance of the scheme. The variance of the speed of the nodes and their initial locations influences the frequency of contact.

In our simulation, we attempt to answer the question of whether the scheme will eventually separate the malicious nodes from the honest nodes between the two network classes and how long it will take for this to happen [9]. Any node-eviction scheme should attempt to optimize the average time, risk, and utility measures under dynamic environment conditions. In our simulation, we study how the evaluation parameters change with respect to the percentage of malicious nodes present in the network. The total number of nodes used in the network was 60, one of which was a police node.

We model node eviction process, as a set of states and transitions. Such a process eventually separates all nodes into two subnets: Subnet I and Subnet II. A node, which is good or bad, initially joins any of the two subnets by convenience. A state transition occurs when a node moves from Subnet I to Subnet II, or vice versa. As the birds of a feather eventually flock together, Subnet I or Subnet II will finally converge into the same kind of nodes, i.e., good or bad only in each subnet. The system is modeled as a network of *who* wants to receive messages from *whom*, controlled by certificates. Each node maintains a List of other nodes Valid Certificates (LVC).

As predicted, the vote class performed the best in terms of average vulnerability time, because every incident triggers segregation, and only half of the population is required to vote a node out by our setting the threshold at 0.5. The police class took second place, since it segregated a bad node once the police catches a node sending a rogue message. The time increases with the percentage of bad nodes because it takes time for the police to arrive in time. The abstinence class performs the worst, since a bad node is moved to Subnet II only if all nodes remove it from their LVC. When the percentage of bad nodes increases, the time dips slightly since the probability of encountering a bad node is higher. Figure 13.2 depicts the time simulation results.

Figures 13.3 and 13.4 summarize the accuracy simulation results. Accuracy was the best category, with the highest unity and lowest risk. Police and abstinence displayed the same unity value of 1, insensitive to the percentage of bad nodes, because their actions depend on first-hand information. No false accusation takes place; hence, good nodes are not mixed with bad nodes. The unity value of the vote class diminishes as the proportion of bad nodes reaches 0.5,

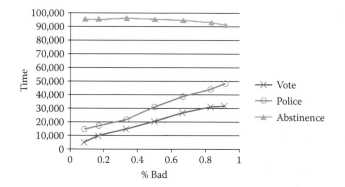

Figure 13.2: Average vulnerability time.

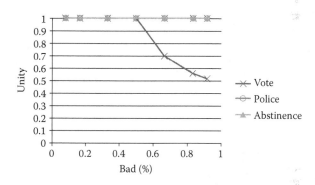

Figure 13.3: Average unity.

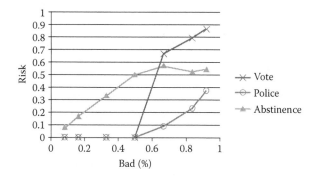

Figure 13.4: Average risk.

its threshold setting, since false accusations by bad nodes move good nodes to Subnet II.

The police class poses the lowest risk among the three, because every detection triggers a bad node being moved from Subnet I to Subnet II. At the end, good and bad nodes are largely segregated, with almost no risk. However, as the percentage of bad nodes increases, it becomes difficult for the single police node to catch all the bad nodes in time, as multiple bad nodes pop up simultaneously at different locations. It is also possible that the police never catch some bad nodes, which demonstrates a rise in risk. The vote class also poses a low risk when the percentage of bad nodes is low but, as the proportion increases beyond 0.5, its threshold setting, the risk rises suddenly due to two reasons: There are fewer good nodes to report and more false accusations by bad nodes; therefore, fewer good nodes remain. As the simulation reaches a state of equilibrium, almost all the nodes, good and bad, end up in Subnet II, returning to the image of the initial state. The abstinence class has the highest risk, since a bad node is moved out of Subnet I only when every other node abstains itself from it. Risk rises steadily as the percentage of bad nodes increases. At some points, the risk fluctuates, since a good node is removed from Subnet I. Notice that, well after the proportion of bad nodes reaches 0.5, the risk value of the abstinence class becomes lower than that of the vote class, due to more bad nodes than good ones distorting the truth.

13.3   Message Authentication: Content Delivery in VANET

The core of VANET applications relies on providing drivers with timely accurate information, namely *content delivery* [10]. However, VANET content delivery poses serious security threats such as confidentiality, integrity, and authentication, due to the distributed, open, and mobile nature of VANET [11]. Various security mechanisms have been proposed; nevertheless, without common metrics to measure their effectiveness, consumer confidence cannot be assured, especially regarding critical road safety concerns [12]. Unfortunately, security measurement is difficult [13] and different from other kinds of measurement such as level of service in transportation [14] or quality of service in wireless multimedia [15].

We propose a security metric to measure the integrity level of security schemes for VANET content delivery, namely, an *asymmetric profit-loss Markov (APLM) model* [16]. With a black-box approach, the model documents the incidents of detecting data corruptions as profits and those of accepting corrupted data as losses. We use a Markov chain to record how the system under assessment self-adjusts its behavior in reaction to profit and loss; where there are more loss states than profit ones, the system is *asymmetric*. We then present how APLM directs the optimization of designing integrity schemes for VANET content delivery, measuring results on a normal VANET content delivery deploying no integrity scheme and four integrity schemes: reputation, voting, voting on reputation (VOR), and random.

When a VANET passes by an RSU, the OBUs on the vehicles deliver to the RSU the traffic status of the upstream road segment. The traffic status could be expressed in traffic density, that is, the number of equivalent passenger cars per mile, with a timestamp attached. Whenever the vehicles are in the vicinity of the RSU, their OBUs respond to repeated requests by the RSU for the traffic status of the upstream road segment. To focus the scope of this chapter, we do not consider other content deliveries such as RSUs exchanging messages for global information, OBUs communicating with each other to avoid car collision, or RSUs advising the OBUs on alternative routes.

As shown in Figure 13.5, an RSU joins a VANET moving in its vicinity. The RSU then establishes concurrent transmission control protocol (TCP) connections with the OBUs on selected vehicles in the VANET. Each OBU contains a subset of fragments of the content. The RSU repeatedly requests each of the OBUs for their fragments over the TCP connections until it successfully assembles all the fragments into the full content. During the process, the RSU decides which fragment to obtain next from which OBU.

Our VANET content delivery application architecture possesses compelling features of scalability, extensibility, and flexibility. Similar to file distribution in peer-to-peer (P2P) architecture, our VANET content delivery self-scales with a bounded delivery time for any number of vehicles in the VANET. Its functionality is extensible to other content deliveries among RSUs and OBUs in duplex directions. The architecture supports flexible applications from collision avoidance to travel efficiency. However, the architecture faces security challenges due to its distributed, open, and mobile nature as discussed previously.

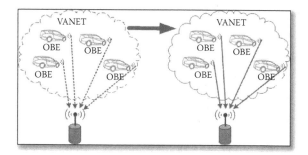

Figure 13.5: Architecture of VANET content delivery.

We propose a new integrity scheme named *voting on reputation for VANET (VOR4VANET)*. The scheme contains two stages: *local reputation calculation*, when an RSU assigns a rating to each of the OBUs based on its own evaluation of past transaction success with that OBU, followed by *voting weighted by reputation*, where a vote weighted by reputations among OBUs, instead of a majority vote, settles content discrepancy.

Local reputation, shown below, is calculated with an exponential weighted moving average over past ratings at the completion of downloading all data fragments needed to assemble the content:

$$R_t = (1 - \alpha) R_{t-1} + \alpha M$$

where:

$R_0 = 0$

$M = 1$ if OBU delivers a good fragment or -1 if it delivers a bad fragment In this trial, α is in [0, 1], with the recommended value of 0.125.

Voting weighted by reputation determines the correct version of a data fragment when its multiple copies from several OBUs carry different values. We adjust the mode calculation of the list with the reputations of the corresponding OBUs:

$$OBU = \text{mode} \left\{ R^h \text{ incidences of } F^h \right\}$$

where R^hs are reputations scaled up to nonnegative integers.

For example, if an RSU receives duplicates of a data fragment from four OBUs, and only one of the four OBUs delivers a "good" fragment while the rest three send "bad" fragments, by a majority vote, the RSU would accept the "bad" fragment. When incorporating their reputations as listed in Table 13.1, the list would equate to 3Gs and 2Bs, resulting in a mode of G; therefore, the RSU would accept the "good" fragment.

When a fresh VANET arrives in the vicinity of an RSU, the RSU checks its reputation base for all the OBUs on the vehicles in the VANET and chooses those OBUs of high reputation having the fragments to cover the entire content. The RSU then establishes concurrent TCP connections with the chosen OBUs and requests each for their fragments. As mentioned before, an OBU may not have the complete set of fragments to cover the entire content. With proper selection of OBUs, the RSU would receive all the fragments needed to assemble the content,

Table 13.1 Majority vote vs. voting weighted by reputation

	H_1	H_2	H_3	H_4	OBU
F^h	B	G	B	B	B
R^h	1	3	0	1	G

most of which would be duplicates. When a discrepancy occurs in the value of a particular fragment due to corruption in some OBUs, the RSU invokes the voting scheme to settle the matter. The verdict will be reached after the RSU receives all the fragments and assembles them into the full content. The RSU then updates its reputation base. If the content fails the integrity check, the RSU repeats its selection process and requests fragments again until either the content delivery succeeds or the VANET passes out of its vicinity.

Our APLM model of content integrity metrics employs *content hosts* such as OBUs in VANET and *content retrievers* like RSUs. The APLM model is based on the idea that an effective integrity scheme would enable content retrievers to avoid "bad" content hosts and request "good" content hosts for fragments needed to assemble a particular content set. The distinct number of content retrievers obtaining at least one corrupted data fragment without detection is represented by each state. Therefore, the state space of Markov chain consists of $(n+1)$ states for content retrievers, a value of 0 denoting that none of the content retrievers have accepted "bad" fragments, 1 if one of them possesses corrupted fragments, (etc.), and n if all of them possess "bad" fragments without them being detected and disregarded. State 0 indicates *profit* while all the other states indicate *loss*; this represents *asymmetric* profit-loss, since there are more loss states than profit states. The heuristic matches the Markov property that the next state depends only on the current state. Through black-box observation, the probabilities of states' transitions can be obtained. In P, the Markov matrix (1), $p_{i,j}$ denotes the probability of transitioning from state i to state j, where the probability of transitioning from state i to each of all the states (itself inclusive) sums to 1, indicated by Constraint 2.

$$P = \begin{bmatrix} p_{0,0} & \cdots & p_{0,n} \\ \vdots & \ddots & \vdots \\ p_{n,0} & \cdots & p_{n,n} \end{bmatrix} \tag{13.1}$$

$$\sum_j p_{i,j} = 1 \tag{13.2}$$

Using the vector π to represent the probabilities of all the steady states, π_i denotes the probability of the network being in state i. Assuming an ergodic property for this Markov process, Equations 13.3 and 13.4 hold true. We can derive the steady-state probabilities, π_i, by solving the linear system of Equation 13.4 and any $(n-1)$ equations taken from Equation 13.3.

$$\pi P = \pi \tag{13.3}$$

$$\sum_i \pi_i = 1 \tag{13.4}$$

Finding the steady-state probability vector π, we can then calculate the integrity score based on profit and loss as in Equation 13.5. The range of $f(\pi)$

is $[-1, 1]$, where "-1" represents the worst, "1" the best, and "0" indicates the system in a state of equilibrium between good and bad. The first term computes profit obtained by remaining in State 0, π_0, normalized to 1 by its coefficient $g(0)$. The second term sums losses at the other states, π_i, normalized to -1 by $g(0)$. Equation 13.5 reflects the asymmetric feature, with only one state carrying profit while the remaining n states cause loss.

$$f(\pi) = g(0)\pi_0 - \left(\sum_{i=1}^{n} g(i) \times \pi_i\right) \qquad (13.5)$$

APLM features a black-box approach to measure an integrity scheme without the need to examine its implementation in detail; it thereby offers feasibility to the measurement process and autonomy without exercising expertise often associated with white-box methods. By utilizing historical statistics recorded as profit and loss, APLM measures integrity levels of five scenarios: normal without deploying any integrity scheme, the two schemes adapted from P2P file distribution, our VOR4VANET, and a random scheme. Let the content hosts of APLM model denote OBUs in VANET and content retrievers for RSUs. We also demonstrate how APLM directs the design of our VOR4VANET.

1. Normal VANET content delivery:

   Under the VANET content delivery architecture illustrated in Figure 13.5, an RSU obtains data fragments from whichever OBUs possess them. Once the RSU receives all the content fragments, it assembles them and checks content integrity. If there is a corruption in a data fragment, which the RSU cannot detect during fragment transmission but will be able to discover only after download completion, the RSU repeats its requests to all OBUs in its vicinity for the missing fragments. Normally, an RSU tends to obtain data fragments from those OBUs that respond faster.

2. Reputation scheme on individual OBU:

   With the reputation scheme, an RSU maintains a local reputation base of all OBUs in a VANET that is passing by. The RSU chooses those OBUs at the top of the reputation list to request the data fragments it needs. Thereby, the level of content integrity increases at the cost of delaying delivery. The idea is borrowed from P2P file distribution, where a reputation base is usually maintained by a trusted central server. Our reputation scheme allows individual RSUs to maintain their own reputation base locally, and doing so in such a distributed fashion lessens the bottleneck effect of centralized schemes. There are various ways for an RSU to rate each OBU based on the OBU's past performance in delivering "good" or "bad" data fragments. We choose the dynamic reputation formula, given in the formula on page 338, which takes the exponential weighted moving average over past ratings to reflect the current status in the system by more recent measurements.

3. Voting scheme on data fragments:

The voting scheme targets the problem which remains in the reputation scheme, where corruptions are detected *after* completion of downloading all fragments. This severely reduces the efficiency of content delivery. Adapted from P2P file distribution, an RSU requests multiple copies of a data fragment from several highly reputable OBUs over concurrent TCP connections. When there is a *discrepancy* (n.b., not *corruption*) among copies, a majority vote takes place to determine which fragment to accept. Obviously, the voting scheme requires more processing overhead. Intuitively, the voting scheme should outperform the reputation scheme in assuring content integrity and delivery efficiency. However, the results from our APLM model surprised us, as indicated by the next scheme, VOR4VANET, where a majority vote under bad influence yields a wrong result. This study demonstrates the effectiveness of our APLM model in directing the optimization of security scheme design.

4. VOR4VANET:

Voting on reputation for VANET integrity (VOR4VANET) contains two stages: local reputation calculation and voting weighted by reputation. The first stage is the same as the reputation scheme on an individual OBU. The second stage differs from the voting scheme on data fragments presented above. Instead of taking a majority vote, VOR4VANET gives greater weight to a more reputable OBU in the voting. In those cases when there are more "bad" OBUs than "good" ones, a majority vote would yield the undesirable result of selecting a corrupted data fragment. Such a situation may be corrected by incorporating reputation into the procedure, giving more reputable OBUs more weight in the voting. The experiments have confirmed our hypothesis. The APLM model in our VOR4VANET directly prevents such design fraud from using the voting scheme on data fragments.

5. Random OBU choice:

In computer science/engineering when optimization relies on heuristics, randomness often works wonders such as caching replacement algorithms. We also propose a scheme to choose an OBU randomly. Out of all the OBUs in a VANET that is passing by, an RSU chooses OBUs at random to request data fragments. Such a scheme involves barely any overhead but improves normal VANET content delivery.

Figure 13.6 shows a result of VANET simulation under a normal setting.

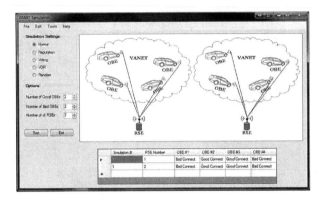

Figure 13.6: VANET simulation.

13.4 Key Management: Physiological Key Agreement in WBAN

Another application domain of the IoT is medical cyberphysical systems (MCPSs) that monitor/control patients' physiological dynamics with embedded/ distributed computing processes and a wireless/wired communication network. MCPSs greatly impact the society with high-quality medical services and low-cost ubiquitous healthcare. The major component that integrates the physical world with cyberspace is the wireless body area network (WBAN) of medical sensors and actuators worn by or implanted into a patient. The life-critical nature of MCPSs mandates safe and effective system design. MCPSs must operate safely under malicious attacks. Authentication ensures that a medical device is what it claims to be and does what it claims to do; the first line of MCPS defense. Traditional authentication mechanisms, reliant on cryptography, are not applicable to MCPSs due to constraints on computing, communication, and energy resources. Recent innovations to secure mobile wireless sensor networks, with multisensor fusion to save power consumption, are not adequate. Despite these challenges, MCPSs present great opportunities, with the unique physical features of WBANs, for noncryptographic authentication and human-aided security. This chapter proposes an authentication framework for MCPSs. By studying medical processes and investigating healthcare adversaries, the novel design crosses the boundary between the physical world and cyberspace. With uneven resource allocation, resource-scarce WBANs utilize no encryption for authentication. Evaluation of this authentication protocol shows promising aspects and ease of adaptability.

An MCPS represents a physiologically closed-loop system, where an automatic controller continuously monitors the patient's vital signs with sensors and administers medication as needed with the aid of an actuator such as an infusion

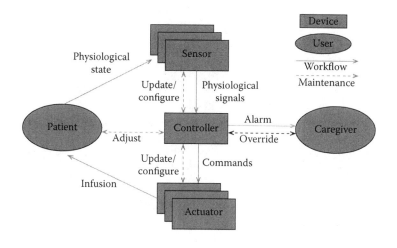

Figure 13.7: MCPS control loop.

pump. Closed-loop control has been applied to the medical device industry, but mostly limited to stand-alone implants. For example, pacemakers deliver electrical impulses by battery-powered electrodes, often combined with defibrillators, to regulate the heartbeat of cardiac patients without human intervention. Some clinical scenarios not based on threshold, however, need coordination of distributed medical devices. Due to patients' different reactions to medications, for instance, seizure detection is deemed ineffective with the current method of threshold-based brain oxygen monitoring. Therefore, physiologically closed-loop control relies on individualized patient modeling and also requires a fail-safe caregiver interface. Figure 13.7 depicts the control loop of a typical MCPS. Boxes represent medical devices, and ovals denote MCPS users. Solid lines indicate the workflow, while dashed lines exemplify the maintenance procedure.

A use case is patient-controlled analgesia (PCA), developed by a team at the University of Pennsylvania jointly with U.S. Food and Drug Administration (FDA) researchers. PCA infusion pumps deliver opioid drugs for postsurgical pain management. A patient can adjust dosage rather than follow a schedule prescribed by a caregiver, because people react differently to the medications. However, overdose causes respiratory failure, leading to death. A PCA closed-loop system solves this safety issue. A pulse oximeter (sensor) continuously monitors two respiratory-related vital signs, heart rate (HR) and blood oxygen saturation (SpO2), and transmits the physiological signals to a controller. The controller, on detecting respiratory depression, commands the infusion pump (actuator), to stop dispensing the pain medication to the patient. The controller also sends an alarm to the caregivers, who have the ability to override the PCA, if an adverse event occurs. The maintenance procedure includes sensors/actuators updating their operational status to the controller and the controller configuring sensors/actuators [18].

Authenticating medical devices in the physical world can avoid resource-intensive cryptography by taking advantages of human biometrics [17]. We adopt a popular noncryptographic authentication scheme, called a physiological signal-based key agreement (PSKA) by Venkatasubramanian et al., to extend our authentication framework to the physical world. The framework is suitable for general WBANs, with any authentication scheme based on biometrics such as electrocardiograms (ECG).

PSKA utilizes photoplethysmogram (PPG) signals to authenticate the sensors worn on a human body utilizing their shared physiological features. The random individuality and universal measurability that vary with time in such features ensure confidence to accept those sensors on the body, while rejecting others not on the body. Therefore, PSKA effectively authenticates medical devices with the aid of a patient themselves, involving neither cryptography nor identification.

PSKA also functions as key distribution to facilitate less computation-intensive symmetric cryptography. By utilizing a fuzzy-vault cryptographic primitive, a sensor locks/hides a secret in a construct called a vault using a set of values A. Another sensor, having only a small subset of values in common with Set A, can unlock/discover the secret. Sharing the same PPG signals, the sensors on the same body can reach agreement of a shared key. Thus, the PSKA provides the apparatus for confidentiality, in addition to authentication, for its communications in the physical world.

We reclassify the on-body medical devices of MCPSs into two: the sensors/actuators as data devices (Ds) and the controller as a single information aggregator (A). Our authentication framework in the physical world contains three stages: physiological feature generation, noncryptographic/nonidentifier authentication, and key agreement. Figure 13.8 illustrates the PSKA process to exemplify our authentication framework [19].

1. Feature generation:

   All Ds and A obtain physiological signal-based features using the four steps below.

   (a) The Ds and A sample (PPG) signals at the same time at a specific rate, irrespective of the parts of the body from which the signals are coming.

   (b) The samples are divided into windows, on each of which a fast Fourier transform (FFT) is performed.

   (c) The peak in each FFT coefficient is detected.

   (d) Each of the peak index-value pairs is quantized into binary strings, which are concatenated to form a feature.

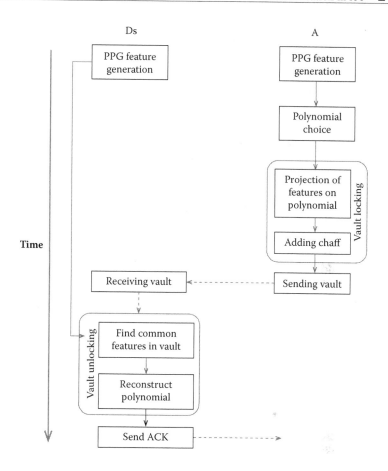

Figure 13.8: PSKA process.

A feature vector is then formed with individual features obtained from a single measurement. The Ds and A in the same WBAN possess feature vectors with a significant number of common values, ready to authenticate all within the group.

2. Group authentication:

The A (denoting the controller in an MCPS as a single information aggregator) initiates authentication for the group containing the A itself and the Ds, representing sensors and actuators as data devices worn on the same body. The following five steps, after the loosely synchronized feature generation described above, complete the group authentication process.

Step 1. Polynomial choice: The A generates a set of random numbers as the coefficients of a polynomial, $\mathrm{p}(x) = \sum_{i=n}^{0} c_i x^i$. The concatenated coefficients form a secret message to be delivered to all the Ds.

Step 2. Vault locking: The A hides the secret message in a vault with two procedures. First, the A projects the features generated from the previous subsection on the polynomial. The A also computes another set of random points as chaff. Then, the A permutes randomly to ensure that the legitimate points and the chaff points are not distinguishable. The result is a vault with the secret message hidden inside.

Step 3. Vault delivery: The A sends the vault to all the Ds.

Step 4. Vault unlocking: On receiving the vault, each D finds the matching features in the vault. It then reconstructs the polynomial, which succeeds in discovering the secret message based on the proven fuzzy-vault cryptographic primitive.

Step 5. Vault acknowledgment: Each D replies back to the A with the secret message.

Thereby, the Ds are authenticated to the A when the A successfully verifies their acknowledgments. This group authentication protocol works because only devices on the same WBAN as the A, measuring the same physiological signals at the same time, could have unlocked the vault and discover the secret message. The success is ensured by the distinctive and temporal variance of certain human biometrics.

3. Key agreement:

In addition to group authentication, the secret message delivered by the A to all Ds can be used as the shared key to facilitate confidential communications among on-body sensors. Figure 13.9 shows this extended PSKA function of group authentication and key agreement.

(a) The A transmits to the Ds the vault, which is a random permutation of the legitimate points and the chaff points, and which hides a secret message:

$$A \rightarrow D: ID_A, V, \text{Nonce}$$
$$MAC(K_S, ID_A\|V\|\text{Nonce})$$

where:

ID_A is the identifier of A

V is the vault

Nonce is a unique random number to ensure transaction freshness

MAC is the message authentication code computed by the shared key K_S, which is locked in the vault V

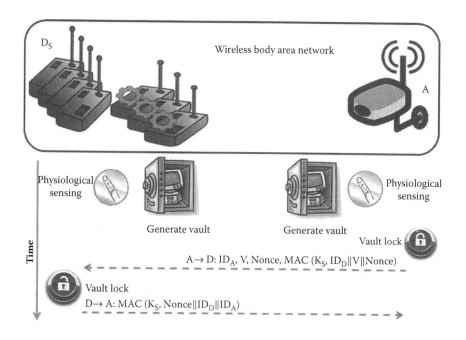

Figure 13.9: On-body group authentication.

(b) Each D, on successfully unlocking the vault V, replies back to the A with the secret message discovered, which is the shared key K_S:

$$D \rightarrow A: MAC(K_S, Nonce \, \| ID_D \, \| \, ID_A)$$

The cyberspace or back-end system authentication for MCPS may apply traditional approaches.

The features of personal distinctiveness, temporal variance, and universal measurability in certain human biometrics present the opportunities to secure MCPS such as one-time key. Furthermore, these features cultivate a new branch of non-cryptographic authentication. We demonstrate group authentication of on-body medical devices without potential deployment of encryption as long as the devices sample certain physiological signals on the same body during the same time. This group authentication extends to key agreement, a kind of key management, which makes symmetric encryption strong with session keys relayed from WBAN to the entire MCPS for a patient.

Bibliography

[1] Alfred J. Menezes, Paul C. van Oorschot, and Scott A. Vanstone, *Handbook of Applied Cryptography*, CRC Press, Taylor & Francis Group, 1996.

[2] Jonathan Andrew Larcom and Hong Liu, Authentication in GPS-directed mobile clouds, in *Proceedings of IEEE Global Communications Conference 2013 (IEEE GLOBECOM 2013)*, pp. 470–475, Atlanta, GA, 9–13 December 2013.

[3] Pietro Michiardi and Refik Molva, Core: A collaborative reputation mechanism to enforce node cooperation in mobile ad hoc networks, in *Proceedings of the IFIP TC6/TC11 Sixth Joint Working Conference on Communications and Multimedia Security: Advanced Communications and Multimedia Security*, pp. 107–121, Portoroz, Slovenia, September 2002.

[4] Maxim Raya, Panos Papadimitratos, Imad Aad, Daniel Jungels, and Jean-Pierre Hubaux, Eviction of misbehaving and faulty nodes in vehicular networks, *IEEE Journal on Selected Areas in Communications*, vol. 25, no. 8, pp. 1–12, 2007.

[5] Arzad Kherani and Ashwin Rao, Performance of node-eviction schemes in vehicular networks, *IEEE Transactions on Vehicular Technology*, vol. 59, no. 2, pp. 550–558, 2010.

[6] Igor Bilogrevic, Mohammad Hossein Manshaei, Maxim Raya, and Jean-Pierre Hubaux, OREN: Optimal revocations in ephemeral networks, *Computer Networks*, vol. 55, pp. 1168–1180, 2011.

[7] CAMP, Vehicle Safety CommunicationsApplications (VSC-A), National Highway Traffic Safety Administration, New Jersey, Final Report DOT HS 811 492AD, 2011.

[8] M.J. North, T.R. Howe, N.T. Collier, and J.R. Vos, A declarative model assembly infrastructure for verification and validation, in *Advancing Social Simulation: The First World Congress*, S Takahashi, D L Sallach, and J Rouchier, Eds. Heidelberg: Springer, 2007, pp. 129–140.

[9] Ikechukwu Kester Azogu and Hong Liu, Performance evaluation of node eviction schemes in inter-vehicle communication, *International Journal of Performability Engineering (IJPE)*, vol. 9, no. 3, pp. 345–351, 2013.

[10] Tamer Nadeem, Sasan Dashtinezhad, Chunyuan Liao, and Liviu Iftode, TrafficView: Traffic data dissemination using car-to-car communication, *ACM Mobile Computing and Communications Review (MC2R)*, vol. 8, no. 3, pp. 6–19, 2004.

[11] Panagiotis Papadimitratos et al., Secure vehiclar communication systems: Design and architecture, *IEEE Communication Magazine*, vol. 46, no. 11, pp. 100–109, 2008.

[12] Daiheng Ni, Hong Liu, Wei Ding, Yuanchang Xie, Honggang Wang, Hossein Pishro-Nik, and Qian Yu, Cyber-physical integration to connect vehicles for transformed transportation safety and efficiency, in *25th International Conference on Industrial, Engineering & Other Applications of Applied Intelligent Systems (IEA-AIE 2012)*, Dalian, China, 9 12 June 2012.

[13] Shari Lawrence Pfleeger and Robert K. Cunningham, Why measuring security is hard, *IEEE Transactions on Security and Privacy*, vol. 8, no. 4, pp. 46–54, 2010.

[14] Jean-Paul Rodrigue, Claude Comtois, and Brian Slack, *The Geography of Transport Systems*, 2nd edn. New York: Routledge, 2009.

[15] Angela YingJun Zhang, SoungChang Liew, and DaRui Chen, Delay analysis for wireless local area networks with multipacket reception under finite load, in *Proceedings of IEEE Global Telecommunications Conference (GLOBECOM 2008)*, pp. 1–6, New Orleans, LA, 30 November–4 December 2008.

[16] Ikechukwu Kester Azogu, Michael Thomas Ferreira, and Hong Liu, A security metric for VANET content delivery, *Proceedings of IEEE Global Communications Conference 2012 (IEEE GLOBECOM 2012)*, pp. 991–996, Anaheim, CA, December 3–7, 2012.

[17] C. C. Poon, Y.-T. Zhang and S.-D. Bao, A novel biometrics method to secure wireless body area sensor networks for telemedicine and m-health, *IEEE Communications Magazine*, vol. 44, no. 4, pp. 73–81, 2006.

[18] K. K. Venkatasubramanian, S. K. Gupta and A. Banerjee, PSKA: Usable and secure key agreement scheme for body area networks, In *IEEE Transactions on Information Technology in Biomedicine*, vol. 14, no. 1, pp. 60–68, 2010.

[19] Mohammed Raza Kanjee and Hong Liu, Authentication and key relay in medical cyber-physical systems, *Security Communication Networks* Special Issue (SCN-SI) on Security and Networking for Cyber-Physical Systems, May 2014.

IOT DATA SECURITY

Chapter 14

Computational Security for the IoT and Beyond

Pavel Loskot

CONTENTS

Although the fundamental processes in society (e.g., the need for travel, business, and entertainment) have not changed for as much as the past 1000 years, the complexities of life and the world have been increasing constantly as these processes are being made ever more efficient [25]. The underlying complex systems are often envisioned as networks of mutually interconnected subunits (so-called

355

structural models, being derived from a physical structure), or as networks capturing interdependencies and relationships (so-called functional models, being derived from a logical structure) [19, 47]. Thus, network models are collections of scalar (often binary) interactions between the pairs of entities. For example, living matter is formed by complex interactions of biomolecules, cells, organs, tissue, individuals, and populations [51]. On the other hand, socioeconomic infrastructures such as telecommunication systems, roads, and distribution of utilities are examples of the largest man-made networks. Notably, the social and biological systems are far more complex than any man-made technology, with the human brain being the most complex structure known in our universe.

From a historical perspective, as the tertiary economic sector of services (established shortly after the second world war) has become saturated, there is a natural pressure to build a new quaternary economic sector to offer new employment opportunities. This new economic sector will benefit from the information revolution of the twenty-first century and from the expanding, knowledge-based digital economy. More importantly, it is expected that the quaternary economy will focus mainly on understanding, controlling, and synthesizing biological systems to improve cognitive and other capabilities of human beings. In other words, as the late twentieth century was about development and deployment of ICT, the beginning of the twenty-first century is about the exploration of active matter and life sciences. For instance, synthetic biology can modify existing organisms, which has many security implications.

The IoT will build bridges between the existing complex systems by extending the reach of the Internet into the physical world. This will allow deeper integration of the human world with nature (down to nanoscale levels) as well as more efficient utilization of resources by intelligent management of flows of people, goods, and assets. The goal is to build pervasive systems and environments that are reliable, unobtrusive, autonomous, and secure. The intelligent systems and smart environments involving the IoT can be considered to be generalizations of the Internet (cf. combinatorial evolution of technology [43]). The controllability of the systems and environments will be enhanced significantly through a network of nested heterogeneous networks with numerous hybrid interfaces, leading to a formation of an extremely complex system of systems. The intelligence will especially concern the interfaces, while the objects and processes will be assigned their unique identifications (IDs). The information flows pertinent to such intelligence must be governed by information security policies including information labeling (classification), modification, ownership, and accountability. The proliferation of the IoT will enable access to information about any environment and about the status of any object, anytime, and anywhere. Establishing these information highways is driven by the deployment of various IoT sensors (physical devices) and markers (logical devices). In addition to ubiquitous sensors, the radiofrequency ID (RFID) tags are another key enabler of the IoT, even though these tags often have very limited computational and memory capabilities (e.g., write-once memory, allowing only for static cryptographic keys). So far, the security of the RFID networks concern the use of so-called blocker tags (to overwhelm the tag reader) and the establishment of privacy zones [26].

Information extracted from the data reported by the IoT is vital to make meaningful decisions to move the system toward a desirable state. Thus, the emergence of the IoT will have profound effects on functionality, dynamics, processes, and activities, including security of many if not all systems on the earth:

- The existing (already complex) systems will become more closely interconnected and immersed.

- The interactions of components within and in between systems will increase.

- The existing services will be modified while the opportunities for new services will emerge.

- Our perception of the environment and the reality we live in will change.

- The scale and scope of security problems (among others) will greatly expand.

For example, the Internet redefined social interactions [3] and is affecting the structure and functions of the human brain [55]. Nanoparticles are now used for sensing the biochemical processes inside biological cells and for drug delivery [16, 17]. The utility grids are enhanced using secure data aggregation to optimize energy consumption [33].

The IoT will also drive machine-to-machine (M2M) communications. Moreover, machine-to-human (M2H) communications are expected to be increasingly more important; for instance, to enhance human brain capabilities, and at the same time, to also enhance machines by exploiting the computational power of the human brain (e.g., to detect, classify, and track multiple objects in arbitrary visual scenes is an overwhelmingly complex task). The IoT networks can be even used to implement brain-to-brain communications [48]. In general, the human brain is the subject of intensive ongoing research [52]. For instance, the brain's complexity has been created in only 4.5 million years as a direct consequence of social interactions and our ability to bypass natural selection (evolution). Unlike very similar biological structures of the body in all human beings, brain structures show enormous variations among individuals. As the human brain is primarily responsible for creating our culture as well as for making decisions, the brain and our mind are now also the subject of serious security concerns. In particular, a new concept of so-called nonlinear or hybrid, network-centered wars involving political, economical, social, psychological, and information contactless encounters as well as conventional military operations, is outlined in a report [10]. This report, frequently debated on the Internet, argues that mankind has entered a new era of permanent war, with the current phase being psychological warfare, primarily targeting human thinking and decision-making. As well as the

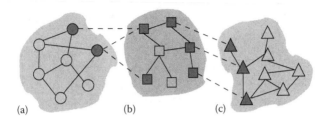

Figure 14.1: An example of three interacting networks (the gateway nodes are filled) where, e.g., (a) is the human brain, (c) represents the surrounding environment, and (b) is the IoT network creating the bridge between the other two networks.

for Internet media, such warfare can exploit new data from mobile phone sensors, and from enhanced personal communications and other ambient technologies [41] to affect our perceptions of reality, and also, in turn, our decisions (cf. ubiquitous advertising). In summary, we may expect emergence of ecosystems of interconnected things deployed in diverse environments with many industries and players involved to make the world we live in more intelligent, predictable, and controllable.

Complex systems are the main focus of many current scientific and technical investigations. These systems can be conveniently modeled as graphs representing interactions of a large number of nodes [14, 47]. They usually require multiple models of different types (structural or physical vs. functional or logical) at different spatiotemporal scales [32]. As an example, Figure 14.1 shows three interacting systems, with Network B acting as a bridge or interface between Networks A and C. For example, Network A is the human brain, Network C is the surrounding environment, and Network B are the IoT sensors and actuators. Even though the security of computer networks and of cybersystems have been studied and understood extensively [50], the security of more general systems having a network-like structure seems to be a new subject [38]. For instance, as the biological and social networks are very complex, defining their security is likely to be rather nontrivial.

In general, security provisioning requires extra resources ("there is no free lunch"), and often, to trade off reliability, availability, and security [4]. The current approaches to security emphasize prevention with pervasive monitoring and control through passive protection, perhaps mimicking security as it evolved in nature. The security of all systems can be described using security policies and procedures. For networks involving technology, security must also account for hardware and software implementations and their updates (due to possibly frequent turnarounds and modernization). When considering the security of complex sociotechnical networks, the main challenges accelerating the demand for their security are:

- Highly fragmented systems with diverse components and hybrid interfaces

- Components with varying levels of security certification, standards compliance and interoperability

- A mixture of components designed with embedded security features and those with security added as an extra feature

- A highly competitive environment with many manufacturers, operators, contractors, suppliers, etc.

- The convergence of information and operation technologies (IT and OT)

- A growing need for remote access and management of subsystems

- A paradigm shift in the motives and targets of the adversaries, fueled by IoT characteristics (e.g., the shift from small to large scale, from ad hoc to well planned, from single domain to concurrent attacks across multiple domains, from material or financial to psychological, etc.)

Ultimately, security provisioning must aim at

- Developing and supporting widely accepted good security practices across IoT industries

- Identifying security monetization opportunities and accounting for underlying costs (e.g., environmental, social, and system downtime costs)

- Developing universal, systematic approaches to holistic security that encompass all complex systems affecting our lives (e.g., embedding security and creating security platforms and concepts, security intelligence, plug-and-play security, etc.)

- Developing automated security threat (risk) assessments and security analytics for arbitrary complex systems or their subsystems

Some of these challenges and aims can be addressed by implementing security at multiple scales, at different segments (creating secure, less secure, and nonsecure zones with the corresponding varying levels of security risks), and at multiple layers (so-called layered security, robust against penetration attacks). Similarly to other networked services and functions, security can be either implemented within the network core or at the network's edges; a viable network security will likely require combination of both these approaches.

One of the main reasons to be concerned about security is that it impacts the sustainability of systems [31]. For instance, malicious behavior, harmful actions, malfunctions, and errors are likely to propagate through the network, and may permanently change the system's internal state [47]. Many real-world network systems and network models are scale-free which makes them very robust against random, ad hoc attacks (i.e., random removal of edges and nodes) [1]. However, these networks are very vulnerable to targeted attacks; for instance, removing the hubs (highly connected nodes) can disrupt the network and its functions very quickly. For example, a phishing attack targeting a specific individual (so-called spear phishing) significantly improves the probability of success [4]. Hence, when considering how to build the secure IoT, the focus should be on targeted, planned attacks. The ad hoc random attacks that prevail in today's computer networks usually cause a temporary service disruption, even though the aggregated cost of damages may be huge. However, a targeted and well-planned attack may cause high-impact and lasting (even permanent) damage in many general network-like systems. For example, a small-scale targeted attack to selected power plants or the electricity distribution grid may cause a long-lasting countrywide blackout.

The bottom line of most security attacks seems to be to identify a vulnerability in the system to bypass its defense mechanisms. Obviously, defense becomes more difficult for more complex systems; as popular wisdom goes: "the system designers have to secure everything, but the attacker has to find only one vulnerability." The most common vulnerability is to make assumptions about system processes, system status, typical behavior of users, expected format of inputs, and so on. The attackers are likely to search for situations when and where these commonly accepted assumptions are violated, and use them to launch an attack. However, making these assumptions can never be entirely avoided due to the complexity of the systems we are dealing with, so no system can ever be made absolutely secure. For example, any process within the system that is predictable can be considered as an assumption that can be exploited by the attacker. Thus, security should be considered to be a dynamic, continuously evolving process rather than a static, one-off solution.

14.1 Characterizing Complex Systems

Many systems in our world can be modeled well as networks of interconnected components. A large number of heterogeneous components and their various spatiotemporal nonlinear interactions make these systems to appear very complex (far beyond complicated). Understanding of these systems is a prerequisite for devising how to make these systems secure. In complex systems, it is, generally, difficult to distinguish causes and effects, how they relate to each other, and how to describe system behavior at all [32]. Locally, the components behave stochastically and predicting their behavior is only possible over short timescales (so-called organized simplicity). However, the compounded behavior of many components becomes a meaningful macroscopic characteristic of the system that is predictable over longer time intervals (so-called organized or unorganized complexity). Predicting the behavior of complex systems is mainly complicated by the nonlinear responses to perturbations (i.e., the whole is not equal to the sum of its parts). In data-driven modeling, it may be straightforward to measure the individual components; however, measuring the interactions (sometimes referred to as protocols) between the components or groups of components is often difficult.

Complex systems have a number of typical intrinsic characteristics: self-organization and adaptation to the environment, emergent macroscopic behavior, and maintaining a dynamic internal state at the boundary between order and chaos [27]. Their self-organization is achieved in a fully distributed manner; centralized control or predictable hierarchy is not possible in complex systems. The adaptation can be described as solving different constrained optimization problems at different spatiotemporal scales (e.g., from continuous homeostasis in cells to habitual behavior of whole populations during evolution). Long-term adaptations are critical for system sustainability and survival. Complex systems usually recover from small perturbations, maintaining stability in an internal (steady) state, but may transition to a new state once the perturbations become large enough. Adaptations may reflect changes in the values of static variables, and even more radical changes of the internal structure. Moreover, complex systems do not have to evolve from scratch. They are often built by reusing the components and subsystems of other complex systems, which can speed up evolutionary developments significantly (cf. the human brain, software, and combinatorial evolution).

Fundamentally, all complex systems can be characterized from different perspectives, domains, contexts, and spatiotemporal scales, as indicated in Figure 14.2. Thus, the services and functions provided by complex systems are observer dependent. The optimization problems defined in different domains

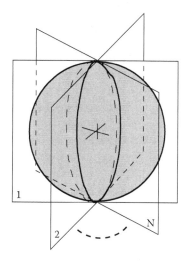

Figure 14.2: The projections (N hyperplanes) of a complex system (the hypersphere).

may have different priorities, so the overall global solution to these problems is also dependent on priorities. For example, society offers education, healthcare, postal deliveries, emergency response services, transportation, supply chains, and other services. The domains may have subdomains, such as those of cultural values and emotions within the domain of society. This has important consequences for the security of complex systems such that a sophisticated attack can evolve from one subsystem (context, service plane, or domain) to another as vulnerabilities are discovered and exploited until the attack reaches its intended objective. Such attacks and, especially, the corresponding defenses are a far more challenging problem than the analogous attacks on computer networks (possibly combined with social engineering as another domain) known as pivoting [4], since the multitude of available domains may help to completely conceal the attack. Thus, detecting and stopping an unfolding attack across multiple different domains may be provably impossible. A good understanding of possible targets and motives of the adversaries (i.e., a good model of them) may significantly increase the chances of their discovery (i.e., to know where, when, and what to look for). Similarly, to make the software environment more secure, it is suggested to minimize the number of concurrently running applications and processes [4]; however, this strategy is not viable, or at least not easy to achieve, for complex systems serving a large number of users with many different services.

Unlike the designers of complex systems who are concerned with the reliability, emerging patterns of behavior, and evolution and adaptation of these systems, the attackers are mainly concerned about not being caught. Thus, the attackers may use a combination of the ad hoc trial-and-error strategy together with computational modeling and planning to plot an attack which gives them an enormous advantage over the defenders. Moreover, as the adaptation of complex systems is usually just good enough (i.e., possibly far from the optimum) to strive and survive, an interesting problem is how to capitalize on this to make complex systems more secure. For instance, complex systems that are more tolerant to perturbations are also likely to be more secure (or easier to be secured).

Introducing the IoT into the existing complex systems will create the intrinsic intelligence needed to enhance the ability of these systems to adapt and self-organize. As the IoT can create interfaces and build bridges among different complex systems, we can expect the controllability of many existing systems to be either significantly improved, or newly created. The resulting sociotechnical (or cybersocial) systems can be then perceived as being built above the information and communication technologies (ICT) and the underlying social networks, and with different scopes of security, as shown in Figure 14.3.

An important class of optimization problems defined in complex systems are so-called wicked problems [27]. These problems are extremely difficult to solve, since they are even difficult to formulate precisely, and in addition, even solving any one of their aspects does not reduce their complexity. The solutions of wicked problems are always only approximations which are even difficult to

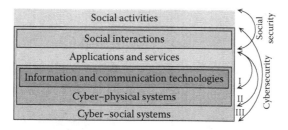

Figure 14.3: The security scopes of cyberphysical and cybersocial systems.

verify. These solutions cannot be obtained iteratively nor by exhaustive search to conquer their complexity. All wicked problems are unique, so solving any one such problem does not help to solve other similar problems. Wicked problems can be described from different perspectives or domains (cf. Figure 14.2) which determines the nature of their solution. The security of complex networks including, for example, cyberwars and global terrorism are good examples of wicked problems.

We can identify some recent trends that play a significant role in the evolution of complex systems. Many of these trends are well known and established in computer science to develop computing systems. For instance, virtualization is a technology to create virtual hardware and software computing platforms [13]. Virtualized computing environments are often used in the education of computer security [9]. More interestingly, we can observe network virtualization tendencies in other types of complex systems: for example, fiat money and derivatives in financial markets (vs. the real economy), virtual friendships on social websites (vs. real human relationships), incremental research results reported in scientific publications (vs. bold and risky research problems that are difficult to publish), manufacturing perceptions and impressions in social networks (vs. going beyond more easily manipulated or even artificially manufactured information labeling using metadata), and so on. Particularly in social networks, ongoing virtualization causes decorrelation of intrinsic processes, changes of (once long-standing) values and shifts in perception; for example, devaluation of the experiences of older generations, and of university education as it no longer guarantees a well-paid job and prospective career, and higher-income activities are no longer more risky nor demanding more resources (e.g., investing in the stock market).

Furthermore, distributing and pooling resources is another example of technology originally devised to build cloud computing platforms [13]. This strategy can be used more generally to build fundamentally new products and services by exploiting multiple types of collaboration and interaction. For instance, functionality can be shared between the smart watch and the smartphone, and a cell phone antenna can be utilized as a wearable element of clothing. A point-of-care medical diagnosis can be performed in a distributed manner in close proximity to the patient rather than in a centralized manner in the laboratory. Laboratory equipment can be centralized and accessed remotely via defined interfaces [40], and so on.

The change of the internal state of a complex system is indicative of prior or ongoing perturbations, and possibly divergence from the normal operating conditions; for example, as the result of a security attack. Such changes can be often inferred using so-called markers which are either observable substances or measurable quantities. For example, a biomarker or biostamp indicates the presence of a living form in the environment, or enables one to distinguish between normal and pathogenic processes occurring in living matter. A genetic marker is a piece of DNA identifying the specific biological species. The decorrelation of selected system processes can serve as a general-purpose marker to quantify system stability and sustainability. Other markers, such as the rate of failure or the amount of flow, are often used to monitor the quality of the services provided.

In the following sections, we will review some representative examples of complex systems and discuss their security aspects.

14.1.1 Wireless networks

Wireless access is fundamental for building modern telecommunication networks, including the sensor networks for the IoT. The nature of wireless transmissions, at the lowest (physical) layer of the protocol stack, creates unique challenges as well as opportunities. The main security challenges of wireless transmissions are jamming and eavesdropping [56]. The jamming station transmits intentionally or accidentally concurrently in the same frequency band as the legitimate station, and the resulting electromagnetic interference normally exhausts the capabilities of the receiving station to recover the transmitted information. The optimum jamming strategy requires knowledge of the legitimate transmission schedule; this can be achieved, for example, by hijacking a legitimate station and altering its transmission schedules and protocols. Jamming efficiency, as well as resistance, can be improved by a group of collaborating stations. Jamming is part of the broader electronic warfare to gain control of the electromagnetic spectrum. In general, the stations in a wireless network can monitor each other's actions to learn (in a distributed, cooperative fashion) and also to suppress (e.g., to penalize) any suspicious or unusual behaviors by rogue stations.

A traditional protection against eavesdropping is based on cryptography [56]. However, particularly for the lightweight wireless sensor nodes to be deployed in the IoT networks, the use of cryptography is severely limited, although not impossible [29]. In general, cryptography is used to implement users' authentication and authorization as well as to create confidentiality of data and of information flows; for example, to restrict multimedia content distribution to only the paying customers.

Recently, information-theoretic approaches to security gained considerable attention [42]. These methods guarantee a secure transmission that is unbreakable (no matter how computationally powerful the eavesdropper may be) and even quantifiable as the maximum secure transmission rate. However, whereas the assumptions about computational power and knowledge of the transmission schedules by the eavesdropper are relieved, all the wireless physical layer security schemes considered seem to rely (in one way or another) on more favorable transmission conditions for the legitimate station than those for the eavesdropper; this can never be permanently guaranteed in practice. For example, time-varying and unpredictable propagation conditions are known approximately only to the end stations of the particular wireless link. More importantly, an unknown number of eavesdroppers can collaboratively bypass the information-theoretic guarantees. On the other hand, it is possible to show that using multiple transmitting and receiving antennas does improve information-theoretic security [42].

14.1.2   *Biological networks*

Many functional as well as structural network models have been devised to study biological systems [19]. Examples of such models are gene regulatory networks, gene coexpression networks, protein residue networks, protein–protein interaction networks, biochemical reaction networks, metabolic networks, intercellular networks, vascular networks, brain networks, and many others. To capture the complexity of biological systems, it is often important to consider multiple, possibly hierarchical models representing different spatiotemporal scales. The network models of biological systems can be used to devise various "hacks" to modify certain functions of these systems; for example, to define personalized medicine [23], to synthesize artificial biological components [5], or to disrupt biological functions using a new generation of the DNA-based biological weapons possibly disguised as genetically modified food or medical vaccinations [45, 53]. Nanotechnology and nanoscale networks exploiting biomarkers will play a key role in bridging the gap to control biological functions at the cellular and subcellular level. Nanotoxicology is concerned with the safety of nanoscale substances and devices which can be extended to cover the issues of (nano)security also. Moreover, the market for innovative healthcare products supported by IoT devices is growing rapidly, with applications mainly in fitness, long-term medical conditions, and preventative medicine.

Biological immunity is a well-known example of the natural security system defending organisms against infection and invasion by foreign substances and attacks by viruses, bacteria, and parasites. The key feature of the immune system is the capability to differentiate between the self and the nonself [57]. The simpler organisms have immune systems composed of the discrete, general-purpose effector cells and molecules. More complex organisms also developed so-called specific immune responses which recognize billions of foreign pathogens. The former subsystem is known as innate immunity, and it is found in most living organisms. It also includes the cellular boundaries such as tissues and a skin as a natural security barrier against invasion (cf. a firewall in computer networks). The latter subsystem greatly benefits from the adaptivity and learning to launch more sophisticated counterattacks against invasion. Moreover, the immune system is fully distributed (no centralized control), tolerant to small errors (malfunctions), and, in normal conditions, it protects itself. The adaptive part of the immune system also exploits diversity combining to build a large number of antibody receptors.

Among well-known and understood examples of bacteriophages attacking susceptible bacterial cells is a T7-phage infection of the *Escherichia coli* cell [18]. In this process, several layers of the defense mechanisms of the bacterium are overcome by the phage. Briefly, the phage attaches to the bacterium and injects into it its viral DNA, including the proteins needed to halt the DNA replication of the host. The host cellular machinery is then used to begin replication of the viral DNA and the supporting proteins.

14.1.3   Social networks

Social networks are the main product of brain activity. They are as vulnerable to attacks and hacking attempts as any other networks [38]. A simple example of the hacking of social systems is making and breaking promises. The resources pertinent to social networks are usually of an abstract nature: social status, ideas, happiness, motivation, freedom, free time, and many others. As these abstract resources can be taken away (stolen), so they are the subject of competition as well as security concerns. The most common attacks to and within social networks are various types of psychological manipulation, with the strategies referred to as pretexting, diversion theft, phishing, and others [4]. While these attacks are well-defined criminal activities, the activities of, for instance, psychopaths can be much more damaging to society, and yet they rarely result in any criminal convictions. The actions of psychopaths can bring down whole companies and even state economies (depending on the social status of the psychopath), and thus, they may affect lives of many more people, unlike computer hackers who usually cause only limited financial damage. In fact, social networks are likely to be much more susceptible to attacks than computer networks.

Psychopathic activities are now much better understood [6]. They are also very illustrative in defining the security of social networks. In particular, psychopaths exploit the vulnerabilities of social networks, similarly to hackers in computer networks. Psychopathy has been recently recognized as a personal survival strategy rather than a personality disorder. This has not only many legal implications, but also implies that psychopathy may propagate through society as an epidemic, as suggested by empirical data as well as our everyday experiences [59]. Sociologists warn about the recent outburst of pathological behaviors in society which may threaten the sustainability of society and social structures. Psychopathy is more likely to be detected in open societies (cultures and institutions) rewarding individualistic (selfish) behaviors than in more traditional, closely interconnected communities. The mind of a psychopath appears to be "shapeless"; an important trait that allows them to take advantage by quickly adapting to diverse everyday situations to maximize personal profit (whereas nonpsychopaths appear to be unable to make such adaptation). At the same time, such flexibility of mind appears to an outside observer as a pattern of random decisions and unpredictable behaviors lacking any long-term goals. The primary objective of all psychopathic efforts is to acquire power so as to gain full control of other people's lives. Psychopaths are prone to take high risks in order to reach their objectives, and are programmed to win at "any cost."

Psychopaths are masters of mind games. Specifically, they seem to have the innate superior psychological skills to decipher other peoples' minds. They use these to uncover the strengths and weaknesses of other people, even during short encounters. They use such knowledge to devise methods of social manipulation to gain power while disguising their intentions and remaining undetected by the system's defense mechanisms (e.g., by important decision-makers in an organization). In a social network, psychopaths quickly map the social structure and categorize the players whom they encounter as: can be manipulated and taken advantage of; have no value for gaining more power, so can be ignored; can be a threat, so have to be eliminated; represent a good opportunity for advancing career and power, so have to be groomed; and so on. This way, they are able to gain genuine support and admiration from the psychologically manipulated individuals while eliminating those who may stop or slow down their advancement to higher social status with more power. Consequently, the discrepancies between their self-presentation, actions, and thinking are significantly larger in psychopaths than in nonpsychopaths. Interestingly, the only people in the system who are able to clearly recognize the ongoing psychopathic attacks are those who were considered to have no value to the psychopaths, so were ignored (and thus, not psychologically manipulated). Furthermore, psychopaths are capable of identifying each other to form short-term (rarely long-term) coalitions to increase the efficiency of their attacks against social networks.

A simple model of a large-scale social network (civilization) is to classify people as: free riders (excessive consumption of the resources compared to little contribution to society), the majority of users (their consumption and contributions balance), and contributors (their contributions exceed their consumption). A balance among these three groups affects the stability and sustainability of social networks. An unprecedented growth of the proportion of free riders in the post–second world war era should be a serious (security) concern. It has been proposed that the contribution level of subjects (e.g., people, things, and even processes) to the sustainability of systems or networks should be assessed by simply considering whether to add or remove particular subjects from the system. Another sociological theory claims that the stability of large-scale social networks require that these two conditions are satisfied: all members are rewarded (i.e., it pays off) for obeying the commonly accepted rules (the law), and most members are convinced that large rewards accumulated by some members are well deserved.

The Internet, as well as the IoT sensors, leaves traces and digital fingerprints as we live, travel, and get involved in many daily activities. For example, the biometric sensors in wearables and other healthcare technologies can and will be used to collect personal data beyond those that are currently being aggregated from social websites. Such data can be used to build accurate predictive models of individuals and groups (a collective mind). The concern is that these models can be exploited not only to identify and suppress psychopathic (or terrorist) activities, but they can be also used to design powerful computational strategies to disrupt or control large-scale social networks. Because of current intensive studies of the human brain and the mind, the privacy of IoT biometric data of (some) individuals may even become the subject of national security.

14.1.4 Economic networks

Economics studies production and distribution of services and goods [60]. It contributes rich and universal tools which can be readily used to describe the dynamics of other systems. For instance, academic publishing, once purely driven by the advancement of our knowledge, is now a much more complex process [2]. In particular, as research methods improved and vast amounts of knowledge were made available, research productivity increased considerably. The number of researchers in science and technology worldwide have increased exponentially in the past two decades, so following marketing and sales rules are now very important for survival in the very competitive world of academia [11]. A study [12] on the outcomes of globalization made the following three (among others) crucial observations: First, globalization revealed that scientific and technical knowledge is very liquid, so it flows to geographical areas with sufficient financial resources. Second, worldwide competition in research created strong pressure to deliver research results at the lowest prices possible (a so-called Dutch auction) while forming an (unsustainable) "winner takes all" competition. Third, "hard work" is no longer a winning strategy once it has been adopted by most players in the system.

As globalization has tremendously increased competition for resources, many networks are forced to operate in a low-resource regime (e.g., many systems have been made more green) which is very different from a regime with abundant resources. Hence, the economic wars in today's world are intensifying as the means of achieving geopolitical objectives. For instance, we can recognize ongoing monetary wars (e.g., quantitative easing, competitive currency devaluations), financial wars (accumulating exports to improve the trade balance, manipulating the prices of commodities such as precious metals and oil, producing suspicious credit and other ratings, using government obligations as debt collateral, etc.), economic sanctions (artificially limiting international trade), as well as intellectual property wars (often concerning the patent portfolios of large pharmaceutical and high-technology companies). The key to understanding economic warfare is that, due to globalization, national economies are now much more tightly interconnected, so any negative consequences are likely to spread through a global economic network [20]. Moreover, exploiting economic asymmetries and creating structures of mutually protective economic elements are some of the (general) tactics used in the current economic wars.

The IoT will improve existing and enable new economic processes such as tracking and managing the inventories of goods, delivering parcels, supporting e-commerce activities (online shopping), optimizing supply chains and manufacturing, creating smart environments for assisted living, personalized healthcare, and so on. Unfortunately, introducing intelligence into these economic processes will also create opportunities for more sophisticated small-scale as well as large-scale attacks and exploitation as part of economic warfare.

14.1.5   Computer networks

Computer security has been the subject of extensive investigations, so most of the research on security exists in this area. The most valuable outcome of these efforts is that the security principles discovered in computer networks can be transferred (possibly with some modifications) to other systems that can be modeled as networks. Thus, all networks are prone to hacking and hijacking and other types of attacks. For example, the service flows in the network can be disrupted, and rogue actions can spontaneously propagate through the network. To better illustrate attacks on computer systems, we describe the principles of a piece of malicious software (malware) known as the rootkit.

A basic idea of the rootkit software is the installation of a small program near or at the core of the operating system (on cell phones, on the IoT middleware, etc.). Such low-level deployment allows the rootkit to hide its presence from most other programs and processes, so it can operate in a stealth mode and remain undetected for very long periods of time [36]. Methods to detect the rootkit involve behavioral-based methods, signature and difference scanning, and a memory dump analysis. The rootkit typically opens the back doors for other malicious software, and can enable to gain the system access with administrator's privileges. Some strategies for rootkit deployment include social engineering to obtain initial administrator-level access to the system, or compromising the core update to be distributed to the server. For instance, the successful hacking of a company's internal computer network was demonstrated several times by first gaining unauthorized access to a less secure personal computer of a close family member of a key employee or executive of the company. Similarly, to secure the network of IoT devices, it is not sufficient to deploy secure gateways with firewalls, as the hackers will search for vulnerable nodes to bypass these defenses. The IoT network will be inevitably less secure than the traditional computer network due to the presence of nodes with limited computing power, so one has to be concerned with attacks initiated at these nodes of the IoT network, since they can then escalate into a conventional attack on the whole computer network (the Internet).

Finally, the cyberwars over computer (and soon also over IoT) networks [15] are becoming the primary objective in the global competition for resources rather than being only a traditionally supporting element of conventional war games [10]. Hence, these wars are large-scale politically and economically motivated hacking attempts. Recently, several governments publicly admitted that they are developing cyberattack strategies in addition to their existing cyberdefenses. The cyberwars are likely to be combined with other types of modern warfare strategies, as explained in [10]. Unfortunately, the IoT will enhance modern cyberwars by, for example, providing more accurate information about remotely located objects and environments (cities, buildings, individuals, weather, energy distribution grids, goods supply grids, etc.).

14.2   Computational Tools for Complex Systems

Empirical data are central to computational engineering for the creation of meaningful models of complex systems, and for the accelerating of the product development cycle and shortening of the time to market. There are two distinct approaches to data-driven modeling: so-called reverse modeling devises mathematical models to fit measured data, whereas forward modeling devises experiments to obtain the data that are the most useful for a given modeling strategy. The first approach has been used in computational science for many years. The newer, second approach aims to develop computational vision systems by attempting to directly reconstruct the characteristics of real-world systems. The second approach also has provenly better power in making predictions about system properties and uncovering unobserved relationships. However, modeling of dynamic systems is, in general, very challenging, as it is often limited to selected processes that are deemed to be the most important. It is very likely that modern-day hackers will exploit these computational approaches extensively to devise sophisticated and possibly multiscale and multidomain attacks against increasingly more complex systems while evaluating and limiting their chances of being detected.

In general, data can be collected from the (IoT) sensors, generated as inputs at human–machine interfaces, or be already stored in databases and remotely accessed via the World Wide Web. There are significant privacy and ethical issues concerning any sources of data, whether considering the sites of their generation or storage. In addition, there are still uncertainties about

- What data to collect and

- How to use information extracted from data

which gives rise to many open issues. For example, just as we do not store every packet passing through the Internet, we should not store all data from every sensor in the IoT. There is a trade-off between the real-time (online) learning from data and the accuracy of extracted information. Distributed data must be aggregated before applying data analytics and visualization. Since "sensing data without knowing the location is meaningless," the utility of the IoT is improved significantly by exploiting the spatiotemporal contexts. Such so-called geospatial analytics are inspired by the long-existing Geographic Information Systems (GIS) [37]. More importantly, structured as well as unstructured data are increasingly labeled by metadata to aid processing (mining) for the extraction of knowledge. It is likely that securing such metadata is more critical than securing actual data.

In general, computational methods are now being introduced into the traditional experiment-driven disciplines in life sciences and humanities such as biology, medicine, psychology, sociology, and even history. The main objective of these efforts is to recreate these disciplines on more rigorous mathematical foundations. It is then only a matter of time before computational security emerges to allow more systematic and rigorous study of security of complex systems. Thus, all major hacking attempts are likely to move away from random ad hoc discoveries and exploitation of vulnerabilities to the use of more scientific approaches. Computational hacking will strive to achieve similar goals, but more systematically and at different (i.e., very large) scales and possibly across different domains, well outside traditional computer networks. Obviously, computational security is concerned about scientific approaches to security rather than the security of computing.

In the following section, we will review some of the most promising modeling methodologies that can be used for computational security analysis of complex systems. A sophisticated targeted attack (i.e., the target is set a priori, in advance) can be constructed analogously to other engineering design work flows. For instance, a computationally aided attack may evolve following these steps:

1. Identifying and gathering relevant data from existing sources, and possibly also actively probing the system to solicit additional useful data.

2. Data evaluation and model building for the targeted system (multiple models at different scales and in different domains likely required).

3. Security assessment of the model using computer simulations (mathematical analysis likely to be intractable due to model complexity).

4. Exploiting the identified vulnerabilities to create an initial strategy of the attack.

5. Refining the attack strategy and devising its implementation under the concealment, available timescales and resources, and other required constraints.

These steps can be iterated in the course of the attack to adaptively increase chances of success and of concealing the attack.

14.2.1   Signal processing tools

Due to uncertainties in system models (parameter values and model structure) and the random behavior of actors often observed in many complex systems, statistical description and statistical signal processing must be used [30]. Many statistical signal processing problems rely on the ergodicity (i.e., the statistical averages are not time varying) and stationarity (i.e., the time averages are non-random) of the underlying random processes in the models. The main idea is that these signal processing methods work well on average, for the vast majority of inputs and system internal states. More recently, statistical signal processing approaches are considering the probability intervals in addition to the first- and second-order statistics corresponding to the statistical mean and variance, respectively [24, 28].

Statistical inferences are the basis of estimation theory, which focuses on the problems of finding the values of model parameters. These parameters are typically arranged into a discrete finite-dimensional vector, or they may be continuous-time signals. Good inference strategies are strongly dependent on how much statistical information is known a priori about the parameters. On the other hand, testing of hypotheses is the main task of detection theory. In this case, the parameters of interest are discrete random variables, and we want to know how likely (how probable) their different outcomes are upon observing some data a posteriori. Estimation and detection theory are both built from the first principles of probability theory. However, for more complex problems—for instance, involving high-dimensional and structured data—more practical methods beyond the first principles have been developed such as machine learning, pattern recognition, and fuzzy logic [46]. For instance, an adversary may use machine learning to identify the predictable patterns of the system processes to devise a powerful attack and avoid detection.

Deep learning attempts to learn efficient representations of unlabeled data, and then to follow similar principles as neural networks with multiple layers of nonlinear processing [22].

Game theory studies mathematical models of cooperative and competing strategies among interacting intelligent players. For example, it can be used to devise unpredictable schedules of security checks under minimum resource constraints [58].

To simulate the collective dynamics of complex systems and solve difficult problems, multiagent models and multiagent systems, respectively, have been developed [21]. The latter involves the intelligent agents within complex networks. The reasoning of these agents can have a form of algorithmic search, function, or reinforcement learning. Prior to multiagent simulations, the dynamics of complex systems were typically modeled by a set of time-dependent differential equations expressing the internal states of the system. These models usually lead to emerging or cyclical system behavior. However, their descriptive power is often limited to highly aggregated scenarios, since they do not account for time-varying relationships among the agents as they exploit their intelligence.

Consensus learning over networks is concerned with the analysis and algorithms for information diffusion in complex systems [49]. It generalizes centralized data fusion which does not scale well and has a single point of failure. It also generalizes highly vulnerable incremental linear learning as shown in Figure 14.4. Distributed learning is robust against link and node failures, and it has a good speed of convergence for small-world type of networks. Using the results of graph theory and control theory, performance guarantees can be given as a function of the network structure [49]. Model (c) in Figure 14.4 has diverse applications, including synchronization of coupled oscillators, flocking, gossiping, belief propagation, and load balancing in networks.

Finally, the algorithms are a crucial step of the implementation of signal processing methods. Their design is especially important for large-scale problems and time-critical applications such as online learning from large numbers of data sources. Algorithm design is also challenging when computing resources are constrained; for example, in IoT sensor nodes, using cryptography for securing information is difficult. Evolutionary algorithms are popular for simulations of large-scale complex systems [7]; they are trial-and-error stochastic optimization methods that are inspired by the principles of Darwinian evolution.

14.2.2 Network science tools

Network science is a rapidly emerging field developing mathematical tools for studying complex networks [47]. It capitalizes on results from many other disciplines such as graph theory, statistical mechanics, and data visualization and algorithms. Initial efforts were focused on describing the structure of

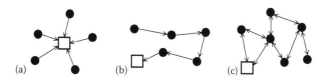

Figure 14.4: Information consensus as (a) centralized fusion, (b) incremental learning, and (c) fully distributed cooperative learning (circles: sources; squares: sinks).

networks [54], since the ultimate goal is to predict system properties from their structure. Further efforts in network science considered processes and phenomena in networks. Current research is concerned with dynamic networks that evolve over time. Some of the most important network (complex system) properties studied by network science are: connectivity, autonomy, emergence, nonequilibrium steady states, self-organization, and evolution. Connectivity is an integral quantity: deciding whether the two nodes are connected is only possible over a nonzero time interval. The autonomy of nodes is a necessary condition to allow their intelligent decisions. Even though the emergent macroscopic behavior from local interactions is nonrandom, it is so complex that it is unpredictable. The existence of states close to an equilibrium (being nonstable) is a crucial condition for the system to keep evolving. Self-organization is a form of structural adaptivity in response to actual or perceived (anticipated) external perturbations or events. Evolution itself is a long-term, large-scale adaptivity to the external environment.

Network models are usually derived from available data, and they are often only approximations or subgraphs of the whole system. The literature refers to the study of network structures as social network analysis (SNA) [54]. SNA offers different types of metrics to evaluate network connectivity, centrality, transitivity (e.g., clustering), similarity, searchability, routing, partitioning (e.g., communities), and other properties. For instance, centrality metrics assess the importance of nodes (or edges) within the network structure; they are predictive of a node's (edge's) influence on some phenomena and events such as malfunction and failure, disease spreading, and information flows.

In general, the network metrics can assume network nodes or edges. The metrics can be defined locally for every node or edge, or for a group of nodes or edges, while possibly taking into account whether the network represents a directed graph or not. Network metrics assuming the unit-weight edges are widely accepted. Redefining the metrics for weighted and, thus, more realistic network models is not straightforward, so many such metrics have been proposed in the literature. Another topic of significant practical interest is to specify the procedures for generating artificial random and nonrandom large-scale network models with the desired structural properties. Except for a simple, purely random network generator, more realistic scale-free and small-world network constructions utilize preferential attachment and random rewirings.

The network robustness against failures, the spreading of epidemics, information cascades, and searching and routing phenomena are of particular interest to computational security. Network robustness is evaluated as a change in the network metrics when nodes or edges are being removed or added. Alternatively, network resilience is its ability to resist a change due to external disturbances. Network resilience is the speed at which the network returns to normal functioning after external perturbations. The spreading of epidemics and information cascades predict the autonomous distribution of material objects (e.g., viruses and mechanical malfunctions) and information (e.g., know-how and news) through the network, respectively. These phenomena can be exploited by the adversary to forecast and, thus, to plan the distribution of malicious objects (e.g., malware) and misleading information (propaganda). Searching networks aims to find a source–destination path in a reasonable amount of time, whereas network routing finds such a path with minimum cost. Since many networks have a small-world property (cf. six degrees of separation), an adversary can (theoretically) reach any node within these networks in only a small number of steps, which makes these networks more vulnerable.

In general, many network models seem to exhibit a threshold robustness against attacks. The attacks below the threshold are normally absorbed by the network, whereas they may cause great damage once they exceed the threshold. This threshold is a function of the network structure (scale-free, small-world) and the defense mechanisms used within the network for its protection. In other words, the defense mechanisms employed within the network affect how the network structure is perceived by the attackers. Depending on knowledge about the network structure, the attacks may target highly centralized (i.e., highly connected) nodes (referred to as hubs). Such targeted attacks are very effective in disrupting scale-free networks (many real-world networks are scale-free). Thus, the knowledgeable attacker is very powerful and can cause significant damage to the system (or, for the same reasons, a disease can be cured more effectively). However, even purely random attacks without any knowledge can be effective in some types of networks, and if they are allowed to accumulate over time.

14.2.3   *Controllability and observability of networks*

Network controllability and observability are important topics in network science. They are also fundamental for the computational security of complex systems. A smart attacker may be motivated to gain at least partial control of the system and to get access to additional resources rather than attempting to cause any damage. Optimum controllability and observability can be derived for a static directed network where every node and edge is assigned a scalar value [34, 35]. The node values represent the system's internal state, and the edge values are attenuations of the node states. Controllability is defined as the ability to drive the system from an arbitrary state to any other state. The task is to find the minimum number of driver nodes to become external inputs to the network to achieve its full controllability.

A brute-force search for the driver nodes is an NP-hard problem. Moreover, the mathematical conditions of controllability are numerically difficult to evaluate for large networks and, also, the edge weights are usually unknown in many practical scenarios. Using a graph-matching technique, it was shown in [34] that the minimum number of driver nodes is strongly degree distribution dependent; surprisingly, the driver nodes are usually the nodes with a small degree connectivity (i.e., not highly connected hubs). Therefore, sparse and heterogeneous networks (likely to be representative of the IoT) are more difficult to control than dense and homogeneous ones.

Similarly, observability of complex networks is the task of estimating their internal state from a finite number of observations. Assuming a linear network, considered above to illustrate network controllability, one may define a dual problem to immediately identify the observation (sensor) nodes of network knowing the driver nodes [35]. To overcome similar computational issues that were mentioned above for network controllability, observability can be approximated by decomposing the network into a set of strongly connected components; typically, it is sufficient to select one sensor node within each of these components. Similarly, partial observability identifies the minimum number of sensor nodes to reconstruct some (but not all) state variables. This is analogous to the problem of defining optimum markers for selected processes in a complex system (cf. Figure 14.2).

The use of conventional trial-and-error random attacks to discover network vulnerabilities may greatly increase the chances of detecting (and stopping) these attacks. Hence, the more sophisticated approaches of computational security using techniques derived from network science can be particularly attractive. For instance, effective attacks against networks may target their controllability as well as observability. Such attacks are dependent on the knowledge the adversary has about the network (e.g., the topology and weightings). In many scenarios, it is fair to assume that the adversary has at least some partial knowledge (e.g., some of the driver and sensor nodes are known) which can be represented as a subnetwork of the original network. The adversary may then employ the outlined computational procedures to devise an optimum attack, including to first attempt to gain additional information about the targeted network.

14.2.4 Network tomography

Network monitoring is a generalization of network observability as defined in the previous subsection. Monitoring of complex systems is essential; for example, for the allocation of resources, ensuring a certain quality of offered services, and for detecting abnormal activities and behaviors to guarantee the network reliability and security [8]. Since separate monitoring of individual nodes and edges is impractical, either active network probing or passive observation is used instead (e.g., end-to-end measurements). This leads to an inverse problem of either reconstructing the network's internal state, or testing hypotheses (e.g., to decide whether the network behaves abnormally) from a finite number of observations. From the implementation perspective, these problems can be interpreted as distributed or collaborative sensing, inference, or decision-making.

Conventional mathematical tomography exploits either sectional or projection imagining and the subsequent computational reconstruction. However, the majority of the methods developed so far for network tomography are straightforward applications of statistical inference and hypothesis testing, with limited or no considerations for the network structure and other network properties. Furthermore, whereas conventional mathematical tomography provides reconstruction guarantees, the uniqueness of network tomography reconstruction has not been shown, but for several specific network instances. Ref. [8] suggests assuming graph embedding in higher-dimensional hyperbolic spaces to obtain proof of reconstruction for more general network structures.

It is clear that network tomography can be vital to computational security. In many scenarios, the network tomography procedures of identifying the network's vulnerabilities can directly lead to efficient attack strategies against the key network infrastructure. For instance, in [8], an adaptive iterative network tomography reconstruction is investigated, where new observation nodes are identified as the network is being partially reconstructed at each step. Such iterative exploration of the network can be readily combined with malicious activities while it is progressing.

14.2.5 Lessons from communications engineering

Continuous information exchanges glue networks together, albeit that these exchanges may take on many diverse forms. Over 60 years of communications engineering revealed some recurring patterns and strategies of how the information flows are efficiently implemented [43, 61]. In particular, information is a measure of uncertainty, and since this uncertainty varies in space and time, information is a function of spatiotemporal coordinates [39]. Additional uncertainty distorting information during its transmission arises due to uncertainties contained in the transmission medium. Hence, successful information transmission can only be achieved statistically (i.e., on average).

Information is always embedded into some form of matter for its transmission over a physical medium in the process referred to as modulation. In complex systems, we observe modulated patterns, but we may not know what information they represent, nor the exact spatiotemporal location where the information originated, nor where it will be extracted, and how it will be used. On the other hand, man-made telecommunication networks exploit many simplifying assumptions, such as: information does not vary over space and time, the information sources and destinations are known exactly, and the use of information outside the telecommunication network is never considered.

The only method available to achieve reliable information transmission appears to be diversity, which requires that the same information is transmitted more than once. The destination is then much more likely to recover transmitted information. Another form of diversity is to adapt the modulation patterns and formats to the transmission medium.

Information transmission experiences many trade-offs in the using of resources. One such fundamental law is to trade off energy for information; it is unclear if information can be converted back to energy (in macroscopic, not quantum systems). It is also unclear how information transmission is optimized in various complex systems; for example, to minimize energy, or to maximize reliability or other objectives, depending on the system considered. As complex systems tend to maintain their internal state to achieve stability, external perturbations are suppressed, and so does the speed of change of the internal state. Since energy is normally minimized if the state transitions to a stable state more quickly, there exists a minimum energy required to deliver information, and thus, to keep the network (and complex systems) together. In fact, the resources required to just keep the network together are so large that, in general, the overall efficiency of providing some utility through the network is very low.

The main implications of these lessons to design secure networks is that security should be embedded into the network and considered from the outset. There seems to be a trade-off between security and reliability; that is, more reliable information transmissions are less secure, and no system can be made absolutely secure nor absolutely reliable. Moreover, security attacks can be concealed more easily if they preserve the system's internal state or do not change it significantly.

14.3 Perspective Research Directions

The IoT builds bridges between different complex systems in our world. The IoT networks span distances across many orders of magnitude, down to nanoscale levels. Hence, these networks allow the environment and systems to be explored and affected to an extent not previously reachable. The Internet was created over two decades ago with no security in mind, so it was soon exploited by rogue users who are now causing billions of dollars' worth of damages to the global economy annually. Since then, the field of computer security has been well established and is evolving continuously.

Our perception of the world is changing as IoT networks are introduced. The complexity of innovative products and services has increased significantly, while the boundaries between diverse systems are becoming blurred. This drives the need for new, holistic approaches to security to reflect the enormous growth in complexity of the ambient world. In particular, security must consider complex systems at different timescales, and span distances across many orders of magnitude while being geographically distributed around the globe. This also requires that the design, deployment, and monitoring of complex systems is performed statistically, and that security features are embedded in systems from their early design whenever possible. More importantly, security now seems to be increasingly more concerned with large-scale attacks representing new kinds of warfare.

Advances in computational modeling, computational engineering, computational sociology, and so on, naturally motivate a new field of computational security. Combining methods from several disciplines such as computer security, network science, network theory, network tomography, and the humanities (sociology, psychology) can allow security provisions that are much more systematic as well as scientific. Even though these approaches are highly mathematical, they are well justified by their potential benefits in building complex systems that are provably secure, stable, and reliable.

In this chapter, some of the trends that the IoT networks are going to bring about were outlined. The perspectives of complex systems and their representation as networks were utilized. Specifically, we discussed the security aspects of some man-made complex systems such as wireless telecommunication networks and computer networks, and also of several other ubiquitous networks such as biological, social, and economic networks. We identified the need to define universal security principles for all these networks and systems. Computational security may emerge as the mainstream approach to design sophisticated attacks, as well as to devise broadly efficient defenses and countermeasures. Thus, computational methods will aid the transition from reactive (suspected) to proactive (suspicious) security considerations.

In the following list, we highlight some perspective research directions in the area of complex systems and networks and their security.

- Multiscale and multidomain modeling of complex dynamic systems with the correct level of granularity is a very challenging problem which is fundamental to the security of pervasive IoT systems. In general, devising metrics to assess the usefulness of models (e.g., their accuracy) is important. Even though the general predictive limits of such models are unclear, they may greatly enhance our understanding of complex IoT systems.

- As all physical systems are, in general, trying to maintain their internal steady state (cf. Newton's first law and low-pass filter analogies), security attacks that are slower than systems' (steady-state) dynamics or responses may be provably undetectable. On the other hand, systems may be unable to respond to attacks that are much more rapid than such dynamics.

- The universal laws governing complex systems including general networks are yet to be discovered. This also includes the design of networked and distributed systems with defined trade-offs between security and other network characteristics; for example, the reliability.

- Even linear networks do not scale linearly in a low-resource regime. Security attacks within nonlinear (complex) systems such as biological and social networks are more serious, as small targeted perturbations may have a large effect on these systems.

■ The networks appear to be inherently extremely inefficient in their utility provisioning. For example, most of the energy seems to be consumed on sustaining the network structure rather than expended on delivering services such as security.

■ It is unclear whether network evolution is an open-ended process, or whether all networks mature and disintegrate in some finite time period, and whether this network lifetime is shortened due to security breaches.

■ There is an enormous need to devise markers to forecast various events within the networks and other complex systems. The markers can be used as proactive security measures. For example, the decorrelation of processes that were once correlated may indicate an upcoming systemic change.

■ Many complex systems are forming hierarchical boundaries between submerged networks. These boundaries could be used to naturally enhance the security of the whole network.

■ The interactions between virtualized systems and the underlying physical systems have not been considered when trying to understand the corresponding security implications. For instance, securing social networks does not secure the underlying biological systems.

■ The soldiers of the future will be experts highly trained in technology, life sciences, humanities, and other interdisciplinary experts, as there is a shift from conventional warfare into other spaces, domains, and interfaces such as the Internet (cyberwars), social networks (psychological wars), economic networks (currency wars), and biological networks (synthetic organisms wars).

Bibliography

[1] R. Albert, H. Jeong, and A.-L. Barabási, "Error and attack tolerance of complex networks," *Nature*, vol. 406, pp. 378–382, Jul. 2000.

[2] B. Alberts, "Impact factor distortions," *Science*, vol. 340, no. 6134, p. 787, May 2013.

[3] Y. Amichai-Hamburger, *The Social Net: Understanding Our Online Behavior*. Oxford, UK: OUP Oxford, 2013.

[4] J. R. Anderson, *Security Engineering: A Guide to Building Dependable Distributed Systems*. New York, USA: Wiley, 2008.

[5] E. Andrianantoandro, S. Basu, D. K. Karig, and R. Weiss, "Synthetic Biology: New engineering rules for an emerging discipline," *Mol. Systems Biol.*, vol. 2, no. 1, pp. 1–14, 2006.

[6] P. Babiak and R. D. Hare, *Snakes in Suits: When Psychopaths Go to Work.* New York, NY: Harper Business, 2007.

[7] T. Back, *Evolutionary Algorithms in Theory and Practice.* New York, NY: OUP USA, 1996.

[8] J. S. Baras, "Network Tomography: New rigorous approaches for discrete and continuous problems," in *Proc. ISCCSP*, 2014, pp. 611–614.

[9] S. M. Bellovin, "Virtual machines, virtual security?" *Commun. ACM*, vol. 49, no. 10, p. 104, Oct. 2006.

[10] J. Bērziņš, "Russia's new generation warfare in Ukraine: Implications for Latvian defence policy," Apr. 2014, nat. Defence Academy of Latvia.

[11] L. Bornmann and H.-D. Daniel, "The usefulness of peer review for selecting manuscripts for publication: A utility analysis taking as an example a high-impact journal," *PLoS One*, vol. 5, no. 6, 2010, doi:10.1371/journal.pone.0011344.

[12] P. Brown, H. Lauder, and D. Ashton, *The Global Auction.* New York, NY: OUP USA, 2011.

[13] R. Buyya, C. S. Yeo, S. Venugopal, J. Broberg, and I. Brandic, "Cloud computing and emerging IT platforms: Vision, hype, and reality for delivering computing as the 5th utility," *Future Gen. Comp. Sys.*, vol. 25, no. 6, pp. 599–616, 2009.

[14] G. Caldarelli, *Networks: A Very Short Introduction.* Oxford, UK: OUP Oxford, 2012.

[15] R. A. Clarke and R. Knake, *Cyber War.* New York, NY: Tantor Media Inc., 2014.

[16] M.-C. Daniel and D. Astruc, "Gold nanoparticles: Assembly, supramolecular chemistry, quantum-size-related properties, and applications toward biology, catalysis, and nanotechnology," *Chem. Rev.*, vol. 104, no. 1, pp. 293–346, 2004.

[17] W. H. De Jong and P. J. Borm, "Drug delivery and nanoparticles: Applications and hazards," *Int. J. Nanomed.*, vol. 3, no. 2, pp. 133–149, Jun. 2008.

[18] M. Demerec and U. Fano, "Bacteriophage-resistant mutants in Escherichia Coli," *Genetics*, vol. 30, no. 2, pp. 119–136, 1945.

[19] E. Estrada, *The Structure of Complex Networks*. New York, NY: OUP USA, 2011.

[20] D. J. Fenn, M. A. Porter, M. McDonald, S. Williams, N. F. Johnson, and N. S. Jones, "Dynamic communities in multichannel data: An application to the foreign exchange market during the 2007–2008 credit crisis," *Chaos*, vol. 19, no. 033119, 2009.

[21] J. Ferber, *Multi-Agent System: An Introduction to Distributed Artificial Intelligence*. Upper Saddle River, NJ: Addison Wesley Longman, 1999.

[22] L. Gomes, "Machine-Learning maestro Michael Jordan on the delusions of Big Data and other huge engineering efforts," *IEEE Spectrum*, 20 October 2014, online: spectrum.ieee.org.

[23] M. A. Hamburg and F. S. Collins, "The path to personalized medicine," *New England J. of Med.*, vol. 363, no. 4, pp. 301–304, 2010.

[24] M. A. M. Hassanien, "Non-ergodic error rate analysis of finite length received sequences," *IEEE Tr. Vehicular Tech.*, vol. 62, no. 7, pp. 3452–3457, Sep. 2013.

[25] C. S. Holling, "Understanding the complexity of economic, ecological, and social systems," *Ecosystems*, vol. 4, no. 5, pp. 390–405, 2001.

[26] A. Juels, "RFID security and privacy: A research survey," *IEEE JSAC*, vol. 24, no. 2, pp. 381–394, Feb. 2006.

[27] S. H. Kaisler and G. Madey, "Complex adaptive systems: Emergence and self-organization," 5 January 2009, online: www3.nd.edu/~gmadey.

[28] O. Kallenberg, *Foundations of Modern Probability*. New York, NY: Springer, 1997.

[29] C. Karlof, N. Sastry, and D. Wagner, "Tinysec: A link layer security architecture for wireless sensor networks," in *Proc. SenSys*, 2004, pp. 162–175.

[30] S. Kay, *Fundamentals of Statistical Signal Processing: Estimation and Detection Theory*. Upper Saddle River, NJ: Prentice Hall, 2001.

[31] D. Korowicz, "Catastrophic shocks through complex socio-economic systems: A pandemic perspective," *FEASTA*, pp. 1–10, Jul. 2013, online: feasta.org.

[32] J. S. Lansing, "Complex adaptive systems," *Annual Rev. Anthropology*, vol. 32, pp. 183–204, 2003.

[33] F. Li, B. Luo, and P. Liu, "Secure information aggregation for Smart Grids using homomorphic encryption," in *Proc. SmartGridComm*, 2010, pp. 327–332.

[34] Y.-Y. Liu, J.-J. Slotine, and A.-L. Barabási, "Control centrality and hierarchical structure in complex networks," *PLoS ONE*, vol. 7, no. 9, p. e44459, 2012.

[35] Y.-Y. Liu, J.-J. Slotine, and A.-L. Barabási, "Observability of complex systems," *Proc. Natl Acad. Sci. USA*, vol. 110, no. 7, pp. 2460–2465, 2013.

[36] A. Lockhart, *Network Security Hacks*, 2nd ed. Sebastopol, CA: O'Reilly, 2007.

[37] P. A. Longley, M. F. Goodchild, D. J. Maguire, and D. W. Rhind, *Geographic Information Systems and Science*, 2nd ed. Chichester, UK: J. Wiley & Sons, 2005.

[38] P. Loskot, "Security aspects of general networks," in *Proc. MIC-Electrical*, 2014, pp. 1–6.

[39] P. Loskot, "Why networking is way ahead to pool knowledge," 7 January 2013, Online: walesonline.co.uk.

[40] P. Loskot, B. Badic, and T. O'Farrell, "Development of advanced physical layer solutions using a wireless MIMO testbed," *Intel Tech. J.*, vol. 18, no. 3, pp. 162–181, 2014.

[41] P. Loskot, M. A. M. Hassanien, F. Farjady, M. Ruffini, and D. Payne, "Long-term drivers of broadband traffic in next-generation networks," *Ann. of Telecom.*, vol. 70, no. 1–2, pp. 1–10, Feb. 2015.

[42] D. Lun, H. Zhu, A. P. Petropulu, and H. V. Poor, "Improving wireless physical layer security via cooperating relays," *IEEE Tran. Sig. Processing*, vol. 58, no. 3, pp. 1875–1888, Mar. 2010.

[43] D. J. C. MacKay, *Information Theory, Inference, and Learning Algorithms*. Cambridge, UK: CUP Cambridge, 2003.

[44] M. Mazzucato, *The Entrepreneurial State: Debunking Public vs. Private Sector Myths*. London, UK: Anthem Press, 2013.

[45] M. Meselson, J. Guillemin, and M. Hugh-Jones, "Public health assessment of potential biological terrorism agents," *J. Emerg. Infect. Diseases*, vol. 8, no. 2, pp. 225–230, Feb. 2002.

[46] K. Murphy, *Machine Learning: A Probabilistic Perspective*. Cambridge, MA: MIT Press, 2012.

[47] M. Newman, *Networks: An Introduction*. Oxford, UK: OUP Oxford, 2010.

[48] M. Nicolelis, "Brain-to-brain communication has arrived. How we did it." Oct. 2014, Online: ted.com/talks.

[49] R. Olfati-Saber, J. A. Fax, and R. M. Murray, "Consensus and cooperation in networked multi-agent systems," *Proceedings IEEE*, vol. 95, no. 1, pp. 215–233, Jan. 2007.

[50] R. E. Pino (Ed.), *Network Science and Cybersecurity*. New York, NY: Springer, 2014.

[51] S. Ramaswamy, "The mechanics and statistics of active matter," *Ann. Rev. Cond. Matter Physics*, vol. 1, pp. 323–345, 2010.

[52] S. Saveljev, "Controlling the human brain (in Russian)," 15 February 2015, Online: youtube.com/watch?v=Oy5YQ-L2pSc.

[53] Q. Schiermeier, "Russian secret service to vet research papers," *Nature News*, no. 526, p. 486, Oct. 2015, doi:10.1038/526486a.

[54] J. Scott, *Social Network Analysis*, 3rd ed. London, UK: SAGE, 2013.

[55] D. Siegel, "How social media is rewiring our brains," 15 January 2015, Online: youtube.com/watch?v=CkMh6xdJNeM.

[56] N. Sklavos and X. Z. (Eds.), *Wireless Security and Cryptography*. Boca Raton, FL: CRC Press, 2007.

[57] L. M. Sompayrac, *How the Immune System Works*, 4th ed. Chichester, UK: Wiley-Blackwell, 2012.

[58] M. Tambe, *Security and Game Theory: Algorithms, Deployed Systems, Lessons Learned*. New York, NY: CUP USA, 2011.

[59] S. Wasserman and K. Faust, *Social Network Analysis*. Cambridge, UK: CUP Cambridge, 1994.

[60] C. Wheelan and B. G. Malkiel, *Naked Economics: Undressing the Dismal Science*, 2nd ed. New York, NY: W. Norton & Comp., 2010.

[61] J. M. Wozencraft and I. M. Jacobs, *Principles of Communication Engineering*. Prospect Heights, IL: Wiley New York, 1965.

Chapter 15

Privacy-Preserving Time Series Data Aggregation for Internet of Things

Rongxing Lu

Xiaodong Lin

Cheng Huang

Haiyong Bao

CONTENTS

15.1   Introduction

In recent years, the networking and collaboration among various devices has experienced tremendous growth. To adapt to the trend, the concept of the Internet of Things (IoT) has been paid great attention, not only from academia but also from industry. Essentially, the IoT is characterized by a large number of intelligent devices sharing information and making collaborative decisions [1]. Due to its potential to support a large number of ubiquitous characteristics and achieve better cost efficiency, the IoT can find many applications in the real world, including eHealthcare systems, smart homes, environmental monitoring, industrial automation, and smart grids, as shown in Table 15.1.

The IoT has attracted a lot of attention; and yet, despite all the attention, many security and privacy challenges have remained. Since most devices in the IoT are often deployed in unattended areas, they are vulnerable to physical attacks that are not detected immediately; and the nature of broadcasting using

Table 15.1   Typical applications and benefits of the IoT

Typical Applications	Benefits
eHealthcare system	Remote patient monitoring for better healthcare
Smart home	Real-time remote security and surveillance
Environmental monitoring	Effective monitoring with lower costs
Industrial automation	Remote equipment management for cost savings
Smart grid	Smart meters, sensors for real-time monitoring power grid

wireless communication also makes it easy for an attacker to launch an eavesdropping attack. As many research efforts have been made about IoT security challenges, in this chapter, we mainly focus on addressing privacy challenges in the IoT.

To address privacy challenges, that is, to protect an individual device's data privacy in the IoT, many privacy-preserving data aggregation schemes have been proposed [2, 3, 5, 6, 10, 27–30]. However, most of them only support one-dimensional data aggregation, which sometimes cannot meet the accuracy requirements of IoT scenarios. Although our previous work, EPPA deals with multidimensional data aggregation [10], it may not support large-space data aggregation very well. Therefore, aiming at the above challenges, we propose a novel privacy-preserving time series aggregation scheme for the IoT, which is characterized by exploiting the properties of group $\mathbb{Z}^*_{p^2}$ to support data aggregation for both small plaintext space and large plaintext space at the same time, which is thus more efficient than traditional data aggregation. Concretely, the main contributions are threefold.

- Firstly, we propose a novel privacy-preserving time series aggregation scheme based on the group $\mathbb{Z}^*_{p^2}$. The proposed scheme can use one single aggregated piece of data to achieve both small plaintext space aggregation and large plaintext space aggregation in a privacy-preserving way at the same time.

- Secondly, with a formal security-proof technique, we show that our proposed scheme can achieve each individual node's data privacy preservation.

- Finally, we implement our proposed scheme in Java and run extensive experiments to validate its efficiency in terms of low computational cost and communication overheads, and discuss the trade-off between utility and differential privacy levels.

The remainder of this chapter is organized as follows. In Section 15.2 and we formalize the system model and security model and identify our research goal. We present the detailed design of our privacy-preserving set aggregation scheme in section 15.4, followed by the security analysis and performance evaluation in sections 15.5 and 15.6, respectively. Section 15.7 reviews some related works and section 15.8 closes the chapter with a summary.

15.2 Models and Design Goals

In this section, we formalize our system model and security model and identify our design goal on time series aggregation in the IoT.

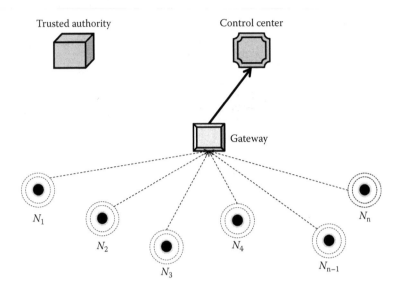

Figure 15.1: System model under consideration

15.2.1 System model

In our system model, we focus on a typical stationary IoT scenario, which mainly includes the following entities: one trusted authority, one control center, one gateway, and a set of nodes $\mathcal{N} = \{N_1, N_2, \ldots, N_n\}$, as shown in Figure 15.1, where n indicates the number of elements in the set $\mathcal{N}$, and its maximal value is denoted as n_{max}.

- Trusted authority (TA): This is a fully trustable entity, whose duty is to manage and distribute key materials to other entities in the system. In general, after key distribution, it will not participate in the subsequent data aggregation process.

- Control center: This is the core entity in the system, which is responsible for data collecting, processing, and analyzing the time series data from $\mathcal{N}$ for monitoring IoT scenarios.

- Gateway: This serves as a relay and aggregator role in the system; that is, it relays information from the control center to $\mathcal{N}$, and at the same time collects and aggregates data from $\mathcal{N}$ and forwards the aggregated data to the control center.

- Nodes $\mathcal{N} = \{N_1, N_2, \ldots, N_n\}$: Each node $N_i \in \mathcal{N}$ is equipped with sensors, which collect and report the time series data $M_i = (m_i, x_i)$, where m_i is a large value while x_i is a smaller value, to the control center via the gateway.

Differing from those previously reported data aggregation schemes [2, 3, 5, 6, 10, 27–30], the proposed time series aggregation in IoT will enable the control center to obtain not only small plaintext space aggregation, that is, $\sum_{i=1}^{n} x_i$, but also large plaintext space aggregation, that is, $\sum_{i=1}^{n} m_i$, which enables the control center to carry out more accurate data analytics for the monitoring and controlling of the IoT.

15.2.2 Security model

In our security model, we consider a generic adversary $\mathcal{A}$ who may compromise the privacy of nodes by eavesdropping on the communication data from the nodes to the gateway and those from the gateway to the control center. We also consider that the protocol participants, the control center and the gateway, are *honest-but-curious*. That is, they are supposed to follow the aggregation protocol appropriately ("honest"); meanwhile, they also try all sorts of measures to seek and infer knowledge of others ("curious"). In our IoT scenario, honest-but-curious participants will not tamper with the aggregation protocols; they do not maliciously distort or drop any received values and intermediate results, and they keep the system running normally. However, by analyzing messages and values routed through them, they try to infer each individual node's data. In addition, nodes $\mathcal{N} = \{N_1, N_2, \ldots, N_n\}$ are also honest, that is, no N_i will report false data to the control center or collude with the control center to obtain other nodes' individual data. Note that other types of attack are possible in IoT scenarios; for example, bad data injection attacks [11], DDoS attacks. Since our focus is on privacy-preserving time series aggregation, those attacks are currently beyond the scope of this research.

15.2.3 Design goal

Our design goal is to develop an efficient and privacy-preserving time series data aggregation scheme for IoT, such that the control center can obtain more varied and abundant information from one single aggregated piece of data. Specifically, the following two desirable goals should be satisfied.

■ **The proposed scheme should be privacy preserving.** Only the control center can read the aggregation results in the proposed scheme, and no one (including the control center) can read each individual user's data.

■ **The proposed scheme should be efficient.** Not only encryption at node side, aggregation at the gateway, but also decryption at the control center should be efficient in terms of computational cost. In addition, the proposed scheme, like other data aggregation schemes [6, 12, 13, 27], should use one single aggregated piece of data for transmission so as to achieve communication efficiency.

15.3 Preliminaries

In this section, we first review Shi et al.'s time series data aggregation scheme [6] and then recall the properties of group $\mathbb{Z}_{p^2}^*$, which will serve as the basis of our proposed aggregation scheme.

15.3.1 Shi et al.'s privacy-preserving time series data aggregation scheme

To enable an untrusted data aggregator to achieve some desirable statistics over multiple participants' data while without compromising each individual's privacy, Shi et al. [6] present an efficient and privacy-preserving time series data aggregation scheme and its enhanced version with the inclusion of a differential privacy technique. Here, the basic construction of Shi et al.'s scheme will be reviewed, which includes three parts: Setup, NoisyEnc, and AggreDec.

■ Setup: Given the parameter λ, a cyclic group $\mathbb{G}$ of prime order p is first chosen, where $|p| = \lambda$ and the decisional Diffie–Hellman problem is hard in $\mathbb{G}$. Then, a trusted dealer chooses a random generator $g \in \mathbb{G}$, and $n + 1$ random secrets $s_0, s_1, \ldots, s_n \in Z_p$, such that

$$s_0 + s_1 + \cdots + s_n = 0 \bmod p \qquad (15.1)$$

After that, a trusted dealer sets the public parameters *param* := $(\mathbb{G}, g, p, H)$, where H is a cryptographic hash function, that is, $H : \mathbb{Z} \to \mathbb{G}$, and assigns the secret key $\mathsf{sk}_0 = s_0$ to the data aggregator, and the secret key $\mathsf{sk}_i = s_i$ to each participant i.

■ NoisyEnc: Let $\hat{x}_i$ be the noisy data of participant i at time step t. Then, participant i computes the ciphertext as follows,

$$c_i = g^{\hat{x}_i} \cdot H(t)^{s_i} \qquad (15.2)$$

■ AggreDec: After receiving all ciphertexts $(c_1, c_2, \ldots, c_n)$ from the participants, the data aggregator computes

$$V = H(t)^{s_0} \cdot \prod_{i=1}^{n} c_i \qquad (15.3)$$

Obviously, because

$$V = H(t)^{s_0} \cdot \prod_{i=1}^{n} c_i = g^{\sum_{i=1}^{n} \hat{x}_i} \cdot H(t)^{\sum_{i=1}^{n} s_i}$$

$$= g^{\sum_{i=1}^{n} \hat{x}_i} \cdot H(t)^0 = g^{\sum_{i=1}^{n} \hat{x}_i} \tag{15.4}$$

and $\sum_{i=1}^{n} \hat{x}_i$ is in a small plaintext space, a brute-force search can be applied to decrypt $\sum_{i=1}^{n} \hat{x}_i$ from $g^{\sum_{i=1}^{n} \hat{x}_i}$. Assuming that each participant's data is in the range of $\{0, 1, \dots, \Delta\}$, the sum of the participants would be within the range of $\{0, 1, \dots, n\Delta\}$. Then, with Pollard's method [14], the decryption of $g^{\sum_{i=1}^{n} \hat{x}_i}$ can only require $O(\sqrt{n\Delta})$.

15.3.2 Properties of group $\mathbb{Z}_{p^2}^*$

In Shi et al.'s aggregation scheme [6], the group $\mathbb{G}$ is an abstract cyclic group of order p, which thus only supports messages in a small plaintext space. In the following, we discuss a concrete group $\mathbb{Z}_{p^2}^*$, and exploit its properties to enable us to support both small plaintext spaces and large plaintext spaces at the same time.

Given the security parameter λ, we choose a safe prime $p = 2q + 1$, where $|p| = \lambda$ and q is also a prime. Then, we can calculate Euler's totient function $\phi(p^2)$ as

$$\phi(p^2) = p^2(1 - \frac{1}{p}) = p(p-1) = 2pq \tag{15.5}$$

which shows that there are a total of $\phi(p^2) = p(p-1) = 2pq$ elements in the group $\mathbb{Z}_{p^2}^*$. Let $x \in \mathbb{Z}_p^*$ be an integer less than p; then, according to Fermat's Little Theorem, we have $x^{p-1} \equiv 1 \bmod p$. That is,

$$x^{p-1} = 1 + k \cdot p \tag{15.6}$$

for some integer k. We raise both sides of Equation 15.6 to the power of p and with the modulus p^2, we have

$$x^{p(p-1)} = (1 + k \cdot p)^p = 1 + \sum_{i=1}^{p} \binom{p}{i} (k \cdot p)^i = 1 \bmod p^2 \tag{15.7}$$

From Equation 15.7, we can see it still holds when $k = 1$. Therefore, let $y = p + 1$; we then have $\gcd(y, p^2) = 1$, and

$$y^p = (p+1)^p = 1 + \sum_{i=1}^{p} \binom{p}{i} p^i = 1 \bmod p^2 \tag{15.8}$$

Summarizing the above, we have the following two properties in group $\mathbb{Z}_{p^2}^*$, which can provide us with more flexible data aggregation.

1. $\forall x \in \mathbb{Z}_p^*$, we have $x^{p(p-1)} = 1 \bmod p^2$.

2. For $y = p + 1$, we have $y^p = 1 \bmod p^2$.

15.4 Proposed Time Series Data Aggregation Scheme

In this section, we present our new privacy-preserving time series data aggregation scheme, which is mainly comprised of four parts: system settings, data encryption at nodes, data aggregation at the gateway, and aggregated decryption at the control center.

15.4.1 System settings

Given the security parameter λ, a safe prime $p = 2q + 1$ is chosen, where $|p| = \lambda$ and q is a prime as well. In addition, a random number $g \in \mathbb{Z}_p^*$ is chosen as a generator of $\mathbb{Z}_{p^2}^*$, $h = g^p \bmod p^2$ is computed, and a secure cryptographic hash function $H : \{0,1\}^* \rightarrow \mathbb{Z}_p^*$ is also selected. Then, the public parameters are $param := (p, g, h, H)$.

The TA chooses n random numbers $s_i \in \mathbb{Z}_{p(p-1)}$, $i = 1, 2, \ldots, n$, and computes $s_c, s_g \in \mathbb{Z}_{p(p-1)}$ such that

$$s_c + s_g + \sum_{i=1}^{n} s_i = 0 \bmod p(p-1) \tag{15.9}$$

Finally, the TA sends s_c as the secret key to the control center, s_g as the secret key to the gateway, and s_i as a secret key to each corresponding node $N_i \in \mathcal{N} = \{N_1, N_2, \ldots, N_n\}$ via secure channels.

15.4.2 Data encryption at nodes

At every time interval t, each node $N_i \in \mathcal{N}$ will report two types of data (m_i, x_i), where the data m_i lies in a large plaintext space, that is, $m_i \in \{0, 1, 2, \ldots, \lfloor \frac{p}{n_{\max}+1} \rfloor\}$, where $n_{\max}$ is the maximal value of the number of nodes n, and the piece of data x_i is within a small plaintext space $\{0, 1, 2, \ldots, \Delta\}$. Concretely, each node N_i uses its secret key s_i to compute

$$c_i = (p+1)^{m_i} \cdot g^{x_i} \cdot H(t)^{s_i} \bmod p^2 \tag{15.10}$$

and reports c_i to the gateway.

15.4.3 Data aggregation at gateway

After receiving all ciphertexts c_i, $i = 1, 2, \ldots, n$, from nodes $\mathcal{N} = \{N_1, N_2, \ldots, N_n\}$, the gateway uses its secret key s_g to perform the following aggregation operation,

$$C = \left(\prod_{i=1}^{n} c_i \right) \cdot H(t)^{s_g} = \left(\prod_{i=1}^{n} (p+1)^{m_i} \cdot g^{x_i} \cdot H(t)^{s_i} \right) \cdot H(t)^{s_g} \bmod p^2$$

$$= (p+1)^{\sum_{i=1}^{n} m_i} \cdot g^{\sum_{i=1}^{n} x_i} \cdot H(t)^{\sum_{i=1}^{n} s_i + s_g} \bmod p^2 \qquad (15.11)$$

and sends the result C to the control center.

15.4.4 Aggregated data decryption at control center

After receiving the aggregated ciphertext C, the control center performs the following steps to recover the aggregated data.

■ Step 1: The control center uses its secret key s_c to compute

$$D = C \cdot H(t)^{s_c} \bmod p^2$$

$$= (p+1)^{\sum_{i=1}^{n} m_i} \cdot g^{\sum_{i=1}^{n} x_i} \cdot H(t)^{\sum_{i=1}^{n} s_i + s_g} H(t)^{s_c} \bmod p^2$$

$$\xrightarrow{\because \sum_{i=1}^{n} s_i + s_g + s_c = 0 \bmod p(p-1)} \qquad (15.12)$$

$$= (p+1)^{\sum_{i=1}^{n} m_i} \cdot g^{\sum_{i=1}^{n} x_i} \bmod p^2$$

■ Step 2: The control center continues to use p to compute

$$\bar{D} = D^p$$

$$= \left((p+1)^{\sum_{i=1}^{n} m_i} \cdot g^{\sum_{i=1}^{n} x_i} \right)^p \bmod p^2$$

$$\xrightarrow{\because (p+1)^p = 1 \bmod p^2} \qquad (15.13)$$

$$= g^{p \sum_{i=1}^{n} x_i} \bmod p^2 = h^{\sum_{i=1}^{n} x_i} \bmod p^2$$

Because $\sum_{i=1}^{n} x_i$ is still within a small plaintext space $\{0, 1, 2, \ldots n\Delta\}$, similarly to Shi et al.'s scheme [6], we can also use Pollard's method to recover $\sum_{i=1}^{n} x_i$ with the computational complexity $O(\sqrt{n\Delta})$.

■ Step 3: After obtaining $\sum_{i=1}^{n} x_i$, the control center computes

$$\hat{D} = \frac{D}{g^{\sum_{i=1}^{n} x_i}} = \frac{(p+1)^{\sum_{i=1}^{n} m_i} \cdot g^{\sum_{i=1}^{n} x_i}}{g^{\sum_{i=1}^{n} x_i}} = (p+1)^{\sum_{i=1}^{n} m_i} \bmod p^2 \quad (15.14)$$

Because $m_i \in \{0, 1, 2, \ldots, \lfloor \frac{p}{n_{\max}+1} \rfloor\}$, we have $\sum_{i=1}^{n} m_i < p$. Therefore, we have

$$\hat{D} = (p+1)^{\sum_{i=1}^{n} m_i} = 1 + p \cdot \sum_{i=1}^{n} m_i + \sum_{i=2}^{\sum_{i=1}^{n} m_i} p^i \cdot \binom{\sum_{i=1}^{n} m_i}{i}$$

$$= 1 + p \cdot \sum_{i=1}^{n} m_i \bmod p^2 \tag{15.15}$$

and thus $\sum_{i=1}^{n} m_i$ can be recovered by computing

$$\sum_{i=1}^{n} m_i = \frac{\hat{D} - 1}{p} \tag{15.16}$$

As a result, two types of aggregated data $\sum_{i=1}^{n} m_i$, $\sum_{i=1}^{n} x_i$, respectively in large plaintext space and small plaintext space, can be obtained by the control center.

15.4.4.1 Discussion on privacy enhancement with differential privacy

Differential privacy is a popular privacy-enhancing technique [15], which has been widely discussed in privacy-preserving data statistics. With the differential privacy technique, proper noises, for example, noises extracted from symmetrical geometric distribution, Laplace distribution, and so on, will be added to the aggregation result, which can make the outputs from similar inputs indistinguishable. Formally, we say that a randomized algorithm A satisfies ε-differential privacy, if for any two data sets D_1 and D_2 differing by a single element, for all $S \subset Range(A)$, $\Pr[A(D_1) \in S] \leq \exp(\varepsilon) \cdot \Pr[A(D_2) \in S]$ holds. The adding of noises is crucial for differential privacy. As the aggregation data are discrete in the proposed scheme, noises extracted from geometric distribution are applied. The noises, generation by the use of geometric distribution was first introduced by Ghosh et al. [16], where the noise is chosen from a symmetric geometric distribution $Geom(\alpha)$ with $0 \leq \alpha \leq 1$. Then, the $Geom(\alpha)$ can be viewed as a discrete approximation of Laplace distribution $Lap(\lambda)$, where $\alpha \approx \exp(-\frac{1}{\lambda})$. The probability density function (PDF) of geometric distribution $Geom(\alpha)$ is

$$\Pr[X = x] = \frac{1 - \alpha}{1 + \alpha} \cdot \alpha^{|x|} \tag{15.17}$$

When the sensitivity of some aggregation function $A(D)$ is $\Delta A = \max_{D_1, D_2} ||A(D_1) - A(D_2)||_1$ for all the data sets D_1 and D_2 differing by at most one element, then, by adding geometric noise r randomly chosen from

$Geom(\exp(-\frac{\varepsilon}{\Delta A}))$ to the original aggregation, the perturbed results can achieve ε-differential privacy, that is, for any integer $k \in Range(A)$, $\Pr[A(D_1)+r=k] \leq \exp(\varepsilon) \cdot \Pr[A(D_2)+r=k]$.

To enhance privacy in our proposed time series data aggregation, after the gateway aggregates all ciphertexts $c_1, c_2, \ldots, c_n$, it runs the following steps:

■ As the sensitivity of the aggregation $\sum_{i=1}^{n} x_i$ is Δ, to achieve ε-differential privacy in the scheme, the gateway first extracts a noise $\tilde{x}$ from the geometric distribution $Geom(\exp(-\frac{\varepsilon}{\Delta}))$.

■ Although m_i can support the space $[0, \lfloor\frac{p}{n_{max}+1}\rfloor]$, we still reasonably consider the sensitivity of the aggregation $\sum_{i=1}^{n} m_i$ to be Δ' in some real application scenarios, which is larger than Δ, but still far less than $\lfloor\frac{p}{n_{max}+1}\rfloor]$. Then, similarly to $\sum_{i=1}^{n} m_i$, the gateway also extracts a noise $\tilde{m}$ from the geometric distribution $Geom(\exp(-\frac{\varepsilon}{\Delta'}))$.

■ Finally, the gateway performs the following aggregation

$$
\begin{aligned}
C &= \left(\prod_{i=1}^{n} c_i\right) \cdot (p+1)^{\tilde{m}} \cdot g^{\tilde{x}} \cdot H(t)^{s_g} \\
&= \left(\prod_{i=1}^{n} (p+1)^{m_i} \cdot g^{x_i} \cdot H(t)^{s_i}\right) \cdot (p+1)^{\tilde{m}} \cdot g^{\tilde{x}} \cdot H(t)^{s_g} \bmod p^2 \quad (15.18) \\
&= (p+1)^{\sum_{i=1}^{n} m_i + \tilde{m}} \cdot g^{\sum_{i=1}^{n} x_i + \tilde{x}} \cdot H(t)^{\sum_{i=1}^{n} s_i + s_g} \bmod p^2
\end{aligned}
$$

and sends C to the control center.

In the end, at control-center side, the aggregated data $\sum_{i=1}^{n} m_i + \tilde{m}$ and $\sum_{i=1}^{n} x_i + \tilde{x}$ can be recovered, which may further enhance each individual node's privacy.

15.4.4.2 Discussion on dynamic node joining and leaving

In IoT scenarios, it is very common for nodes to join and leave frequently. Therefore, to deal with this dynamic environment, the following dynamic key management strategy can be applied by the TA.

■ Node joining: When a node N_j joins, the TA randomly chooses a subset of nodes $\{N_{a1}, N_{a2}, \ldots, N_{az}\}$ of $\mathcal{N}$, where each node N_{ai} has its secret key s_{ai}. Then, the TA assigns a random secret key s_j to the joining node N_j, and a new secret key $\bar{s}_{ai}$ to each N_{ai} such that

$$
s_j + \sum_{i=1}^{z} \bar{s}_{ai} = \sum_{i=1}^{z} s_{ai} \bmod p(p-1) \quad (15.19)
$$

With this strategy, the aggregation at the gateway and the decryption at the control center will not be affected.

■ Node leaving: Similarly, when a node N_j with the secret key s_j leaves, the TA also randomly chooses a subset of nodes $\{N_{a1}, N_{a2}, \ldots, N_{az}\}$ of $\mathcal{N}$, where each node N_{ai} has its secret key s_{ai}. Then, the TA assigns a new secret key $\bar{s}_{ai}$ to each N_{ai} such that

$$\sum_{i=1}^{z} \bar{s}_{ai} = \sum_{i=1}^{z} s_{ai} + s_j \bmod p(p-1) \qquad (15.20)$$

Note that the above dynamic key management can also be suitable for multiple users' joining and leaving cases.

15.5 Security Analysis

In this section, we analyze the privacy properties of the proposed time series aggregation scheme. Specifically, following the security model discussed earlier, we will show that each individual node's data privacy can be preserved.

In the proposed aggregation scheme, each node's data are encrypted in the form of $c_i = (p+1)^{m_i} \cdot g^{x_i} \cdot H(t)^{s_i} \bmod p^2$. Without the masking from $H(t)^{s_i}$, it is obvious that the plaintext, denoted as $M_i = (m_i, x_i)$, can be easily derived from $(p+1)^{m_i} \cdot g^{x_i}$ by using the same procedure in the decryption at the control center. Therefore, the security of $M_i = (m_i, x_i)$ is highly dependent on $H(t)^{s_i}$. In the following, we formally prove that M_i is indistinguishable during a chosen plaintext attack, even though an adversary $\mathcal{A}$ knows the public key $Y_i = g^{s_i}$ corresponding to the secret key s_i of the node N_i.

Definition 15.1 Computational Diffie–Hellman problem in $\mathbb{Z}_{p^2}^*$ *Given a generator g of group $\mathbb{Z}_{p^2}^*$, and g^a, g^b for unknown $a, b \in \mathbb{Z}_{p(p-1)}$, to compute $g^{ab} \in \mathbb{Z}_{p^2}^*$.*

Definition 15.2 Decisional Diffie–Hellman problem in $\mathbb{Z}_{p^2}^*$ *There are two distributions*

$$
\begin{aligned}
DH &= \{(A, B, C) = (g^a, g^b, g^{ab}) | g \in \mathbb{Z}_{p^2}^*, a, b \in \mathbb{Z}_{p(p-1)}\} \\
Rand &= \{(A, B, C) = (g^a, g^b, R) | g, R \in \mathbb{Z}_{p^2}^*, a, b \in \mathbb{Z}_{p(p-1)}\}
\end{aligned}
\qquad (15.21)
$$

The decisional Diffie–Hellman (DDH) problem states that, for given $(A, B, C) \in \mathbb{Z}_{p^2}^*$, to decide $(A, B, C) \in (g^a, g^b, g^{ab})$ or $(A, B, C) \in (g^a, g^b, R)$. The advantage of a distinguisher D, denoted by $Adv(D)$, is defined by

$$Adv(D) = \left| \Pr_{DH}[D(A, B, C) = 1] - \Pr_{Rand}[D(A, B, C) = 1] \right| \qquad (15.22)$$

For the DDH assumption, we assume there is no probabilistic polynomial time distinguisher D running in time τ that has a nonnegligible advantage $Adv(D) = \varepsilon$.

Theorem 15.1

Let $\mathcal{A}$ be an adversary against the node N_i's ciphertext $c_i = M_i \cdot H(t)^{s_i}$ with time τ. After q_h queries to the random oracles, its advantage is a nonnegligible ε. Then, the DDH problem in $\mathbb{Z}^*_{p^2}$ can be solved with another probability ε' with time τ', where

$$\varepsilon' = \frac{\varepsilon}{2}, \quad \tau' \leq \tau + q_h \cdot T_h$$

where T_h denotes the time cost for each hash query.

Proof. Assuming that there is an adversary $\mathcal{A}$ which runs in polynomial time and has a nonneglible advantage ε to break the semantic security of the ciphertext $c_i = M_i \cdot H(t)^{s_i}$ in the proposed scheme, then we can construct another algorithm $\mathcal{B}$ which has access to $\mathcal{A}$ and achieves a nonneglible advantage ε' to break a DDH problem instance $(A = g^a, B = g^b, C)$.

Let $A = g^a$ be the public key of node N_i corresponding to the secret key $s_i = a$, though the value of a is unknown. We make $A = g^a$ available to the adversary $\mathcal{A}$, and allow $\mathcal{A}$ to make q_h times hash oracle $H()$ queries, where $H()$ is modeled as a random oracle [17] on a different time point t_i.

Each time $\mathcal{A}$ queries on t_i, $\mathcal{B}$ randomly chooses a number $r_i \in \mathbb{Z}^*_{p(p-1)}$, stores (t_i, r_i, B^{r_i}) in a hash list, and returns $H(t_i) = B^{r_i}$ to $\mathcal{A}$. Obviously, because $H()$ is modeled as a random oracle, the hash query is indistinguishable from the real world.

At some time point, $\mathcal{A}$ chooses two messages $M_0, M_1 \in \mathbb{Z}^*_{p^2}$ for a ciphertext query in time period t^*, and sends them to $\mathcal{B}$. At this moment, $\mathcal{B}$ first retrieves (t_*, r_*, B^{r_*}) in a hash list with the search condition t^*, flips a bit $\beta \in \{0,1\}$ and generates a ciphertext $c_i = M_\beta \cdot H(t^*) = M_\beta \cdot B^{r_*} \bmod p^2$. Finally, $\mathcal{B}$ sends c_i to $\mathcal{A}$. After receiving c_i, $\mathcal{A}$ returns $\mathcal{B}$ a bit β' as its guess for β. $\mathcal{B}$ then returns 1 for $\beta' = \beta$, and returns 0 otherwise.

On one hand, if $(A = g^a, B = g^b, C)$ comes from the random distribution *Rand*, the ciphertext $C_i = M_\beta \cdot B^{r_*} \bmod p^2$ is uniformly distributed, hence independently of β. As a result,

$$\Pr_{Rand}[\mathcal{B}(A,B,C) = 1|\beta = \beta'] = \frac{1}{2} \tag{15.23}$$

On the other hand, when $(A = g^a, B = g^b, C = g^{ab})$ comes from the Diffie–Hellman (DH) distribution, one may remark that $C_i = M_\beta \cdot B^{r_*} \bmod p^2$ is a valid ciphertext of M_β, following a uniform distribution among the possible ciphertexts. Then,

$$\Pr_{DH}[\mathcal{B}(A,B,C) = 1|\beta = \beta'] = \frac{1}{2} + \frac{\varepsilon}{2} \tag{15.24}$$

The advantage of $\mathcal{B}$ in distinguishing the *DH* and *Rand* distributions is

$$Adv(\mathcal{B}) = \varepsilon' = \left| \Pr_{DH}[\mathcal{B}(A,B,C) = 1] - \Pr_{Rand}[\mathcal{B}(A,B,C) = 1] \right|$$

$$= \left| \frac{1}{2} + \frac{\varepsilon}{2} - \frac{1}{2} \right| = \frac{\varepsilon}{2} \tag{15.25}$$

By a simple computation, we can also obtain the claimed bound for $\tau' \leq \tau + q_h \cdot T_h$. Thus, the proof is completed. □

From the above theorem, we can see, under the DDH assumption, each individual node N_i's data is privacy preserving, even though the public key $Y_i = g^{s_i}$ is available to the adversary. Next, we show that our proposed scheme, once enhanced with the differential privacy technique, is also secure against differential attacks.

Theorem 15.2
Each node N_i's data is also secure against differential attacks in the enhanced aggregation.

Proof. For a given privacy level ε, the gateway perturbs the aggregation without recovery but adding appropriate geometric noises in the form of ciphertext. In such a way, ε-differential privacy can be achieved. Specifically, for data in a small plaintext space, the gateway adds the noise $\bar{x}$, which is chosen from $Geom(\exp(-\frac{\varepsilon}{\Delta}))$ to the exact aggregation to obtain the perturbed one. We assume that an adversary is able to gain two perturbed pieces of aggregation data $s + \bar{x}_s$ and $t + \bar{x}_t$, where s and t are aggregations of the two data sets differing by at most one element, while $\bar{x}_s$ and $\bar{x}_t$ are two corresponding geometric noises. Similarly, in [12], since $|s - t| \leq \Delta$, for any integer k, we have

$$\tau = \frac{\Pr[s + \bar{x}_s = k]}{\Pr[t + \bar{x}_t = k]} = \frac{\Pr[\bar{x}_s = k - s]}{\Pr[\bar{x}_t = k - t]}$$

$$= \frac{\frac{1-\alpha}{1+\alpha} \alpha^{|k-s|}}{\frac{1-\alpha}{1+\alpha} \alpha^{|k-t|}} = \alpha^{|k-s|-|k-t|} \qquad \text{where } \alpha = e^{-\frac{\varepsilon}{\Delta}} \tag{15.26}$$

Since

$$-|s - t| \leq |k - s| - |k - t| \leq |s - t| \qquad \text{and } 0 < \alpha < 1 \tag{15.27}$$

we have

$$e^{-\varepsilon} = (e^{-\frac{\varepsilon}{\Delta}})^{\Delta} = \alpha^{\Delta} \leq \alpha^{|s-t|} \leq \tau \leq \alpha^{-|s-t|} \leq \alpha^{-\Delta} = (e^{-\frac{\varepsilon}{\Delta}})^{-\Delta} = e^{\varepsilon} \tag{15.28}$$

Similarly, we can also prove that the data in a large plaintext space can also achieve ε-differential privacy, when a noise is chosen from the distribution $Geom(\exp(-\frac{\varepsilon}{\Delta'}))$. This completes the proof. □

Table 15.2 **Parameter settings**

Parameter	Value
λ	$\lambda = 1024$
$\mathbb{Z}_{p^2}^*$	$\mathbb{Z}_{p^2}^*$ is a group order $\phi(p^2) = p(p-1)$, where $|p| = \lambda$
n_{max}	$n_{max} = 1000$
n	$n = 200, 400, 600, 800, 1000$
Δ	$\Delta = 20$
ε	Differential privacy level $\varepsilon = 1, 2, 3$

15.6   Performance Evaluation

In this section, we evaluate our proposed privacy-preserving time series aggregation scheme in terms of computational cost and communications overheads, and analyze the utility in a differential privacy enhanced version as well. Concretely, we implement our scheme by Java (JDK 1.8) and run our experiments on a laptop with a 3.1 GHz processor, 8GB RAM, and Windows 7 platform. The detailed parameter settings are shown in Table 15.2.

Although the decryption complexity of the small plaintext space data $\sum_{i=1}^{n} x_i$ in the proposed scheme is $O(n\Delta)$, and may be accepted by the powerful control center in the IoT, we still build a hash table (stored in a zip file of around 167 KB) to accelerate the decryption lookup process in decryption in our experiment, where each entry in the hash table is the hash value of h^j with $0 \le j \le (n_{max} + 1) \cdot \Delta$. We ran our experiments 10 times, and the average results are reported below.

15.6.1   Computational costs

From the experiments, the average encryption at the node only takes 35 ms, which is very efficient for IoT scenarios. Figure 15.2 shows that the computational costs of aggregation at the gateway and decryption at the control center vary with the number of nodes n from 200 to 1000, with an increment of 200. From the figure, we can see that both of them are efficient, and the number of nodes n has little effect on aggregation and decryption, due to the direct aggregation over ciphertexts and a hash table to look up the small plaintext space decryption, built in advance.

15.6.2   Communication costs

When $|p| = 1024$, any ciphertext (including c_i and C) in group $\mathbb{Z}_{p^2}^*$ is less than or equal to 2048 bits.

15.6.3   Utility in differential privacy enhanced version

The novelty of our proposed scheme for supporting two types of data aggregation makes it suitable for many potential practical scenarios in the IoT. In the

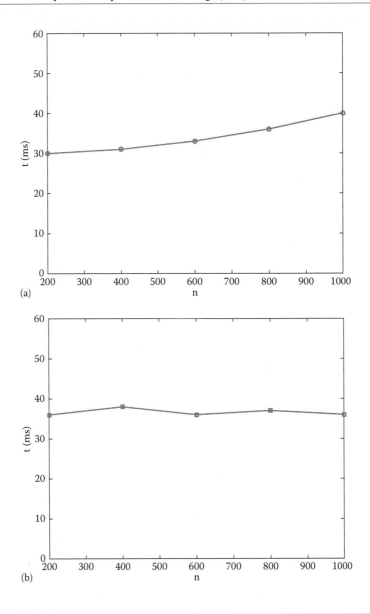

Figure 15.2: Computational costs of (a) aggregation at the gateway and (b) decryption at the control center varying with n.

following, we put emphasis on the evaluation of utility of differential privacy enhanced version. Concretely, we take a smart grid as an example here to elaborate the advantages and effectiveness of our enhanced version. Differing from

Table 15.3 Parameter settings

Description	Parameter	Value
Number of users	n	10000
User measurement	$x_i.m_i$	{0.000, 0.001, 0.002,..., 29.999, 30.000}
Differential privacy level	ε	1,2,3
– Sensitivity of small plaintext space data	Δ	30
– Sensitivity of large plaintext space data	Δ'	999

previously reported aggregation schemes for smart grids, our scheme can support data aggregation of user measurements including not only the integer part (small plaintext data $x_i \in [0,30]$) but also the decimal part (large plaintext data $m_i \in [0,999]$). The detailed parameter settings are listed in Table 15.3.

Similar to the aggregation scheme in [18], we implement an electricity consumption simulator having the ability to generate realistic 1 min consumption traces synthetically, which is extended from the basic simulator presented in [19]. Based on the simulator, we produce traces for 10,000 households, and the distribution of residents in each household follows the U.K. statistics on household sizes in 2011 [20].

Figures 15.3 and 15.4 illustrate the traces of actual total measurements and noisy total consumption for small and large plaintext space, respectively. We also set ε, the differential privacy level, to $1,2,3$ for each of the two scenarios. As can be seen from the figures, the larger ε is, the smaller the noise that will be added, and then the utility is higher; while the smaller ε is, the larger the noise that will be included, and then a higher level of the privacy can be guaranteed. Compared with the case of $\varepsilon = 3$, the utility at $\varepsilon = 1$ is lower, but it is still acceptable. Therefore, in real scenarios, there is a trade-off between privacy and utility.

15.7 Related Works

In this section, we briefly discuss some other research studies [6, 12, 13, 18, 21–24, 27] that are closely related to our scheme. Based on BGN homomorphic encryption techniques [25], some data aggregation schemes, for example, [6, 12], have been proposed, which focus on protecting individual users' privacy. The schemes in both [6] and [12] are secure against differential attack. In addition, multifunction data aggregation is also researched in [12]. However, these two schemes can only support one-dimensional data aggregation. Furthermore, since BGN-based [25] aggregation schemes depend on brute-force search techniques [14] to be able to decrypt the sum of the plaintext, all of the existing similar schemes have the disadvantage of limiting each user's reported measurement in the small plaintext space. Based on Paillier's homomorphic encryption techniques [26], some privacy-preserving data aggregation schemes [21, 24, 27] have been proposed, which eliminates the small plaintext space limitation. However, these proposed Paillier homomorphic-based schemes can only support one-dimensional data aggregation as ever. In addition, some data aggregation schemes based on other techniques have been designed. For example, based on modular addition-based encryption, some privacy-preserving aggregation schemes for smart grid communications, for example, [18, 23], have been proposed, which are secure against differential attack by adding Laplace noise to the real measurement. In [22], Jia et al. proposed a privacy-preserving data aggregation scheme in which coefficients of the polynomial are used to hide users' individual measurements. However, none of the aforementioned schemes can support more than one-dimensional data aggregation simultaneously, which greatly hinders practical applications.

Focusing on improving efficiency, we previously proposed an efficient EPPA protocol [13], which supports multidimensional data aggregation. EPPA significantly reduces the computation and communication overheads by encrypting multidimensional data into one single ciphertext. However, because the superincreasing sequence is the key requirement and characteristic of EPPA to support encryption and aggregation of users' structured data by homomorphic cryptosystem techniques, EPPA may still not support multidimensional large size data aggregation very well.

Although our proposed scheme here addresses similar issues, that is, providing efficient, privacy-preserving, and differentially private aggregation in the IoT, in contrast to the above studies, the emphases of our research still have some differences: (1) Our proposed scheme supports data aggregation of both small and large plaintext space messages (of arbitrary length comparative to the length of the system security parameter) and (2) our proposed scheme is secure against differential attack for both types of data aggregation; thus, it greatly enhances security, and improves efficiency and practicability.

15.8 Summary

In this chapter, we have proposed a novel privacy-preserving time series data aggregation scheme for the IoT. The proposed scheme is characterized by exploiting the properties of group $\mathbb{Z}_{p^2}^*$ to support data aggregation for both small and large plaintext spaces at the same time, which thus is more efficient than the traditional one-dimensional data aggregation. Detailed security analysis shows that the proposed scheme is privacy preserving, that is, no one can read each individual node's data, and only the control center can read the aggregation results. Furthermore, when the differential privacy technique is applied, the proposed scheme is also secure against differential attacks. Through extensive performance

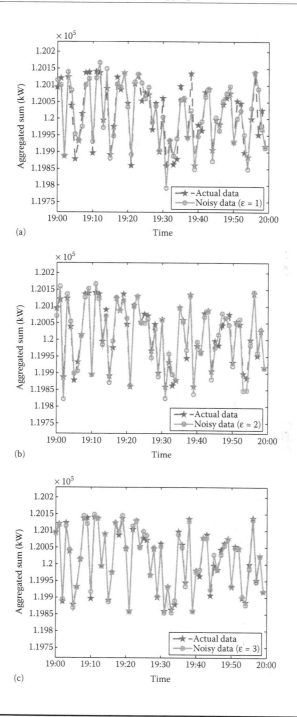

Figure 15.3: Differential privacy for small plaintext space data aggregation; (a) $\varepsilon = 1$; **(b)** $\varepsilon = 2$; **(c)** $\varepsilon = 3$.

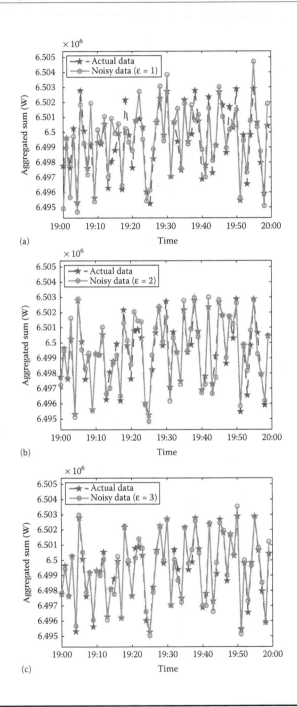

Figure 15.4: Differential privacy for large plaintext space data aggregation; (a) $\varepsilon = 1$; **(b)** $\varepsilon = 2$; **(c)** $\varepsilon = 3$.

evaluation, we have also demonstrated that our proposed scheme is efficient in terms of computational costs and communication overheads. Therefore, our proposed scheme can be applied in various IoT scenarios.

Bibliography

[1] R. Lu, X. Li, X. Liang, X. Shen, and X. Lin, "GRS: the green, reliability, and security of emerging machine to machine communications," *IEEE Communications Magazine*, vol. 49, no. 4, pp. 28–35, 2011.

[2] F. D. Garcia and B. Jacobs, "Privacy-friendly energy-metering via homomorphic encryption," in *Security and Trust Management - 6th International Workshop, STM 2010, Athens, Greece, September 23-24, 2010, Revised Selected Papers*, 2010, pp. 226–238.

[3] Z. Erkin and G. Tsudik, "Private computation of spatial and temporal power consumption with smart meters," in *Applied Cryptography and Network Security - 10th International Conference, ACNS 2012, Singapore, June 26-29, 2012. Proceedings*, 2012, pp. 561–577.

[4] L. Chen, R. Lu, and Z. Cao, "Pdaft: A privacy-preserving data aggregation scheme with fault tolerance for smart grid communications," *Peer-to-Peer Networking and Applications*, vol. 8, no. 6, pp. 1122–1132, 2015.

[5] K. Alharbi and X. Lin, "LPDA: A lightweight privacy-preserving data aggregation scheme for smart grid," in *International Conference on Wireless Communications and Signal Processing, WCSP 2012, Huangshan, China, October 25-27, 2012*, 2012, pp. 1–6. [Online]. Available: http://dx.doi.org/10.1109/WCSP.2012.6542936

[6] E. Shi, T. H. Chan, E. G. Rieffel, R. Chow, and D. Song, "Privacy-preserving aggregation of time-series data," in *Proceedings of the Network and Distributed System Security Symposium, NDSS 2011, San Diego, California, USA, 6th February-9th February 2011*, 2011.

[7] Haiyong Bao and R. Lu, "A new differentially private data aggregation with fault tolerance for smart grid communications," *IEEE Internet of Things Journal*, vol. 2, no. 3, pp. 248–258, 2015.

[8] L. Chen, R. Lu, Z. Cao, K. Alharbi, and X. Lin, "Muda: Multifunctional data aggregation in privacy-preserving smart grid communications," *Peer-to-Peer Networking and Applications*, vol. 8, no. 5, pp. 777–792, 2015.

[9] C. Li, R. Lu, H. Li, L. Chen, and J. Chen, "PDA: a privacy-preserving dual-functional aggregation scheme for smart grid communications," *Security and Communication Networks*, vol. 8, no. 15, pp. 2494–2506, 2015.

[10] R. Lu, X. Liang, X. Li, X. Lin, and X. Shen, "EPPA: an efficient and privacy-preserving aggregation scheme for secure smart grid communications," *IEEE Trans. Parallel Distrib. Syst.*, vol. 23, no. 9, pp. 1621–1631, 2012. [Online]. Available: http://dx.doi.org/10.1109/TPDS.2012.86

[11] Y. Liu, M. K. Reiter, and P. Ning, "False data injection attacks against state estimation in electric power grids," in *Proceedings of the 2009 ACM Conference on Computer and Communications Security, CCS 2009, Chicago, Illinois, USA, November 9-13, 2009*, 2009, pp. 21–32. [Online]. Available: http://doi.acm.org/10.1145/1653662.1653666

[12] L. Chen, R. Lu, Z. Cao, K. AlHarbi, and X. Lin, "Muda: Multifunctional data aggregation in privacy-preserving smart grid communications," *Peer-to-Peer Networking and Applications*, pp. 1–16, 2014.

[13] R. Lu, X. Liang, X. Li, X. Lin, and X. Shen, "EPPA: An efficient and privacy-preserving aggregation scheme for secure smart grid communications," *IEEE Transactions on Parallel and Distributed Systems*, vol. 23, no. 9, pp. 1621–1631, 2012.

[14] A. J. Menezes, P. Van Oorschot, and S. Vanstone, *Handbook of Applied Cryptography*, CRC Press, 1996, p. 12.

[15] C. Dwork, "Differential privacy," in *Automata, Languages and Programming, 33rd International Colloquium, ICALP 2006, Venice, Italy, July 10-14, 2006, Proceedings, Part II*, 2006, pp. 1–12.

[16] A. Ghosh, T. Roughgarden, and M. Sundararajan, "Universally utility-maximizing privacy mechanisms," *SIAM J. Comput.*, vol. 41, no. 6, pp. 1673–1693, 2012. [Online]. Available: http://dx.doi.org/10.1137/09076828X

[17] M. Bellare and P. Rogaway, "Random oracles are practical: A paradigm for designing efficient protocols," in *CCS '93, Proceedings of the 1st ACM Conference on Computer and Communications Security, Fairfax, Virginia, USA, November 3-5, 1993.*, 1993, pp. 62–73. [Online]. Available: http://doi.acm.org/10.1145/168588.168596

[18] J. Won, C. Y. Ma, D. K. Yau, and N. S. Rao, "Proactive fault-tolerant aggregation protocol for privacy-assured smart metering," in *INFOCOM 2014*. IEEE, 2014, pp. 2804–2812.

[19] I. Richardson, M. Thomson, D. Infield, and C. Clifford, "Domestic electricity use: A high-resolution energy demand model," *Energy and Buildings*, vol. 42, no. 10, pp. 1878–1887, 2010.

[20] Office for National Statistics, "Families and households, 2001 to 2011," Jan. 2012, http://www.ons.gov.uk/ons/rel/family-demography/families-and-households/2011/rft-tables-1-to-8.xls.

[21] V. Rastogi and S. Nath, "Differentially private aggregation of distributed time-series with transformation and encryption," in *Proceedings of the 2010 ACM SIGMOD International Conference on Management of data.* ACM, 2010, pp. 735–746.

[22] W. Jia, H. Zhu, Z. Cao, X. Dong, and C. Xiao, "Human-factor-aware privacy-preserving aggregation in smart grid," *IEEE System Journal*, pp. 1–10, 2013.

[23] G. Acs and C. Castelluccia, "I have a dream! (differentially private smart metering)," in *Information Hiding.* Springer, 2011, pp. 118–132.

[24] F. Li, B. Luo, and P. Liu, "Secure information aggregation for smart grids using homomorphic encryption," in *2010 First IEEE International Conference on Smart Grid Communications (SmartGridComm).* IEEE, 2010, pp. 327–332.

[25] D. Boneh, E.-J. Goh, and K. Nissim, "Evaluating 2-DNF formulas on ciphertexts," in *Theory of cryptography.* Springer, 2005, pp. 325–341.

[26] P. Paillier, "Public-key cryptosystems based on composite degree residuosity classes," in *Advances in cryptology - EUROCRYPT'99.* Springer, 1999, pp. 223–238.

[27] L. Chen, R. Lu, and Z. Cao, "Pdaft: A privacy-preserving data aggregation scheme with fault tolerance for smart grid communications," *Peer-to-Peer Networking and Applications*, vol. 8, no. 6, pp. 1122–1132, 2015.

[28] HaiyongBao and R. Lu, "A new differentially private data aggregation with fault tolerance for smart grid communications," *IEEE Internet of Things Journal*, vol. 2, no. 3, pp. 248–258, 2015.

[29] L. Chen, R. Lu, Z. Cao, K. Alharbi, and X. Lin, "Muda: Multifunctional data aggregation in privacy-preserving smart grid communications," *Peer-to-Peer Networking and Applications*, vol. 8, no. 5, pp. 777–792, 2015.

[30] C. Li, R. Lu, H. Li, L. Chen, and J. Chen, "PDA: a privacy-preserving dual-functional aggregation scheme for smart grid communications," *Security and Communication Networks*, vol. 8, no. 15, pp. 2494–2506, 2015.

Chapter 16

A Secure Path Generation Scheme for Real-Time Green Internet of Things

Yung-Feng Lu

Chin-Fu Kuo

CONTENTS

The Internet of Things (IoT) is expected to offer promising solutions to transform the operation and role of many existing systems such as transportation systems and manufacturing systems, and enables applications in many domains. The IoT aims to connect different things over networks. The goal of the IoT is to provide a good and efficient service for many applications. A real-time IoT application must react to stimuli from its environment within time intervals dictated by its environment. The instant when a result must be produced is called a deadline.

Wireless sensor networks (WSNs) have recently been in the limelight for many domains. The IoT can be explained as a general-purpose sensor network. WSNs will constitute an integral part of the IoT paradigm, spanning many different application areas. Since sensor nodes usually are developed by low-cost hardware, one major challenge in the development of many sensor network applications is to provide high-security features with limited resources. In this chapter, we first present a path generation framework with deadline considerations for real-time query processing. To meet the deadline, the framework will assign the time budget to the routing path and then derive a feasible path with the assigned time budget. Then, we present a novel key establishment scheme, named a half-key scheme, that is based on the well-known random key predistribution scheme and DDHV-D deployment knowledge, to provide resource-efficient key management in wireless sensor nodes with reduced memory space requirements and better security enforcement. The capability of the proposed approach is evaluated by an analytical model and a series of experiments.

16.1   Introduction

16.1.1   *Data gathering of IoT*

The next wave in the era of computing will be outside the realm of the traditional desktop. The Internet of Things (IoT) is a novel networking paradigm which allows the communication among all sorts of physical objects over the Internet [22]. In the IoT paradigm, many of the objects that surround us will be on the network in one form or another [7, 41, 73]. Ubiquitous sensing enabled by WSN technologies cuts across many areas of modern-day living. This offers the ability to measure, infer, and understand environmental indicators, from delicate ecologies and natural resources to urban environments [3, 25, 54].

Recent technological advances have enabled the development of low-cost, low-power, and multifunctional sensor devices. These nodes are devices with integrated sensing, processing, and communication capabilities. Sensor technology has enabled a broad range of ubiquitous computing applications, such as agricultural, industrial, and environmental monitoring [6, 49, 50, 60, 76]. As shown in Figure 16.1, WSN can work as part of the IoT; the collection and processing of such data leads to unprecedented challenges in mining and processing such data. Such data needs to be processed in real time and the processing may be highly distributed in nature [1, 39]. However, sensor networks are different from traditional networking. The sensor network has some physical resource constraints and special properties, thus contributing to the green IoT concept

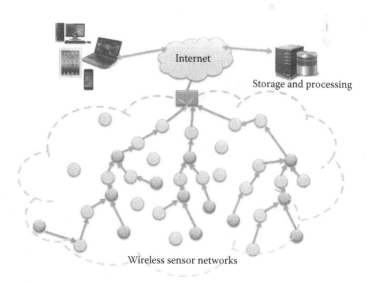

Figure 16.1: System model with multihop communications in the green internet of things.

[11, 87, 98]. We need to redesign the management methodology for it. The physical resource constraints of the sensor network include limited bandwidth and quality of service (QoS), limited computation power, limited memory size, and a limited supply of energy. The effective lifetime of the sensor is determined by its power supply. Energy conservation is one of the main system design issues. Ref. [28] shows that the power consumption of each sensor node is determined by the cost of transmission. For example, it requires 5000 nJ of energy to transmit a bit in a sensor node, and it requires 5 nJ of energy to process a single instruction.

In scientific settings, WSNs can act as intelligent data collection instruments; one might task the relevant subset of nodes to sense the physical world and transmit the sensed values, using multihop communication paths, toward a base station where all the processing takes place [32]. Since the energy cost of processing data is one order of magnitude smaller than the energy cost of transmitting the same data [13, 35, 37], it is more energy efficient to carry out as much processing as possible inside the WSN, as this is likely to reduce the number of bytes that are transmitted to the base station. From the viewpoint of this work, one approach to in-WSN processing construes the WSN as a distributed database, and the processing task injected into nodes for execution is the evaluation of a query evaluation plan (QEP). To optimize QEPs, many mechanisms [24, 52] aim to develop sensor network query processors (SNQPs) that drastically reduce the need for bespoke development while ensuring sufficient low levels of energy consumption as to deliver deployments of great longevity.

To support QoS requirement, SPEED [27] and MMSPEED [20] are QoS-based routing protocols that provide soft end-to-end deadline guarantees of packets for WSN. In SPEED, each node keeps information only about its immediate neighbors and utilizes geographic location information to make localized routing decisions. The MMSPEED protocol is an extension of the SPEED protocol. It is designed to provide probabilistic QoS differentiation with respect to timeliness and reliability domains. MMSPEED provides multiple delivery speed options for each incoming packet, and each incoming packet is placed into appropriate queues according to its speed class. A multiconstrained QoS multipath routing (MCMP) protocol [31] uses braided routes to deliver packets to the sink according to certain QoS requirements that are expressed in terms of reliability and delay. A message-initiated constraint-based routing (MCBR) protocol [95] is composed of explicit specifications of constraint-based destinations, route constraints and QoS requirements for messages, and QoS aware metastrategies. Building on a previously proposed QoS provisioning benchmark model, the energy-constrained multipath routing (ECMP) protocol [4] extends the MCMP protocol by formulating the QoS routing problem as an energy optimization problem that is constrained by reliability, play-back delay, and geospatial path selection constraints.

Moreover, we also need to consider properties of sensor networks: Sensor networks have a large number of sensor nodes. Individual sensor nodes are connected to other nodes in their vicinity via a wireless communication interface. Thus, some researchers aim to reduce the impact of interference. The interference-minimized multipath routing protocol (I2MR) [23] aims to support high-rate streaming in low-power WSNs by considering the recent advances in the design of high-bandwidth backbone networks. I2MR tries to construct zone-disjoint paths and distributes network traffic over the paths discovered by assuming a special network structure and the availability of particular hardware components. The low-interference energy-efficient multipath routing protocol (LIEMRO) [58, 59] improves the performance demands of event-driven sensor networks (e.g., delay, data delivery ratio, throughput, and lifetime) through construction of an adequate number of interference-minimized paths. LIEMRO utilizes an adaptive iterative approach to construct a sufficient number of node-disjoint paths with minimum interference from each event area toward the sink node. It improves the performance demands of event-driven applications by distributing network traffic over high-quality paths with minimum interference.

16.1.2   Key management of wireless embedded systems

In recent years, wireless embedded systems (WEBs) have attracted wide attention due to their suitability for monitoring complex physical-world phenomena [64, 70, 79, 88, 90]. WEBs have enabled various applications in many domains such as environment monitoring [53], home and industrial automation [78, 81], cyberphysical systems [61, 71], ubiquitous computing [63], security enforcement and surveillance [83], and military systems [2]. The play a vital role in sensing, gathering, and disseminating information about environmental phenomena. An instance of a WEB, a WSN, usually consists of a large number of battery-powered wireless embedded sensor systems and some base stations. To secure WEBs, the data transmission must be encrypted and authenticated. Since WEBs are usually developed by low-cost hardware, one major challenge in the deployment of many WEB applications is to provide high-security features with limited resources.

The provision of good protection for WEBs is an important issue; there are a lot of research results based on key encryption and management that have been proposed to enhance the security of wireless embedded sensor systems. Besides some excellent works on key encryption [48, 72, 91, 96], key renewal schemes [47, 82], and lightweight authentic bootstrapping [33], many researchers have also proposed effective key management schemes, such as those in [8, 16–18, 44, 46, 62, 92–94], in the past few years. Tseng [77] proposed an authenticated group key agreement protocol for resource-limited mobile devices. A detailed survey of such schemes is provided by Xiao et al. in [86].

Eschenauer and Glicor proposed a random key predistribution scheme (RKPS) as an effective solution, in which each wireless embedded sensor node is randomly assigned a subset of keys from a key pool before deployment [18]. The RKPS consists of three phases: key predistribution, shared-key discovery, and path-key establishment. If two neighboring nodes share one key, then a direct link may be established. A q-composite random key predistribution scheme [8] extends the random key predistribution scheme by requiring two adjacent communicating sensor nodes to share at least q keys. The rationale behind the extension is to provide a higher resilience for key management. Liu and Ning [44] take advantage of the location information to improve network connectivity. To reduce the storage requirements of wireless embedded sensor systems and resolve the scalability issue, researchers have proposed group-based or deployment-information-based methodologies; for example, [17, 45, 94, 97]. Although the probability of two wireless embedded sensor systems sharing common keying information increased, a significant amount of keying information had to be preloaded to each wireless embedded sensor node, regardless of whether a particular piece of information would be used in the future. Perrig et al. [57] considered a secure architecture in which each node shares a secret key with the base station. Two sensor nodes must use the base station as a trusted third party to set up a new key. Lai et al. [42] proposed a session key negotiation protocol based on a signal master key predeployed at sensor nodes. Wong and Chen [84] considered key exchanging for low-power computing devices, where one of the participants must be a powerful server.

16.2 Real-Time Query Processing in the Green Internet of Things

In this chapter, we propose a real-time query-processing framework for the green IoT. Assuming the sensors have a query plan, we then need to derive a feasible query propagation plan to transmit data packets. In real-time applications, the main challenge is to guarantee that the data packet meets its deadline. In such applications, such as a natural disaster monitoring system, the energy consumption is of secondary importance. To support real-time queries in the green IoT, the routing path must be adjusted to the packet deadlines and try to minimize energy consumption.

16.2.1 Real-time query processing in the green internet of things

The IoT is expected include billions of connected devices communicating in a machine-to-machine (M2M) fashion [12]. As for the definition of the IoT: The IoT allows people and things to be connected anytime, anyplace, with anything and anyone, ideally using any path/network and any service [80]. It is expected to offer promising solutions to transform the operation and role of many existing systems such as transportation systems and manufacturing systems, and enables many applications in many domains. The IoT aims to connect different things over its network. The goal of the IoT is to provide a good and efficient service for many applications.

A real-time application (RTA) is an application program that functions within a time frame that the user senses as immediate or current [9, 10, 85]. Correct system behavior depends not only on the logical results of the computations, but also on the physical time when these results are produced. By system behavior, we mean the sequence of outputs in time of a system. A real-time IoT application must react to stimuli from its environment within time intervals dictated by its environment. The instant when a result must be produced is called a deadline. Many applications are time-critical, such as healthcare, traffic control, and alarm monitoring. As a result, the IoT must collect the data from its things (or sensor networks) before the given deadline. Thus, real-time query processing is needed to provide time-efficient results.

Due to advances in sensor technology, sensors are becoming more power-ful, cheaper, and smaller in size. Thus, it has simulated large-scale deployments. From their origin, WSNs were designed, developed, and used for specific appli-cation purposes. In contrast, the IoT is not focused on specific applications. The IoT can be explained as a general-purpose sensor network [21]. The IoT would not be targeted to collect specific types of sensor data; rather it would deploy sensors where they can be used for various application domains [55, 56]. Thus, WSNs will constitute an integral part of the IoT paradigm, spanning many differ-ent application areas. Note that WSNs can exit without the IoT, but the IoT can-not exist without WSNs. This is because WSNs provide the majority of hardware infrastructure support, through providing access to sensor nodes. In any case, the sensor nodes of WSNs are usually operating with limited battery power, hence the need for energy-efficient techniques to reduce the power consumed, thus con-tributing the green IoT concept [87].

In general, the topology of the WSNs can vary from a simple star network to an advanced wireless mesh network. In this work, we focus on the multi-hop wireless mesh network type. Energy efficiency in WSNs has been studied [14, 36, 74]. Sensors with data to transmit should relay this data to a single source using multihop. Nodes that do not have data to transmit or that are not relaying the data of other nodes can be put to sleep. Energy efficiency is achieved by reducing the number of active nodes. Without considering the deadline, some researchers proposed a designated-path scheme for energy-balanced data aggre-gation in WSNs [38]. The proposed scheme predetermines a set of paths and runs them in round-robin fashion so that all the nodes can participate in the workload of gathering data and transferring them to the sink node. In this chapter, we have taken the time requirements further. Note that data gathered on WSNs would not transmit directly in the query plan. The realistic paths for gathering data (i.e., the query propagation plan) must meet the deadline and try to minimize the overall energy consumption.

16.2.2 Query processing in the green internet of things

16.2.2.1 Query plan in wireless sensor networks

In WSNs, sensors collect and transmit information under limited power and radio bandwidth. Traditional approaches for deploying these applications require months of design and engineering time [75]. Sensor query-processing architecture using database technology can, however, facilitate deployment of sensor networks, greatly reducing programming effort and time-to-deployment for many such applications. Some query-processing systems such as TinyDB [52], Directed Diffusion [34], and Cougar [89] provide users of WSN applications with a high-level interface to perform queries. Users specify the data of interest through simple, declarative queries, just as in a database system, and the infrastructure efficiently collects and processes the data within the sensor network.

Queries in TinyDB are disseminated through the entire network and collected via a routing tree. The root node of the routing tree is end point of the query, which is generally where the user that issued the query is located. Nodes within the routing tree maintain a parent–child relationship to properly propagate results to the root.

Recalling the traditional database, a query plan (or QEP) is a set of steps used to access information in an SQL relational database management system [40, 65–68]. This is a specific case of the relational model concept of access plans. Since SQL is declarative, there are typically a large number of alternative ways to execute a given query, with widely varying performances [19]. When a query is submitted to the database, a query plan is usually generated by the query optimization module of a database system; it consists of a partial order of the physical operators, such as join, sort, and table scan, of a query for the manipulation of database data.

In WSNs, the role of query plan is different form the traditional database. Due to the characteristics of sensor networks, queries need to transfer data to all sensors, and gather data from all sensors. We need to consider the power consumption issues of WSNs. Data gathering on a sensor network should not transmit to the query plan directly. We need to generate a more realistic query plan, a so-called propagation plan that has considered sensor characteristics, to gather sensor data. As shown in Figure 16.2, there are three phases of query data gathering. The first phase is data dissemination from the sink to the sensor nodes. This is usually done through multicasting. Then, the sensor senses objects, and then, the information is retrieved. Finally, the sensor nodes transmit data packets to the sink node based on a query plan. In this phase, sensor nodes usually transmit packets by unicasting.

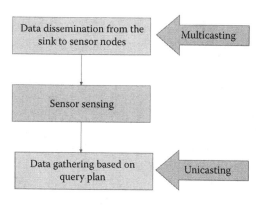

Figure 16.2: Three phases of query data gathering.

16.2.3   Network model and problem definition

Consider an example where a user asks the sink an urgent question via a declarative query, just as in a database system, and then the sink generates a high-level query plan for the query. Finally, the infrastructure conducts a query propagation plan from the query plan to efficiently collect and process the data within the sensor network.

In this chapter [51], we study a power-efficient networking problem to generate a data-gathering path that does not violate the given deadline. In this section, we present a network model for research on real-time query processing. Related terminologies and a problem definition are also specified.

16.2.3.1   Query processing and network model

The purpose of this research is to generate a data-gathering path of a query plan that does not violate the given deadline D and so that the total energy consumption is minimized. We consider a homogeneous WSN. A WSN consists a of set of sensor nodes $S = \{s_1, s_2, \ldots, s_M\}$. There is a set of K discrete power levels $P = \{P_1, P_2, \ldots, P_K\}$ given for sensor nodes, where $P_i > P_j$ if $i > j$ such that a higher power level results in a larger range to send data to another sensor node in the WSN. Each power level P_i is associated with a fixed range R_i for signal transmissions and an energy consumption amount $C(P_i)$, where $R_i > R_j$ and $C(P_i) > C(P_j)$ if $P_i > P_j$.

We assume that the sensor nodes can change their power level during the runtime. The set of edges, denoted as E^s, includes all possible edges. An edge $e_{i,j} \in E^s$ represents the connection of sensor nodes s_i and s_j. Furthermore, each edge $e_{i,j} \in E^s$ may be associated with a weight $e_{i,j}$ that is equal to $C(P_i)$, where a weight denotes the energy consumption of the signal transmission. As pointed out in [28], the energy consumption for a node to transmit signals to another node increases as the nth power of the distance between the two nodes, for $n \geq 2$. Assuming that node s_i transmits k bits of data and the distance to node s_j is d, the energy consumption function is as follows:

$$\mathbf{C_{i,j}}(k,d) = \varepsilon_{amp} \times k \times d^2 \qquad (16.1)$$

A query plan (QP) in the sensor network consists of a collection of physical operators (such as select, join, data acquisition, and aggregate) and a partial order of them. There could be more than one QP pending or executing in the system simultaneously, as shown in Figure 16.3. In many database systems, a query optimizer may have a joint consideration of pending or executing QPs such that the same nodes or subtrees in several QPs are merged, as shown in Figure 16.3 (without the dashed lines and the virtual root node). As a result, a collection of QPs after merging could be considered as a directed acyclic graph (DAG). In addition, we always add a virtual root node for pending/executing QPs such that the input format could be general, as shown in Figure 16.3 (with the dashed lines and the virtual root node).

As shown in Figure 16.4, the plan for data gathering could be viewed as a query plan $QP = (V, E^q)$; each edge $e_{m,n} \in E^q$ represents a need to transfer $d(e_{m,n})$ bytes of data from v_m to v_n; for example: $d(e_{C,D}) = 200$. The query plan QP has a partial order among E^q; for instance, the data transfer of $e_{i,j}$ must be processed before that of $e_{j,k}$. The data-gathering path at runtime is defined as a query propagation plan $EP = (V, E^{ep})$. A subset of EP at runtime is shown in Figure 16.5. Each edge $e_{i,j} \in E^{ep}$ represents $data_{i,j}$ units of data

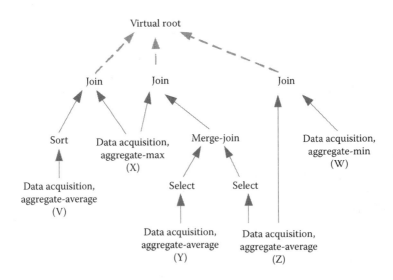

Figure 16.3: DAG-structured query plan.

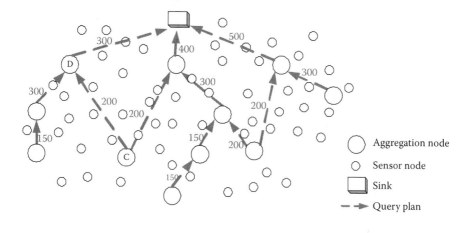

Figure 16.4: A sensor network with a DAG-structured query plan.

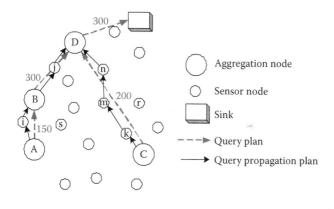

Figure 16.5: The data gathering path of a query plan.

from v_i to v_j. Let $X \to Y$ be a path, for example, $X-i-j-k-Y$ be a sequence of paths such that $EP = \{ep_i, ep_i \in E^q\}$. The function $st(e_{i,j})$ represents the start time of the path $e_{i,j}$, and $ft(e_{i,j})$ represents the finish time of the path $e_{i,j}$. As shown in Figure 16.5, $A \to D$ is a path. Both sequence of paths $A-B-D$ and $A-i-B-j-D$ are possible for $E^{ep}_{A,D}$. Because the relationship between energy consumption and distance is exponential (about 2), it is easy to discover that the energy consumption of the sequence of path $A-B-D$, with higher power levels, is greater than that of path $A-i-B-j-D$. However, the transmission time of path $A-i-B-j-D$ greater than that of path $A-B-D$. In a real-time environment, the path $A-i-B-j-D$ might not meet the time constraint. As a result, we need to adjust the power level of the nodes and set up the path to meet the time constraint.

In this chapter, we study the path generation methodology for a query propagation plan that we formulate as a real-time query-processing (RTQP) problem, to find a data-gathering path of the given query plan that does not violate the given deadline D and where the total energy consumption is minimized.

16.2.3.2 Problem definition

We formulate the real-time query-processing (RTQP) problem explored in this chapter as follows:

Problem 16.1

Input: *A given sensor network $SN = (V, E^s)$ with a query plan $QP = (V, E^q)$ and a given deadline D.*

Output: *A query propagation plan $EP = (V, E^{ep})$.*

Goal: *To find a sequence for execution path $EP_{s,t}$ which starts from v_s and ends at v_t, ($EP_{s,t}$ transfers data before $EP_{s'=t,t'}$) such that $ft(sink) \leq D$ and $\sum_{i,j} \sum_{e_{j,j} \in EP} p(e_{i,j}) d(e_{i,j}^q)$ is minimized.*

This problem is very hard to solve directly. We must convert this problem into two subproblems to reduce complexity. The first one is the time budget assignment problem that we summarize in Problem 16.2, and the other is one is the RSP problem that we summarize in Problem 16.3.

Problem 16.2

Input: *DAG $G = (V, E)$ and a budget B are given. Each edge $e_{m,n} \in E$ represents a need to transfer $k_{m,n} = tr(e_{m,n})$ bytes of data form v_m to v_n with distance $d_{m,n} = dis(e_{m,n})$. Edge $e_{m,n}$ is associated with a power consumption function $p_{m,n} = p(k_{m,n}, b_{m,n}, d_{m,n})$.*

Output: *DAG $GB = (V, E^b)$. Each edge $b_{m,n} \in E^b$ represents the budget of G.*

Goal: *The critical path of GB is $CP = (V', E^{b'}) \in GB$. The problem is to find an assignment of B such that $\sum E^{b'} \leq GB$, and power consumption $\sum_{e \in G} p(k_{m,n}, b_{m,n}, d_{m,n})$ is minimized.*

Problem 16.3

Input: *Given a sensor network $SN = (V, E)$, with a start node s, a target node t, and a deadline D, each edge $e_{ij} \in E$ has an associated positive integral cost $c_{i,j}$ and a positive integral delay $d_{i,j}$.*

Output: *Find a path s to t. The cost (respectively, delay) of a path is defined as the summation of the costs (respectively, delays) along all of its edges.*

Goal: *Find the minimum cost s-t path in SN such that the delay along this path does not exceed a given bound D.*

The real-time query-processing problem is NP-hard; we prove this as Theorem 16.1:

Theorem 16.1 NP-hardness.
The RTQP problem is NP-hard.

Proof. This arises directly from a special case when we only consider one segment the of query plan and a deadline D. It reduces to an RSP [26] problem.

16.2.4   A path generation framework

In this section, we explain the proposed framework. Figure 16.6 shows the framework of our mechanism. It contains four major procedures: discovery of the minimal-cost path, discovery of the critical path, budget reassignment, and path regeneration. The first of these is the discovery of the minimal-energy path; we set the nodes on a minimal power level and then apply Dijkstra's algorithm to find the minimal-energy path of the segments. The second one is the discovery of the critical path; we adapt the above to search for the critical path. The third component is the subdeadline; we apply the above to assign the assignment of a subdeadline of the segments. We assign the subdeadline to each segment with the proportion of the transmission time of the critical path segments. An RSP method is then adopted to derive the new paths.

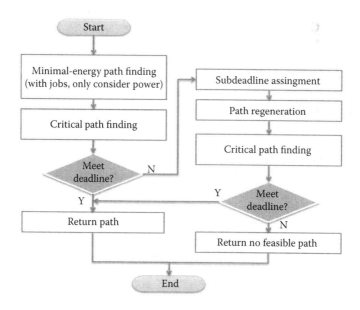

Figure 16.6: The framework of our mechanism.

The main function of propagation plan generation, referred to as Algorithm 16.1, is the major framework of our proposed mechanism. At the beginning, the algorithm assumes that all nodes are set with minimal transmission range. This algorithm will invoke a further algorithm, called *MinimalCost − PathFinding_Dijkstra*() to derive the paths of the query plan with minimal cost (Line 2). This algorithm explores the minimal-cost paths of the query plan using Dijkstra's algorithm. To assign a feasible time budget, it is necessary to know the longest transmission path. Then, the algorithm invokes Algorithm 16.3 to determine which is the critical path (Line 3). If the derived path violated the deadline, this algorithm would invoke Algorithm 16.4 to reassign the time budget and then generate a new path using Algorithm 16.4.

Algorithm 16.2 derives the minimal-cost propagation plan of the given query plan. It is revised from a revision of the well-known Dijkstra algorithm [15]. For a given sensor network $SN = (V, E^s)$ and a query plan $Q = (V, E^q)$, the algorithm finds the path with the lowest cost between that sensor node and every other sensor node. It can also be used to find costs of the shortest paths from a single sensor node to a single destination sensor node by stopping the calculation once the minimal-cost propagation plan to the destination sensor node has been determined.

Algorithm 16.3 shows how to find the critical path under the PERT algorithm (Lines 2–8) [69]. In this algorithm, we calculate the longest sensor node path to the end of the sink node, and the earliest and latest that each activity can start and finish without making the transmission time longer. This algorithm determines which paths are "critical" (i.e., on the longest path) and which have "total float" (i.e., can be delayed without making the transmission longer).

Algorithm 16.4 shows the budget reassignment procedure. As shown in Figure 16.7, a sensor network can derive the critical path via PERT. To save energy, one could reduce the transmission range of some sensor nodes. Thus, we need to know the slack of in time that we can achieve and we can thus assign to the transmission path using Algorithm 16.4 (Line 2). To explore the overall time budget of a path, we need first to explore the paths. In Line 8, Algorithm 16.3 explores the paths via the function *ExplorePath*() (Algorithm 16.5) that checks paths from parent nodes to leaf nodes. Then, it invokes Algorithm 16.8 (Line 9) to assign the time budget of nodes. After that, it invokes a function that reassigns the segment budget by the budget of the nodes (Line 10).

Algorithm 16.5 derives the slack times of query plan segments. Given a sensor network $SN = (V, E^s)$, a query plan $Q = (V, E^q)$, and a deadline D, Algorithm 16.5 derives the slack times of nodes from leaf to sink (Line 5). Algorithm 16.6 checks the paths from parent nodes to leaf nodes. Algorithm 16.7 reassigns node budgets; it assigns the new subdeadline of each query plan node, according to the proportion of a segment in the critical path. The new budget of a segment will be reassigned as

$$NodeTB_{currentNode} = NodeTB_{child} - \frac{P_{i,j}}{|Path|} \times TB_{subsegment}$$

Finally, we adopt Algorithms 16.9 and 16.10 to generate the execution path.

Algorithm 16.1: Propagation Plan Generation

input : Sensor network $SN = (V, E^s)$, query plan $Q = (V, E^q)$, deadline D
output: Find a set of sequences of execution path $EP=\{EP_{i,j}\}$
1 **PROCEDURE: PropagationPlanGeneration(SN, Q)**
2 **begin**
3     $EP \leftarrow MinimalCostPathFinding(SN, Q)$;
4     $CP \leftarrow CriticalPathFinding\_PERT(SN, EP)$;
5   **if** *Finish time of CP violate dead line* **then**
6        $BudgetReassignment(Q)$;
7        $PathGenerating(SN, Q)$;
8   Return path EP;
9 **end**

Let us use a QEP derivation example to illustrate the proposed scheme. Consider a query plan $QP = (V, E^q)$ on a sensor network $WSN = (V, E^s)$ as shown in Figure 16.8a, where each edge in E^q is a query plan edge. The orange nodes are the query plan nodes. Other nodes are relay sensor nodes. Suppose that the objective is to derive a feasible query propagation plan $EP = (V, E^{ep})$. Let wireless sensor network WSN be taken as an example for the explanation of the algorithm. At the beginning, all sensor nodes are set to the minimal power level at which they can be connected. Then, we apply the Dijkstra algorithm (Algorithm 16.2) to derive the minimal-energy paths. After this process, we can derive a query propagation plan QEP as shown in Figure 16.8b. The derived QEP is not equal to QP. This is because the Dijkstra algorithm has discovered some paths consuming smaller amounts of energy than those in the query plan. Then, we can compute the transmission time of a segment of the query plan. The variable ti=the transmission time of a segment of the query plan. For example, $t5$ is the segment EB. The variable tjk= the transmission time between node j and node k. For example, $t5$ is the transmission time from node E to node B. It is $tEu + tuv + tvB$, as shown in Figure 16.8c. Finally, as shown in Figure 16.8d, we can derive all transmission times of QP with the derived QEP.

Then, we adopt the PERT algorithm (Algorithm 16.4) to derive the critical path. We set all nodes in a topological order (Line 3, Algorithm 16.3) and follow the PERT process (Lines 4–7) to find the path with the largest transmission time of all paths. After that, we can determine the critical path, as shown in Figure 16.9a. The process then returns, the critical path to Algorithm 16.1.

In Line 4 of Algorithm 16.1, we check the transmission time of the critical path. In this example, $t_{CP} = t1 + t3 + t5$. If the time of the critical path t_{CP} is smaller or equal to deadline D, the derived query propagation plan can satisfy all requirements and the plan is returned. Otherwise, it is necessary to derive a new propagation plan for each segment associated with the critical path. We need

---

Algorithm 16.2: MinimalCostPathFinding_Dijkstra

---

input : Sensor network $SN = (V, E^s)$, query plan $Q = (V, E^q)$
output: Find a set of sequence of minimal cost execution path $EP = \{EP_{i,j}\}$

1   *PROCEDURE: **MinimalCostPathFinding(SN, Q)***
2   **begin**
3   **forall** $E^q \in Q$ **do**
4         $MinimalCostPathFinding\_Dijkstra(SN, E_{ij}^q)$;
5   Return path EP;
6   **end**
7   *PROCEDURE: **MinimalCostPathFinding_Dijk-stra(SN, E_{ij}^q)***
8   **begin**
9      $initialize\_signle\_source(SN, S)$;
10     $S \leftarrow \varnothing$;
11     $N \leftarrow V[SN]$;
12   **while** $N \neq \varnothing$ **do**
13        $u \leftarrow ExTract\_Min(SN)$;
14        $S \leftarrow S \bigcup \{u\}$;
15   **foreach** *vertex* $V \in Adj[u]$ **do** $RELAX(u,v,w,t)$;
16   Return path EP_{ij};
17   **end**
18   *PROCEDURE: **ExTract_Min(N)***
19   **begin**
20   Return the node that requires minimal transmit distance in the node set of minimal cost;
21   **end**
22   *PROCEDURE: **RELAX(u,v,w,t)***
23   **begin**
24   **if** $d[v] > d[u] + w(u,v)$ **then**
25        $d[u] \leftarrow d[u] + w(u,v)$;
26        $\Pi[v] \leftarrow u$;
27   **if** $d[v] = d[u] + w(u,v)$ **then**
28   **if** $R[v] > R[u] + r(u,v)$ **then**
29           $\Pi[v] \leftarrow u$;
30   **end**

Algorithm 16.3: Critical Path Finding

---

input : Sensor network $SN = (V, E^s)$, query plan $Q = (V, E^q)$
output: Critical path CP

1 *PROCEDURE: CriticalPathFinding_PERT(SN, Q)*
2 **begin**
3   Initialize fin[v] ← 0;
4   **forall** *vertex $v_j \in V$, Consider vertices v in topological order* **do**
5   **foreach** *edge $v - w$* **do**
6   set $fin[w] = \max(fin[w], fin[v] + time[w])$;
7   set $DistanceMax[w] = \max(DistanceMax[w], DistanceMax[v] + distance[w][v])$;
8      $CP \leftarrow Report\_PERT\_Critical\_Path()$;
9 **end**

Algorithm 16.4: Budget Reassignment

---

input : Sensor network $SN = (V, E^s)$, query plan $Q = (V, E^q)$, deadline D
output: Time budget of segments TB=$\{TB_{i,j}\}$

1 *PROCEDURE: BudgetReassignment(Q)*
2 **begin**
3      $SlackComputing(Q)$;
4      $NodeTB_{Sink} \leftarrow DeadLine$;
5      $NodeTB_{VirtualLeaf} \leftarrow 0$;
6   **repeat**
7   Select the node set of smallest slack into setA;
8   node t ← select the biggest node from setA;
9          $P \leftarrow ExplorePath(t)$;
10          $NodeBudgetReassign(P)$;
11   **until** *all of path have been reassign* ;
12      $SegmentBudgetReassign()$;
13 **end**

---

to assign subdeadlines to the segments. For example, in this figure, the critical path is $E - B - A - Sink$. We will apply a two-phase mechanism to deal with the segments, EB, DB, BA, CA, and $A - Sink$.

In our example, the transmission time of the critical path is greater than the deadline. Thus, we follow a two-phase mechanism to derive the query propagation plan. First, we assign the subdeadlines for the segments of the plan. Then, we generate the path for each segment of on the critical path to meet the assigned subdeadline with the RSP algorithm.

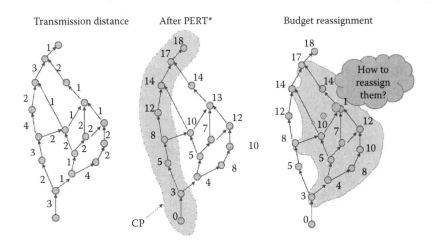

Figure 16.7: Budget reassignment of non-critical-path nodes.

Algorithm 16.5: Slack Computing (to decide the time budget reassignment sequence)

input : Sensor network $SN = (V, E^s)$, query plan $Q = (V, E^q)$, deadline D
output: Slack of each segment SP=$\{SP_{i,j}\}$
1 *PROCEDURE: SlackComputing(Q)*
2 **begin**
3   **forall** *vertex* $v_j \in V$ **do**
4          $slack_{node}[j] \leftarrow \infty$;
5      $slack_{node}[\text{virtual Leaf}] \leftarrow 0$;
6      $SlackNodeComputing(Sink)$;
7 **end**
8 *PROCEDURE: SlackNodeComputing(node v)*
9 **begin**
10   **if** *node v = virtual Leaf* **then**
11         $slack_{node}[v] \leftarrow 0$;
12   **else**
13   **foreach** *edge v − w,* **do** $slack_{node}[v] = min(SlackComputing(w) + SlackEdgeComputing(v, w))$;
14   return $slack_{node}[v]$
15 **end**
16 *PROCEDURE: SlackEdgeComputing(node v, node w)*
17 **begin**
18   return $DistanceMax[w] - DistanceMax[v] - distance[w][v]$;
19 **end**

Algorithm 16.6: Explore Path

---

input : Query plan $Q = (V, E^q)$, basis node
output: Find a set of sequences of transmission path TP=$\{TP_{i,j}\}$

1 *PROCEDURE: ExplorePath(t)*
2 **begin**
3   $\quad c \leftarrow$ child(t);
4   $\quad TP \leftarrow Q^q_{c,t}$;
5   $\quad$ **repeat**
6   $\quad\quad p \leftarrow$ parent(t);
7   $\quad\quad TP \leftarrow Q^q_{t,p}$;
8   $\quad$ **until** *parent(t) have already computed* ;
9   $\quad$ return TP;
10 **end**

Algorithm 16.7: Node Budget Reassign

---

input : Query plan segment P, time budget basis $TB_{segment}$
output: Time budget of nodes in segment NodeTB=$\{NodeTB_i\}$

1 *PROCEDURE: NodeBudgetReassign(P)*
2 **begin**
3   $\quad$ **repeat**
4   $\quad\quad Basis \leftarrow TB_{subsegment}$;
5   $\quad\quad proportion \leftarrow \frac{P_{i,j}}{|Path|}$;
6   $\quad\quad NodeTB_{currentNode} \leftarrow NodeTB_{child} - proportion \times Basis$;
7   $\quad$ **until** *all TB of nodes in this subpath have been reassign* ;
8 **end**

Algorithm 16.8: Segment Budget Reassign

---

input : Query plan segment P, time budget of nodes TB
output: Time budget of segment TB=$\{TB_{i,j}\}$

1 *PROCEDURE: SegmentBudgetReassign(Q)*
2 **begin**
3   $\quad$ **forall** $Q^p_{u,v}$ **do**
4   $\quad\quad TB_{u,v} \leftarrow NodeTB_v - NodeTB_u$;
5 **end**

Algorithm 16.9: Path Generating

    input : Sensor network $SN = (V, E^s)$, query plan $Q = (V, E^q)$, time budget
              TB, deadline D

    output: Find a set of sequences of execution path EP=$\{EP_{i,j}\}$

1 *PROCEDURE: PathGenerating(SN, Q)*

2 **begin**

3   **forall** *vertex* $v_j \in V$ **do**

4          $status[j] \leftarrow$ WAITING;

5          $edge\_status[j] \leftarrow$ WAITING;

6   **repeat**

7   chose the leaf node CP_j from CP;

8          $EdgeGenerating(j)$;

9   remove CP_j from CP;

10   **until** *CP has no element* ;

11 **end**

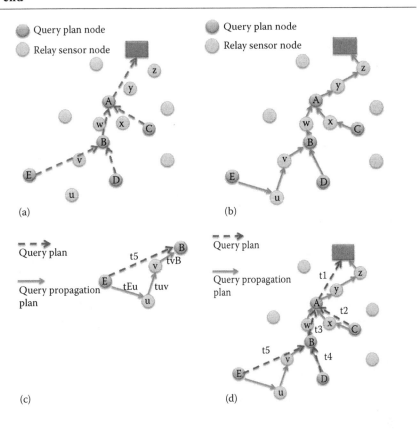

Figure 16.8: An example construction process of a query propagation plan (Part I).

Algorithm 16.10: Edge Generating

```
/* note: this version of algorithm did not consider the
   scenario where two paths use the same edge e_{i,j}.        */
```
input : Sensor network $SN = (V, E^s)$, edge $j \in Q = (V, E^q)$, time budget
 TB, deadline D
output: Find a set of sequences of execution path $EP_j = \{EP_{i,j}\}$
1   *PROCEDURE: EdgeGenerating(j)*
2   **begin**
3      $status[j] \leftarrow COMPUTING$;
4   **forall** *vertex* $v_i \in Adj[j]$ **do**
5   **if** $status[i] = WAITING$ **then**
6            $EdgeGenerating(i)$;

7   **forall** *vertex* $v_i \in Adj[j]$ **do**
8   **if** $edge\_status[i][j] = WAITING$ **then**
9   **if** $RS\_P[i] = \infty$ **then**
10               $RS\_P[i] \leftarrow 0$;
11            $RS[i][j] \leftarrow RS\_P[i]$;
12            $Budget \leftarrow B[i][j] + S[i][j] + RS[i][j]$;
13            $EP_{i,j} \leftarrow edgeRouting(P_{i,j}, Budget)$;
```
                /* Find path for P_{i,j} using RSP algorithm or
                   using run time determining algorithm        */
```
14            $RS\_P[i] \leftarrow Min(RS\_P[i], remaining\ budget)$;
15            $edge\_status[i][j] \leftarrow COMPUTED$;

16      $status[j] \leftarrow COMPUTED$;
17   **end**

We invoke Algorithm 16.4 to assign the subdeadlines. We denote *Di* as the subdeadline of a segment *i* in the query plan. As shown in Figure 16.9b, the basic concept is to assign the subdeadline according to the proportion of the transmission time of each segment in the critical path. For example,

$$D1 = D \times \frac{t1}{t1 + t3 + t5}, \; D5 = D \times \frac{t5}{t1 + t3 + t5}.$$

After we assign the subdeadline, then as shown in Figure 16.9c, d, and e, we use the RSP algorithm to derive a new path for each segment to meet its subdeadline. For example, the segment *EB*, because it has a subdeadline of *D5*, will generate the new path *EVB* to replace the path *EuvB*. Since *D4* is large enough, the segment *DB* transmits data via path *DB*. Finally, the returned query propagation plan is as shown in Figure 16.9f.

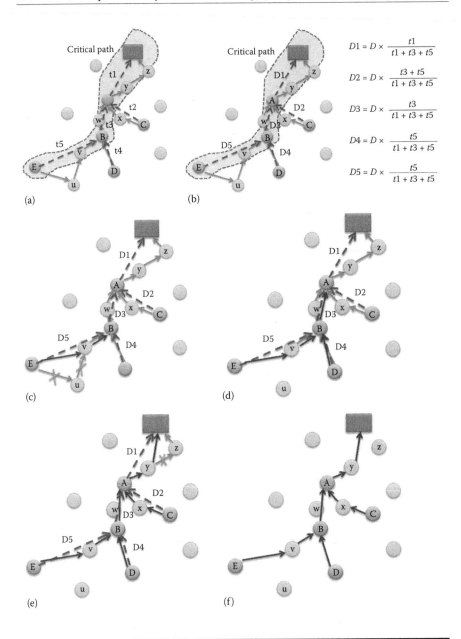

$$D1 = D \times \frac{t1}{t1 + t3 + t5}$$

$$D2 = D \times \frac{t3 + t5}{t1 + t3 + t5}$$

$$D3 = D \times \frac{t3}{t1 + t3 + t5}$$

$$D4 = D \times \frac{t5}{t1 + t3 + t5}$$

$$D5 = D \times \frac{t5}{t1 + t3 + t5}$$

Figure 16.9: An example construction process of a query propagation plan (Part II).

16.2.5 Properties

In this section, we discuss the properties of the proposed framework: (1) relation of transmission time and transmission distance, (2) setup of transmission range, (3) function for time budget and energy consumption, (4) time budget assignment policy, (5) completeness of the budget reassignment algorithm, and (6) completeness of the path generation algorithm.

Lemma 16.1

Relation of transmission time and transmission distance. *Let the transmission range of sensor nodes is be r. An amount of energy Tu is required to transmit one unit of packet of data over one unit of distance. The transmission time Tb of distance Distance$_{s,t}$ is:* $T = \lceil \frac{Distance_{s,t}}{R} \rceil \times Tu.$

Proof. The total distance is *Distance$_{s,t}$*. It requires *Tu* of energy to transmit one unit of packet of data over one unit of distance. This would be require transmission $\lceil \frac{Distance_{s,t}}{R} \rceil$ times. Thus, it requires $T = \lceil \frac{Distance_{s,t}}{R} \rceil \times Tu$ to transmit the packet.

Lemma 16.2

Setup of transmission range. *Let one unit of packet data be transmitted from s to t, the distance over which to transmit is Distance$_{s,t}$, which requires Tu of energy to transmit one unit of packet data over one unit of distance, and the time budget for this transmission is Tb. The minimal transmission range R required for the sensor nodes is*

$$R = \frac{1}{\lceil \frac{Tb}{Tu} \rceil} \times Distance_{s,t}.$$

Proof. This packet can be transmitted at most $\lceil \frac{Tb}{Tu} \rceil$ times. Thus, the minimal range of the sensor node requires

$$R = \frac{1}{\lceil \frac{Tb}{Tu} \rceil} \times Distance_{s,t}.$$

Corollary 16.1

Function for time budget and energy consumption. *Assuming node s$_j$ receives k bits of data transmitted from s$_i$, which requires Tu to transmit one unit of packet data over one unit of distance, the time budget of this transmission is T$_{i,j}$, and the distance transmitted from s$_i$ to node s$_j$ is Dis$_{i,j}$, the energy consumption function is as follows:*

$$\mathbf{C}_{(i,j)}(k, T_{i,j}, Dis_{i,j}) = k\{\varepsilon_{amp}(\frac{(Dis_{i,j})^2}{\lceil \frac{T_{i,j}}{Tu} \rceil})\} \tag{16.2}$$

Proof. We can easily derive this from Lemma 16.2 and Equation 16.1.

Property 16.1. Time budget assignment policy. *The time budget of query plan segment is assigned by the transmission distance proportion of the critical path.*

Proof. From the algorithm *Budget Reassignment*, we can find that all of the sensor nodes will reassign a new time budget via *NodeBudgetReassign(P)*. From Line 4 of this algorithm, we know our budget will be reassigned by the proportion of the transmission distance of the critical path.

Property 16.2. Completeness. *The BRA algorithm will reassign the time budget of all query plan segments.*

Proof. From Algorithm 16.4, in Line 2, the slack will be computed and then assigned to all the nodes in Line 9. After that, all budgets will be assigned to each path segment in Line 11.

Property 16.3. Completeness. *The PGA algorithm will generate the paths of all query plan segments.*

Proof. From Algorithm 16.1, all query plans will generate a path in Line 6. As the result, the PGA algorithm (Algorithm 16.9) will generate paths of all query plan segments.

16.2.6 Performance evaluation

We use network simulator ns-2 (ns-2.35) with the dynamic transmission power control extension [29, 30] to perform the simulations. We assume the deadlines are $80\%, 50\%, 30\%$ of the transmission time of the critical path. Thus, we need to reroute the transmission path to meet the deadline. We evaluate the power consumption of three methods. They are: our proposed scheme (RTQP), the scheme which has fast transmission with greatest transmission range (Fast), and each segment of the query plan has the same time budget (Avg). Figure 16.10 shows the WSN of our simulations. To further discuss the effect of our proposed scheme, we evaluate nine different query plans based on this WSN of a green IoT in Figure 16.11. After the execution of Algorithm 16.2, we can derive propagation plans of query plan topologies. Note that the critical path of these topologies is $A-B-C-D$. The segment $B-C$ has the longest transmission time in the critical path.

Figure 16.12 shows the evaluation results. Each subfigure represents the results of the corresponding topology; for example, Figure 16.12a represents the evaluation results of the topology in Figure 16.11a. The x-axis shows the evaluated schemes. The y-axis shows the power consumption of the query plans. Since the fast scheme always uses the greatest transmission range, it always consumes the greatest amount of energy. As shown in the figures, the RTQP and average schemes display similar results while the deadline is set as 80% of the transmission time of the critical path. This is because wireless sensor nodes can transmit data over a small distance. Since the RTQP scheme has good subdeadline assignment for query plan segments, it can always outperform the average scheme

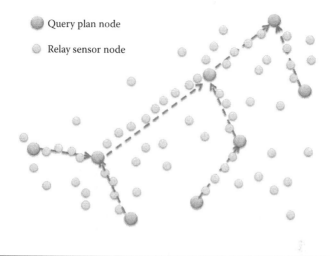

Figure 16.10: Wireless sensor networks in the IoT of our experiments.

while the deadline is only 50% amount of critical path. In the case of the 30% deadline, we also find that the RTQP scheme outperforms the average scheme.

In Figure 16.12a, the energy consumption of the RTQP scheme is lower than that of the other algorithms because the time budget of the nodes has been reassigned according to the proportion of the critical path. Thus, the power level of the nodes in the RTQP scheme is in better balance than the other algorithms. As a result, the RTQP scheme consumes less energy than the other schemas. The results of Figure 16.12b–i are also similar to the results of Figure 16.12a. The same behaviors are also invoked by each scenario.

16.2.7 Summary

A WSN can act as an intelligent data collection instrument; we can use it as a database of the IoT. We assume the sink node can generate a feasible query plan and the sensor nodes can then derive a feasible query propagation plan in run time. This research proposes a path generation methodology for a query propagation plan, which we formulate as a real-time query-processing problem, to find a data-gathering path of the given query plan that does not violate the given deadline and minimizes the total energy consumption. The properties of the proposed algorithms are also discussed. To evaluate the performance of the proposed RTQP scheme, we constructed a simulation model using ns-2.35. The performance of the RTQP scheme is compared with that of other related mechanisms. We have very encouraging results for the RTQP scheme in comparison to the other schemes.

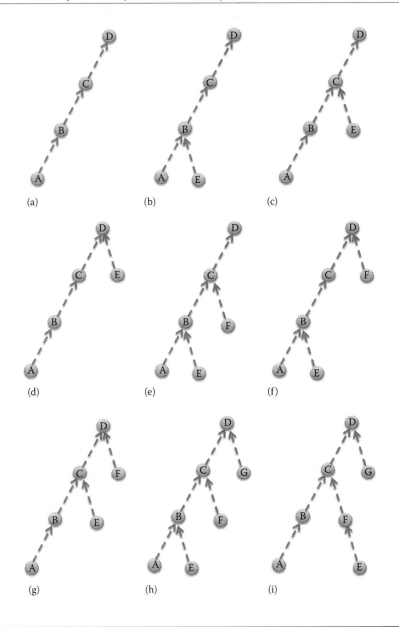

Figure 16.11: Evaluated query plan topologies.

16.3 Half-Key Key Management

While many previous studies and implementations have been carried out on network connectivity and compromised links after node capture, this chapter focuses its research on the reduction of memory space requirements of a sensor

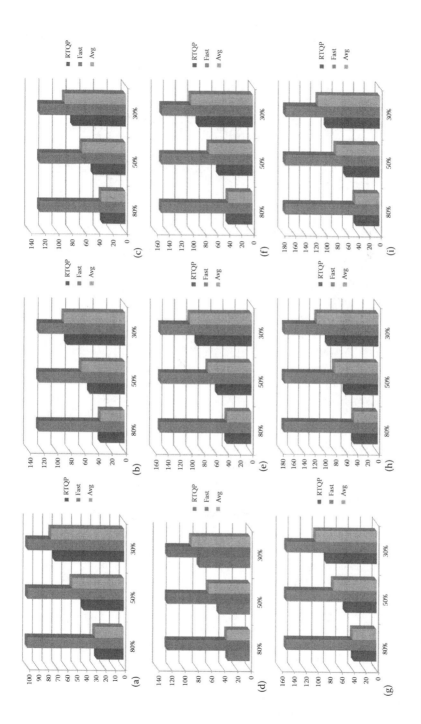

Figure 16.12: The evaluation results.

network on key predistribution and the provision of better security enforcement in sensor data delivery. In this section, we propose a half-key approach based on the well-known DDHV-D scheme [17] to provide resource-effective key management. We present an analytical model to exploit the properties of the proposed approach on connectivity, the number of session key candidates, and compromised links after node capture. The capability of the proposed approach is evaluated by a series of experiments, for which we have very encouraging results.

16.3.1 Preliminaries

16.3.1.1 Background schemes

In practice, it is quite common that a group of wireless embedded sensor nodes are deployed around a single deployment point. Such a group-based deployment model is modeled as follows: $\aleph$ nodes are deployed into an arbitrary target field S_f, and the location of each node $i(i = 1,...,\aleph)$ follows some distribution of probability density function (pdf) $f_i(x,y)$, where $(x,y) \in S_f$ are the coordinates of the node.

The sensor nodes that are to be deployed are divided into $t \times n$ equal-sized groups $G_{i,j}$ for $(i = 1,...,t$ and $j = 1,...,n)$, where (i,j) is called the group (or grid) index. The center (x_i, y_j) of each grid $G_{i,j}$ is called deployment point, which is the desired location of the nodes of the corresponding group. Because of the randomness in the deployment process, a group of nodes may spread into a local area around the deployment point to which the group of nodes should be deployed. Hence, DDHV-D assumes that the real location of each group of nodes follows some distribution $f_{i,j}(x,y) = f(x,y,\mu_x,\mu_y)$, where $(\mu_x,\mu_y) \in S_f$ are the coordinates of the deployment point of the group.

If we consider a normal distribution of deployed nodes—let us consider nodes dropping from a helicopter—we assume the deployment distribution for any node u in group $G_{i,j}$ follows a two-dimensional Gaussian distribution. When the deployment point of group $G_{i,j}$ is at (x_i, y_j), the pdf for node u in group $G_{i,j}$ is as follows [43]:

$$f(x,y|u \in G_{i,j}) = \frac{1}{2\pi\sigma^2}e^{\frac{-[(x-\mu_{x_i})^2+(y-\mu_{y_j})^2]}{2\sigma^2}} \tag{16.3}$$

where σ^2 is the variance of distribution.

Given the global key-space pool S and the overlapping factors a and b, let us use $|S_c|$ to represent the size of $S_{i,j}$ (for the sake of simplicity, we let all $S_{i,j}$ have the same size in this example). DDHV-D uses the following procedure to select key spaces for each key-space pool $S_{i,j}$. First, key spaces for the first group $S_{1,1}$ are selected from S; then, key spaces for the groups in the first row are selected from S and their left-hand neighbors. After that, key spaces for the groups from the second row to the last row are selected from S and their left-, upper-left-, upper-, and upper-right-hand neighbors. For each row, DDHV-D conducts the process from left to right.

After the deployment of nodes, let $\xi(ij,i'j')$ represent the number of shared key spaces between the deployment groups $G_{i,j}$ and $G_{i',j'}$. A key predistribution scheme using DDHV-D deployment knowledge $\xi(ij,i'j')$ can be computed as follows:

$$
\xi(ij,i'j') = \begin{cases}
|S_c|, & \text{when } (i,j) = (i',j'); \\
a|S_c|, & \text{when } (i,j) \text{ and } (i',j') \text{ are} \\
& \text{horizontal or vertical neighbors;} \\
b|S_c|, & \text{when } (i,j) \text{ and } (i',j') \text{ are} \\
& \text{diagonal neighbors;} \\
0, & \text{otherwise.}
\end{cases}
$$

In addition, to provide a better resilience, the DDHV scheme combines with the Blom scheme [5] to provide an important threshold property, which is called $\lambda - secure$: as long as no more λ nodes are compromised, all communication of links of noncompromised nodes remain secure. The DDHV [16] scheme assumes an agreed $(\lambda + 1) \times \aleph$ Vandermonde matrix G over a finite field $GF(\bar{h})$, where $\aleph$ is the size of the network and $\bar{h} > \aleph$. This matrix G is public information and may be shared by different systems; even adversaries are allowed to know G. During the key generation phase, the sink creates a random $(\lambda + 1) \times (\lambda + 1)$ symmetric matrix D over $GF(\bar{h})$, and computes an $\aleph \times (\lambda + 1)$ matrix $A = (D \cdot G)^T$, where $(D \cdot G)^T$ is the transpose of $D \cdot G$. Matrix D must be kept secret, and should not be disclosed to adversaries or to any sensor nodes. Because D is symmetric, it is easy to see that

$$
A \cdot G = (D \cdot G)^T \cdot G = G^T \cdot D^T \cdot G = G^T \cdot D \cdot G = (A \cdot G)^T \tag{16.4}
$$

that is, $A \cdot G$ is a symmetric matrix. If we let $K = A \cdot G$, we know that $K_{ij} = K_{ji}$, where K_{ij} is the element in the ith row and jth column of K. The idea is to use K_{ij} (or K_{ji}) as the pairwise key between node i and node j. Each node u only needs to (1) store the uth row of matrix A and (2) store the uth column of matrix G. Nodes i and j would be able to compute K_{ij} and K_{ji} independently in the following manner by each node independently:

$$
K_{ij} = A_c(i) \cdot G(j) = A_c(j) \cdot G(i) = K_{ji}.
$$

16.3.1.2 Motivation

The storage unit of keys or key space in related work usually is a full-sized key or a full-sized of key space. For example, in DDHV [16], the size of a key-space matrix A_i would be $(\lambda + 1) \times 64$ bits if the size of a common key is 64 bits. This is because the key-space matrix G is defined over a finite field $GF(2^{64})$.

This research [50] is motivated by the memory limitation in key maintenance of WSNs. We are interested in half-key-space predistribution scheme so that nodes do not need to store full-sized keys or key spaces. Instead, in the proposed scheme, the sensor nodes store half-sized keys or key spaces and the common session key may be concatenated using two half keys when they need to communicate with each other.

16.3.2   The half-key-space predistribution scheme

The goal of this scheme is to allow sensor nodes to derive common session keys with each of their neighboring nodes after deployment. Our scheme consists of two phases: key-space predistribution and session key establishment. The behaviors of key-space predistribution are similar to the DDHV-D scheme [17], but, because a session key is derived via two half keys, this phase is considerably different.

16.3.2.1   Off-line phase: Key-space predistribution

In this phase, we first generate a global key-space pool with $|S|$ half-key spaces. Since one session key could be concatenated using two half keys, the size of each half-key-space is only half that of a full key space. This phase is conducted offline and before the sensor nodes are deployed. Similar to DDHV-D, we firstly divide the key-space pool S into $t \times n$ key-space pools $S_{i,j}$ (for $i = 1, \ldots, t$ and $j = 1, \ldots, n$), with $S_{i,j}$ corresponding to the deployment groups deployed in neighboring (or nearby) locations.

16.3.2.1.1   Setting up of key-space pools

Before we describe our proposed scheme, we define a *key space* as a matrix D as defined in the previous section. The size of the elements of that matrix are one-half of a full key. We say a node i holds key space D if i stores the secret information; for instance, the corresponding row $A(i)$ of key-space matrix A, generated from (D, G) using Blom's scheme. Note that two nodes can calculate a half key if they hold a common key space.

Basically, the proposed scheme is to follow DDHV-D to set up the key-space pools. We generate $|S|$ half-key spaces for the global half-key-space pool S. We set the length of a half key at one-half of a full key; for example, a half key is 32 bits if a session key is 64 bits. For the global half-key pool S, we use the following procedure to generate them:

■ **Step 1: Generating matrix G.** We first select a primitive element from a finite field $GF(\overline{h})$, where $|\overline{h}|$ is larger than the desired half-key length (and also $\overline{h} > 2\frac{\aleph}{t \times n}$), to create a generator matrix G of size $(\lambda + 1) \times 2\frac{\aleph}{t \times n}$. Let $G(j)$ represent the jth column of G. We provide $G(j)$ to node j. Note that, since we can make a new key space by changing the value of matrix

D or matrix G, we change matrix G and generate φ (where $\varphi = |S|$) matrices D for all half-key spaces.

■ **Step 2: Generating matrices D.** We generate φ random, symmetric matrices $D_1, ..., D_\varphi$ of size $(\lambda + 1) \times (\lambda + 1)$. We call each tuple $s_i = (D_i, G)$ (for $i = 1, ..., \varphi$), a *key space*. We then compute the matrix $A_i = (D_i \cdot G)^T$. Let $A_i(j)$ represent the jth row of A_i.

Since each half-key space has $2\frac{\aleph}{t \times n}$ rows, it may be shared by two groups of nodes (only). To avoid two different nodes storing the same row of a half-key space, we need a policy to assign these secret pieces of information to nodes. For each half-key space pool, we define a half-key space as a local space if it makes a selection from the global half-key-space pool; otherwise, we define it as a visited space. The following procedure describes how we choose key spaces for each half-key-space pool of groups:

■ For group $S_{1,1}$, select $|S_c|$ half-key spaces as local spaces from the global half-key-space pool S; then, remove these $|S_c'|$ half-key spaces from S.

■ For group $S_{1,j}$, for $j = 2, ..., n$, select $a \cdot |S_c|$ key spaces as visited spaces from the half-key-space pool $S_{1,j-1}$; then, select $\omega = (1 - a) \cdot |S_c|$ half-key spaces as local spaces from the global half-key-space pool S and remove the selected ω half-key spaces from S.

■ For group $S_{i,j}$, for $i = 2, ..., t$ and $j = 1, ..., n$, select $a \cdot |S_c|$ half-key spaces as visited spaces from each of the half-key-space pools $S_{i-1,j}$ and $S_{i,j-1}$ if they exist; select ω (defined below) half-key spaces as local spaces from the global half-key-space pool S and remove these ω half-key spaces from S.

$$\omega = \begin{cases} (1 - (a+b)) \cdot |S_c| & \text{for } j = 1 \\ (1 - 2(a+b)) \cdot |S_c| & \text{for } 2 \le j \le n-1 \\ (1 - (2a+b)) \cdot |S_c| & \text{for } j = n \end{cases}$$

16.3.2.1.2 Key-space predistribution

Algorithm 16.11 shows the key-space predistribution procedure. Suppose the storage size of each sensor node is τ units (a unit is a full key). Since the size of each half-key space is $\frac{\lambda+1}{2}$ units, each sensor node could store $\frac{2\tau}{\lambda+1}$ half-key spaces. So, after the key-space pools are set up, for each sensor node in the deployment group $G_{i,j}$, we randomly select $\frac{2\tau}{\lambda+1}$ key spaces from its corresponding half-key-space pool $S_{i,j}$ (Line 2); then, for each selected key space, we load the corresponding row of its matrix into the memory of the node. Since each half-key space may be shared by two group of nodes, that is, as a local space or a visited space, to avoid collision, for each half-key space, a node loads the row in the top half of the space if this space is a local space (Lines 3–5); otherwise, it loads the row in the bottom half of the space (Lines 6–8).

Algorithm 16.11: Key-Space Predistribution (for Each Node $n_{ij.l}$)

---

1 **for** $m = 1$ *to* $\frac{2\tau}{\lambda+1}$ **do**
2       $n_{ij.l}$ randomly select one new key space A_x from S_{ij};
3   **if** A_x *is a local space* **then**
4           $k = l$;
5   Assign $A(k)$ to $n_{ij.l}$;
6   **else**
7           $k = l + \frac{\aleph}{t \times n}$;
8   Assign $A(k)$ to $n_{ij.l}$;

16.3.2.2 Online phase: Session key establishment

During the initialization of a sensor network, each node must discover all of the half-key spaces shared with its neighboring nodes within its wireless communication range. The discovery of shared half-key spaces may be accomplished by the broadcasting of key-space identifiers among nodes. If the number of half-key spaces shared between two neighboring nodes is more than a user-defined threshold q, then we say that these two nodes could establish a communication link; otherwise, there is no communication link between these two nodes. The rationale behind this setting is to avoid potential attacks, due to an insufficient number of session key candidates.

Algorithm 16.12 shows the establishment of the session key. If two neighboring nodes i and j may have a communication link, and i wishes to send data to j, then the initialization of the communication link may be done as follows:

Node i first randomly picks up one half-key space s_x in the common half-key space set of i and j and sends the identifier id_x of the selected half-key space to j. When j receives the identifier, j also randomly picks up one half-key space s_y in the common half-key space set of i and j and sends the identifier id_y of the selected half key to i. Then, they can compute the selected half keys, that is, $k_{x,ij}$ and $k_{y,ij}$, using Blom's scheme: Initially node i possesses $A_c(i)$ and $G(i)$, and node j possesses $A_x(j)$ and $G_x(j)$. After exchanging $G_x(i)$ and $G_x(j)$, then the shared half key between nodes i and j, $k_{x,ij} = k_{x,ji}$, can be computed in the following manner by each node independently:

$$k_{x,ij} = A_x(i) \cdot G_x(j) = A_x(j) \cdot G_x(i) = k_{x,ji}$$

Since both nodes share half-key spaces s_x and s_y and know their identifiers, n_i and n_j, they could create a session key on each side by having $k_{x,ij}$ and $k_{y,ij}$ as the prefix and suffix of the session key, respectively. The session key would be used for data encryption/decryption until a new session key is generated based on a similar approach. Note that, to avoid cryptanalysis attacks, any two nodes involved in data transmission should change their session keys for every specified time interval. In the next section, we will exploit some security-related properties of the proposed half-key approach.

Algorithm 16.12: Session Key Establishment

---

1 n_i randomly picks up one half-key space s_x in the common half-key space set of n_i and n_j;
2 n_i sends the identifier (seed) id_x of the selected half-key space to n_j;
3 After n_j receives the identifier, n_j randomly picks up one half-key space s_y in the common half-key space set of n_i and n_j;
4 n_j sends the identifier id_y of the selected half-key space to n_i;
5 n_i and n_j compute the prefix half key
   $k_{x,ij} = A_x(i) \cdot G_x(j) = A_x(j) \cdot G_x(i) = k_{x,ji}$;
6 n_i and n_j compute the postfix half key
   $k_{y,ij} = A_y(i) \cdot G_y(j) = A_y(j) \cdot G_y(i) = k_{y,ji}$;
7 n_i and n_j derive a session key: $k_s = k_{x,ij} || k_{y,ij}$;

16.3.3 Analysis study

The purpose of this section is to propose an analytic model to provide insights into the performance evaluation of the proposed approach. In this research, three properties are under consideration: *the number of session key candidates, connectivity*, and *resilience against node capture*. Connectivity is defined as the probability of the number of half keys shared between two neighboring nodes is no less than a given threshold q. The number of session key candidates is defined as the minimum possible number of session keys being created for data encryption/decryption. Resilience against node capture is defined as the percentage of the secure links that are compromised after a certain number of nodes are captured by the adversaries. Note that the analysis is complicated by the possibility of nodes sharing in keys or half keys.

16.3.3.1 Session key candidates analysis

Under the basic key predistribution scheme [18], one single shared key is required for two neighboring nodes in data transmission. Because only one shared key is required, the number of session key candidates for the scheme would be one [18]. The random key predistribution scheme [8] requires any two neighboring nodes to communicate with each other when they share q keys. The number of session key candidates is at least q. Since the half-key scheme proposed in this work creates a session key by merging two shared half keys, the number of session key candidates is at least q^2 when two neighboring nodes communicate with each other when they share q half keys.

16.3.3.2 Connectivity analysis of half-key space predistribution scheme using deployment knowledge

Lemma 16.3 [8]
Let p'_q be the connectivity of a sensor network under the q-composite key predistribution scheme. If the memory size of each sensor node in the sensor network is fixed, the connectivity p'_q of the sensor network may be derived by the following equation:

$$p'_q = 1 - \frac{((P-K')!)^2}{(P-2K')!P!} \times \sum_{i=0}^{q-1} \frac{(P-2K')!(K'!)^2}{i!(P-2K'+i)!((K'-i)!)^2}, \tag{16.5}$$

where K' is the number of keys assigned to each sensor node and P is the size of the key pool.

Lemma 16.4 [17]
Let $A(u,v)$ be the event that u and v are neighbors; let $B'(u,v)$ be the event that u and v share at least one common key spaces. The local connectivity P^1_{local} (i.e., the probability of two neighboring nodes being able to find a common key space) is the following conditional probability:

$$P^1_{local} = Pr(B'(u,v)|A(u,v)) = \frac{Pr(B(u,v)\,and\,A(u,v))}{Pr(A(u,v))} \tag{16.6}$$

$$= \frac{\sum_{i\in\Psi}\sum_{j\in\Psi} Pr(B(n_i,n_j)) \cdot Pr(A(n_i,n_j))}{\sum_{i\in\Psi}\sum_{j\in\Psi} Pr(A(n_i,n_j))} \tag{16.7}$$

Let $A(u,v)$ be the event that u and v are neighbors; let $B^q(u,v)$ be the event that u and v share at least q common key spaces. The local connectivity P^q_{local} (i.e., the probability of two neighboring nodes being able to find a common key space) is the following conditional probability:

$$P_{local} = Pr(B^q(u,v)|A(u,v)) = \frac{Pr(B^q(u,v)\,and\,A(u,v))}{Pr(A(u,v))} \tag{16.8}$$

$$= \frac{\sum_{i\in\Psi}\sum_{j\in\Psi} Pr(B^q(n_i,n_j)) \cdot Pr(A(n_i,n_j))}{\sum_{i\in\Psi}\sum_{j\in\Psi} Pr(A(n_i,n_j))} \tag{16.9}$$

Theorem 16.2
Let p_q be the connectivity of a sensor network under the proposed half-key space predistribution scheme with a user-defined parameter q. If the memory size of each sensor node in the sensor network is fixed, the connectivity p'_q of the sensor network may be derived by the following equation:

$$p_q = 1 - \frac{((P-K)!)^2}{(P-2K)!P!} \times \sum_{i=0}^{q-1} \frac{(P-2K)!(K!)^2}{i!(P-2K+i)!((K-i)!)^2} \tag{16.10}$$

where K is the number of half keys assigned to each sensor node (note that $K = 2K'$) and P is the size of the key pool.

Proof. The correctness of this theorem follows from Lemma 16.3. ■

Lemma 16.5 [17]
Let $g_i = g(z_i|n_i \in G_i)$ represent the probability that a sensor node n_i from group G_i resides within the attack circle. The probability g_i may be derived by the following equation:

$$g_i = 1\{z < R\}[1 - e^{-\frac{(R-z)^2}{2\sigma^2}}] + \int_{|z-R|}^{z+R} 2\ell \cos^{-1}(\frac{\ell^2 + z^2 - R^2}{2\ell z}) f_R(\ell|n_i \in G_i)d\ell \quad (16.11)$$

where $\ell\{.\}$ is the set indicator function and $f_R(\ell|n_i \in G_i)$ is given by Equation 16.2.

Lemma 16.6 [17]
Let $Pr(A(n_i, n_j))$ be the event that node n_i and node n_j are neighbors. If the deployment model is of normal distribution, the probability $Pr(A(n_i, n_j))$ of two sensor nodes n_i and n_j being neighbors may be derived by the following equation:

$$Pr(A(n_i, n_j)) = \int_{y=0}^{Y} \int_{x=0}^{X} f_R(d_{j\theta}|v \in G_j) \cdot g(d_{i\theta}|u \in G_i) \cdot dxdy \quad (16.12)$$

where $d_{j\theta}$ is the distance between θ and the deployment point of group j, and $g(d_{i\theta}|u \in G_i)$ is the probability that the sensor node n_i from group G_i resides within the θ-circle.

Theorem 16.3
If the memory size of each sensor node in the sensor network is fixed, let $B^q(n_i, n_j)$ be the probability of any two nodes n_i and n_j sharing at least q key spaces to form secure communications. $B^q(n_i, n_j)$ may be derived by the following equation: $B^q(n_i, n_j) = 1 - (p(i, j, 0) + p(i, j, 1) + \ldots + p(i, j, q - 1))$

$$\text{where } p(i, j, x) = \frac{\sum_{k=0}^{min(\tau, \xi(i,j))} \binom{\xi(i,j)}{k} \binom{|S_c| - \xi(i,j)}{\tau - k} \binom{k}{x} \binom{|S_c| - k}{\tau - x}}{\binom{|S_c|}{\tau}^2}.$$

Proof. Let $p(i, j, x)$ be the probability that any two nodes n_i, n_j have exactly x key spaces in common. Any given node has $\binom{|S_c|}{\tau}$ different ways of choosing its τ key space from the key pool of size $|S_c|$. Hence, the total number of ways for both nodes to pick τ key spaces each is $\binom{|S_c|}{\tau}^2$. Suppose the two nodes have x keys in common. The first node selects k key spaces from the ξ shared key spaces; it then selects the remaining $\tau - k$ key spaces from the non-shared key spaces. For the second node, there are $\binom{k}{x}$ ways to choose the x common key spaces. Since the second node only shares x key spaces with the first node, it has to select $\tau - x$ key spaces from the remaining $|S_c| - k$ key spaces from its key-space pool. Hence, we have

$$p(i,j,x) = \frac{\sum_{k=0}^{min(\tau,\xi(i,j))} \binom{\xi(i,j)}{k}\binom{|S_c|-\xi(i,j)}{\tau-k}\binom{k}{x}\binom{|S_c|-k}{\tau-x}}{\binom{|S_c|}{\tau}^2}$$

Let $B^q(n_i, n_j)$ be the probability of any two nodes n_i and n_j sharing sufficient key spaces to form secure communications. $B^q(n_i, n_j) = 1 - $ (probability that two nodes share insufficient keys to form connections), hence $B^q(n_i, n_j) = 1 - (p(i,j,0) + p(i,j,1) + \cdots + p(i,j,q-1))$. ■

16.3.3.3 Resilience analysis of half-key space: Predistribution scheme using deployment knowledge

In this research, we consider a realistic scenario in which the adversary intrudes into a region inside the WEBs and randomly captures and compromises x_c wireless embedded sensor nodes within this region. The region is assumed to be a circle at point $Z(x,y)$ with radius R_c. We term such a circle as the *attack circle* and call R_c the *attack radius*.

Before we present our detailed analysis on resilience, we summarize our approach as follows for the benefit of clarity: Based on the above assumptions, we can calculate, among all the sensors in the attack circle, the average number of sensors that are deployed from each specific group. Since the adversary compromises x_c sensors randomly inside the cycle, the average number of compromised sensors that are deployed from the specific group can be derived by Lemmas 16.5 and 16.7. Then, we calculate the fraction of additional communication that an adversary can compromise, based on the information retrieved from the x_c captured nodes in Theorem 16.4.

Lemma 16.7 [17]

Suppose the adversary captures nodes randomly within a region that is assumed to be a circle at point $Z(x,y)$ with radius R_c. With N sensors divided into n groups, each group has N/n sensor nodes. Let $x_i(x,y,R_c,x_c)$ represent the expected number of captured sensor nodes that are deployed from group G_i. Let $X_i(x,y,R_c,x_c)$ represent the weighted sum of the numbers of nodes that have been captured from all these groups. The expected number $X_i(x,y,R_c,x_c)$ may be derived by the following equation:

$$X_i(x,y,R_c,x_c) = \sum_{j \in \Psi_i} \left(\frac{\xi(i,j)}{|S_c|} \cdot x_i(x,y,R_c,x_c) \right)$$

$$= \sum_{j \in \Psi_i} \left(\frac{\xi(i,j)}{|S_c|} \cdot x_c \cdot \frac{g_i}{\sum_{j \in \Psi} g_j} \right) \qquad (16.13)$$

Lemma 16.8

Let c be a link in the key-sharing graph between two nodes that are not compromised, and K be the communication key used for this link. The probability of c being broken given that x nodes are compromised is:

$$Pr(c \text{ is broken}|C_x) = \sum_{j=\lambda+1}^{x} \binom{x}{j} (\frac{\tau}{\omega})^j (1 - \frac{\tau}{\omega})^{x-j} \qquad (16.14)$$

where λ is for $\lambda -$ secure, ω is the size of the group pool, and each node selects τ from the group key pool.

Theorem 16.4

Let x_c be the number of sensor nodes captured by the adversary, and r the resilience of the proposed scheme against node capture. The fraction of compromised links r of x_c nodes being compromised can be derived by the following equation:

$Pr(c \text{ is compromised}|A(u,v) \text{ and } B(u,v))$

$$\leq \frac{1}{XY} \left(\sum_{i \in \Psi} \frac{\sum_{j \in \Psi} p(\xi(i,j)) \cdot Pr(A(n_i, n_j))}{\sum_{i' \in \Psi} \sum_{j \in \Psi} p(\xi(i',j)) Pr(A(n_{i'}, n_j))} \right)^2 \qquad (16.15)$$

$$\cdot \int_{y=0}^{Y} \int_{x=0}^{X} \left(\sum_{a=\lambda+1}^{X_i(x,y,R_c)} \binom{X_i(x,y,R_c)}{a} (\frac{\tau}{|S_c|})^a (1 - \frac{\tau}{|S_c|})^{X_i(x,y,R_c)-a} \right)^2 dxdy \quad (16.16)$$

Proof. Let c be the link between u and v and $C(x,y)$ be the event that the attack circle is centered at (x,y). Let K_i and K_j be the events that c is derived by key spaces in S_i and S_j, respectively. Due to the fact that $C(x,y)$ is independent of $A(u,v)$ and $B(u,v)$, we have

$Pr(c \text{ is compromised}|A(u,v) \text{ and } B(u,v))$

$$\leq \frac{1}{XY} \int_{y=0}^{Y} \int_{x=0}^{X} \sum_{i \in \Psi} \sum_{j \in \Psi} \{Pr_1 \cdot Pr_2$$

$$\times Pr(c \text{ is compromised}|K_j \text{ and } C(x,y) \text{ and } A(u,v) \text{ and } B(u,v))$$
$$\cdot Pr(K_j|A(u,v) \text{ and } B(u,v))\}dxdy \qquad (16.17)$$

where

$$Pr_1 = Pr(c \text{ is compromised}|K_i \text{ and } C(x,y) \text{ and } A(u,v) \text{ and } B(u,v))$$
$$Pr_2 = Pr(K_i|A(u,v) \text{ and } B(u,v))$$

According to the result given by Du et al. [17], for any of the $|S_c|$ keys belonging to group G_i that might be used by any link, we have

$Pr(\text{c is compromised}|K_i \text{ and } C(x,y) \text{ and } A(u,v) \text{ and } B(u,v))$

$$= \sum_{l=\lambda+1}^{X_i(x,y,R_c)} \binom{X_i(x,y,R_c)}{l} \left(\frac{\tau}{|S_c|}\right)^l \left(1 - \frac{\tau}{|S_c|}\right)^{X_i(x,y,R_c)-l} \tag{16.18}$$

According to the result given by DDHV-D [17], we have

$$Pr(K_j|A(u,v) \text{ and } B(u,v)) = \frac{Pr((K_i \text{ and } B(u,v)) \text{ and } A(u,v))}{Pr(A(u,v) \text{ and } B(u,v))} \tag{16.19}$$

where

$$Pr((K_i \text{ and } B(u,v)) \text{ and } A(u,v))$$
$$= \frac{1}{(nt)^2} \sum_{j\in\Psi} p(\xi(i,j)) \cdot Pr(A(u,v)|u \in G_i \text{ and } v \in G_j) \tag{16.20}$$

$$Pr(A(u,v) \text{ and } B(u,v)) = \frac{1}{(nt)^2} \sum_{i\in\Psi} \sum_{j\in\Psi} Pr(A(n_i,n_j))Pr(B(n_i,n_j)) \tag{16.21}$$

So, we have

$$Pr(K_j|A(u,v) \text{ and } B(u,v)) = \frac{\frac{1}{(nt)^2} \sum_{j\in\Psi} p(\xi(i,j)) \cdot Pr(A(n_i,n_j))}{\frac{1}{(nt)^2} \sum_{i\in\Psi} \sum_{j\in\Psi} Pr(A(n_i,n_j))Pr(B(n_i,n_j))}$$
$$= \frac{\sum_{j\in\Psi} p(\xi(i,j)) \cdot Pr(A(n_i,n_j))}{\sum_{i\in\Psi} \sum_{j\in\Psi} Pr(A(n_i,n_j))Pr(B(n_i,n_j))} \tag{16.22}$$

Combining Equation 16.22 and Lemma 16.8, we have

$Pr(\text{c is compromised}|A(u,v) \text{ and } B(u,v))$

$$\leq \frac{1}{XY} \int_{y=0}^{Y} \int_{x=0}^{X} \sum_{i\in\Psi} \sum_{j\in\Psi} \{Pr(\text{c is compromised}|K_i \text{ and } C(x,y) \text{ and }$$

$A(u,v) \text{ and } B(u,v)) \cdot Pr(K_i|A(u,v) \text{ and } B(u,v))$
$$\times Pr(\text{c is compromised}|K_j \text{ and } C(x,y) \text{ and } A(u,v) \text{ and } B(u,v))$$
$$\cdot Pr(K_j|A(u,v) \text{ and } B(u,v))\}dxdy \tag{16.23}$$

Since the events that link c uses the space from K_i and K_j are independent, we have

$Pr(\text{c is compromised}|A(u,v) \text{ and } B(u,v))$

$$\leq \frac{1}{XY} \left(\sum_{i\in\Psi} \frac{\sum_{j\in\Psi} p(\xi(i,j)) \cdot Pr(A(n_i,n_j))}{\sum_{i'\in\Psi} \sum_{j\in\Psi} p(\xi(i',j)) Pr(A(n_{i'},n_j))} \right)^2 \cdot$$

$$\int_{y=0}^{Y} \int_{x=0}^{X} \left(\sum_{a=\lambda+1}^{X_i(x,y,R_c)} \binom{X_i(x,y,R_c)}{a} \left(\frac{\tau}{|S_c|}\right)^a \left(1 - \frac{\tau}{|S_c|}\right)^{X_i(x,y,R_c)-a} \right)^2 dxdy \tag{16.24}$$

■

16.3.4 Performance evaluation

The purpose of this section is to evaluate the performance of the proposed scheme, referred to as the half-key predistribution scheme (HKPS). A simulation model was constructed for performance evaluation, in which nodes and their neighboring relationship were randomly generated. The performance of HKPS was evaluated, compared to the q-composite random key predistribution scheme (RKPS) [8]. Note that each node was assigned a subset of keys randomly chosen from a key pool prior to network deployment. Under HKPS and RKPS, a communication link could be established between two neighboring nodes if they shared at least q keys, where q was a user-defined parameter within the experiments. The performance metrics for the experiments were connectivity, the number of session key candidates, and the resilience against node capture.

16.3.5 Connectivity

Figure 16.13 shows the connectivity observed in the experimental results under HKPS-D and RKPS-D (note that XXX-D denotes the "XXX" scheme with deployment knowledge), where q was set as 1. The y-axis represents connectivity, and the x-axis represents the memory size required to store half keys or full keys in each node under HKPS-D and RKPS-D. It was observed that the larger the memory space was, the better the connectivity. This was because the number of (half) keys stored in each node increased when the memory size did. HKPS-D outperformed RKPS-D because the length of a half key (under HKPS-D) was one-half of a full key (under RKPS-D). Furthermore, when the memory space for each node was equal to 40, that is, a space for 40 half keys, the connectivity of HKPS-D was close to 0.6. However, the connectivity of RKPS-D was about 0.3.

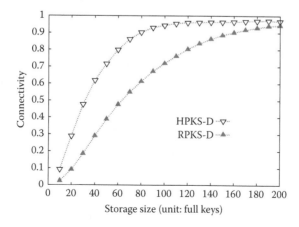

Figure 16.13: Connectivity of HKPS-D and RKPS-D when q = 1.

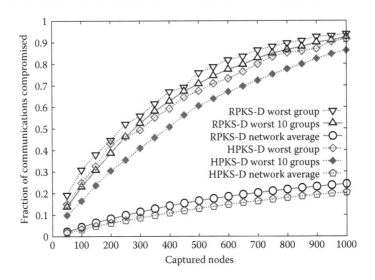

Figure 16.14: Resilience against node capture under HKPS-D and RKPS-D ($m = 200, \lambda = 0, q = 1$).

16.3.5.1 Resilience against node capture

Sensor nodes under experiments were assumed to have the same memory space and the same number of session key candidates under both HKPS and RKPS, that is, one or nine. Connectivity was set at 0.5 and the storage size was set at 200. Note that a larger key pool size, in general, provided better security support. Such an implication could be revealed by the following experiments on resilience against node capture under HKPS and RKPS.

Figure 16.14 shows the resilience against node capture under HKPS-D and RKPS-D when λ was set at 0 and q was set at 1. The x-axis represents the number of captured nodes, and the y-axis shows the ratio of the compromised links against all links, that is, the resilience against node capture. It was shown that our proposed scheme outperforms RKPS-D. Similarly, Figure 16.15 shows the resilience against node capture under HKPS-D and RKPS-D when λ was set at 19 and q was set at 1. Figure 16.16 shows the resilience against node capture under HKPS-D and RKPS-D when λ was set at 19 and q was set at 9. As shown in the experimental results, our proposed HKPS-D scheme also significantly outperformed the RKPS-D scheme.

16.3.6 Summary

This work proposes a half-key approach based on the random key predistribution scheme [8] and DDHV-D deployment knowledge [17] to provide

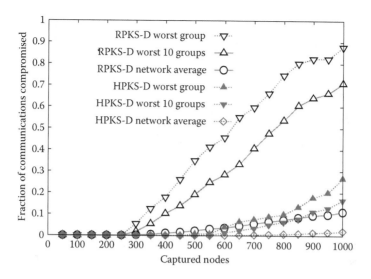

Figure 16.15: Resilience against node capture under HKPS-D and RKPS-D ($m = 200, \lambda = 19, q = 1$).

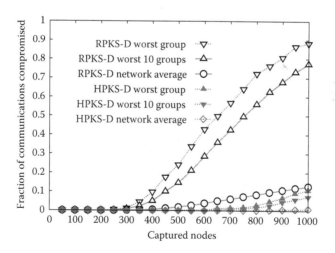

Figure 16.16: Resilience against node capture under HKPS-D and RKPS-D ($m = 1200, \lambda = 19, q = 9$).

resource-efficient key management in wireless embedded systems. Distinct from past research, this research focuses its study on the reduction of the memory space requirements of wireless embedded sensor nodes on key predistribution and the provision of better security enforcement in sensor data delivery. We present an analytical model to exploit the properties of the proposed approach on connectivity and resilience against node capture. The capability of the proposed approach is evaluated by a series of experiments. It was shown that the analytical results matched the experimental results very well. The proposed scheme also provides significant improvements on these two properties, as demonstrated by the experiments.

Bibliography

[1] Charu, C. Aggarwal, Naveen Ashish, and Amit Sheth. The internet of things: A survey from the data-centric perspective. In Charu C. Aggarwal, editor, *Managing and Mining Sensor Data*, 383–428. Springer, 2013.

[2] Isaac Amundson and Xenofon D. Koutsoukos. A survey on localization for mobile wireless sensor networks. In *Proceedings of the 2nd International Conference on Mobile Entity Localization and Tracking in GPS-Less Environments, MELT'09*, 235–254, Berlin, 2009. Springer-Verlag.

[3] Luigi Atzori, Antonio Iera, and Giacomo Morabito. From "smart objects" to "social objects": The next evolutionary step of the internet of things. *Communications Magazine, IEEE*, 52(1):97–105, January 2014.

[4] Antoine B. Bagula and Kuzamunu G. Mazandu. Energy constrained multipath routing in wireless sensor networks. In *Proceedings of the 5th International Conference on Ubiquitous Intelligence and Computing*, UIC '08, 453–467, Berlin, 2008. Springer-Verlag.

[5] Rolf Blom. An optimal class of symmetric key generation systems. In *Proc. of the EUROCRYPT 84 Workshop on Advances in Cryptology: Theory and Application of Cryptographic Techniques*, 335–338, New York, 1985. Springer-Verlag.

[6] Chiara Buratti, Andrea Conti, Davide Dardari, and Roberto Verdone. An overview on wireless sensor networks technology and evolution. *Sensors*, 9(9):6869–6896, 2009.

[7] Jesús Carretero and J. Daniel García. The Internet of Things: Connecting the world. *Personal Ubiquitous Comput.*, 18(2):445–447, February 2014.

[8] Haowen Chan, Adrian Perrig, and Dawn Song. Random key predistribution schemes for sensor networks. In *S&P '03: Proceedings of the 2003 IEEE Symposium on Security and Privacy*, 197, Washington, DC, 2003. IEEE Computer Society.

[9] Yuan-Hao Chang, Ping-Yi Hsu, Yung-Feng Lu, and Tei-Wei Kuo. A driver-layer caching policy for removable storage devices. *Trans. Storage*, 7(1):1:1–1:23, June 2011.

[10] Jian-Jia Chen and Chin-Fu Kuo. Energy-efficient scheduling for real-time systems on dynamic voltage scaling (DVS) platforms. In *Proceedings of the 13th IEEE International Conference on Embedded and Real-Time Computing Systems and Applications, RTCSA '07*, 28–38, Washington, DC, 2007. IEEE Computer Society.

[11] Yan Chen, Feng Han, Yu-Han Yang, Hang Ma, Yi Han, Chunxiao Jiang, Hung-Quoc Lai, et al. Time-reversal wireless paradigm for green internet of things: An overview. *Internet of Things Journal, IEEE*, 1(1):81–98, February 2014.

[12] Yen-Kuang Chen. Challenges and opportunities of internet of things. In *Design Automation Conference (ASP-DAC), 2012 17th Asia and South Pacific*, 383–388, January 2012.

[13] T.S. Chou, S.Y. Chang, Y.F. Lu, Y.C. Wang, M.K. Ouyang, C.S. Shih, T.W. Kuo, J.S. Hu, and J.W.-S. Liu. EMWF for flexible automation and assistive devices. In *Real-Time and Embedded Technology and Applications Symposium, 2009. RTAS 2009. 15th IEEE*, 243–252, April 2009.

[14] Felipe da Rocha Henriques, Lisandro Lovisolo, and Marcelo Goncalves Rubinstein. Algorithms for energy efficient reconstruction of a process with a multihop wireless sensor network. In *Circuits and Systems (LASCAS), 2013 IEEE Fourth Latin American Symposium on*, 1–4, Feb 2013.

[15] Edsger W. Dijkstra. A note on two problems in connexion with graphs. *Numerische Mathematik*, 1:269–271, 1959.

[16] Wenliang Du, Jing Deng, Yunghsiang S. Han, and Pramod K. Varshney. A pairwise key predistribution scheme for wireless sensor networks. In *CCS '03: Proceedings of the 10th ACM conference on Computer and communications security*, 42–51, New York, 2003. ACM Press.

[17] Wenliang Du, Jing Deng, Yunghsiang S. Han, and Pramod K. Varshney. A key predistribution scheme for sensor networks using deployment knowledge. *IEEE Trans. Dependable Secur. Comput.*, 3(1):62, 2006.

[18] Laurent Eschenauer and Virgil D. Gligor. A key-management scheme for distributed sensor networks. In *CCS '02: Proceedings of the 9th ACM conference on Computer and communications security*, 41–47, New York, 2002. ACM.

[19] Hua-Wei Fang, Mi-Yen Yeh, Pei-Lun Suei, and Tei-Wei Kuo. An adaptive endurance-aware b+-tree for flash memory storage systems. *Computers, IEEE Transactions on*, PP(99), 2013.

[20] Emad Felemban, Chang-Gun Lee, and Eylem Ekici. MMSPEED: Multipath multi-speed protocol for QoS guarantee of reliability and timeliness

in wireless sensor networks. *IEEE Transactions on Mobile Computing*, 5(6):738–754, June 2006.

[21] Bernhard Firner, Robert S. Moore, Richard Howard, Richard P. Martin, and Yanyong Zhang. Poster: Smart buildings, sensor networks, and the internet of things. In *Proceedings of the 9th ACM Conference on Embedded Networked Sensor Systems*, SenSys '11, 337–338, New York, 2011. ACM.

[22] Stefan Forsström and Theo Kanter. Continuously changing information on a global scale and its impact for the internet-of-things. *Mob. Netw. Appl.*, 19(1):33–44, February 2014.

[23] Bin Fu, Renfa Li, Xiongren Xiao, Caiping Liu, and Qiuwei Yang. Noninterfering multipath geographic routing for wireless multimedia sensor networks. In *Multimedia Information Networking and Security, 2009. MINES '09. International Conference on*, 1:254–258, 2009.

[24] Ixent Galpin, Christian Y. Brenninkmeijer, Alasdair J. Gray, Farhana Jabeen, Alvaro A. Fernandes, and Norman W. Paton. SNEE: a query processor for wireless sensor networks. *Distrib. Parallel Databases*, 29(1-2):31–85, February 2011.

[25] Jayavardhana Gubbi, Rajkumar Buyya, Slaven Marusic, and Marimuthu Palaniswami. Internet of things (IoT): A vision, architectural elements, and future directions. *Future Gener. Comput. Syst.*, 29(7):1645–1660, September 2013.

[26] Refael Hassin. Approximation schemes for the restricted shortest path problem. *Mathematics of Operations Research*, 17(1):36–42, 1992.

[27] Tian He, John A. Stankovic, Chenyang Lu, and Tarek Abdelzaher. SPEED: A stateless protocol for real-time communication in sensor networks. In *Proceedings of the 23rd International Conference on Distributed Computing Systems*, ICDCS '03, 46–55, Washington, DC, 2003. IEEE Computer Society.

[28] Wendi Rabiner Heinzelman, Anantha Chandrakasan, and Hari Balakrishnan. Energy-efficient communication protocol for wireless microsensor networks. In *HICSS*, 2000.

[29] Pei Huang. Dynamic transmission power control (in ns2). *dekst.award space.com*.

[30] Pei Huang, Hongyang Chen, Guoliang Xing, and Yongdong Tan. SGF: A state-free gradient-based forwarding protocol for wireless sensor networks. *ACM Trans. Sen. Netw.*, 5(2):14:1–14:25, April 2009.

[31] Xiaoxia Huang and Yuguang Fang. Multiconstrained QoS multipath routing in wireless sensor networks. *Wirel. Netw.*, 14(4):465–478, August 2008.

[32] Yu-Kai Huang, Chin-Fu Kuo, Ai-Chun Pang, and Weihua Zhuang. Stochastic delay guarantees in zigbee cluster-tree networks. In *Communications (ICC), 2012 IEEE International Conference on*, 4926–4930, June 2012.

[33] Muhammad Ikram, Aminul Haque Chowdhury, Bilal Zafar, Hyon-Soo Cha, Ki-Hyung Kim, Seung-Wha Yoo, and Dong-Kyoo Kim. A simple lightweight authentic bootstrapping protocol for IPv6-based low rate wireless personal area networks (6LoWPANs). In *IWCMC '09*, 937–941, New York, 2009. ACM.

[34] Chalermek Intanagonwiwat, Ramesh Govindan, and Deborah Estrin. Directed diffusion: a scalable and robust communication paradigm for sensor networks. In *MobiCom '00: Proceedings of the 6th annual international conference on Mobile computing and networking*, 56–67, New York, 2000. ACM Press.

[35] Farhana Jabeen and Alvaro A. A. Fernandes. An algorithmic strategy for in-network distributed spatial analysis in wireless sensor networks. *J. Parallel Distrib. Comput.*, 72(12):1628–1653, December 2012.

[36] Yan Jin, Ling Wang, Ju-Yeon Jo, Yoohwan Kim, Mei Yang, and Yingtao Jiang. Eeccr: An energy-efficient m-coverage and n-connectivity routing algorithm under border effects in heterogeneous sensor networks. *Vehicular Technology, IEEE Transactions on*, 58(3):1429–1442, March 2009.

[37] Holger Karl and Andreas Willig. *Protocols and Architectures for Wireless Sensor Networks*. John Wiley, 2005.

[38] Yong-ki Kim, R. Bista, and Jae-Woo Chang. A designated path scheme for energy-efficient data aggregation in wireless sensor networks. In *Parallel and Distributed Processing with Applications, 2009 IEEE International Symposium on*, 408–415, August 2009.

[39] Chin-Fu Kuo, Lieng-Cheng Chien, and Yung-Feng Lu. Scheduling algorithm with energy-response trade-off considerations for mixed task sets. In *Proceedings of the 2013 Research in Adaptive and Convergent Systems*, RACS '13, 410–415, New York, 2013. ACM.

[40] Kam-Yiu Lam, Jiantao Wang, Yuan-Hao Chang, Jen-Wei Hsieh, Po-Chun Huang, Chung Keung Poon, and Chun Jiang Zhu. Garbage collection for multi-version index on flash memory. In *Proceedings of the Conference on Design, Automation & Test in Europe, DATE '14*, 57:1–57:4, Leuven, Belgium, 2014. European Design and Automation Association.

[41] Ivan Lanese, Luca Bedogni, and Marco Di Felice. Internet of things: A process calculus approach. In *Proceedings of the 28th Annual ACM Symposium on Applied Computing*, SAC '13, 1339–1346, New York, 2013. ACM.

[42] Bocheng Lai, Sungha Kim, and Ingrid Verbauwhede. Scalable session key construction protocol for wireless sensor networks. In *IEEE Workshop on Large Scale Real-Time and Embedded Systems*, 2003.

[43] Alberto Leon-Garcia. *Probability and Random Processes for Electrical Engineering*. Addison-Wesley, 2nd edition, 1994.

[44] Donggang Liu and Peng Ning. Location-based pairwise key establishments for static sensor networks. In *SASN '03: Proceedings of the 1st ACM workshop on Security of ad hoc and sensor networks*, 72–82, New York, 2003. ACM.

[45] Donggang Liu, Peng Ning, and Wenliang Du. Group-based key pre-distribution in wireless sensor networks. In *WiSe '05: Proceedings of the 4th ACM workshop on Wireless security*, 11–20, New York, 2005. ACM.

[46] Donggang Liu, Peng Ning, and Rongfang Li. Establishing pairwise keys in distributed sensor networks. *ACM Trans. Inf. Syst. Secur.*, 8(1):41–77, 2005.

[47] Zhihong Liu, Jianfeng Ma, Qingqi Pei, Liaojun Pang, and YoungHo Park. Key infection, secrecy transfer and key evolution for sensor networks. *IEEE Trans. Wireless. Comm.*, 9:2643–2653, August 2010.

[48] Javier Lopez. Unleashing public-key cryptography in wireless sensor networks. *J. Comput. Secur.*, 14(5):469–482, 2006.

[49] Yung-Feng Lu, Chin-Fu Kuo, and Ai-Chun Pang. A half-key key management scheme for wireless sensor networks. In *Proceedings of the 2011 ACM Symposium on Research in Applied Computation*, RACS '11, 255–260, New York, 2011. ACM.

[50] Yung-Feng Lu, Chin-Fu Kuo, and Ai-Chun Pang. A novel key management scheme for wireless embedded systems. *SIGAPP Appl. Comput. Rev.*, 12(1):50–59, April 2012.

[51] Yung-Feng Lu, Jun Wu, and Chin-Fu Kuo. A path generation scheme for real-time green internet of things. *SIGAPP Appl. Comput. Rev.*, 14(2): 45–58, June 2014.

[52] Samuel R. Madden, Michael J. Franklin, Joseph M. Hellerstein, and Wei Hong. TinyDB: an acquisitional query processing system for sensor networks. *ACM Trans. Database Syst.*, 30(1):122–173, 2005.

[53] Alan McGibney, Antony Guinard, and Dirk Pesch. Wi-Design: A modelling and optimization tool for wireless embedded systems in buildings. In *LCN*, 640–648, 2011.

[54] Wei-Chen Pao, Yung-Fang Chen, and Chia-Yen Chan. Power allocation schemes in OFDM-based femtocell networks. *Wireless Personal Communications*, 69(4):1165–1182, 2013.

[55] Wei-Chen Pao, Yung-Feng Lu, Wen-Bin Wang, Yao-Jen Chang, and Yung-Fang Chen. Improved subcarrier and power allocation schemes for wireless multicast in OFDM systems. In *Vehicular Technology Conference (VTC Fall), 2013 IEEE 78th*, 1–5, September 2013.

[56] Charith Perera, Arkady B. Zaslavsky, Peter Christen, and Dimitrios Georgakopoulos. Context aware computing for the internet of things: A survey. *CoRR*, abs/1305.0982, 2013.

[57] Adrian Perrig, Robert Szewczyk, Victor Wen, David Culler, and J.D. Tygar. SPINS: Security protocols for sensor networks. In *Proceedings of MOBI-COM*, 2001.

[58] Marjan Radi, Behnam Dezfouli, Shukor Abd azak Razak, and Kamalrulnizam Abu Bakar. Liemro: A low-interference energy-efficient multipath routing protocol for improving QoS in event-based wireless sensor networks. In *Sensor Technologies and Applications (SENSORCOMM), 2010 Fourth International Conference on*, 551–557, 2010.

[59] Marjan Radi, Behnam Dezfouli, Kamalrulnizam Abu Bakar, Shukor Abd Razak, and Mohammad Ali Nematbakhsh. Interference-aware multipath routing protocol for QoS improvement in event-driven wireless sensor networks. *Tsinghua Science and Technology*, 16(5):475–490, 2011.

[60] Sutharshan Rajasegarar, Christopher Leckie, and Marimuthu Palaniswami. Hyperspherical cluster based distributed anomaly detection in wireless sensor networks. *J. Parallel Distrib. Comput.*, 74(1):1833–1847, January 2014.

[61] Ragunathan Rajkumar, Insup Lee, Lui Sha, and John A. Stankovic. Cyber-physical systems: the next computing revolution. In *DAC*, 731–736, 2010.

[62] Amar Rasheed and Rabi Mahapatra. Key predistribution schemes for establishing pairwise keys with a mobile sink in sensor networks. *IEEE Trans. Parallel Distrib. Syst.*, 22:176–184, January 2011.

[63] Joel J. P. C. Rodrigues and Paulo A. C. S. Neves. A survey on IP-based wireless sensor network solutions. *International Journal of Communication Systems*, 23(8):963–981, 2010.

[64] Roy Shea, Mani B. Srivastava, and Young Cho. Optimizing bandwidth of call traces for wireless embedded systems. *Embedded Systems Letters*, 1(1):28–32, 2009.

[65] Pei-Lun Suei, Che-Wei Kuo, Ren-Shan Luoh, Tai-Wei Kuo, Chi-Sheng Shih, and Min-Siong Liang. Data compression and query for large scale sensor data on cots dbms. In *Emerging Technologies and Factory Automation (ETFA), 2010 IEEE Conference on*, 1–8, September 2010.

[66] Pei-Lun Suei, Victor C. S. Lee, Shi-Wu Lo, and Tei-Wei Kuo. An efficient b+-tree design for main-memory database systems with strong access locality. *Inf. Sci.*, 232:325–345, May 2013.

[67] Pei-Lun Suei, Yung-Feng Lu, Rong-Jhang Liao, and Shi-Wu Lo. A signature-based grid index design for main-memory RFID database applications. *J. Syst. Softw.*, 85(5):1205–1212, May 2012.

[68] Pei-Lun Suei, Jun Wu, Yung-Feng Lu, Der-Nien Lee, Shih-Chun Chou, and Chuo-Yen Lin. A novel query preprocessing technique for efficient access to XML-relational databases. In *Database Technology and Applications, 2009 First International Workshop on*, 565–569, April 2009.

[69] H. S. Swanson and R. E. D. Woolsey. A PERT-CPM tutorial. *SIGMAP Bull.*, (16):54–62, April 1974.

[70] Matthew Tancreti, Mohammad Sajjad Hossain, Saurabh Bagchi, and Vijay Raghunathan. Aveksha: a hardware-software approach for non-intrusive tracing and profiling of wireless embedded systems. In *Proceedings of the 9th ACM Conference on Embedded Networked Sensor Systems*, SenSys '11, 288–301, New York, 2011. ACM.

[71] Lu An Tang, Xiao Yu, Sangkyum Kim, Jiawei Han, Chih-Chieh Hung, and Wen-Chih Peng. Tru-Alarm: Trustworthiness analysis of sensor networks in cyber-physical systems. In *ICDM*, 1079–1084, 2010.

[72] Somanath Tripathy. Lisa: Lightweight security algorithm for wireless sensor networks. In *ICDCIT*, 129–134, 2007.

[73] Chun-Wei Tsai, Chin-Feng Lai, Ming-Chao Chiang, and L.T. Yang. Data mining for internet of things: A survey. *Communications Surveys Tutorials, IEEE*, 16(1):77–97, 2014.

[74] Kun-Yi Tsai, Yung-Feng Lu, Ai-Chun Pang, and Tei-Wei Kuo. The speech quality analysis of push-to-talk services. In *Wireless Communications and Networking Conference, 2009. WCNC 2009. IEEE*, 1–6, April 2009.

[75] Hsueh-Wen Tseng, Shiann-Tsong Sheu, and Yun-Yen Shih. Rotational listening strategy for IEEE 802.15.4 wireless body networks. *Sensors Journal, IEEE*, 11(9):1841–1855, September 2011.

[76] Hsueh-Wen Tseng, Shan-Chi Yang, Ping-Cheng Yeh, and Ai-Chun Pang. A cross-layer scheme for solving hidden device problem in IEEE 802.15.4 wireless sensor networks. *Sensors Journal, IEEE*, 11(2):493–504, 2011.

[77] Yuh-Min Tseng. A secure authenticated group key agreement protocol for resource-limited mobile devices. *Comput. J.*, 50:41–52, January 2007.

[78] Dan Stefan Tudose, Andrei Voinescu, Madi-Tatiana Petrareanu, Andrei Bucur, Dumitrel Loghin, Adrian Bostan, and Nicolae Tapus. Home automation design using 6LoWPAN wireless sensor networks. In *DCOSS*, 1–6, 2011.

[79] Nikos Tziritas, Thanasis Loukopoulos, Spyros Lalis, and Petros Lampsas. Agent placement in wireless embedded systems: Memory space and energy optimizations. *Parallel and Distributed Processing Workshops and PhD Forum, 2011 IEEE International Symposium on*, 0:1–7, 2010.

[80] Ovidiu Vermesan, Peter Friess, Patrick Guillemin, Sergio Gusmeroli, Harald Sundmaeker, Alessandro Bassi, Ignacio Soler Jubert, Margaretha Mazura, Mark Harrison, Markus Eisenhauer, and Pat Doody. Internet of things strategic research roadmap. *Technical report, Cluster of European Research Projects on the Internet of Things (CERP-IoT)*, 2011.

[81] Berta Carballido Villaverde, Susan Rea, and Dirk Pesch. InRout: a QoS aware route selection algorithm for industrial wireless sensor networks. *Ad Hoc Networks*, 10(3):458–478, 2012.

[82] Gicheol Wang, Deokjai Choi, and Daewook Kang. A lightweight key renewal scheme for clustered sensor networks. In *ICUIMC '09*, 557–565, New York, 2009. ACM.

[83] Xue Wang, Sheng Wang, and Daowei Bi. Distributed visual-target-surveillance system in wireless sensor networks. *Trans. Sys. Man Cyber. Part B*, 39:1134–1146, October 2009.

[84] Duncan S. Wong and Agnes H. Chan. Efficient and mutually authenticated key exchange for low power computing devices. In *Advances in Cryptology, ASIACRYPT 2001*, December 2001.

[85] Jun Wu. CA-SRP: An energy-efficient concurrency control protocol for real-time tasks with abortable critical sections. In *Proceedings of the International C* Conference on Computer Science and Software Engineering*, C3S2E '13, 125–127, New York, 2013. ACM.

[86] Yang Xiao, Venkata Krishna Rayi, Bo Sun, Xiaojiang Du, Fei Hu, and Michael Galloway. A survey of key management schemes in wireless sensor networks. *Comput. Commun.*, 30(11-12):2314–2341, 2007.

[87] Elias Yaacoub, Abdullah Kadri, and Adnan Abu-Dayya. Cooperative wireless sensor networks for green internet of things. In *Proceedings of the 8th ACM Symposium on QoS and Security for Wireless and Mobile Networks*, Q2SWinet '12, 79–80, New York, 2012. ACM.

[88] Poonam Yadav and Julie A. McCann. EBS: decentralised slot synchronisation for broadcast messaging for low-power wireless embedded systems. In *Proceedings of the 5th International Conference on Communication System Software and Middleware*, COMSWARE '11, 9:1–9:6, New York, 2011. ACM.

[89] Yong Yao and Johannes Gehrke. The cougar approach to in-network query processing in sensor networks. *SIGMOD Rec.*, 31(3):9–18, 2002.

[90] In-Su Yoon, Sang-Hwa Chung, and Jeong-Soo Kim. Implementation of lightweight TCP/IP for small, wireless embedded systems. In *Proceedings of the 2009 International Conference on Advanced Information Networking and Applications*, 965–970, Washington, DC, 2009. IEEE Computer Society.

[91] Chia-Mu Yu, Yao-Tung Tsou, Chun-Shien Lu, and Sy-Yen Kuo. Practical and secure multidimensional query framework in tiered sensor networks. *IEEE Transactions on Information Forensics and Security*, 6(2):241–255, 2011.

[92] Chia-Mu Yu, Chun-Shien Lu, and Sy-Yen Kuo. A simple non-interactive pairwise key establishment scheme in sensor networks. In *SECON'09: Proceedings of the 6th Annual IEEE communications society conference on Sensor, Mesh and Ad Hoc Communications and Networks*, 360–368, Piscataway, NJ, 2009. IEEE Press.

[93] Chia-Mu Yu, Chun-Shien Lu, and Sy-Yen Kuo. Noninteractive pairwise key establishment for sensor networks. *IEEE Transactions on Information Forensics and Security*, 5(3):556–569, 2010.

[94] Zhen Yu and Yong Guan. A key management scheme using deployment knowledge for wireless sensor networks. *IEEE Trans. Parallel Distrib. Syst.*, 19(10):1411–1425, 2008.

[95] Ying Zhang and M. Fromherz. Message-initiated constraint-based routing for wireless ad-hoc sensor networks. In *Consumer Communications and Networking Conference, 2004. CCNC 2004. First IEEE*, 648–650, 2004.

[96] Jianliang Zheng, Jie Li, Myung J. Lee, and Michael Anshel. A lightweight encryption and authentication scheme for wireless sensor networks. *IJSN*, 1(3/4):138–146, 2006.

[97] Li Zhou, Jinfeng Ni, and Chinya V. Ravishankar. Efficient key establishment for group-based wireless sensor deployments. In *WiSe '05: Proceedings of the 4th ACM workshop on Wireless security*, 1–10, New York, 2005. ACM.

[98] Liang Zhou. Green service over internet of things: a theoretical analysis paradigm. *Telecommunication Systems*, 52(2):1235–1246, 2013.

Chapter 17

Security Protocols for IoT Access Networks

Romeo Giuliano

Franco Mazzenga

Alessandro Neri

Anna Maria Vegni

CONTENTS

Abstract

Nowadays, we are immersed in a digital world with a huge number of sensors, and devices, connected following a great variety of typologies. Internet Protocol (IP) v6 and the standardization of the novel Internet of Things (IoT) protocols enable new services and applications. Moreover, the heterogeneity of IP and non-IP devices requires novel security techniques, allowing non-IP devices to connect over a short range with a mediator gateway, and then forming a *capillary access network*. Providing security and privacy is hard in the conventional Internet, and is even more challenging in the IoT because of global connectivity and heterogeneous and resource-constrained devices.

In this chapter, we present the background on security algorithms for both uni- and bidirectional terminals, in the context of IoT scenarios. We review the current security and privacy solutions in the IoT, and discuss research challenges for novel IoT security and privacy solutions. Particularly, we deal with security algorithms based on a local key renewal, performed considering only the local clock time. Finally, conclusive remarks and future trends are outlined at the end of the chapter.

17.1   Introduction to IoT

In recent years, a vast number of devices (i.e., objects) have been connected to the Internet than people. In 2020, it is expected that there will be 50 billion things connected to the Internet, and it is estimated that there will be seven devices per person [1]. As a first consequence, there will be a huge amount of data and information generated by these objects.

To exploit the numerous opportunities opening up for the creation of applications in the areas of automation, sensing, and so on, it is necessary to have a standardized and flexible platform to manage the emerging IoT. This is a new paradigm able to manage information generated by "objects" is widely distributed in the environment, including without human presence [2, 3]. Its basic characteristics are the possibility of addressing (i.e., each object should be uniquely identifiable), monitoring (i.e., able to interface with the environment), connecting (i.e., able to inject data into the Internet), analyzing the system (i.e., able to perform complex or simple computations[1]), and reacting (i.e., able to interact with the environment). The huge number of objects, which can produce a stream of information and data from the environment and forward it to the Internet, provides a wide range of applications and services. The main domains in which the IoT can allow the development of innovative applications are the transportation and logistic domain, the healthcare domain, the smart cities

---

[1]In Japan and South Korea, this characteristic is the most important. In fact, the term *ubiquitous computing* is usually adopted rather than IoT.

domain, and the personal and social domain, as well as other futuristic domains such as those related to enhanced gaming. The following list provides examples of the main applications enabled by the IoT:

1. Transportation and logistic domain: inventory, product management, object tracking, parking/traffic

2. Healthcare domain: data collection, person/medicine tracking

3. Smart cities domain: (industrial/business) energy and smart grid, smart metering, industrial plants, infrastructure/utilities, agriculture; (personal) smart home, smart building, environmental monitoring, security (e.g., fire and elevators) and surveillance, heating, ventilating and air-conditioning (i.e., HVAC), lighting, sensors (e.g., temperature, humidity, presence of gases)

4. Personal and social domain: entertainment, social networking, personal objects (losses, thefts), appliances

The evolution of technology also caused a change in the potential of IoT. Nowadays, IoT architecture is based on four main pillars, recalling the main technologies enabling the most common vertical applications related to automation or machine interaction [1].

Radio frequency identification (RFID) [4] is the most diffused technology, with the aim of identifying and tracking objects through tags spared in the environment or attached to an object. Then, the user is able to connect the scanned tag and the central server, where information is contained. The standardization of the electronic product code (EPC) favored its diffusion among industries.

A second pillar is machine-to-machine (M2M) communications. Although they have acquired a wider meaning, at the beginning in 2004 they were restricted to communications between a device/product with a remote (and dedicated) application platform/server through cellular networks or fixed wide area networks.

The third pillar is wireless sensor networks (WSNs), which consist of several sensors widely separated in the environment, able to monitor physical values (e.g., temperature, humidity, motion, pressure, and pollutants) and to communicate wirelessly in a multihop mode. The reference standard in WSN is the IEEE802.15.4 [5], and many devices on the market refer to it. Moreover, the modern WSN can be bidirectional, enabling the sensor node to act locally, even if with non-time-critical characteristics. Full duplexing allows wireless sensor and actuator networks (WSANs).

Finally, the fourth pillar is supervisory control and data acquisition (SCADA), which is an autonomous system able to monitor smart systems (i.e., complex industrial processes), most of all with real-time requirements, through the closed-loop control theory, where human control or interaction is not feasible.

One aspect that cannot be neglected is the management of the vast amount of data that will be generated by tens (or more) of billions of objects from the environment to the Internet. A cloud platform becomes fundamental to store, compute, and visualize data, transforming them into meaningful information. A possible implementation is in [3].

Among the main issues regarding the diffusion of IoT, we cite the following:

■ The lack of a common (and standardized) platform, forcing the software developers to implement vertical (and rigid) architectures to provide specific services

■ The need to address each object

■ The heterogeneity of terminals: the protocol stack is not equal for all the objects, causing different processing capabilities and supported functionalities for each object

■ The need to guarantee the security of data collected by each object and their transmission to the application platform, which is fundamental to the IoT diffusion and standardization process

This chapter is organized as follows. In Section 17.2, we provide a background on security and privacy issues in IoT scenarios, and present the main technologies used to deal with these issues. In Section 17.3, we discuss some algorithms for providing secure connections for unidirectional and bidirectional (non-IP) communications. Finally, conclusions are drawn at the end of the chapter together with a discussion of cognitive security in the context of IoT.

17.2   Related Works on Security Protocols

The aim to provide security and privacy constraints in IoT scenarios remains a challenge, mainly due to the huge number of heterogeneous devices (i.e., around 20 billion devices in 2013, which will increase to 32 billion by 2020), as well as data exchanged via insecure connections. Furthermore, the concept of security is extended not only to device-to-device communications (i.e., end-to-end data confidentiality and integrity), but also to network aspects (i.e., authenticity of devices and access to networks). As an example, many hackers create fake networks (termed *botnets*) to steal data and user privacy information.

Phishing and spam attacks involving IoT devices are becoming an issue. In January 2014, researchers at the security provider Proofpoint discovered an IoT cyberattack where by devices (i.e., home appliances such as home routers, televisions, and refrigerators) sent malicious e-mail spam. Then, thingbots were created to compromise things. Also, privacy issues are proving to be more complicated to fix, since devices in IoT networks are associated with a person, thus resulting in a lack of privacy.

Generally, different security requirements should be addressed to guarantee network, and data security. First, confidentiality is necessary to limit network access and data only to authorized users (i.e., devices). Second, data integrity and authentication should be guaranteed so that messages are successfully transmitted and are reliable to the receiver. Finally, data authentication and availability should be provided, as well as detection of malicious intruders.

In IoT scenarios, a number of technologies have been developed to achieve information privacy and security goals [6], such as transport layer security (TLS), which could also improve the confidentiality and integrity of the IoT, and onion routing, which encrypts and mixes Internet traffic from different sources, and encrypts data into multiple layers, by using public keys on the transmission path. Finally, a recent in-depth review on the security aspects of IoT is provided in [7].

The IoT platform will become a reality due to two main pillars: 6LowPAN [8] and constrained application protocol (CoAP) [9]. 6lowPAN enables embedded nodes to use a restricted subset of IPv6 addresses, while CoAP—a software protocol targeted at small, low-power sensors—allows these devices to offer services to other machines, enabling resource-efficient implementation. In more detail, the idea of 6LoWPAN is a combination of IPv6 and IEEE 802.15.4. The most important difference is the size of the IPv6 packet, so that the Internet Engineering Task Force (IETF) 6LoWPAN working group proposed an adaptation layer that optimizes IPv6 packets through fragmentation and assemblies to be supported by the IEEE 802.15.4 link layer.

A 6LoWPAN network consists of one or more LoWPAN networks connected to the Internet through the edge router, which controls flows incoming and outgoing from the LoWPAN. LoWPAN devices are characterized by their short radio range, low data rate, low power, and low cost. In a LoWPAN, there are two types of devices: (1) the full function devices (FFD) and (2) reduced function devices (RFD) connected to the edge router, responsible for communications with the Internet. Moreover, the LoWPAN supports two types of topologies: star topology, in which nodes communicate with one coordinator responsible for managing communications within the network, and mesh topology, in which nodes can communicate with each other directly. Within LoWPAN, devices do not use the IPv6 address or user datagram protocol (UDP) full header to communicate; it remains at the edge router to communicate with the outside. Finally, routing issues in 6LoWPAN are addressed by the IETF-ROLL (Routing over Low-power and Lossy Network) working group, to seek a proper routing solution for this kind of network. IETF-ROLL has proposed RPL (routing protocol for low-power and lossy networks) [10], which has opened a new area of research and development.

Security issues in 6LoWPAN are analyzed by Rghioui et al. in [11]. 6LoWPAN networks can suffer from several attacks on the security level that aim to cause direct damage to the network or just to spy on the network's confidential information. These attacks can be classified into two types: *internal* attacks provided by malicious nodes and *external* attacks by unauthorized devices. Moreover, these attacks may be *passive*, when the main purpose of the attacker is to spy on the network and capture secret information, or *active*, when interfering directly with the network's performance and causing it to malfunction, such as denial-of-service (DoS) attacks. In [12], Kasinathan et al. present a DoS detection architecture for 6LoWPAN, into which they integrate an intrusion detection system (IDS). Finally, there are several threats, and each layer in the 6LoWPAN stack can undergo specific attacks, occurring at different layers [11]. Surveys on the main protocol stacks for IoT are presented by Palattella et al. in [13] and by Tan and Koo in [14].

17.3 Time-Based Secure Key Generation and Renewal

The time-based secure key generation approach has the aim of efficiently managing (and also renewing) the keys of a secure connection, while guaranteeing the integrity of data transmitted over an insecure channel. The main feature is a local key synchronization and generation by means of the generation of symmetric encryption keys at both sides of the communication channel (i.e., at the transmitter and receiver sides). Specifically, the transmitter (receiver) will encrypt (decrypt) data by means of an encryption (decryption) key extracted from a shared sequence of keys. Moreover, to enhance the security level of the data transmission, the selected key will be changed during transmission.

The key change can be planned on a time or an event basis, and obviously, must be synchronized between the two communication parties. The principle of time-based secure key generation is schematically depicted in Figure 17.1.

In this approach (see Figure 17.1), the key generation process is an operation performed independently by each communication party. In fact, unlike any other key management algorithms, no additional messages are required to be exchanged to agree about a key, and the only requirement is that the key generation function should create the same keys for both communication parties based on the timestamp (TS in Figure 17.1) of the device. The validity of the secure keys is restricted to a time interval, so that reply attacks based on valid messages sent using keys generated in past time intervals are discarded. Leveraging such features, we evince that, as a main advantage of the time-based secure key generation approach, there is no need for a server to manage secure keys. Moreover, the keys are generated locally on both sides of the communication link (i.e., transmitter and receiver) and are not shared along the connectivity link. In Figure 17.1, we have assumed that clocks are locked to a global positioning system (GPS) timescale. This could be difficult to achieve for the IoT, since devices might not be able to receive GPS signals, or they might not be equipped with GPS. However, as shown in the following section, this principle can be extended to the considered heterogeneous IoT scenarios.

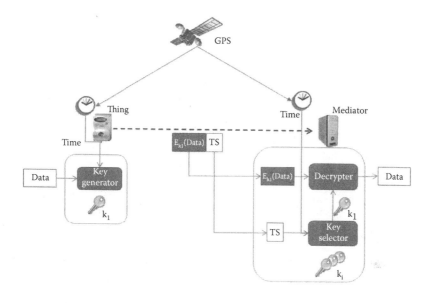

Figure 17.1: Principle of time-based secure key generation.

In the following subsections, we present the security access algorithm in the case of (1) unidirectional and (2) bidirectional data transmissions.

17.3.1 Security access algorithms for unidirectional data transmissions

Due to their simplicity, unidirectional devices cannot perform any secure procedure for secure keys exchange with the mediator. The transmitter just sends a message without any feedback, that is, it fails to receive any signal, and is equipped with an internal clock, which is assumed not to be accurate. Then, a generic non-IP unidirectional terminal executes the following steps to send data to the gateway/mediator in a secure way:

1. It generates the encryption key locally, based on the time measured by a local clock.

2. It creates the message and encrypts it with the generated key; the message includes the payload and (possibly) any other data to be used to enhance security.

3. It computes the hash values using the message text and the generated key and attaches them to the message.

4. It sends the message to the gateway/mediator.

Timestamp	Destination identity	Plain source identity	Security level ID	Hash value	Frame counter	Encrypted source identity	Payload

Plain part   Encrypted part

Figure 17.2: Format of the message sent by a non-IP terminal to the mediator.

The message includes fields that can be grouped into a *plain part* and an *encrypted part*, as shown in Figure 17.2:

1. *Plain part*: The timestamp (obtained by the local clock); the plain part identity (allowing the gateway/mediator to identify it locally for a security procedure such as key generation); the hash (for assessing message integrity; if the hash is calculated using the message texts (see Figure 17.2) and the generated encryption key, then the hash can also be used to verify the identity of the transmitter); a security-level parameter, which is present when several security degrees are allowed at the application level for different types of messages (e.g., simple state data and setting sensor data can be secured differently).

2. *Encrypted part* (optional): The encrypted part identity, which could be used to enhance authentication; the frame counter, which is increased by one at each frame sent; the payload (used to convey information to the related application running on the remote server).

When the mediator receives the ciphered message, it can decipher it by generating the correct decryption key starting from the attached timestamp. In fact, based on the information provided by the timestamp, the mediator can calculate or select the key to decrypt the given message; if the temporal difference between the current time and the timestamp exceeds a predefined threshold, the message is discarded. Values of consecutive timestamps could also be used by the gateway/mediator to estimate the behavior of the clocks of the unidirectional devices in terms of phase and drift. This could allow the gateway/mediator to follow the evolution of the device's clock and then to easily adapt the temporal window in which the timestamp is considered to be valid. We note that being able to verify that the received timestamp time series is monotonically increasing enables replay attacks to be avoided.

The gateway/mediator can organize message reception with all connected terminals in a receiving table. Each table entry is indexed by the *tri*-ple field: <plain identity, timestamp, SLP>. The other fields of each entry contain the key to decrypt the received message. The entry is deleted from the table when the validity related to the timestamp expires. The security-level parameter (SLP) can indicate the security algorithm to be used for decryption (e.g., AES

Table 17.1 **Example of the receiving table**

Plain ID	Timestamp	SLP	Key
ID 51	dd mm yy, 173545	1	$a_0 a_1 \ldots a_{n-1}$
ID 27	dd mm yy, 181011	1	$b_0 b_1 \ldots b_{n-1}$
ID 74	dd mm yy, 174457	2	$c_0 c_1 \ldots c_{n-1}$
...	...	...	...

for confidentiality or SHA for integrity) obviously predefined in the installation phase. The organization allows different parallel communications with a simple IoT terminal, related, for example, to a periodic sensor detection, a setting parameter, or a critical detected datum. In Table 17.1, it is reported an example of the receiving table is reported.

17.3.2 Security access algorithms for bidirectional data transmissions

For bidirectional terminals (i.e., each device can send and receive packets), the mediator can periodically broadcast its clock timing in a dedicated message [15], and its identity in the plain part of the message. Terminals can align their local clocks to the gateway/mediator terminal, and then generate the security keys in accordance with the algorithm previously described.

Since devices are close to the gateway/mediator, propagation delays can be neglected. Furthermore, as for the unidirectional case, the security keys have a validity time interval sufficiently long to transmit one or more packets and to absorb possible retransmissions or any other unwanted delay. In this case, even the key renewal can be performed with a time-based generation algorithm.

Note that each terminal can be served (i.e., being in the coverage area) by more than one mediator gateway. Thus, the mediator gateway identity is fundamental for bidirectional transmissions to distinguish several mediator gateways, which could also have clocks running at (slightly) different times. Thus, the terminal should insert the mediator gateway identity in the sent message; otherwise, the message cannot be correctly decrypted due to possible gateway desynchronization, causing encryption with the wrong key.

Note that the analysis of possible solutions based on public-key cryptography is outside the scope of this chapter. As an example, for gateway-to-IoT device transmissions, the gateway/mediator could broadcast the public key in the area. IoT devices use this key to encrypt their identity and data and to communicate with the gateway/mediator. In this case, the only problem to be solved to guarantee secure reception is preserving the integrity of the transmitted packet. This can be solved by adding a hash to the packet transmitted by the IoT device.

17.4 Cognitive Security

Traditional robust static security can be insufficient, especially for wireless communications (lack of fixed infrastructure), meaning constant surveillance and lack of privacy. Moreover, cooperative wireless protocols are more vulnerable, and dynamic network conditions do not allow normalcy to be distinguished from anomaly.

With the explosive deployment of wireless technologies and the rapid evolution of mobile devices and applications, and then fully distributed control loose security management, mobile devices are subject to security trade-off. Today, we need a new approach to providing security, because even adaptive security is insufficient. This new approach is termed *cognitive security* [16]. As is well known, "cognitive" involves conscious intellectual activity, such as knowing and perceiving, and is based on the possibility of being reduced to empirical factual knowledge. It follows that cognitive security adds cognition by exploiting technologies such as machine learning, knowledge representation, and network control and management, while solving security problems.

Cognitive security authenticates a user through properties, patterns, or knowledge specific to the user that have been continuously learned and updated.

Figure 17.3 shows a principle scheme of cognitive security work applied to the capillary network. The cognitive engine collects all the received data from the terminals in the capillary network at the mediator. Possible parameters to be collected are the transmission–reception time difference of frames for each terminal, the transmission frequencies, the packet lengths, the queue lengths, and so on. In the case of unidirectional terminals, the timestamp difference related to received frames provides information about the emission rate of the source,

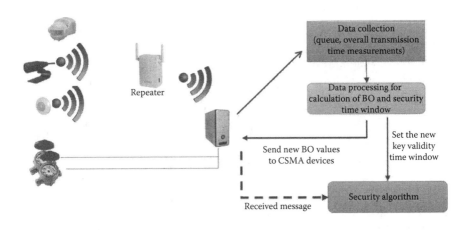

Figure 17.3: Principle scheme of cognitive security work in the capillary network. BO, backoff time; CSMA, carrier sense multiple access.

which should be compared with its target emission rate. For bidirectional terminals, their timestamp difference measured at the mediator should be compared with the set value.

Based on these parameters and on comparison with historical data, a cognitive security-based algorithm should be able to adapt security thresholds to counteract possible intruders/disturbers or terminals that are not correctly working. As an example, the cognitive security engine can modify the backoff time (BO) of the same terminals to increase their possibility of accessing shared channel and transmitted frames. When a traffic anomaly at a certain terminal is detected, the mediator analyzes the identity, that is, the ID parameter, of this terminal, which is considered as a potential disturber. If the disturber is declared nontrustworthy (i.e., secure), the mediator modifies the transmission parameters of terminals of the capillary network, to increase the bidirectional sent frames, and also notifies the ID disturber about the management entity of the capillary network. On the contrary, if the anomalous terminal is trustworthy, the mediator notifies the terminal ID to management entity that the terminal has been compromised.

Thus, possible countermeasures are

- The mediator modifies access parameters to a set of terminals, based on the information of the application level. Possible access parameters that can be modified are: (1) the generation rate of the frames; (2) the reduction of the backoff time to repeat a new access to the channel; (3) the reduction the measured time to detect the presence of the transmission of another terminal (e.g., acting on the Short InterFrame Space).

- The terminal can perform packet aggregation to improve its performance.

- The mediator modifies the time validity of the security keys to avoid replay attacks or clock desynchronization.

Thanks to a collection of data in the capillary network, the mediator is able to toughen the security in the network by properly modifying some parameters related to (1) the channel access, (2) the security techniques, or (3) the transmission characteristics of the traffic: in short, by applying the cognitive security paradigm.

17.5 Conclusions

In this chapter we investigated security aspects related to IoT network access. Specifically, we distinguished communications from a gateway/mediator to non-IP uni- and bidirectional IoT devices.

The IoT security issue has been presented by means of a time-based solution that generates and renews the keys for a secure transaction, for the case of both uni- and bidirectional non-IP devices. The concept is based on the use of the timestamp of the local transmitter (inserted in the plain part of the sent frame) to determine the key for encryption. Then, it is exploited by the receiver to select the proper security key for decryption without any key exchange or extra messages on air. This technique drastically reduces security attacks and greatly simplifies the device capabilities, which is fundamental for IoT environments.

Finally, the concept of cognitive security is introduced and applied to the time-based security solution, highlighting the main parameters that could be monitored and measured by actors (i.e., the mediator in a capillary network) to enforce and toughen the security in a variegated and variable scenario such as the IoT.

References

[1] H. Zhou, *The Internet of Things in the Cloud: A Middleware Perspective*, CRC Press, Boca Raton, FL, 2012.

[2] L. Atzori, A. Iera, and G. Morabito, "The Internet of Things: A survey", *Computer Networks*, Vol. 54, 2010, pp. 2787–2805.

[3] J. Gubbi, R. Buyya, S. Marusic, and M. Palaniswami, "Internet of Things (IoT): A vision, architectural elements and future direction", *Future Generation Computer Systems*, Vol. 29, 2013, pp. 1645–1660.

[4] RFID guide. http://www3.nd.edu/g̃madey/Activities/CAS-Briefing.pdf.

[5] IEEE Standard for Local and metropolitan area networks–Part 15.4: Low-Rate Wireless Personal Area Networks (LR-WPANs), available online at https://standards.ieee.org/getieee802/download/802.15.4-2011.pdf.

[6] R.H. Weber, "Internet of Things: New security and privacy challenges," *Computer Law & Security Review*, Vol. 26, No. 1, 2010, pp. 23–30.

[7] H. Suo, J. Wan, C. Zou, and J. Liu, "Security in the Internet of Things: A review," in *Proc. of Intl. Conf. on Computer Science and Electronics Engineering (ICCSEE)*, vol. 3, pp. 648–651, March 2012.

[8] G. Mulligan, "The 6LoWPAN architecture," in *Proc. 4th ACM workshop on Embedded Networked Sensors (EmNets '07)*, pp. 78–82, 2007.

[9] Z. Shelby, K. Hartke, C. Bormann, and B. Frank, "Constrained Application Protocol (CoAP)," IETF draft, January 2012.

[10] T. Winter, P. Thubert, A. Brandt, J. Hui, R. Kelsey, P. Levis, K. Pister, R. Struik, J.P. Vasseur, and R. Alexander, "RPL: IPv6 routing protocol for low-power and lossy networks," Request for Comments (RFC): 6550, March 2012.

[11] A. Rghioui, M. Bouhorma, and A. Benslimane, "Analytical study of security aspects in 6LoWPAN networks," in *Proc. of 5th Intl. Conf. on Information and Communication Technology for the Muslim World*, 2013.

[12] P. Kasinathan, C. Pastrone, M.A. Spirito, and M. Vinkovits, "Denial-of-Service detection in 6LoWPAN based Internet of Things," in *Proc. of IEEE 9th Intl. Conf. on Wireless and Mobile Computing, Networking and Communications (WiMob)*, 2013, pp. 600–607, 7–9 October 2013.

[13] M.R. Palattella, N. Accettura, X. Vilajosana, T. Watteyne, L.A. Grieco, G. Boggia, and M. Dohler, "Standardized protocol stack for the Internet of (important) Things," *IEEE Communications Surveys & Tutorials*, vol. 15, no. 3, pp. 1389–1406, 2013.

[14] J. Tan, and S.G.M. Koo, "A survey of technologies in Internet of Things," in *Proc. of IEEE Intl. Conf. on Distributed Computing in Sensor Systems (DCOSS)*, 2014, vol., no., pp. 269–274, 26–28 May 2014.

[15] R. Giuliano, A. Neri, and D. Valletta, "End-to-end secure connection in heterogeneous networks for critical scenarios", *WIFS 2012, Proc. of the 2012 IEEE Intl. Workshop on Information Forensics and Security*, pp. 264–269, Tenerife, Spain.

[16] K. Witold, "Towards cognitive security systems", in *Proc. of Cognitive Informatics Cognitive Computing (ICCI*CC), 2012 IEEE 11th International Conference on*, pp. 539–539, August 2012.

Bibliography

daCosta, F. *Rethinking the Internet of Things: A Scalable Approach to Connecting Everything*, Apress Open, 2013.

Evans, D. "The Internet of Things: How the next evolution of the Internet is changing everything," White Paper, April 2011, available online: http:// www.iotsworldcongress.com/documents/4643185/3e968a44-2d12-4b73-9691- 17ec508ff67b.

Giuliano, R., F. Mazzenga, A. Neri, and A.M. Vegni, "Security access protocols in IoT networks with heterogenous non-IP terminals," in *Proc. of IEEE Intl. Conf. on Distributed Computing in Sensor Systems (DCOSS)*, pp. 257–262, 18–26 May 2014, Marina Del Rey.

Giuliano, R., F. Mazzenga, A. Neri, A.M. Vegni, and D. Valletta, "Security implementation in heterogeneous networks with long delay channel," in *Proc. of 2012 IEEE 1st AESS European Conference on Satellite Telecommunications, ESTEL 2012*, pp. 1–5, Rome.

Giuliano, R., F. Mazzenga, and M. Petracca, "Consumed power analysis for mobile radio system dimensioning", *IEEE International Conference on Communications (ICC 2013)*, June 2013, Budapest, Hungary.

Giuliano, R., F. Mazzenga, M. Petracca, and R. Pomposini, "Performance evaluation of an opportunistic distributed power control procedure for wireless multiple access", in *Proc. of 5th Intl. Symp. on Communications Control and Signal Processing, ISCCSP 2012*, May 2012, Rome.

Giusto, D., A. Iera, G. Morabito, and L. Atzori (Eds.), *The Internet of Things*, Springer, 2010. ISBN: 978-1-4419-1673-0.

Inzerilli, T., A.M. Vegni, A. Neri, and R. Cusani, "A location-based vertical handover algorithm for limitation of the ping-pong effect," in *Proc. of 4th IEEE Intl. Conf. on Wireless and Mobile Computing, Networking and Communications (WiMob 2008)*, pp. 385–389, 12–14 October 2008, Avignon, France.

Mionardi, D., S. Sicari, F. De Pellegrini, and I. Chlamtac, "Internet of Things: Vision, application and research challenges", *Ad Hoc Networks*, Vol. 10, 2012, pp. 1497–1516.

Palma, V. and A.M. Vegni, "On the optimal design of a broadcast data dissemination system over VANET providing V2V and V2I communications: The vision of Rome as a smart city," *Journal of Telecommunications and Information Technology (JTIT)*, no.1, 2013, p.4148.

Petracca, M., R. Giuliano, and F. Mazzenga, "Application of UWB technology for underlay signaling in cognitive radio networks", *Recent Patents on Computer Science* 2012, vol. 5, no. 2, pp. 109–116.

Spiess, P., S. Karnouskos, D. Guinard, D. Savio, O. Baecker, L. Souza, and V. Trifa, "SOA-based integration of the Internet of Things in enterprise services", *Proceedings of IEEE ICWS 2009*, July 2009, Los Angeles, CA.

Su, K., J. Li, and H. Fu, "Smart city and the applications," in *Proc. of International Conference on Electronics, Communications and Control (ICECC)*, pp. 1028–1031, 9–11 September 2011.

Vienna University of Technology. European Smart Cities. http://www.smart-cities.eu/.

SOCIAL
AWARENESS

Chapter 18

A User-Centric Decentralised Governance Framework for Privacy and Trust in IoT

Jorge Bernal Bernabe

Jose Luis Hernandez

Mara Victoria Moreno

Antonio Skarmeta

Niklas Palaghias

Michele Nati

Klaus Moessner

CONTENTS

The Internet of Things (IoT) is changing the way that people share information and communicate with their surrounding environment, enabling a constant, and sometimes unconscious, data exchange between things and people. This situation demands new security and privacy-preserving solutions to cope with the dynamic and ubiquitous nature of IoT environments.

Chapter 18 gives an overview of the main security and privacy-enhanced technologies that are being developed in the scope of the SocIoTal EU project, such as attribute-based cryptography for secure data sharing, anonymous credential systems for minimal disclosure of personal data, and access control mechanisms based on capability tokens. These mechanisms are encompassed in a new IoT security framework, which is based on the Architecture Reference Model (ARM) and puts strong emphasis on context management as the cornerstone aspect to drive security decisions. The context can be obtained and inferred by different device-centric enablers, such as the face-to-face enabler and the indoor localization enabler.

18.1   Introduction

Recently, the number of Internet-connected objects and devices has exceeded the number of humans on Earth, marking the dawn of a new era of the Internet of Things (IoT). The initial roll out of IoT devices has been fueled primarily by industrial- and enterprise-centric use cases. There are some end-user applications and the "quantified-self" has seen significant pick up despite most users not being aware of the privacy and data-ownership implications. However, the real exploitation potential for smart services to address the needs of individual citizens, user communities, or society at large is limited at this stage and not obvious to many people.

Unleashing the full potential of the IoT means going beyond the enterprise-centric systems and moving toward a citizen-inclusive IoT in which IoT devices and the information flows provided by people are encouraged. This will allow the unlocking of a wealth of new citizen-centric IoT information based on which a new generation of services of high societal value can be built.

The move toward a citizen-inclusive IoT in which citizens provide IoT devices and contribute information flows are encouraged will have a significant impact on people and societies in general. A variety of technological socioeconomic barriers will have to be overcome to enable such inclusive IoT solutions. In particular, the human perception of the IoT is critical for a successful uptake of the IoT in all areas of society. The perceived level of trust and confidence in the technology are crucial in forming public opinion on the IoT and as such are extremely important challenges that have to be addressed. This is a real challenge with IoT solutions, which are expected to behave seamlessly and act in the background, invisible to their users.

To ensure large-scale uptake of the IoT in all areas of society, IoT architecture and the protocols of an inclusive IoT ecosystem must be simple and must provide motivation for every citizen to contribute an increasing number of IoT devices and information flows in their households and make them available to their immediate community and to the IoT at large. In addition to the simplicity in terms of how the system is used and the immediate and clear benefits provided by the system to each individual user, implementation must be done in such a way to ensure adequate control and transparency. This is needed to grow confidence and to allow a better understanding of what is happening with the information and devices contributed. If transparency and user control are not treated adequately in such a community- grown IoT system, there is a real danger delete for and insert that the systems delete to and insert will be perceived with suspicion and mistrust by users, which may result in opposition and refusal of such technology, thus hindering its widespread deployment.

The SocIoTal[1] project investigates, designs, and provides key enablers for a reliable, secure, and trusted IoT environment that will enable the creation of a socially aware citizen-centric IoT. The approach is based on incentivizing and encouraging people to contribute their IoT devices and information flows on one side and to be able to gain from having themselves access to information provided by other users. SocIoTal will provide the techno-social foundations to unlock billions of new IoT information streams taking a citizen-centric IoT approach toward the creation of large-scale IoT solutions of interest to society. By equipping communities with secure and trusted tools that increase user confidence in the IoT environment, SocIoTal will enable their transition to smart neighborhoods, communities, and cities.

The goal is to establish an IoT ecosystem that puts trust, user control, and transparency at its heart in order to gain the confidence of everyday users and citizens. Providing adequate, socially aware tools and mechanisms that simplify complexity and lower the barriers of to entry will encourage citizen participation in the IoT. Since the majority of these barriers are related to security and privacy concerns, this chapter focuses on the main mechanisms devised in the scope of SocIoTal to deal with these issues. Namely, the privacy-preserving solution that allows more flexible secure sharing models, the identity management (IdM) mechanism that supports minimal disclosure of private information as well as the access control mechanism. These security and privacy solutions are driven by the context in order to cope with the pervasive and ubiquitous nature of the IoT. Thus, this chapter also describes how the context can be provided and inferred by two of the main enablers addressed in SocIoTal, that is, the enabler and the indoor localization enabler. The context-aware security solutions are framed in the scope of a novelty security framework, which is being designed and implemented in the scope of the project.

18.2   Background and State of the Art

Several European initiatives have defined different architectures in order to design IoT services and applications under a common view. Typically, these approaches have been tailored to specific domains addressing a small subset of requirements regardless of the global nature of the IoT. This was identified as one of the main barriers to for a broad adoption of the IoT. In this direction, the EU FP7 IoT-A[1] project represents the most remarkable initiative for the creation of a harmonized vision of the IoT in Europe, by optimizing the interoperability among isolated IoT applications to create a global ecosystem of services under a common understanding. The main result of IoT-A was the definition of an ARM [4] for IoT systems, to promote a common understanding at a high abstraction level, through the description of essential building blocks. The results from IoT-A promoted the emergence of additional initiatives adopting the ARM as the starting point of design activities, such as the EU FP7 IoT6 [28] or BUTLER [3] projects. However, a common feature of the resulting architectures of

---

[1] http://iot-a.eu

these efforts is that they are not focused on the definition of suitable security and privacy mechanisms for IoT scenarios, supporting aspects such as privacy by design, and data minimization principles.

In this sense, although some non-IoT, related projects such as Daidalos[2], SWIFT[3] and Primelife[4] have undertaken the application of user-centric, privacy-preserving IdM schemas, there is a lack of clarity regarding the definition of an integral architecture that is able to tackle these security and privacy issues for IoT scenarios. The IoT security framework presented in Section 18.3 is based on the ARM architecture, extending it with novel mechanisms regarding security and privacy preserving.

The current security and privacy-preserving solutions need to be adapted to the envisioned IoT scenarios, allowing more flexible sharing models (beyond the classic request/response approach), as well as fleeting and dynamic associations between entities, while preserving privacy. Privacy-enhancing technologies help to deal with this problem, providing the means to achieve anonymity, pseudonymity, data minimization and unlikability as well as other techniques to provide confidentiality and integrity of sensitive data. Individuals should be able to control which of their personal data are being collected, under which circumstances as well as who is collecting such information. At the same time, the IoT paradigm implies device limitations such us memory, computation, storage, and energy capacity, which means that usability aspects also need to be taken into account when designing the security and privacy framework for the IoT.

In this context, the IdM solution adopted in our IoT framework is based on the usage of partial identities [16] as an identity-preserving mechanism that allows users to define a subset of their personal attributes, from their real identity, in order to identify them in a given context. The idea is to avoid using the whole credential (e.g., the user X.509 certificate) when using a service since probably only a small set of the credential attributes are really needed. Anonymous credential systems (such as Idemix [7] or Uprove[21]) allow the user to send cryptographic proofs, instead of the whole credential, stating that he or she is in possession of certain attributes or claims. In this kind of anonymous credential system, an entity firstly obtains the credential from a credential authority (issuer), and then generates a customized cryptographic proof, which is sent to the other part (e.g., a service) with the aim of convincing him or her that he or she is in possession of the credential. The anonymous credential functionalities are adopted in our framework as part of the IdM functional group. Users can take advantage of the usage of partial identities to authenticate themselves against IoT services in order to access them securely while disclosing, at the same time, the minimum amount of personal data. Anonymous credentials and partial identities can be applicable

---

[2]Daidalos: http://www.ist-daidalos.org/
[3]SWIFT: http://ist-swift.sit.fraunhofer.de/
[4]Primelife: http://primelife.ercim.eu/

to the information exchange between the user and an IoT service following either a synchronous or an asynchronous way.

Additionally, providing confidentiality and integrity to information exchange is a paramount security issue that also still needs to be addressed in the IoT environment. To this aim, the IoT framework should allow the use of different partial identities according to the context, for each data transaction carried out from the data producer to the consumer. An attribute base encryption (ABE) [14] mechanism can be employed to ensure confidentiality during data exchange. ABE can rely on anonymous credential systems that one adopted in the IoT framework presented in Section 18.3, to obtain private keys associated with certain user's attributes, after demonstrating the possession of such attributes in the partial identity. Then, these private keys can be used by the consumer to decrypt the information ciphered by the producer, as long as the consumer complies with the attribute sharing policies. This secure data dissemination approach based on ABE is presented in Section 18.3.4.

The realization of IoT scenarios imposes significant restrictions on privacy and access control, since everyday physical objects are being seamlessly integrated into the Internet infrastructure. Current access control mechanisms need to consider efficient and proper IdM schemes to be able to cope with scenarios with billions of objects while end-to-end security is preserved. Additionally, IoT scenarios are intended to manage particularly sensitive data as any information leakage could seriously damage the user's privacy. This problem is exacerbated in the IoT, since any entity connected to the Internet will be able to create new information and communicate it to any other entity. Traditional access control approaches solutions were not designed with these aspects in mind and, in most cases, they are not able to meet the needs of these incipient ecosystems regarding scalability, interoperability, and flexibility. These challenges have attracted increasing attention from the community and recently several efforts have started to emerge in this direction. The authors in [27] present an abstraction of the Usage Control (UCON) model [24] in the IoT. The proposal is based on a trust management center, which is responsible for updating the trust values of devices and services in each usage request.

Under the main foundations of ZBAC and SPKI Certificate Theory the [11], the application of capability-based access control (CapBAC) on IoT scenarios is considered in [15], which is based on the work carried out in the EU FP7 IoT@Work project. The proposed approach is based on policy decision points (PDPs) that are queried by services to get authorization decisions. Therefore, when the subject tries to access the data of a particular resource, such user attaches the capability token to the access request. Then, the PDP is responsible for deciding whether the entity is authorized or not. The decision is based on the received capability and the internal rules defined for such a resource. CapBAC is also considered by [22] for secure access to services. Once the capability is verified by the service, a protected session is established for subsequent communications. Based on the main foundations of these works, distributed capability-based access control (DCapBAC) [17] has recently been introduced as a feasible access control approach for deployment on the IoT, even when constrained devices are used. DCapBAC allows a distributed approach in which constrained devices are enabled with authorization logic by adapting the communication technologies and data-interchange format. The access control system proposed herein and explained in Section 18.3.3 is based on the usage of DCapBAC, along additional access control and security features in order to provide a holistic IoT access control system.

18.3 SocIoTal Security Framework

The SocIoTal security framework is, to a certain extent, a realization of the security functional group of the IoT ARM architecture devised within the scope of the IoT-A EU project. Nonetheless, our IoT security framework, in contrast to ARM, gives special attention to privacy-preserving mechanisms, as well as secure data sharing. Thus, the framework includes some other modules beyond the five classical ones defined in IoT-A. Namely, the framework extends the ARM with a context manager, as a transversal component that enables the rest of the components in the framework to cope with the pervasive and ubiquitous nature of IoT. In addition, the framework includes a privacy-preserving IdM system that endows users with the means to achieve anonymity, data minimization, and unlinkability based on anonymous credential systems. Furthermore, the security framework introduces a group manager component to deal with more flexible secure sharing models within bubbles of users or smart objects or both. Figure 18.1 shows the main components of the SocIoTal security framework. As can be seen, it describes seven main groups, that is, authentication, authorization, identity management, trust and reputation, context manager, and group manager.

The *authentication* component enables the authentication of users and smart objects based on the provided credentials. It allows binding a real identity to a subject. As a result of the authentication process, an assertion is generated to be used afterward in the authorization process, to declare that a specific subject was authenticated successfully. In this sense, the SAML protocol[5] is used in our framework for handling the authentication tokens. Traditional authentication mechanisms based on, for instance, login-password or electronic IDs have been addressed and solved even in the emerging IoT paradigm. Our framework also addresses more sophisticated ways of performing authentication by ensuring, at the same time, privacy and minimal disclosure of attributes. Thus, this kind of alternative privacy-preserving way of authentication is handled in the framework by the IdM component. The IdM system is described in detail in Section 18.3.2.

---

[5]SAML: http://docs.oasis-open.org/security/saml/v2.0

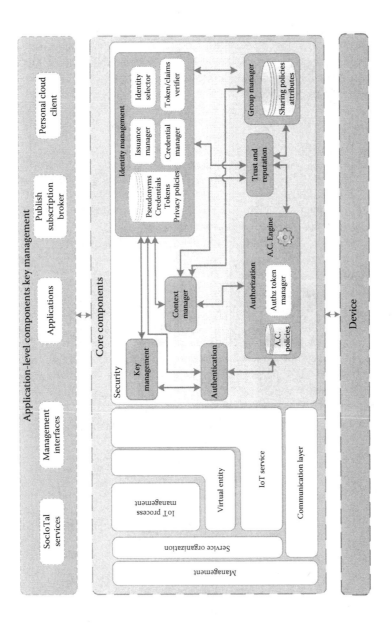

Figure 18.1: ARM-based Security Framework for the IoT.

The *key exchange and management (KEM)* component assists peers involved in a communication in the process of establish the a security context, such as setting up tunnels for a security communication. It involves cryptographic key exchange and provides interoperability between peers to reach an agreement regarding the security functions to use for the communication. Our framework focuses on the part of the KEM component that deals with the keys management in the privacy-preserving IdM system as well as the group manager by means of the CP-ABE ciphering scheme.

The *trust and reputation* component enables the establishment of a trusted and reliable IoT environment where users can safely interact with the IoT service. There are different situations where trust and reputation scores that are worked out by the trust and reputation module can be useful in IoT scenarios. Thus, the trust and reputation component allows other security components of the framework to take security and privacy decisions according to the quantified trust scores. The goal of the scores is twofold. On the hand, they can be used to interact securely within IoT services. Thus, for producers, they allow authorization decisions to be taken on sharing data according to trust scores, while they enable consumers to obtain to obtain data from producers in a reliable way, in order to obtain data only from those services that satisfy certain trust scores On the other hand, the scores can be used to manage bubbles (circle of trust) and share data in the bubble according to the context and trust values. The trust and reputation component could evaluate the degree of social interaction between users involved in a bubble, as perceived by the interactions of their devices. The trust and reputation component usually follows a process with four main operations:

1. The trust and reputation component continuously gathers information about the entities in the system in order to obtain behavioral information. An entity can refer to a user, smart objects, or even communities (or bubbles) as a whole. trust and reputation component gathers the information mainly from the CM component of the framework.

2. After gathering the information, different kinds of algorithms and techniques can be used to compute the trustworthiness of a given entity. computational, energy, and storage restrictions of IoT devices need to be taken into account. As a result, the trust and reputation component up with a score about the entity.

3. Based on the scores as well as other useful information that may be needed, the entity chooses the best entity to interact with or just declines an interaction.

4. For future interactions, once the communication between the entities is done, the trust and reputation component updates the score of the target entity, rewarding or punishing such an interaction.

The remaining four components of the security framework, that is, context management, identity management, authorization, and group management, are designed and implemented with innovative security techniques specifically intended for the IoT, and therefore, they are given more importance in this book chapter. The following four subsections detail these four security components of the framework.

18.3.1 Context-driven security and privacy

As shown in Figure 18.1, the role of context is central in the SocIoTal security framework. The following components will require access to the context manager:

- Identity manager: To create and manage multiple identities associated with a given user/device and to load and expose them to other devices and architecture components depending on the device context.

- Authorization manager: To enable capability-based access to the data and services provide by a user's devices according to both the provider and consumer context.

- Group manager: To define and identify groups (e.g., bubbles) according to the context of the devices and to securely share data among them, by distributing private keys within specific groups.

- Trust & reputation manager: To compute device reputation scores, according to the device and user context and to define the trustworthiness level of a given device with respect to another.

Figure 18.2 shows the architecture and the functionalities provided by the SocIoTal CM. As SocIoTal is dealing with distributed architecture and a mixture of embedded and mobile/static devices, part or all the different modules and functionalities of the CM could be hosted on the devices, associated gateways, or cloud/back-end infrastructure or both.

To expose the SocIoTal-generated context to external and third-party components and developers, and depending on the considered scenarios and involved devices, that is, static versus mobile ones, sensor motes instead of mobile smartphones, two different mechanisms for context communication have been envisioned:

- Query based: The component that requires the context (context consumer) makes a request in terms of a query, so that the CM, which acts as a context repository, can use that query to produce the requested results.

- Pub/sub based: In this case, the context consumer subscribes with a CM by describing its context requirements. The CM acting as a context broker

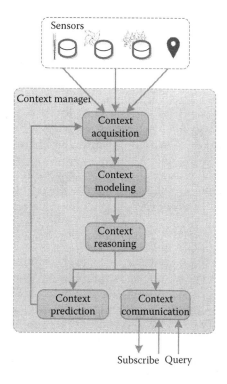

Figure 18.2: Context manager architecture.

will then return the results periodically or when an event occurs (threshold violation).

To support such functionalities, the SocIoTal context broker is implemented based on the FI-WARE Orion context broker [13] module, thus providing the back-end/cloud functionalities for sharing and accessing SocIoTal context information from selected components. According to this, each device producing context will implement its Context communication module as an NGSI context provider registered to the back-end SocIoTal context broker. By using standardized interfaces, that is, the NGSI-9 registerContext, a specific context notion can be registered according to a well-defined model. As soon as a specific context value is extracted by the context reasoning module, this can be pushed, using the NGSI-10 update-context method, to the SocIoTal context broker and either distributed to subscribed modules or stored for further access using a query-based approach.

It should be noted that the communication with the context broker is subject to security mechanisms (e.g., authorization and encryption) to ensure that only trustworthy devices interact with it.

In order to share context in a simple and highly transferable format, JSON message to communicate the context notion to interested modules has been defined according to the OMA context management specification [12], which defines every context as a particular *context entity* made of a collection of *context elements*, each one represented by attributes and metadata. A collection of attributes, based on the SensorML attributes specification [8], will be used to define each specific SocIoTal context.

It is the role of the context reasoning module, which, depending on the device features, can run directly on SocIoTal devices or on more powerful gateways or cloud/back-end services, and according to the definition provided by the context modeling, which extracts the required context element based on the different data stream source provided by the context acquisition module. In the case of specific SocIoTal context, context reasoning module is implemented by the two device enablers defined in Section 18.4, namely, the F2F enabler and the indoor localization enabler. The two enablers allow the specification of the context for user mobile phone devices in which a new group or bubble can be based on the indoor position of devices and their users and detect social relations among them. An example of the extracted and shared SocIoTal context model providing such information can include the following attributes:

■ Location: Providing the position of the device according to specific indoor localization coordinates.

■ Relation: Providing a list (eventually empty) of pseudonyms representing the ID of devices for which an F2F relation along with its type is detected.

■ DateTimeStamp: A temporal reference for when the previous two context attribute are obtained.

Finally, in order to manage the scarce resources available on envisioned SocIoTal devices, such as smart cities sensors or to preserve the user experience of other devices, such as user smartphones, a context prediction module is envisioned to optimize the operation of context reasoning by leveraging on the periodicity of observed contexts. This will allow a reduction in reduce the burden of continuously acquiring sensing data and extract in context information. In case of social relations and indoor positions, it is expected that a periodicity can be observed in the user behavior while carrying around his or her SocIoTal device (e.g., mobile smartphone producing and sharing data).

18.3.2 *Privacy-preserving identity management*

IdM encompasses the technologies and processes that are aimed at controlling and managing private and secure access to information and resources while at the same time, protecting the user or smart object profiles. The IdM should provide the means to storing the information of entities such identifiers, credentials, and pseudonyms. It is also responsible for defining, managing, and issuing the identities and credentials of entities, taking into account that in the IoT environment an entity can refer to both persons and smart objects.

IdM systems usually provide interfaces to make identity information and management accessible for both users and administrators. Traditional IdM systems lack the proper means to deal with privacy preserving and usually do not provides the means for their users to deal with minimal disclosure of private information. In traditional IdM systems that use common credentials, the service provider can usually store all the tokens and users credentials (e.g., X.509 certificates) that are presented to it. The problem is that the service provider can then link them together. Thus, users should be given full control over their data in order to determine which private data are disclosed in which context. The IdM should provide a proper mechanism to manage their partial identities in a private way according to the context, that is, providing anonymity and unlinkability.

The IdM component of the security framework is an anonymous credential system that ensures user privacy and minimal disclosure of personal information when accessing IoT services. It is based on the Idemix [7] anonymous credential system, but it is adapted to IoT scenarios. Namely, in order to address the SocIoTal uses cases, where mobile smartphones are usually employed, the IdM is firstly implemented to deal with deployments in smartphones based on Android.

Having part of the IdM deployed in end users, smartphones allows the control and management of personal data in the smartphone, defining partial identities and describing rules defining the way its personal information is disclosed according to the context. In this kind of scenario, users could interact directly with other peers, members of communities, and bubbles to share information and access each other's IoT services, so that user devices could act as consumers and producers of information.

The SocIoTal IdM, unlike traditional IdMs such as Fi-ware [13], addresses a small set of functionalities, focusing on the authentication process and the privacy-preserving mechanism that enable users to use different partial identities to access target devices according to the context. Other IdM functionalities used in traditional web contexts, such as user profile management and single sign on (SSO), are left in SocIoTal to open existing solutions, which already successfully provide those functionalities. The privacy-preserving IdM system of the SocIoTal security framework relies on two main operations, the credential issuance and the credential presentation processes, which are detailed next.

The *credential issuance* process is one of the main protocols, along with the presentation process, that is required in the anonymous credential system. Figure 18.3 shows the main interactions of the credential issuance operation.

1. Firstly, the subject IoT device requests a credential to the issuer entity. In case it is the first time that the subject asks for a credential (and it does not present another credential or proof), the issuer must identify the subject with an out-of-band authentication process or any other bootstrap electronic authentication.

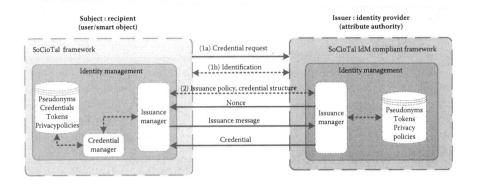

Figure 18.3: IdM credential issuance.

2. With the aim of obtaining a credential, the subject needs a credential structure definition. This credential structure, which defines the attribute structure of the credential, can be provided by the issuer, or be already known by the subject. Depending on the implementation, the issuer can also provide the subject with a issuance policy to indicate which existing credentials the subject must possess in order to be issued with a new credential. Optionally, in case a credential is based in another existing credential, the credential structure should describe which attribute will be reused in the new credential.

3. Once both parties have finished the initialization and share the same credential definition, the issuer computes a random value called a *nonce*, which that is sent to the subject. The subject computes a cryptograph message (also known as *token*), which includes the attributes to be incorporated into the credential, following the credential structure, and optionally satisfying the issuance policy. detailed mathematical description of signatures and encryption schemes is omitted since they depend on the underlying crypto engine implementation.

4. The issuance message with the token is sent to the issuer. Depending on the implementation, in case that in Step 2 an Issuance policy were requested by the Issuer, it verifies that the token satisfies such a policy. Then, the Issuer creates the cryptographic part of the credential, signing the attributes with his secret key. It also creates a proof of correctness. The issuer can save the pseudonym and the context for accountability purposes.

5. The Issuer replies, sending the subject a cryptographic message with the proof of correctness, and the attributes signature. The subject verifies receipt of the cryptographic material, generates the credential based on this message, and stores the credential.

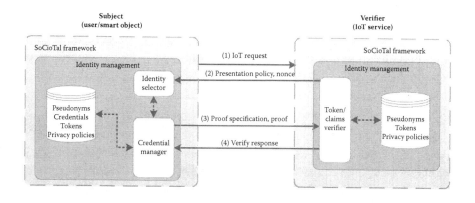

Figure 18.4: IdM presentation process.

The *presentation process* is the main operation provided by the privacy-preserving IdM system, based on anonymous credentials. Figure 18.4 shows the presentation process with the interactions between a subject that wants to prove possession of certain attributes in its partial identity (i.e., in this credential).

1. The subject makes a request to an IoT Service, that is, the Verifier, which requires the Subject to present certain cryptographic proof of possessing a credential or certain attributes.

2. The verifier computes a random value called *nonce* that is sent to the subject. Based on the actual context, the Identity Selector module of the Subject makes use of the credential manager to select the best credential (the partial identity) or pseudonym to be used against the Verifier among the ones it already has available in its database. Optionally, if supported by the underlying crypto engine, in case the Subject does know the proof specification required by the IoT service, the Verifier can send the Subject a Presentation policy stating which data a user has to reveal to the Verifier to gain access to the requested IoT service. In other words, the presentation policy defines which credentials and attributes are required or which conditions have to be fulfilled by the attributes or both.

3. The Subject (acting as Prover) defines the proof specification from the selected credential(s) to be used against the Verifier. This proof includes the nonce, the attributes as well as statements about attributes. Then, the Prover builds a cryptographic object as proof and sends the proof along with the specification to the Verifier.

4. The Verifier validates the incoming proof specification using the cryptographic proof. It computes the verifying protocols to check that the attributes' statements, and pseudonyms are valid.

5. The Verifier, depending to the result of the validation, can send an affirmative or negative response to the subject. In case of a successful identity validation, the IoT Service can then redirect the subject to the authorization component to make an authorization decision based on authorization policies.

The Presentation process can be used to authenticate users and smart objects anonymously with minimal attribute disclosure of private information. When a subject wants to access an IoT service, and both parties are endowed with the SocIoTal privacy preserving IdM system, the user can provide proof of credential following the Credential Presentation process as a means of authentication to gain access to the IoT service.

18.3.3 Capability-based access control for IoT

DCapBAC has been postulated as a feasible approach to be deployed in IoT scenarios [17] even in the presence of devices with tight resource constraints. It features a lightweight and flexible design that allows the authorization functionality to be embedded onto IoT devices, providing the advantages of a distributed security approach for the IoT in terms of scalability, interoperability, and end-to-end security. The key element of this approach is the concept of capability, which was originally introduced by [10] as a "token, ticket, or key that gives the possessor permission to access an entity or object in a computer system." This token is usually composed of a set of privileges that are granted to the entity holding the token. Additionally, the token must be tamperproof and unequivocally identified in order to be considered in a real environment. Therefore, it is necessary to consider suitable cryptographic mechanisms to be used even on resource-constrained devices, which enable an end-to-end secure access control mechanism. This concept is applied to IoT environments and extended by defining conditions which that are locally verified on the constrained device. This feature enhances the flexibility of DCapBAC since any parameter that is read by the smart object could be used in the authorization process. DCapBAC is based on JavaScript object notation (JSON) [9] as the representation format for the token, the use of emerging communication protocols such as the constrained application protocol (CoAP) [25] and 6LoWPAN, as well as a set of cryptographic optimizations for elliptic curve crytography (ECC). DCapBAC along with a policy-based mechanism based on XACML [23] is used in SocIoTal an access control system to infer the access control privileges to be embedded into the capability token.

18.3.3.1 Capability token

The format of the capability token is based on JSON. Compared with more traditional formats such as XML, JSON is getting more attention from academia and industry in IoT scenarios, since it is able to provide a simple, lightweight, efficient, and expressive data representation, which is suitable for use on constrained networks and devices.

Figure 18.1 shows a capability token example that allows a subject device (Smart Object A) to perform the *Get* action over the resource *position* in an (Smart Object B). The capability token also indicates the target device, which enforces the authorization decision, to authorize access to the subject device only in case the trust index about the subject device is over 5. Notice that this feature requires the target device to have deployed the Trust Manager component of the framework, in charge of quantifying trustworthiness.

```
{"id": "Jd93_jZ8Ls5VOqP",
 "ii": 1412941013,
 "is": "coap://tokenManager.um.es",
 "su": "aB4wSICIXC1pm2pkW9YMPQyFudc=CPhYdgOAQwcOYgURwP1q02WSv=",
 "de": "coap://smartObjectB.um.es",
 "si": "TqZaXuxZ5dmZU6k3PtiWwI3NrjH=7u5By5OHzlOOtq4TmkrZU2JPd=",
 "ar": [
 {"ac": "GET"
  "re": "position"
  "co": [{
   "t": 5,
   "u": trust,
   "v": 0.7}]}
 "nb": 1412941013,
 "na": 1412941456
} Legend: "id"-> identifier "ii"-> issued time "is"->issuer "su"->
subject "de"-> device "si"-> signature "ar"-> accessRights "ac"->
action "re"-> resource "co"-> condition "t"-> type "u"-> unit "v"->
value "nb"-> not before "na"-> not after
```

Listing 18.1: Trust-aware capability token example.

18.3.3.2   DCapBAC scenario

In a typical DCapBAC scenario, an entity (subject) tries to access the resource of another entity (target). Usually, a third party (issuer) generates a token for the subject the specifying which privileges it has. Thus, when the subject attempts to access a resource hosted in the target, it attaches the token that was generated by the issuer. Then, the target evaluates the token granting or denying access to the resource. Therefore, a subject that wishes to access certain information from a target, needs to send the token together with request. Thus, the target device that receives such a token can know the privileges (contained in the token) that the subject has and it can act as a Policy Enforcement Point (PEP). This simplifies the access control mechanism, and it is a relevant feature of IoT scenarios since complex access control policies are not required to be deployed on end devices.

The basic operation of DCapBAC is shown in Figure 18.5. The initial step, the Issuer entity, which could be instantiated by the device owner or another entity in charge of the smart object, issues a capability token to the Subject to be able to access such device. Additionally, in order to avoid security breaches, such

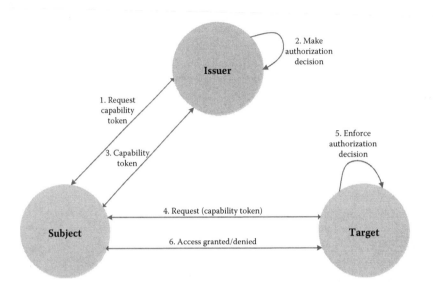

Figure 18.5: Authorization process based on DCapBAC.

token is signed by the Issuer. In the SocIoTal access control system, this process is based on the use of XACML policies. Therefore, in the case of a "Permit" decision, a capability token is generated with that specific privilege. In addition, XACML Obligations can be used to embed contextual conditions to be locally verified by the target device. Once the Subject has received the capability token, it attempts to access the device data. For this purpose, a request is generated (e.g., by using CoAP), in which the token is attached. According to Figure 4, this request does not have to be read by any intermediate entity. When the Target receives the access request, the authorization process is carried out. First, the application checks the validity of the token (i.e., if it has expired) as well as the rights and conditions to be verified. Then, the Issuer signature is verified with the corresponding public key. Depending on the specific scenario, this key can be delivered to smart objects during the commissioning or manufacturing process, or it can be recovered from a predefined location. Finally, once the authorization process has been completed, the Target generates a response based on the authorization decision.

Additionally, this approach provides support for advanced features, such as access delegation. In this case, a subject S (acting as a delegator) with a capability token CT can generate another token CT' for S' (acting as a delegated), in which a subset of the privileges of CT are embedded. Consequently, CT' can be used by S' to get access to a resource in a target smart object. Furthermore, S can grant the right to S' for additional delegations. This feature is valuable to address the dynamic and pervasive nature of IoT scenarios and everyday life. For example, elderly people can provide temporary privileges or delegate them to home help personnel to get access to their homes in case of an emergency situation. In the case of delegation, it is necessary to sign each new capability token with the corresponding subset of privileges, in order to allow a full auditability of access and avoid security breaches.

18.3.4 Secure group data sharing

The realization of scenarios with entities composing dynamic communities requires the definition of appropriate mechanisms to design a scalable and distributed security solution for the envisioned use cases. Unlike the current Internet, in such dynamic coalitions, IoT interaction patterns are often based on short and volatile associations between entities without a previously established trust link. Providing basic security properties to such data exchange is a paramount security issue that also needs to be properly addressed by allowing more flexible sharing models (beyond the classic request/response approach), as well as fleeting and dynamic associations between entities, while the privacy of involved entities is still preserved.

Because of the usage of resource-constrained devices, symmetric-key Cryptography (SKC) has been widely used on the IoT, requiring that the producer and the consumer share a specific key. Nevertheless, this approach is not able to provide a suitable level of scalability and interoperability in a future with billions of heterogeneous smart objects. These issues are tackled by public key cryptography (PKC), but present significantly higher computing and memory requirement as well as the need to manage the corresponding certificates. A common feature with SKC is that PKC allows a producer to encrypt information to be accessed only by a specific consumer. However, given the pervasive, dynamic, and distributed nature of the IoT, it is necessary to consider different scenarios in which some information can be shared with a group of consumers or a set of unknown receivers and, therefore, is not addressable *a priori*.

In that sense, identity-based encryption (IBE) [6] was designed as an alternative without certificates for PKC, in which the identity of an entity is not determined by a public key, but a string. Consequently, it enables more advanced sharing schemes since a data producer could share data with a set of consumers whose identity is described by a specific string. In this direction, attribute-based encryption (ABE) [14] represents the generalization of IBE, in which the identity of the participants is not represented by a single string, but by a set of attributes related to their identity. Just as IBE, it does not use certificates, while cryptographic credentials are managed by an entity usually called attribute authority (AA). In this way, ABE provides a high level of flexibility and expressiveness, compared with previous schemes. In ABE, a piece of information can be made accessible to a set of entities whose real, probably unknown identity is based on a certain set of attributes.

Based on ABE, in a CP-ABE scheme [5], a ciphertext is encrypted under a policy of attributes, while keys of participants are associated with sets of attributes. In this way, a data producer can exert full control over how the information is disseminated to other entities, while a consumer's identity can be intuitively reflected by a certain private key. Moreover, to enable the application of CP-ABE on constrained environments, the scheme could be used in combination with SKC. Thus, a message would be protected with a symmetric key, which, in turn, would be encrypted with CP-ABE under a specific policy. In the case of smart objects, which cannot apply CP-ABE directly, the encryption and decryption functionality could be realized by more powerful devices, such as trustworthy gateways. In addition, CP-ABE can rely on IdM systems (e.g., anonymous credentials systems) to obtain private keys associated with a certain user's attributes from a specific AA, after demonstrating the possession of such attributes in the partial identity. Then, these private keys can be used by consumers to decrypt data, which is disseminated by producers, as long as the consumer satisfies the policy that was used to encrypt.

18.3.4.1 Secure data-sharing scenario

The realization of IoT scenarios with entities composing dynamic communities requires the definition of appropriate mechanisms to design a scalable and distributed security solution for the envisioned use cases. Unlike the current Internet, in such dynamic coalitions, IoT interaction patterns are often based on short and volatile associations between entities without a previously established trust link. Providing basic security properties to such data exchange is a paramount security issue that also needs to be properly addressed by allowing more flexible sharing models (beyond the classic request/response approach), as well as fleeting and dynamic associations between entities, while the privacy of involved entities is still preserved.

Figure 18.6 shows the scenario in which a specific smart object disseminates information to make it visible only to a specific set of entities. This process is based on the CP-ABE cryptographic scheme, which is used to allow secure communication between objects belonging to the same bubble. In this case, a Smart Object A (from Bubble A) tries to get access to data being shared in Bubble B. It is assumed that smart objects in a Bubble X maintain at least one CP-ABE key associated with the attribute "bubbleX" that allows them to exchange information in a secure way. Thus, the Smart Object A needs to obtain a CP-ABE key associated with the same attribute in order to access data being shared among objects of Bubble B.

According to Figure 18.6, each bubble has a group manager, as an entity responsible for generating CP-ABE keys to allow secure sharing transactions. Therefore, during an off-line stage, the Smart Object A contacts the group manager from Bubble B to get a CP-ABE key to get access to the information being

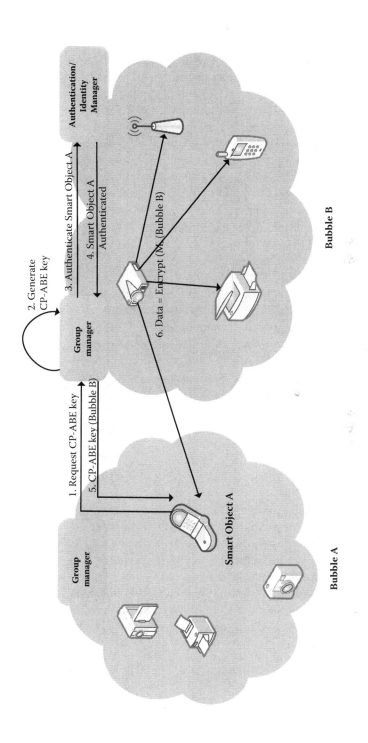

Figure 18.6: Secure data sharing in bubbles.

disseminated among smart objects i Bubble B. Likewise for the access control process, before the key generation process is carried out, the group manager verifies that the requester smart object is who it claims to be. This process can be based on traditional authentication mechanisms (e.g., based on login/password or X.509 certificates). Optionally, anonymous credential systems (e.g., Idemix) could be used in order to preserve the privacy of the smart object. Once Smart Object A is successfully authenticated, the group manager generates and delivers a CP-ABE key, which is associated with the attribute "bubbleB." In addition, this key could be associated with other identity attributes that were proved during the authentication process (e.g., attributes in an Idemix proof), enabling the composition of sub-bubbles according to different combinations of identity attribute values. After Smart Object A has received the corresponding cryptographic key, during an online stage, it can makes use of it in order to decrypt the information which is disseminated by smart objects in bubble B

In addition to bubbles or communities, which can be statically defined, given the pervasive and dynamic nature of the scenarios envisioned by SocIoTal, it is necessary to consider the application of security mechanisms to cope with the requirements of the so-called opportunistic bubbles. An opportunistic bubble is a kind of dynamic sharing group that is not registered as a static community anywhere. Unlike the previous approach, this kind of bubble leverages opportunistic contact and ad hoc connection among devices, simulating the way that people communicate in the physical world. Opportunistic bubbles are formed spontaneously, particularly based on physical proximity, and using short-range communications technologies without infrastructure. Due to the inherently mobile nature of smart objects (such as mobile phones), this model has an important interest to be exploited in the IoT. For example, in a real-life scenario, a user can create an opportunistic network of mobile phones when he or she goes into a restaurant to share information with other people who satisfy a specific combination of identity attributes.

This creation of opportunistic groups of entities is given by the fact that an entity can encrypt some data using a combination of attributes to make them visible only to a set of entities whose keys satisfy this combination. Unlike traditional cryptographic schemes (e.g., based on symmetric group keys), in this approach there is no need to generate new keys to enable a secure sharing between subgroups of entities. Indeed, information can be encrypted under different CP-ABE policies and decrypted by the same CP-ABE key. In this way, during the sharing process, any third party is involved, enabling a secure and ad hoc communication between smart objects.

18.4 Device-Centric Enablers for Privacy and Trust

This section introduces two of the main device-centric enablers that are being developed in the scope of SocIoTal EU project. On the one hand, the F2F enabler allows the measurement of social interactions opportunistically based on off-the-shelf smartphones. On the other hand, the indoor localization enabler allows the determination of the position of devices inside buildings by means of magnetic field measurements. The enablers are the baseline to infer context information that is used to make security decisions such as quantifying trustworthiness or to drive the access control process. This section also includes a first evaluation of each enabler to demonstrate their accuracy and feasibility.

18.4.1 Face-to-face enabler, from context to trust

An F2F enabler is an accurate and reliable system for opportunistically measuring social interactions based on off-the-shelf smartphones, without the need of any external hardware. First, it consists of a novel, hierarchical machine-learning-based methodology for estimating the interpersonal distance between users with only 6 Bluetooth (BT) RSSI samples. It showcases two models for detecting interaction zones and for inferring if users are within proximity or not. Second, it incorporates into the social interaction detection process, a method for computing the relative orientation of a user that allows estimations to be performed regardless of the on-body wearing position. Third, it introduces a collaborative sensing mechanism, allowing devices to exchange sensed information such as the user's facing direction and BT RSSI measurements. These components are incorporated into a coherent system, enabling accurate and pervasive sensing of real-world social interactions.

In light of its nature of representing an enabling technology, the SocIoTal F2F enabler will provide a software component implementing a dedicated IoT service. It is envisioned that such a service will be deployed on devices that are part of the SocIoTal platform. Because the information extracted by the F2F IoT service relates to the user's personal sphere, it appears clear how mobile smartphones are the target devices that will benefit from such information, when participating in the SocIoTal platform. The information provided by the F2F IoT service will be exploited in different ways, in order that the participation and information shared by SocIoTal devices are more secure and privacy preserving, thus satisfying the envisioned platform requirements. More details are provided in the following sections.

18.4.1.1 From face-to-face to context

As described above, the face-to-face (F2F) F2F enabler will extract information about the nature of the social relations incurring occurring among people, by classifying them accordingly to the users' orientation and interpersonal distance. Such information can be used in a number of different ways. First of all, by extracting the F2F information periodically or according to well-defined events, a more accurate characterization of the device context can be provided. Such characterization will allow from one side to better classify the environment surrounding a given SocIoTal device (e.g., mobile smartphones) in terms of discovered surrounding devices and relations with them. As SocIoTal devices, smartphones are seen in terms of their capability to provide additional IoT services, related to the production and consumption of the information generated by their embedded sensors and users in supporting the creation of citizen-centric services. The extracted context will then be used from one side to locally decide which information the device could share, in terms of available IoT services and according to the devices detected nearby and the nature of the discovered relations. On the other side, the extracted context information can be used to annotate all the information generated, thus guaranteeing access to it according to specific authorization policies, globally applied by the SocIoTal authorization component. For instance, specific information cannot be shared if the originating device is detected to be in a context in which it is surrounded only by a device in an intimate relation (i.e., home context).

18.4.1.2 From context to trust

On the other side, the information about F2F relations, generated by each SocIoTal-enabled smartphone devices, can be used and analyzed to establish relations about different devices and their users. Such information will be fed to the SocIoTal trust and reputation manager (T&RM) to allow the computation of reputation scores for classifying the recurrent social relations occurring between two given users and their devices. By relying on such information, each SocIoTal device will be able to create its user social graph, by relying on the type and frequency of real social encounters. According to the type and frequency of the relations, different reputations can be associated with different devices and evolve over time, while new information is acquired. By using the information extracted by the F2F enabler and consuming it locally, each device will be able to extract this reputation score and share it accordingly with the T&RM or the shared context information can be consumed by a dedicated infrastructure module to extract and update similar knowledge. According to this, the request for and sharing of information generated by a given device can be allowed with only specific other devices that comply with a defined trust and reputation score. For instance, specific information cannot be shared when asked by device, that never previously shared an intimate social relation with the considered one (i.e., device owned by office colleagues).

18.4.1.3 Related IoT service

According to the SocIoTal architecture, the F2F enabler provides IoT services, generating and exposing a well-defined type of information, accessible through specified application programming interfaces (APIs). As above this was described, the authorization and trust & management blocks consume this information. The enabler data available for creation, retrieval, modification, and deletion are briefly summarized next (see also Figure 18.7).

The F2F interaction detection enabler incorporates two concrete IoT services:

■ *DirectionData.* This IoT service retrieves the facing direction of the users, that is, the direction of the front part of the user's torso, inferred by the walking locomotion of the user. The knowledge of the user's facing direction will lead to the computation of the user's relative orientation. The relative orientation is one of the key parameters of F2F interaction detection.

■ *NearbyDevicesData.* This IoT service retrieves information about the nearby devices of a user, by utilizing a communication mean such as BT discovery, which acquires data including signal strength, device details, and so on. Given the data collected, the enabler is capable of estimating the interpersonal distance of the users, thus inferring the occurrence of a F2F interaction or not.

These two IoT services convey data to a virtual entity (VE) service for F2F interaction detection and several other components of the system. The combined information provided by the VE as well as the simple atomic information provided by the implemented IoT services can be accessed either directly (using dedicated APIs) or redistributed in the form of context information shared through the CM.

18.4.1.4 *Interfaces with authorization, trust and reputation management blocks*

The F2F enabler will extract information about social relations with surrounding devices. Such information can provide contextual information to be used by the trust manager to verify the trustworthiness of a subject Device B with respect to the considered Device A. This information will be used when services (e.g., data sharing) are requested to Device A. In addition this information can be used to verify rules requested by the authorization manager, that is, based on the device-detected context (e.g., the device is in a public environment, surrounded by many untrusted [non]SocIoTal devices).

The context information that needs to be extracted (e.g., using the context inference module) and shared (e.g., using the context communication module) can be modeled as shown in Figure 18.8.

The F2F enabler is able to recognize ongoing real-world interactions. Further, by inferring the interaction zone in which the social interaction is taking place, it is able to estimate the social relationship among people. Information about the entities involved in the social relation is provided using the pseudonyms associated with corresponding SocIoTal devices and annotated with information about

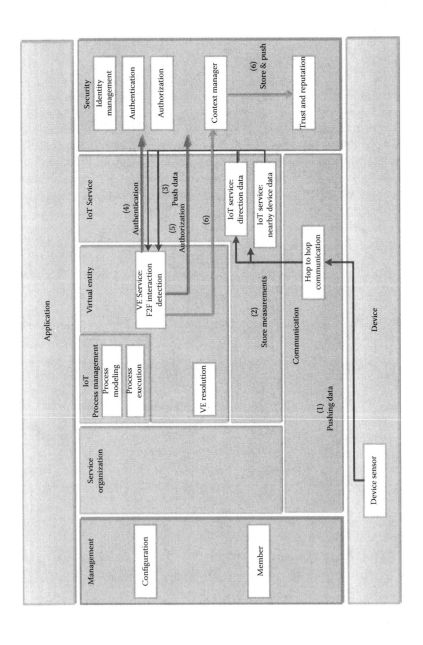

Figure 18.7: Face-to-face enabler architecture compliant with IoT-A (functional/information view).

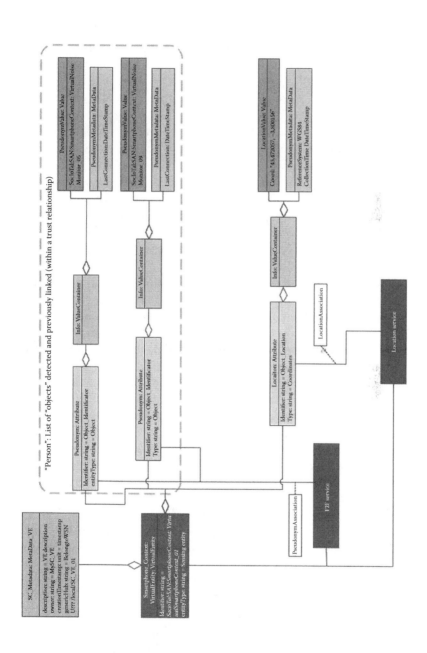

Figure 18.8: Context information and model for the F2F enabler.

the type of relation (e.g., personal, social, or public) and the location where it is taking place.

Two ways are envisioned in the SocIoTal architecture to allow access to services, such as data sharing, provided by SocIoTal devices. In both cases, privacy-preserving data access is regulated according to the type and number of detected social relations with target and other surrounding devices. In the first way, services can be accessed directly with a peer-to-peer (P2P) communication among devices. However, this functionality is not exposed to components outside the F2F enabler. These IoT services constitute the internal functionality required for the correct F2F interaction detection inference. In the second way, services are accessed through a centralized architecture, making use of the SocIoTal context broker. In this situation, the IoT service provides information about an occurrence of an F2F interaction and contextual data about the particular inference.

Figure 18.9 shows how the information provided by the F2F enabler and exposed as an IoT service running on a mobile phone is generated and integrated with other SocIoTal components. When two SocIoTal devices, namely A and B, come into proximity, their type of F2F relation is measured by making use of their BT radio and other mobile sensors to estimate the distance and facing direction of the devices. The extracted raw information is fed to the context inference module and knowledge about social relations is extracted according to a defined SocIoTal context model (Figure 18.8). Such information can be shared with direct communication among the involved devices, for real-time and decentralized operations, but it is also shared with other central SocIoTal components for remote access and use. By making use of the provided NSGI-9 (e.g., registerContext) interface, sharing of the extracted context is pushed by the context communication module on the device through the centralized SocIoTal context broker.

The shared social relationship estimation is accessed by the central trust & reputation manager to infer the trust relationship among people and compute the reputation score. The trustworthiness of a device/user with respect to another can be initially considered as a weighted average (between 0 and 1) of the number and type of observed relations between two well-defined devices. As an example, personal relations can be weighted as 0.5, while social can weight as 0.3 and public as 0.2. Two given device/user couples tend to trust each other if the number of personal relations they experience is higher. This allows a social graph derived from real social relations to build, which can be used alone or in conjunction with existing ones (e.g., Facebook, Twitter, LinkedIn) to assess the degree of relations between two different devices/users. On the other hand, by combining all the social relations incurred by a device, a reputation score can also be built. The reputation score can be a weighted average (between 0 and 1) of all the numbers and types of relations observed by a given device, including also relations with (non)SocIoTal devices. As an example, personal relations can be weighted as 0.3, while social can weight as 0.2, public as 0.1, and those with (non)SocIoTal devices can be averaged as 0.4. By doing so, devices that tend to be in contact

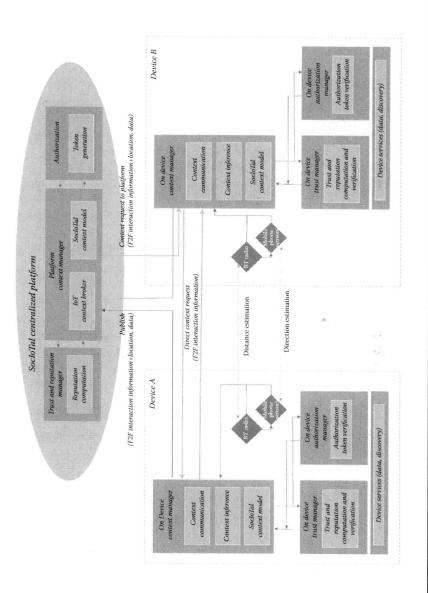

Figure 18.9: F2F enabler integration.

with many (non)SocIoTal devices receive a lower reputation scores as they can be in situation where they are exposed to more security threats. When a change in the reputation score is detected, such information is pushed back to the context broker using the NSGI-10 updateContext interface and accessed from interested components that can subscribe to the broker by using the provided NSGI-10 sub-scribeContext interface.

Similarly, the same information can be locally accessed by the on device trust manager to internally compute reputation scores thus avoiding sharing any information with external components and preserving privacy. In addition, the on device trust manager can request reputation score of a given device (e.g., Device B) the requiring access to device-provided services (e.g., data sharing) through the context broker and the central trust & reputation manager and use it to verify the suitability of the device to access the required functionalities.

Additionally, the F2F enabler information generated by Device A is com-municated to the authorization component through the SocIoTal context broker. The provided information is used to define the context surrounding the device, in terms of devices and observed relations. Such information is used by the authorization component to apply various rules about the services Device B (which eventually was previously in contact with Device A) can request and access onto Device A. Such information can be used alone or together with the one provided by the trust & reputation manager in order to issue a capa-bility token. Such token is provided to Device B using the context broker or dedicated direct interfaces and presented to device A, which verifies it using its on device authorization manager before granting access or not to the required services.

While this approach assumes that data are consumed by directly accessing the intended devices that provide the data, in case the shared information is accessed through the context broker, the F2F context information is used as an attribute to annotate any sensing data provided by a given device, and access to the context broker from requesting Device B should be regulated by the centralized autho-rization manager in the same way.

18.4.1.5 Initial evaluation of the F2F enabler

A real-world experiment serves the purpose of evaluating and proving the via-bility and robustness of the F2F enabler for the real-time detection of social interactions. We also benchmarked against accurate active radiofrequency iden-tification (RFID) approach [26], to understand if similar accuracy could be achieved. It should be noted that the whole interaction detection process is per-formed online on users' devices.

For this set of experiments, eight participants were recruited. All partici-pants were PhD students between 25 and 30 years old. The experiments took place in an indoor typically, furnished office room. We performed three sets of experiments. In each set, five random participants interacted with each other. Each participant was given a mobile device (HTC One S), which had a deployed SocIoTal application and an active RFID tag from [26]. Further, the ground truth was established by a human observer, as the number of participants was not too large.

The participants placed the mobile device in one of their trousers pocket (user's choice) with arbitrary orientation and walked for a few seconds (unspecified directions) until uDirect [19] converged. The participants then entered the office room for the experiment. An RFID tag was also attached to participants' chest. An RFID reader was deployed in the room and connected to a laptop to log RFID tag–detected interactions.

Through our empirical evaluations, we collected 756 social interaction inferences from the F2F enabler and 40,000 from the RFID [26]. Table 18.1 summarises the results of the experiments. The RFID-based technique only provides detected interactions and does not give any evidence about undetected interactions. Our initial evaluation confirms that our prototype system has been able to correctly identify 81.4% of interactions with only six RSSI samples. Benefiting from collaborative sensing, the F2F enabler acquires RSSI samples faster than smartphone-based, state-of-the-art solutions. This implies that the F2F enabler detects even short-term interactions that were hidden from previous solutions, with a reasonable accuracy.

Wearable techniques such as RFID provide more frequent estimations, which in turn improve the granularity of the data. One notable observation from our experiments was the role of the facing direction in reducing false-positive errors. Table 18.1 also confirms that the ratio between false positives and true positives in our approach was considerably less than that of the RFID solution (more than 9%). However, our approach shows an increase in the number of false-negative errors. Overall, it can be inferred from these observations that our proposed solution has a more conservative approach than the RFID-based solution. The inference is made with more concrete evidence, that is, both matched the facing direction and proximity, resulting in more false negatives and less false-positive errors. In contrast, the RFID-based approach shows more liberal decision-making on the identification of the interactions, which has resulted in more false positives.

Table 18.1   Overall accuracy for social interaction detection

F2F Enabler			RFID Approach [26]	
	Positive	Negative	Positive	Negative
Positive	**117**	47	**25,950**	—
Negative	93	**499**	12,846	—

18.4.1.6 The F2F enabler as a tool

As described in Section 18.4.1.3, the F2F enabler will implement an IoT service suitable for smartphones that will extract the type of social relations incurred by the device with other enabled SocIoTal devices. Such an enabler will be implemented as a black box and provided as a framework that can be either accessed through a software development kit (SDK) implementing APIs in order to extend other applications and integrate other components. In a simpler way, the information generated will initially be exposed and made available to other infrastructure components and developers through the SocIoTal context broker, thus simplifying and standardizing integration. The information about F2F relations, generated by each enabled smartphone device, can be used and analyzed to establish relations about different devices and their users. Such information will be fed to the trust manager component of the framework to allow computation of the reputation score for classifying the recurrent social relation incurring between two given users and their devices. By relying on such information, each SocIoTal device will be able to create its user social graph, by relying on the type and frequency of real social encounters.

18.4.2 Indoor localization enabler: from context to access control

This section presents a novel approach for indoor localization based on the magnetometers that are integrated in common smartphones. Unlike most of current phone-based proposals for localization [18], our system does not rely on an additional support infrastructure. Our solution only requires a personal smartphone that is able to sense the magnetic field available inside buildings. During the first stage of our system, we generate maps containing the magnetic field profile of the building where the localization problem needs to be solved. This represents the off-line training phase of the system. Then, during the online phase, users provide the system with the measurements of the magnetic field vectors sensed by their phone, and using these measurements, our system is able to provide accurate localization data of such users. We evaluate the proposed mechanism based on data samples collected and compare the performance with other existing phone solutions such as Wi-Fi.

18.4.2.1 From context to access control

The location data obtained from the indoor localization enabler are used to provide distributed access control to smart objects. This access control mechanism is built on top of distributed CapBAC [17]. This system makes use of an in-application programming (IAP)-based communications architecture with emerging protocols, which have been designed for constrained environments, such as 6LoWPAN or the CoAP.

The basic operation of our access control mechanism is as follows. As an initial step, the issuer entity of the system, which could be the device's owner or manager, issues a capability token to the subject granting permissions on the device. Furthermore, such issuer signs this token to prevent security breaches. Once the subject has received the capability token, he or she tries to make use of the smart object. To do this, when he or she is close to the geographical area of the target device, he or she generates a request including the magnetic field values and the capability token. In addition, this request must be signed in order to get access to the smart object. For this purpose, the CoAP request format has been extended with three headers: the capability, the signature, and the magnetic field vector.

The first task to be performed by the authorization engine is an assessment of whether the subject is inside the same building's zone where the smart object is placed. We base our assessment on the magnetic field characterization associated with the landmark identified in such a zone. Such landmarks' centroid is represented through mean and deviation values associated with each magnetic field feature. Deviation is the parameter that indicates the zone's extension covered by each landmark in terms of the magnetic field.

Therefore, given a device located in a building's zone where the magnetic field landmark lj with centroid Cj has been identified, the required device must assess if the distance between the mean values of the landmark centroid and the vector of the magnetic field features extracted from the measurements sent by the user, is smaller than the deviation associated with such landmark's centroid. If it is smaller, this means that the subject is inside the same building zone as the device. Otherwise, the authorization process is aborted and the service is denied. If the previous requirement is satisfied, the second evaluation task is carried out. It consists of evaluating the capability token, which is attached to the access request. In the case that the capability token is successfully evaluated, the last task involved in our authorization engine is launched. During this step, it is evaluated if the subject is inside the security zone defined for the required service (which can be denoted as SZ). For this evaluation, it is necessary to first estimate the subject's position by using the radial basis function (RBF) defined for the associated landmark. Once the subject's location is estimated, Zk, the distance between the subject and the device is calculated, and then, it is evaluated if such a distance is smaller than SZ. For this last evaluation, the mean accuracy value (z) associated with the RBF utilized is considered to estimate the subject's position.

A similar approach can be considered for other security mechanisms, such as for trust and reputation computation. Furthermore, this indoor localization mechanism is able to provide different levels of accuracy in its data, depending on the granularity of the clustering applied during its off-line phase. In this way, different levels of computational cost and time consumption can be considered according to the final requirements of the security mechanism implemented.

18.4.2.2   Related IoT service: Indoor localization

From an architectural point of view, Figure 18.10 shows the functional view for the indoor location enabler. A description of this view follows.

The magnetic device sensor pushes data to the MagneticMeasurement IoT service. Here, the following security consideration should be taken the device makes a push operation over an IoT service. The data are pushed into the indoor location detection VE service, which calculates the indoor location of Device A. (Notice that this VE Service may be placed in Device B different from where the MagnecticMeasurement IoT service is placed.) In the case where two devices are involved, the following security consideration should be taken: the IoT service tries to access a VE Service in the same way that a user or an application tries to access a service. The quantified location is sent back to Device A, and the VE service updates the location position in the IoT LocationPosition service. In the case where two devices are involved, the following security consideration should be taken: the VE service tries to access an IoT service in the same way that a user or application tries to access a service. Afterward, the indoor localization service can be used in different scenarios to feed the CM component of the framework and then make security decisions according to the localization context.

18.4.2.3   Interfaces with identity, trust and reputation management blocks

The main interaction of the localization enabler is within the CM component of the SocIoTal security framework. The indoor location enabler feeds the CM with the obtained localization information. From a security perspective, the localization data obtained from the enabler are then delivered to the CM. Then, the context is used mainly by the authorization component of the framework to make authorization decisions accordingly.

The role of the CM within the SocIoTal security framework is to provide the context notion to the different architecture modules in order to support their activities. The device context manager supports NGSI standard compliance and deals with the local context and is able to obtain context from the global context broker. The localization service can interact with the CM by means of the NGSI interface.

The context can be managed globally in the back end and locally in devices or gateways. SocIoTal security components are able to make security decisions taking into account their local context along with the context coming from the back end. The CM can publish raw context events (e.g., coming from the sensor) or elaborated events (as a result of the events processed by the CM engine) to the back end context broker. The context broker is usually placed in the cloud or in a data center in order to maintain and process context events coming from different devices. In addition to published events, the CM can also accept events

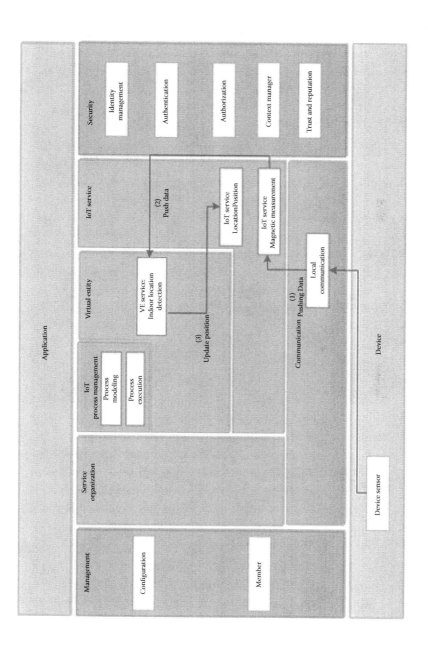

Figure 18.10: Indoor location enabler architecture compliant with IoT-A (functional/information view).

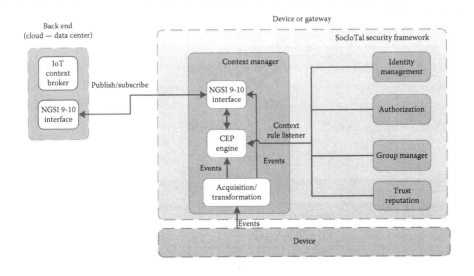

Figure 18.11: Context manager main interactions with security components.

from any NGSI compliant event producer. Thus, the devices could be notified by the context broker with new context events needed to take security decisions.

Different security components, namely, the IdM, authorization component, group manager, and the trust and reputation can be subscribed to the context engine and take security decisions depending on the inference of the rules defined in the CM inference engine. Thus, each time a rule is activated, the security components can be notified with the consequent information derived in the rule. Figure 18.11 shows the way the different security components of the security framework can access the CM to make security decisions accordingly.

The device CM can be registered as an NGSI context provider in the ensure hyphenated back-end IoT context broker, which can be done by calling the NGSI-9 registerContext operation. After context registration, the CM can send events by calling the NGSI-10 updateContext method of the NGSI interface in the back end. The context broker acts as an event consumer.

Additionally, the device CM can be subscribed to receive context from the IoT context broker. It should be noted that the communication with the context broker could be subject to security mechanisms to ensure that only trustworthy devices interact with it. Thus, the device can take security decisions locally based on the context when necessary. In this case, the context broker acts as a context provider. The main security components can access the inferred context data in the local context engine.

The authorization functional component of the SocIoTal security framework is based on a combination of access control models and techniques. To accomplish the main features of the proposed system, contextual information is a key aspect to be considered when making access control decisions. According to the

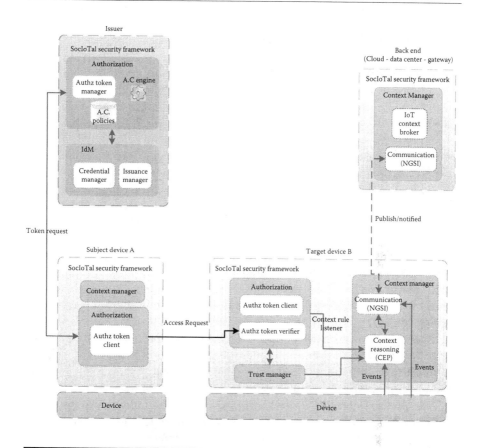

Figure 18.12: SocIoTal context-aware access control.

SocIoTal access control system, a subject entity gets a capability token in order to obtain access data from a target device. This token is usually generated by an issuer entity, which makes access control decisions that are embedded into the token. Therefore, when the subject entity tries to get access to a resource being hosted in a target entity, it provides the capability token previously obtained. Such a token can contain contextual restrictions to be locally verified when the token is evaluated by the target device. At that moment, the target device can use contextual information from its local CM, as well as other data stemming from the IoT context broker deployed in the back end. This process is shown in Figure 18.12, in which context information from the CM is used by the target device when verifying the capability token.

18.4.2.4   The indoor location enabler as a tool

The magnetic field localization enabler will implement the indoor localization service. This service, as described at the beginning of Section 18.5, is able to determine the location of a subject device analyzing the magnetic field measurements. The service can be accessible by any device connected to the Internet. Additionally, the localization measurements will be exposed and make them made available to the context broker, which will make localization even more accessible for other infrastructure components. Additionally, to simplify the integration process and deal with the indoor localization access control scenario, an android app will be implemented to enable smartphones to access the indoor localization service to obtain the location of the device being analyzed.

18.4.2.5   Initial evaluation of the indoor localization enabler

In order to evaluate our indoor localization mechanism, firstly it is necessary to choose the optimum parameters from the design of the different techniques that compound the mechanism. For this, in the target building where the localization problem needs to be solved, it is necessary to collect data about the magnetic field distribution throughout the building space of interest. Using the data collected, an optimum configuration for each technique involved can be obtained after an analysis of the results of localization errors. Therefore, in this section, we describe the experiments carried out in the Computer Science Faculty of the University of Murcia to get the optimum parameters to implement the localization mechanism proposed as well as validation of the mechanism with 10-fold cross- validation over the training data set.

We have developed a sensing application on an Android that has been deployed in a HTC One smartphone. This phone is equipped with a Hall-effect geomagnetic sensor3 in three axes. The sensor implements a dynamic offset estimation (DOE) algorithm to automatically compensate for the magnetic offset fluctuations, thereby making it more resilient to magnetic field variations within the device [56]. In addition, the effect of high frequency, ambient noise is mitigated by averaging the measurements prior to phone calibration. Our application is able to gather magnetometer signals with a frequency of 25 Hz and record them into a database allocated in an external platform (in Dropbox in our case).

Ten subjects were selected from the Information and Communications Engineering Department of the University of Murcia to perform the experiments for which the data were collected. In this way, the data collected cover different user paths inside the building at the same moment of a day and on different days. In the University of Murcia there is no ethics requirement for experiments with humans, but the experiments performed respected every aspect related to the privacy and confidentiality of the participants. During the data collection, the subjects were asked to walk on predefined trajectories along the first floor of the Computer Science Faculty. The participants walked along these baseline trajectories while carrying their phones in their hand and in a fixed orientation with respect to a reference system of coordinates. All participants carried their phone in the same position, considering the same phone orientation.

Since data collection was performed on different days and at different moments during the same day, variability in the context conditions was included in our base data sets. Therefore, the size of the final data set considered for testing was 1065 measurements. Then, using this data set, the data processing techniques presented in Section 6.2.2 were analyzed considering different values for their implementation in Matlab^{TM}.

To provide a detailed analysis of the results achieved by our localization mechanism, we focus on the localization results obtained for the first floor of the Computer Science Faculty depicted in Figure 18.3. The magnetic field distribution along this floor does not present high variability compared with the zones where the lifts are located. Therefore, the location results achieved in this floor cover the cases of buildings where there is not the best contextual conditions to apply a localization solution following the approach of using the magnetic field sensed inside. As a result of the off-line training phase of our localization system, a two-dimensional (2-D) map containing the magnetic field features of the building and the RBFs was obtained. This map was associated with a predefined phone orientation and the position in which the participants carried their phone.

By considering the map of the building containing the magnetic field profile resulting from the training phase, the classification mechanism charged with

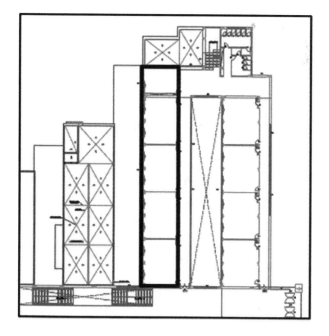

Figure 18.13: Magnetic field landmarks identified in a corridor.

assigning to each new measurement the zone where it belongs was evaluated. Firstly, we obtained the mean and deviation values of the accuracy achieved by each RBF implemented for estimating the user position. Table 18.2 shows the results. As can be seen, it is possible to achieve very accurate localization results, with a mean value of 3.9 m and deviation of 2.7 m. However, note that the number of different sources of magnetic field perturbation in the scenario under analysis is not high, which is related to the main drawback of the solutions that follow the approach of using magnetic field measurements for indoor localization.

To validate the results achieved by our mechanism, we take as reference the most similar work presented in the literature [20]. In this work, the nearest neighbor (NN) algorithm [2] considering Manhattan distance is used for estimating user positions, that is, as a regression technique. We compare the results obtained considering such a regression technique with the results associated with our proposal, in which we use RBFs as the regression mechanism. As in our approach, the authors also propose using the three elements of the magnetic field measurements sensed inside the building to solve the indoor localization problem. Now we focus on the localization results obtained for the corridor depicted in Figure 18.13. This corridor presents a high human activity level due to the numerous laboratories allocated there. Besides, it is one of the longer zones of the floor selected for the tests (28 m).

Based on our data set of the magnetic field sensed in this corridor, we apply the NN technique to estimate user positions once the building zone where the user is located is known. We present the results of this comparison in terms of accuracy. As a result, the RBF achieves a mean accuracy of 3.9 m, whereas a value of 5.7 m is achieved by NN. Therefore, our approach applying RBFs to solve localization estimates improves on the results provided by related studies.

Based on the assessments of the proposed indoor localization system, we conclude that the magnetic field measured by the magnetometers integrated in smartphones represents a feasible and accurate solution for solving localization

Table 18.2 Accuracy in location estimation and accuracy deviation

Accuracy in location estimation (m)	Accuracy deviation (m)
5.3	4.0
1.5	1.2
0.4	0.2
5.5	3.3
4.1	3.0
3.8	4.3
2.1	1.5
2.1	2.1

problems in buildings containing perturbation sources of the earth's magnetic field, such as lifts, electronic devices, machines, and so forth. Of note is the fact that the reference building that was used for testing presents a medium level of magnetic field perturbation, so that the results provided in this context are applicable to similar types of building. Furthermore, the mechanism proposed to generate magnetic field profile maps, the classifier design, and the estimation process is totally reproducible for any building.

18.5   Conclusion

This chapter has introduced the IoT security framework based on the ARM that has been designed in the scope of the SocIoTal EU project. The framework extends the original ARM security functional group, putting strong emphasis on security and privacy concerns in order to cope with more dynamic sharing models required by the pervasive IoT scenarios.

The chapter has shown the main security and privacy-preserving solutions that are being designed and implemented in the scope of the SocIoTal project. Namely, the secure group data-sharing solution that allows more flexible sharing models, the IdM mechanism that supports minimal disclosure of private information as well as the access control mechanism based on capability tokens.

Moreover, this work has provided an overview about how context information can be used by the security components of the framework with the aim of developing adaptive security mechanisms for the IoT. In this sense, two of the main context enablers of SocIoTal, that is, the F2F enabler and the indoor localization enabler, have been described and evaluated showing how they can be used as baseline to drive the security behavior within the framework.

Bibliography

[1] SOCIOTAL. Creating a socially aware citizen-centric Internet of Things. EU FP7 SocIoTal Project, 2013.

[2] David W. Aha, Dennis Kibler, and Marc K. Albert. Instance-based learning algorithms. *Machine Learning*, 6(1):37–66, 1991.

[3] Deliverable 3. Integrated system architecture and initial Pervasive BUTLER proof of concept. EU FP7 Butler Project, 2013.

[4] Alessandro Bassi, Martin Bauer, Martin Fiedler, Thorsten Kramp, Rob van Kranenburg, Sebastian Lange, and Stefan Meissner. Springer, Berlin, 2013.

[5] John Bethencourt, Amit Sahai, and Brent Waters. Ciphertext-policy attribute-based encryption. In J. Kilian (ed.) *Security and Privacy, 2007. SP'07. IEEE Symposium on*, pages 321–334. IEEE, 2007.

[6] Dan Boneh and Matt Franklin. Identity-based encryption from the Weil pairing. In *Advances in Cryptology—CRYPTO 2001*, pages 213–229. Springer, Berlin, 2001.

[7] Jan Camenisch and Els Van Herreweghen. Design and implementation of the idemix anonymous credential system. In *Proceedings of the 9th ACM Conference on Computer and Communications Security*, pages 21–30. ACM, New York, 2002.

[8] Open Geospatial Consortium. Sensor Model Language (SensorML), 2015.

[9] D. Crockford. RFC 4627: The application/JSON media type for javascript object notation (JSON). IETF RFC 4627, July 2006. http://www.ietf.org/rfc/rfc4627.txt.

[10] Jack B. Dennis and Earl C. Van Horn. Programming semantics for multiprogrammed computations. *Communications of the ACM*, 9(3):143–155, 1966.

[11] C. Ellison, B. Frantz, B. Lampson, R. Rivest, B. Thomas, and T. Ylonen. SPKI Certificate Theory. RFC 2693 (Experimental), September 1999.

[12] Fi-WARE. NGSI-9/NGSI-10 information model, 2014.

[13] Fi-WARE. Publish/subscribe context broker — orion context broker, 2015.

[14] Vipul Goyal, Omkant Pandey, Amit Sahai, and Brent Waters. Attribute-based encryption for fine-grained access control of encrypted data. In *Proceedings of the 13th ACM Conference on Computer and Communications Security*, pages 89–98. ACM, New York, 2006.

[15] Sergio Gusmeroli, Salvatore Piccione, and Domenico Rotondi. A capability-based security approach to manage access control in the Internet of Things. *Mathematical and Computer Modelling*, 58(5):1189–1205, 2013.

[16] Marit Hansen, Peter Berlich, Jan Camenisch, Sebastian Clauß, Andreas Pfitzmann, and Michael Waidner. Privacy-enhancing identity management. *Information Security Technical Report*, 9(1):35–44, 2004.

[17] José L. Hernández-Ramos, Antonio J. Jara, Leandro Marín, and Antonio F. Skarmeta. DCapBAC: Embedding authorization logic into smart things through ECC optimizations. *International Journal of Computer Mathematics*, 1–22, 2014.

[18] Jeffrey Hightower and Gaetano Borriello. Location systems for ubiquitous computing. *Computer*, 34(8):57–66, 2001.

[19] Seyed Amir Hoseinitabatabaei, Alexander Gluhak, Rahim Tafazolli, and W. Headley. Design, Realization, and Evaluation of uDirect: An approach for pervasive observation of user facing direction on mobile phones. *IEEE Transactions on Mobile Computing*, 13(8):1981–1994, 2014.

[20] Binghao Li, Thomas Gallagher, Andrew G. Dempster, and Chris Rizos. How feasible is the use of magnetic field alone for indoor positioning? In *International Conference on Indoor Positioning and Indoor Navigation*, Sydney, volume 13, page 1–9. IEEE, New York, 2012.

[21] Wojciech Mostowski and Pim Vullers. Efficient u-prove implementation for anonymous credentials on smart cards. In M. Rajarajan, F. Piper, H. Wang, and G. Kesidis (eds) *Security and Privacy in Communication Networks*, pages 243–260. Springer, Berlin, 2012.

[22] Martin Naedele. An access control protocol for embedded devices. In *Industrial Informatics, 2006 IEEE International Conference on*, pages 565–569. IEEE, New York, 2006.

[23] E Rissanen. Extensible access control markup language (XACML) version 3.0 oasis standard, 2012.

[24] Ravi Sandhu and Jaehong Park. Usage control: A vision for next generation access control. In V. Gorodetsky, L. Popyack, and V. Skormin (eds), *Computer Network Security*, pages 17–31. Springer, Berlin, 2003.

[25] Z. Shelby, K. Hartke, and C. Bormann. The constrained application protocol (COAP). *IETF RFC 7252*, 10, June 2014.

[26] Juliette Stehlé, Nicolas Voirin, Alain Barrat, Ciro Cattuto, Lorenzo Isella, Jean-François Pinton, Marco Quaggiotto, Wouter Van den Broeck, Corinne Régis, Bruno Lina, and Philippe Vanhems. High-resolution measurements of face-to-face contact patterns in a primary school. *PLoS ONE*, 6(8):e23176, 2011.

[27] Guoping Zhang and Wentao Gong. The research of access control based on UCON in the Internet of Things. *Journal of Software*, 6(4):724–731, 2011.

[28] Sébastien Ziegler, Cedric Crettaz, Latif Ladid, Srdjan Krco, Boris Pokric, Antonio F. Skarmeta, Antonio Jara, Wolfgang Kastner, and Markus Jung. Lecture Notes in Computer Science, 7858:161–172, 2013. Springer, 2013.

Chapter 19

A Policy-Based Approach for Informed Consent in Internet of Things

Ricardo Neisse

Bertrand Copigneaux

Abdur Rahim Biswas

Ranga Rao Venkatesha Prasad

Gianmarco Baldini

CONTENTS

Abstract

Informed consent is an essential element of data protection for information and communication technology (ICT) systems as the consent of a data subject (e.g., the citizen) is often necessary for a third party to legitimately process personal data. To provide informed consent regarding the use of personal data, the citizen must have a clear understanding of how his/her personal data will be used by ICT applications. This may not be an easy task, especially for a citizen with a limited understanding of the complexities of ICT systems, as End User License Agreements (EULAs) are often either too complex or too generic to be easily understood. This issue is likely to become more critical in the Internet of Things (IoT) where the collection of personal data can happen in various ways, which are often not evident to the user. There is a need to define new models of informed consent that (a) address the different capabilities and features of the user of IoT systems and applications and (b) make the provision of informed consent easier. In this chapter, we describe an approach to informed consent founded on a policy-based framework whereby policies that are more suited to the complexities of IoT and that can be refined on the basis of the specific features of the user or categories of users can be used to implement EULAs or more sophisticated forms of informed consent.

19.1   Introduction

The term *informed consent* originates in the medical community and describes the process for obtaining permission from a patient to perform a medical procedure on the basis that he/she has been fully informed about the benefits and risks of the procedure, and has agreed to the procedure being undertaken. Informed consent may only be given by patients who have adequate reasoning faculties and are aware and in possession of all relevant facts at the time the informed consent is given.

The informed consent process has now been adopted to regulate the interactions of citizens within the digital world. From a legal perspective, the notion of informed consent is essential for the data protection of information and communication technology (ICT) systems as the consent of a data subject (e.g., the citizen) is often necessary for a third party to legitimately process personal data. Within the European Union, the data protection directive [1], which defines conditions under which personal data can be processed, specifies that the consent must be "freely given, specific and informed" and "unambiguous." The foreseen evolution of this regulation [2] further strengthens this definition of consent by narrowing it to "explicit, clear affirmative action," thus excluding the possibility of implicit content.

To provide informed consent regarding the use of personal data, the citizen must have a clear understanding on how his/her personal data will be used by the ICT systems and applications. This may not be an easy task, especially for a citizen with a limited understanding of the complexities of ICT. On the other hand, informed consent must be collected before ICT applications can be used. This has led to the development of End User License Agreements (EULAs), which are often too complex or too generic for most users of ICT applications. The complexity and length (e.g., often tens of pages) of current instances of EULAs has developed a *consent fatigue*, whereby most users accept the licensing agreement by default, often without reading it [3]. The generic nature of EULAs is evident in the fact that they are often the same for all users regardless of the user's role or proficiency in the use of the ICT application.

These issues notwithstanding, EULAs are generally used because there are no widely available alternatives. For the user then, the choice is between (a) not having access to the ICT application or (b) having some potential risks to his/her privacy in a distant and vague future. In recent times, point (b) has become less vague and more menacing as private photos and information are now routinely distributed to the public.

In addition, EULAs are often static artifacts and are not related to a specific context or domain. For example, in many cases the EULA is the same regardless of whether the ICT application is used for personal or business reasons. Moreover, EULAs are often not changed by the ICT application developers, even if the usage of personal data may change due to technology trends (e.g., cloud computing), or interaction with other applications.

There is a need for a more sophisticated tool for informed consent, which would provide the following features at a minimum:

1. Support different types of users across the full spectrum of users in the digital divide (i.e., from the most ICT literate to the least) and/or support different user roles.

2. Be customizable so that the user can change settings if he/she wishes to within preestablished parameters, as defined by the regulations or the application developer.

3. Support different type of contexts or changes in the environment.

Beyond the ICT domain, the issue of providing a tool for "informed consent" with these features is further complicated by the evolution of the IoT. The definition of EULAs for end users may be further complicated by the limited processing capabilities of IoT devices, the distributed nature of the IoT, and the integration of the digital with the real world. The numbers of potential data operations in a fully deployed IoT make the adoption of EULA less practical. In addition, the nature of the informed consent required would vary depending on the data provided by the IoT device and the related data flow.

In other words, IoT device manufacturers should provide more decentralized control over the processing of personal data in the new data-driven environment IoT so that users could gain a better understanding of what data of theirs is collected and how it is used. This should be reflected in the definition of a new approach to providing informed consent in the IoT.

In this chapter, we propose a new approach and the related tools to support the features and challenges identified earlier. The approach is based on the definition, deployment, and adoption of policies through a policy-based framework.

The policies consist of authorizations and obligations specified as Event-Condition-Action (ECA) enforcement rules. These rules use as a reference a set of interrelated design models representing different aspects of the IoT system and the related data flows.

The policy framework consists of a collection of metamodels for the specification of a computer system structure, information, behavior, context, identities, organizational roles, and security rules. The policy framework adopts a generic design language to represent the architecture of a distributed system across application domains and levels of abstraction, including the refinement of relations inspired in the Interaction System Design Language (ISDL) [4].

As described in the rest of the chapter, sets of policies can be used to define profiles for informed consent. The policy framework provides the flexibility to adapt the informed consent definition for different types of users and different types of IoT contexts or conditions.

This chapter is organized as follows: Section 19.2 provides an analysis of the issues and challenges in implementing an effective informed consent process in the Internet of Things. Section 19.3 provides an overview of the current results from the research domain on the potential approaches for Informed Consent with a specific focus on the IoT. Section 19.4 describes the design of the informed consent based on the policy-based framework. The system describes the generic framework and how it can be applied to provide informed consent in the IoT systems and devices. Finally, Section 19.5 concludes the book.

19.2 Problems Defining Informed Consent in the Internet of Things

Informed consent is a term which originates in the medical research community and describes the process by which a person—such as a patient or a participant in a research study—has been fully informed about the benefits and risks of a medical procedure and has agreed on the medical procedure being undertaken on them. Informed consent should be given based on a clear appreciation and understanding of the facts, implications, and future consequences of the action of providing the consent. In order to give informed consent, the individual concerned must have adequate capabilities and be in possession of all relevant information at the time that the informed consent is given.

From a legal perspective, the notion of consent is essential in data protection as the consent of a data subject is often necessary for a third party to legitimately process personal data. Within the European Union, the data protection directive [3], which defines conditions under which personal data can be processed, specifies that the consent must be "freely given, specific and informed" and "unambiguous." The foreseen evolution of this regulation [4] further strengthens this definition of consent by narrowing it to "explicit, clear affirmative action" excluding the possibility of implicit content.

In Europe, the Article 29 Working Party issued an opinion on privacy issues with regard to mobile applications [5], which could also be extended to the future development of the IoT. The Working Party provided specific suggestions to the industry players (e.g., application developers, application stores, application and service providers, manufacturers of mobile devices). The suggestions included the provision of tools for free, specific, informed consent, a "readable, understandable and easily accessible" privacy policy. The Article 29 Working Party emphasized that the notice should be provided "at the point in time when it matters to consumers, just prior to the collection of such information by applications." It required that the initial notice contain the minimum information required by the EU legal framework, and that further information be made available through links to the whole privacy policy. The Working Party also defined the minimum information that should be included: (a) identity and contact details (b) precise categories of data to be processed by the application which requires the informed consent (c) information as to whether the data would be disclosed to third parties, and (d) the rights of users in terms of withdrawal of consent and deletion of data.

The Working Party's analysis identified the following areas to be addressed by a new approach to informed consent:

- Lack of control and information asymmetry. As mentioned earlier in this chapter, the large amount of data generated by IoT systems and devices can be difficult to control using conventional systems. In other words, the generation of data flows can hardly be managed with the classical tools used to ensure the adequate protection of the data subjects' interests and rights.

- Quality of the user's consent. Users may not be aware of the data processing carried out by specific objects. They also may not know precisely when and if an object is connected or not.

■ Inferences derived from data and repurposing of original processing. Modern analytical techniques can cross-correlate data from different sources to extract information that may point to personal data, even if the single data flows do not include personal data.

■ Security risks: security versus efficiency. As pointed out before in this chapter, there is a trade-off between the need to design and implement confidentiality, integrity, and availability measures and the need to optimize the use of computational resources—and energy—by objects and sensors.

Ensuring this level of "informed consent" can already be an issue in itself for traditional ICT applications, while the technical and legal complexity of the problem can be an obstacle to informing potential end users. This has led to the development of so-called End User License Agreements (EULAs). A EULA is usually presented as a text box in an ICT application, containing a large amount of legal text, with a scroll function to enable reading of the entire document. The text box has a process (e.g., a check box) that requires the approval of the EULA by the user on the basis that the user has read and understood the content of the EULA. If the response is positive, the ICT application grants access to the requested services. There are various problems with EULAs, when they are applied to ICT and in extension to the IoT domain. The set of issues includes

1. The EULA text is the same for all users. There is no effort to tailor the text of the EULA to specific categories of users or a specific context (e.g., home or office). From one point of view this is understandable, as the legal framework which generates the EULA is usually generic. Still, informed consent and the treatment of personal data in ICT are slightly different from the healthcare domain where the informed consent concept originated.

2. The EULA text is usually very long and complex and often difficult for the generic user to read and understand. Therefore, there is the risk that the condition identified earlier, that "a clear appreciation and understanding of the facts, implications, and future consequences of the action," is not satisfied.

In the IoT, informed consent is even more complex than in ICT systems because the following challenges may also arise:

1. The "things" collecting data, may not be fully evident as this data could be embedded in other devices. For example, future wearable sensors for healthcare could collect and transmit information for healthcare purposes that the user is not aware of.

2. The difficulty in requesting informed consent because of the lack of identifiable places and times to implement consent mechanisms (e.g., when

a driver uses an intelligent car, which provides information to a remote server).

3. The difficulty in opting out for specific services. It may be difficult to understand when and how the user can opt out from some services, and maintain the consent from other services, if an IoT device supports various applications.

4. While the informed consent for a specific ICT application in a defined place or context (e.g., at home) could be relatively easy to obtain, a change in the context (e.g., from home to office, or if a person's role changes) may require a new informed consent but this may not be easy to achieve in the current situation.

In addition, the collection and processing of data by the IoT can be more complex and pervasive than the current model of a specific website or application, whereby the human–computer interface (HCI) is well defined and specific to the Web portal or the HCI application. Data can be collected and distributed from a variety of IoT devices, which can be used by the individual at the same time and place. For example, an individual may wear a wearable device for healthcare, may use a mobile phone or drive a connected vehicle all at the same time. In addition, all of these devices may independently transmit to external servers or remote applications. In this situation, there would also be different interfaces, data collections points, and data flows for the personal data of the individual. Even if the informed consent is provided separately for each specific device or application, remote analytical tools could be used to exploit the combination of collected and processed data to identify correlated information, which could impact on the privacy of the individual. For example, even if the position of a car is obfuscated by decreasing the accuracy of information obtained from a GNSS system, other events (e.g., road charging or the position from a cellular base station) could be used to pinpoint the position of the individual in a more accurate way. This example shows that there is the need for a more sophisticated process for informed consent, which addresses the increased complexity of IoT. We propose a policy-based framework to address this complexity and the issues presented above.

19.3   State of Art

19.3.1   *Dynamic and context-aware approach*

To face the dynamic and distributed nature of the IoT, it has been proposed to take into account the context in which an application operates to influence the authorization [6] and disclosure of information [7]. In particular, [6] addresses the problem of highly dynamic environments where standard access control may not work in an effective way, as authorization is most often static or controlled by applications, which can lead to users being considered authorized even after a change of the context. With context changes, we cannot assume that a user is authorized for the duration of their use of an application, even if that user is still authenticated. This paper then proposes an approach based on dynamic authorization, which could also be used for informed consent, even though authorization is only one of the features. In a similar way, [7] presents an approach called the Dynamic Disclosure-Control Method (DDCM), which offers protection of location privacy through an agent. The agent provides a location privacy method that adapts to variations in contexts by using an efficient context analysis process. In addition, this method takes into consideration the privacy preferences of the users and the object operators. The approach proposed in this paper is based on these existing trends toward dynamic and context-aware disclosure of information.

19.3.2   Semiautonomous agent

The use of a user-centric, semiautonomous agent to negotiate the consent of the user with third-party applications, based on user-defined rules and preferences, has been discussed in [8].

This approach has the advantage of keeping the end user in control (by clearly defining the rules) while allowing for the scaling up to a large number of data authorization operations (matching the needs of a fully deployed IoT). However, the scope of the privacy coach proposed in [8] was limited to a single technology (RFID) and did not take into account any context information.

19.3.3   Reputation systems

A complementary and promising approach to further increase end-user information on privacy and data protection in IoT would be to rely on reputation systems. Initial examples of reputation-based systems for trustworthy communications are provided in [9,10].

We propose to extend and generalize them to both become visible to end users and integrate user feedback and perceptions on IoT applications' respect for privacy. Such ranking could not only provide information for end users but also be taken into account in the semiautomatic selection of which nodes are trustworthy for information sharing.

19.3.4   Behavior modeling

Advanced techniques have been developed to model and analyze the behavior of humans and their interactions with each other and with ICT technologies, as presented in [11–13]. These models can be used to create user adaptive systems as in [14], which describe services tuned to the individual preferences of the users.

Although this profiling can in itself lead to privacy issues, we argue that it could, with the necessary safeguards [15], be used to better understand individual user privacy requirements. We propose here to use profiling to automatically propose data operation authorization decisions to the user that match his/her previous decisions.

19.3.5 Analysis of the EULA

The use of standardized license agreements can be made available in three different but coherent version: full legal text (enforceable legal text), simplified text (short and understandable by most user), and machine-readable versions (enabling automated processing). All these forms have been experimented with some degree of success in the copyright domain, as described in [16].

In this section, we describe new technologies and tools for improving the basic EULA model. A very good analysis on this topic is provided in [17], where various technologies are identified. An example is the EULAlyzer [18], which analyzes the text of the EULA and tries to identify words that could hint to a specific use of the personal data (e.g., advertising or mobile commerce). The tool then notifies the user about these words and shows the context (e.g., the portion of the EULA text) where the words have been identified. In this way, the user is alerted to the potential misuse of their personal data by the application. The EULAlyzer is a good example of a category of tools that analyze EULAs and aim to protect the user from misuse by web applications. A potential issue with this category of tools is that the web application can modify the EULA against the analyzer and obfuscate the use of personal data using other words or a specific jargon. Another potential issue of this category of tools is that it does not address some of the basic challenges of the EULA described in the introduction, regarding the customization of the informed consent for different classes of users when the EULA is still the same for all users. As outlined earlier in this chapter, EULAs have too many limitations for IoT.

19.4 Overview of the System

The policy-based approach that we propose in this paper builds on and combines these different trends to offer a solution to the informed consent problem in the IoT.

As described in Figure 19.1, the system is user-centric. A graphic user interface enables the user to define a set of rules embedded in policies that should be both simple enough for the user to comprehend and complex enough to enable advanced users to fine-tune them if necessary. The user can also define

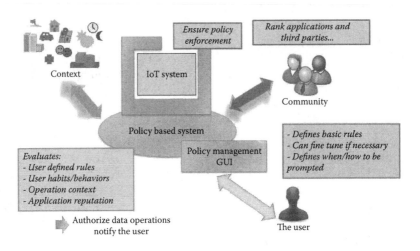

Figure 19.1: Overview of the system.

how and when to be contacted by the system and notified about a change in the context.

The policy-based system is a semiautonomous agent in itself, whose main role is to authorize or deny data operations on behalf of the user. In making each decision, the agent evaluates the rules/policies defined and chosen by the user but also takes into account context elements, and eventually information relating to user behavior and the reputations of third parties.

To handle the reputation system, the user is able to participate in communities which evaluate and rank IoT applications and third parties (e.g., service providers and application developers).

To ensure the policy enforcement, the whole system is built on an IoT platform that embeds policy enforcement components and the policy framework, as described in the rest of this chapter. To be implemented successfully, the system must address the following requirements:

1. Support different types of users across the full spectrum of users in the digital divide (i.e., from the most ICT literate to the less) and/or the different roles. This includes the necessity of providing the user with easily understood information in a simple GUI, and also the setup of mechanisms to train and motivate the user to define policies (i.e., to ensure regular use of the system).

2. Be customizable so that the user can change settings if he/she wishes. One of the challenges of customization is to adapt the GUI to follow the user proficiency.

3. Support different type of contexts or changes in the IoT environment and ensuring the enforcement of the policies chosen by the user.

A description on how the proposed policy-based framework is able to address these challenges is provided in Section 19.4.1.

19.4.1 Policy-based framework

We propose to use the Model-based Security Toolkit (SecKit) for the specification and enforcement of informed consent rules [19]. The SecKit includes a collection of metamodels, runtime components, and technology-specific Policy Enforcement Points (PEPs) that are used as the foundation for the security engineering process. Models in the SecKit are specified to represent the IoT system data, identities, behavior, structure, context, roles, trust relationships, threat scenarios for risk analysis, and security policy rules as reactive or preventive security countermeasures for the identified threats.

From a methodology point of view, the first step for the specification of policy rules is to model the target IoT system, which is done in the SecKit using a generic design language to represent the architecture of a distributed system across application domains and levels of abstraction. The system design is divided into two domains named *entity domain* and *behavior domain*, with an assignment relationship between entities and behaviors. In the entity domain the designer specifies the entities and interaction points between entities representing communication mechanisms. In the behavior domain the behavior of each entity is detailed including actions, interactions, causality relations, data, and identity information attributes. Activities in the behavior domain may handle user data or identities. For example, an IoT weather station may provide the current indoor home temperature (data) of a specific person (identity).

The second step is to specify the supporting models necessary, depending on the security requirements including business roles, context information and/or situations, and trust relationships. The context model specifies types of *Context Information* and *Context Situations*. Context Information is a simple type of information about an entity that is acquired at a particular moment in time, and Context Situations are a complex type that models a specific condition that begins and finishes at specific moments in time [20]. For example, the "GPS location" is a Context Information type, while "at home" and "at work" are examples of context situations where a person or employee (target entity) is at their home or work environment. The person that is at home or at work is assigned a role in that specific situation.

The results of the context situation monitoring are events generated when the situation begins and ends. These events contain references to the entities that participate in the situation and can be used to support the specification of the policy rules. *Policy rules can be specified to represent authorizations to be granted when a situation begins and data protection obligations that should be fulfilled when the situation ends*. For example, access to the patient data can be allowed when an emergency situation starts with the obligation that all data is deleted when the emergency ends. For example, a security policy may be specified to allow access to data when the situation starts and to trigger the deletion of the data when the situation ends. Existing policy language standards like XACML [21] only support the specification of context as attributes and of textual obligations to be fulfilled when the access to data is granted and not in the future.

The security policies have to be disseminated to the device(s) that are gathering the data under consideration in a secure way. Depending on the security policy, the device has to trigger and apply the appropriate mechanism for transmitting the data in the exact format needed by the application. This includes a two-step process: first the device has to map the policies for the application to specific data-gathering policies, and second, it should identify the encryption/security level of the data to identify the proper transmission mechanisms, considering also the energy efficiency requirements of the devices (i.e., using an adaptive encryption scheme). For example, in a traffic-monitoring scenario, users in cars may be sending information regarding traffic in an application server. The application should know only how much traffic there is at every street segment. The users' phone has the ability to send various types of traffic-related data, that is, the exact location every second, speed every second, direction of movement, etc. If the application wants to estimate the traffic, the related policies should be considered by the devices of the users, so that only an average speed per time period and street segment is sent, in order to avoid disclosing the exact location of the user at each point of time (ensuring privacy by design). Actually, intermediate nodes (i.e., the gateway) should also consider these policies and send to the application server only aggregated/average data so that the location of the users will be hidden from the application point of view. Other applications that need to know the exact location of the user (depending on their access control policies) will indeed be identified as such by the devices, which will transmit the exact location (i.e., for a person to track his car if it is stolen).

The security rules model consists of the *security rule templates* (aka *policy rules*) specified to be enforced and the configuration rules for these templates. The security rule templates are Event-Condition-Action rules, with the Action part being an enforcement action of Allowing, Denying, Modifying, or Delaying a service or data in the IoT device or system. Furthermore, the Action part may also trigger the execution of additional actions to be enforced, or specify trust management policies to increase/decrease the trust evidence for a specific trust aspect. For the purposes of informed consent, we also support the execution of an *Ask for User Consent* abstract activity, which may be instantiated differently depending on the current user situation (busy, available, in a meeting, etc.) and the previously specified user preferences. From an informed consent perspective, users have two alternatives: (1) to specify *a priori* the consent rules to allow, deny, modify, or delay an activity or (2), to specify rules that declare when the users should be explicitly asked for consent in an interactive way. Considering the second alternative our policy language is very expressive and allows users to specify temporal and cardinality constraints for the informed consent rules, for example, that consent should be explicitly asked once per hour, or once per day, if the data access requests are not more than 10 per day.

The security rule semantics is based on temporal logic and is evaluated using a configurable discrete time-step window of observed events, for example, 30 s. Details about the security rule model are described in previously published research papers [19]. Examples of security policy rules are provided for our scenario implementation in the following section.

From an architectural perspective, policy rules representing users' informed consent requirements can be exchanged between administrative domains that collaborate in an IoT scenario. For example, when a smart device exchanges data with a user mobile phone, the smart device can exchange the informed consent policies that regulate the authorizations and obligations associated to the exchanged data that should be enforced by the mobile phone. This delegation of sticky flow policies must be supported by trust management mechanisms in order to guarantee or increase the level of assurance with respect to the enforcement of the policy rules by the mobile phone.

19.4.2 Enforcement

In this section, we show how the SecKit is applied to an IoT architecture already defined in the FP7 iCore project and based on the concept of Virtual Objects and Composite Virtual Objects to represent IoT devices and IoT systems and applications. The architecture is described in [22] and the main concepts of the architecture are described here.

The main concept of the iCore project is to enable each IoT node with multiple functionalities based on its capability. For this, three key communication abstraction layers are identified, besides the necessary connectivity layers such as PHY, MAC and Network layers. They are (a) Virtual Object (VO) layer, (b) Composite Virtual Object (CVO) layer, and (c) Service layer. VO abstraction is applied to each node/device, which makes it easy to reuse the IoT devices. Figure 19.2 is a pictorial description of the architecture.

For example, the ambient light control in a smart building could indeed use the projector VO to realize that there is a movie or slide projected in a particular room and therefore lights can be turned off. The idea is to reuse IoT devices in multiple applications. The CVO layer enables the IoT devices to interact with other devices, and can mashup multiple VOs to offer smart applications. For example, a smart home has strict requirements regarding energy reduction, light control, climate control, and security. By combining multiple VOs, these

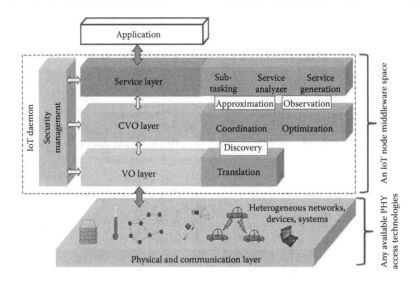

Figure 19.2: FP7 iCore architecture.

requirements could be served. At the service layer, multiple application requirements can be addressed. Going along with the previous example, the service
layer enables an ambient light control application to use information from the
projector by querying IoT devices (or services) in the vicinity, learning from
the obtained information, and making intelligent decisions. Of course, this also
requires semantic interoperability on all respective layers.

Figure 19.1 shows the SecKit enforcement components. In our enforcement
architecture, the IoT Framework and platform are monitored by a technology-
specific Policy Enforcement Point (PEP), which observes and intercepts service,
CVO, and VO invocations, taking into account event subscriptions of a Policy
Decision Point (PDP). The PEP component signals these events to the PDP, and
receives enforcement actions in case a tentative event is signaled. If required
for policy evaluation the PDP may implement custom actions to retrieve status
information of VOs and CVOs and subscribe to context information and situation
events with the Context Manager component, both using existing functionality
provided by the IoT Framework.

In order to concrete implementation scenarios, the SecKit must be extended
with technology-specific runtime monitoring components. In the iCore project,
we provide one extension to support monitoring and enforcement of policies
through a MQTT broker, which is the technology adopted by most of the project
partners to support communication between VOs, and CVOs. The SecKit may be
used in a hospital scenario where VOs and CVOs represent the staff and medical
devices being used that communicate using a MQTT middleware. Policies are

specified to control access to the hospital staff information (e.g., location) and to control the access to medical devices represented as VOs.

19.4.3 Application of the SecKit framework to internet of things for informed consent

Figure 19.3 shows the design of the IoT system behavior and the *Consent Manager* behavior type, which instantiates the action type *Ask For User Consent*. This action represents an abstract activity that interacts with the user and requests the consent for a specific operation. Interactive consent policy rules instantiate this behavior type to request the user consent.

Figure 19.4 shows the design of context information and situation types. In this figure we highlight the context situation *Working*, which models a *person* that is currently performing this activity.

Figure 19.5 shows the approach we adopt for specification of trust beliefs that are used in the policy rule language. Trust relationships are specified for a specific trust aspect, for example, to enforce privacy preferences, and a trust value is assigned to a specific entity for this aspect. In this example, the European Commission Joint Research Center (JRC) is considered to be very trustworthy for the aspect. The measurement of trust beliefs is implemented using the Subjective Logic opinion triangle, which assigns belief, disbelief, and uncertainty values to an opinion. A complete description of the approach adopted by us is already provided in [23].

Figure 19.6 shows a sample interactive consent policy rule template. In this rule, whenever an access to user data is detected by an event, and the relevant

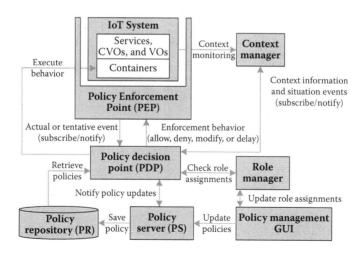

Figure 19.3: IoT system behavior model.

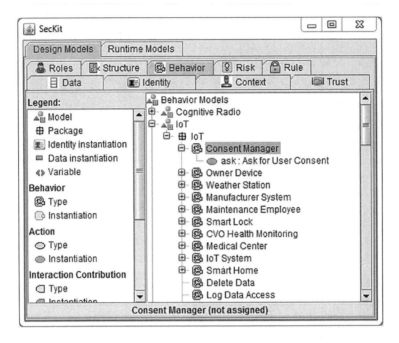

Figure 19.4: Context model.

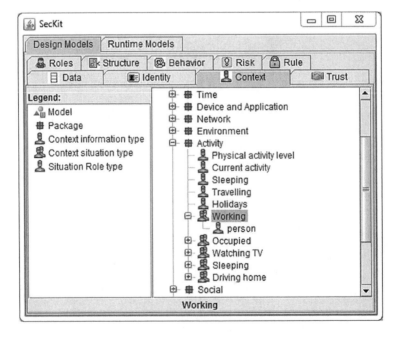

Figure 19.5: Visualization of trust beliefs and reputation values.

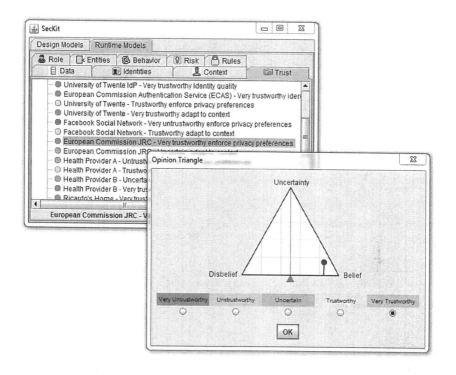

Figure 19.6: Interactive informed consent rule.

person is currently not working, a consent manager is instantiated to interactively request consent to allow, deny, modify, or delay the data being access. We illustrate the specification also for a variable in this rule template, to represent the specific person for whom the template should be instantiated and enforced. Variables can be also used to parameterize the event being considered and to specify generic consent rules that apply for all activities of a particular subtype. For example, a set of rules could be specified to all access to personal information, access to photos, and so on.

Figure 19.7 illustrates an informed consent rule template that is defined *a priori* by users and allows all data access but anonymizes the identity of the users. By anonymization in this context, we mean simply replacing the *identity* attribute by the string *anonymous*. Depending on the specific requirements, anonymization could also include the replacement/modification by a service specific user pseudonym.

Figure 19.8 illustrates a policy rule that allows any operation to be performed by trustworthy entities. This rule is generic and does not specify in the event part the specific activity that should be allowed, for example, any data access

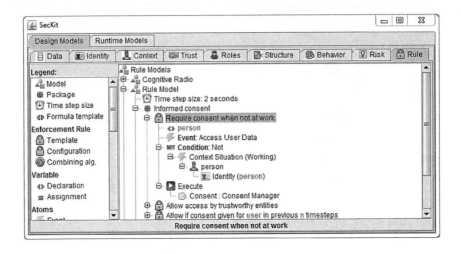

Figure 19.7: Off-line informed consent anonymization rule.

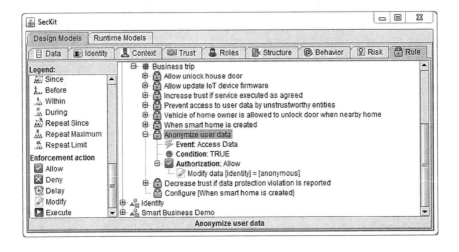

Figure 19.8: General-purpose trust-based access rule.

request would be allowed. This type of generic rule allows more generality in the specification of policy rules.

Figure 19.9 illustrates a policy rule template that allows access in the case of a specific user identified by a variable who has given consent (result = true) in the previous *n* time steps. It is up to each user that instantiates this template to specify how often the system should require his consent, for example, every day

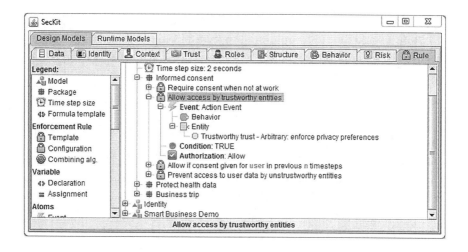

Figure 19.9: Consent check rule.

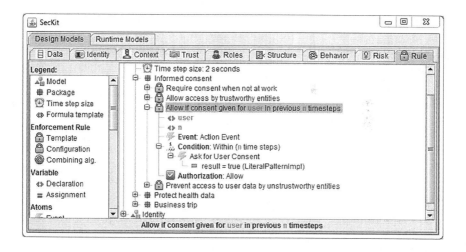

Figure 19.10: Composite rule with combining algorithm conflict resolution.

or every hour. An additional variable could be specified in this rule template to identify the specific activity where consent is required more often than for others.

Figure 19.10 shows a policy rule template that instantiates the previously described policy rules in a more complex template, which demonstrates the conflict resolution approach we adopt in our policy rule language. In this example, access is denied by default to all untrustworthy entities that try to access user data. However, in case a trustworthy entity tries to access the data or if consent was explicitly given by the user in the previous *n* time steps, the access is allowed. The strategy for the combining algorithm adopted here is named *Allow overrides*, meaning that if any of the nested rules allows the access, this is the final result chosen by the container rule template.

19.5 Conclusion and Future Developments

In this paper we introduced a novel approach to handle the authorization of data operation in the IoT, by combining a number of previously introduced different approaches, to produce a semiautonomous, rule based agent, which integrates context-awareness and enforcement through the "SecKit" mechanisms.

We believe that such an approach can significantly improve the way the informed consent question is handled. The definition of rule can be very specific (taking into account the type of data, type of operation, identity of the third party requesting the data operation and context of the operation) enabling detailed control by the end user. Further developments can now be envisioned to further refine the system and increase its impact on the informed consent of IoT end users: (a) providing information, training, and motivation of the user and (b) facilitating the definition of rules and policy implementation, which are described next.

19.5.1 Information, training, and motivation of the user

To actually achieve the "informed consent" of data operations by a large number of end users, some of which have low digital literacy, the ability to define data policy rules and strongly customize their application to the context of their use is not by itself sufficient. Dedicated efforts are needed to ensure that the policy-based system helps to inform, train, and motivate the user to actually care about his privacy and to define policies. Thus, the following additions to the system have to be considered:

■ The development and embedding of contextual help mechanisms in the Policy Management GUI to inform the user about the behavior of the system and the potential impacts of his decisions.

■ Progressive complexity of the user interface: one option could be to have different levels of complexity in the Policy Management GUI to accommodate users from beginner (that need a simple GUI to start defining policy rules) to expert levels (who need to be able to define complex custom rules). Tutorials can be created to help train the user in defining their first rules and to present the challenges of data protection in the IoT context.

19.5.2 Facilitating definition of rules and policy implementation

The definition of policy rules (made available by the system we propose) can take significant time and effort. To achieve large-scale acceptance and further increase the proposed system efficiency, additional dedicated efforts should target the facilitation of policy rules definitions. The following possibilities should be considered:

- GUI mechanisms to easily regroup and visualize rules applying to similar situations, similar actors (e.g., third parties accessing data), data operations, or type of data, could help the user in defining his rule set and validating that it covers what he cares about.

- To rapidly populate the policy repository for new users, the system could be initially populated by rules defined through a user-friendly questionnaire that introduces the user to the issue of privacy in the IoT.

- The addition of a rule suggestion engine to automatically propose rules to the user when none of the rules in the policy repository apply could significantly help the end user to define policy rules. Mechanisms based on behavior modeling [4] and on community systems were identified in the FP7 BUTLER project [24] to analyze the individual decisions made by the user. By gathering knowledge on the decision, the system is eventually able to propose decisions and new rules to the user. When confronted with a situation not taken into account by a user-defined rule the "policy-based system" would therefore first examine the behavior modeling component, and then, the community system in order to gather insights on a potential decision to propose to the end user. The end user is given the opportunity to either enforce this decision only once, or to define it as a policy rule in his Policy Repository for all future actions matching this signature.

19.6 Acknowledgment

This work was funded by the EC through the FP7 projects iCore (287708) and BUTLER (287901) and H2020 project FESTIVAL (643275).

Bibliography

[1] European Union, Directive 95/46/EC of the European Parliament and of the Council of 24 October 1995 on the protection of individuals with regard to

the processing of personal data and on the free movement of such data. 1995, http://eur-lex.europa.eu/ (last accessed 27 March 2015).

[2] European Union, Proposal for a directive of the European Parliament and of the Council on the protection of individuals with regard to the processing of personal data by competent authorities for the purposes of prevention, investigation, detection or prosecution of criminal offences or the execution of criminal penalties, and the free movement of such data. 2013, http://eur-lex.europa.eu/ (last accessed 27 March 2015).

[3] T. Ploug and S. Holm, Public health ethics: Informed consent and routinisation. *Journal of Medical Ethics*, 39(4), 214–218, 2013.

[4] D. Quartel, Action relations—Basic design concepts for behaviour modelling and refinement. PhD Thesis, 1998, University of Twente, the Netherlands.

[5] European Union, Article 29 Data Protection Working Party. 2013, http://ec.europa.eu/justice/data-protection/article-29/documentation/opinion-rec ommendation/files/2013/wp202_en.pdf (last accessed 27 March 2015).

[6] J.-Y. Tigli, S. Lavirotte, G. Rey, V. Hourdin and M. Riveill, Context-aware authorization in highly dynamic environments. *IJCSI International Journal of Computer Science Issues*, 4(1), 2013.

[7] M. Elkhodr, S.Shahrestani and H. Cheung, A contextual-adaptive location disclosure agent for general devices in the Internet of Things. *Proceeding of the 38th IEEE Conference on Local Computer Networks (LCN)*, October 2013, Sydney, Australia.

[8] G. Broenink, J.H. Hoepman, C. van't Hof, R. Kranenburg, D. Smits and T. Wisman, The privacy coach: Supporting customer privacy in the internet of things. In Michahelles, F. (Ed.), *What Can the Internet of Things Do for the Citizen (CIOT)*, 2010, pp. 72–81, Radboud University Nijmegen, the Netherlands.

[9] A. Boukerche and X. Li, An agent-based trust and reputation management scheme for wireless sensor networks. *Global Telecommunications Conference 2005, GLOBECOM'05*. Volume 3, 2005, IEEE, New York.

[10] D. Chen, G. Chang, D. Sun, J. Li, J. Jia and X. Wang, TRM-IoT: A trust management model based on fuzzy reputation for internet of things. *Computer Science and Information Systems*, 8(4),1207–1228, 2013.

[11] T. Henderson and S. Bhatti, Modelling user behaviour in networked games. *Proceeding Multimedia '01, Proceedings of the Ninth ACM International conference on Multimedia*, 2001, pp. 212–220, ACM, New York.

[12] C. A. Yeung and T. Iwata, Modelling user behaviour and interactions: augmented cognition on the social web. *Foundations of Augmented Cognition. Directing the Future of Adaptive Systems—Sixth International Conference,* Lecture Notes in Computer Science, 2011, Volume 6780, pp. 277–287, Orlando, FL.

[13] S. Angeletou, M. Rowe and H. Alani, Modelling and analysis of user behaviour in online communities. *The Semantic Web—ISWC 2011,* Lecture Notes in Computer Science, 2011, Volume 7031, pp. 35–50.

[14] R. Rawat, R. Nayak and Y. Li, Individual user behaviour modelling for effective web recommendation. *2nd International Conference on E-Education, E-Business, E-Management and E-Learning (IC4E 2011),* 2011, IEEE, Mumbai, India.

[15] A. Kobsa and J. Schreck, Privacy through pseudonymity in user-adaptive systems.*Journal ACM Transactions on Internet Technology (TOIT) TOIT Homepage,* 3(2), 149–183, 2003.

[16] H. Abelson, B. Adida, M. Linksvayer and N. Yergler, The Creative Commons Rights Expression Language Technical Report. 2008, Creative Commons, http://wiki.creativecommons.org/Image: Ccrel-1.0.pdf (last accessed 27 March 2015).

[17] C. Flick, Informed consent in information technology: Improving end user licence agreements. *Professionalism in the Information and Communication Technology Industry,* 3, 127, 2013.

[18] Brightfort, Eulalyzer. https://www.brightfort.com/eulalyzer.html (last accessed 30 April 2015).

[19] R. Neisse,I. Nai Fovino,G. Baldini, V. Stavroulaki, P. Vlacheas and R. Giaffreda, A Model-based security toolkit for the Internet of Things. *International Conference on Availability, Reliability and Security (ARES),* 2014, University of Fribourg, Switzerland.

[20] P.D. Costa,I. T. Mielke, I. Pereira,and J. P. A. Almeida,A model-driven approach to situations: Situation modeling and rule-based situation detection. *Enterprise Distributed Object Computing Conference (EDOC), 2012 IEEE 16th International,* 2012, pp. 154, 163, Tsinghua University, Beijing, China.

[21] E. Rissanen, XACML: Extensible Access Control Markup Language v3.0, 2010, http://docs.oasis-open.org.

[22] C. Sarkar, S. N. Akshay Uttama Nambi, R. Venkatesha Prasad, A. Rahim, R. Neisse, and G. Baldini, DIAT: A scalable distributed architecture for IoT. *IEEE Internet of Things Journal,* 2(3), 230–239, 2015.

[23] R. Neisse,M. Wegdam,and M. van Sinderen, Trust management support for context-aware service platforms. In A. Aldini and A. Bogliolo (Eds.), *User-Centric Networking (Lecture Notes in Social Networks)* 2014, pp. 75–106, Springer.

[24] Butler FP7 European Union collaborative project. Requirements and exploitation strategy. 2012, http://www.iot-butler.eu/ (last accessed 27 March 2015).

Chapter 20

Security and Impact of the Internet of Things (IoT) on Mobile Networks

Roger Piqueras Jover

CONTENTS

The ongoing evolution of wireless cellular networks is creating a new ecosystem with the pervasive presence of a great variety of network-enabled objects, which, based on unique addressing schemes, are able to interact with each other. Cellular connectivity is reaching beyond smartphones and tablets, providing access to data networks for connected home appliances, machinery, and vehicles. The rapid evolution of mobile networking technologies and the transition toward Internet Protocol (IP) v6 might drive this trend in to an ecosystem in which every single consumer item could be reachable through the cellular network. This convergence of the Internet and cellular mobility networks is breeding new machine-to-machine (M2M) communication systems, which are the enabling platform for the IoT [1].

Cellular-based IoT applications are experiencing dramatic growth, backed up by large investments from network operators [2]. Current studies predict the cellular IoT to be 1000 times more profitable than mobile data and as lucrative for operators as the short messaging service (SMS) [3]. This is an attractive new market for cellular operators, which are currently dealing with a heavily competitive market and declining revenues. Consequently, IoT applications are among the common denominator of some of the largest investments in mobile and cellular technology innovation. Cellular operators are seeking valuable partnerships in markets such as connected cars [4] and remote health care systems [5]. Indeed, forecasts predict the health industry to be one of the main drivers of the IoT market over the next few years [6].

Consensus exists in the industry that great growth in mobile cellular connectivity from M2M and embedded mobile applications will be experienced. More than 50 billion nonpersonal data-only mobile devices are expected to join existing mobile networks, supporting this plethora of emerging applications [7]. Consequently, to provide ubiquitous broadband connectivity to the IoT, *massive device connectivity* (billions of connected devices) is one of the main goals for the design and planning of future 5G mobile systems [8].

The fourth generation of mobile networks, Long-Term Evolution (LTE), has been designed for greatly enhanced capacity to provide support to a large number of connected devices. As such, LTE introduces significant improvements at the radio access network (RAN) and a more flexible IP-only architecture at the evolved packet core (EPC). Although a substantial percentage of current M2M systems operate over legacy second- and third-generation (2G and 3G) mobile networks, LTE is expected to be the main driver of the emergence of the IoT on cellular networks [9].

This massive deployment of IoT applications and their underlying M2M devices on mobile networks presents a great challenge for network operators. The traffic characteristics of many IoT applications, substantially different than user traffic from smartphones and tablets, is known to result in network resource utilization inefficiencies [10]. There is an increasing concern about the potential impact the IoT could have on LTE mobility networks, specifically regarding the surge in both traffic and control plane load [11]. Moreover, known security vulnerabilities in legacy mobile networks have been exploited to eavesdrop communications between M2M embedded devices and reverse engineer IoT systems.

Fueled by the great challenges of securely deploying and scaling IoT systems over cellular networks, both the research and the standardization community are leading several efforts to enhance the security of the IoT in the context of mobile cellular networks.

20.1   Security Threats against IoT Embedded Devices and Systems

Security and privacy are two of the main challenges of the IoT, particularly due to the emerging threats embedded devices face as a result of their unique limitations in terms of connectivity, computational power, and energy budget. Providing secure communications among M2M devices over cellular networks is an emerging research area, with different approaches being adopted. On the one hand, efforts aim to secure the device itself [12], and, on the other hand, network/provider-based architectures that benefit from the existing authentication methods of a cellular telecom operator are being proposed [13]. In parallel, privacy is increasingly becoming one of the major concerns in these kinds of systems, especially given the surge of applications handling critical information. This is a particularly crucial area in certain IoT system categories, such as the case of network-enabled medical environments [14].

Despite the increasing focus on securing M2M communications, efforts still have to be made to design more effective security architectures and transfer them into actual system deployments. An alarming lack of basic security features in certain applications has allowed researchers to discover new vulnerabilities and attack vectors against IoT systems, such as allowing remote ignition of a car's engine [15] and getting root access on a home-automation connectivity hub [16]. Other basic security flaws have been highlighted by the media, such as different types of connected devices being remotely accessible over the Internet with default or no access credentials [17].

IoT service providers often rely on the implicit wireless network encryption and authentication to protect traffic from eavesdropping and man-in-the-middle (MitM) attacks. Most deployments leverage legacy cellular 2G links, generally considered to be insecure and with outdated encryption schemes. This allows decryption and eavesdropping communications over the air [18]. For example, a recent study identified a popular geolocation platform that transmitted application information as plain text in the body of text messages. This allowed security researchers to reverse engineer the entire M2M application [19].

Cellular M2M systems should be designed with the assumption that any attacker can eavesdrop the traffic over the air, so extra layers of encryption should be encouraged. Nevertheless, certain low-power embedded devices with limited computing resources might not be capable of strong extra encryption. Overall, new M2M system implementations should be designed to leverage LTE mobile networks, with a state-of-the-art encryption and authentication scheme, preserving traffic privacy and security. Moreover, although studies have shown the feasibility of launching attacks and gaining control over network-enabled devices by means of deploying rogue base stations [20], this would not be possible over a mutually authenticated LTE access link.

20.2 IoT Security Impacts against Mobile Networks

Aside from the security and privacy of IoT-connected devices, the deployment of M2M systems on wireless mobile networks also has important security implications for the network itself. Resource allocation to millions of embedded devices is a big challenge for the heavily used mobile infrastructures of cellular network providers [7]. Beyond the challenge of network operation under such a load of IoT traffic, M2M traffic is considered to be one of the main factors within the overall LTE network security framework [21]. Industry and standardization forums defining the main security threats and requirements for mobile network security are indeed highlighting the IoT and its potential impact.

The traffic characteristics of many IoT applications, substantially different from user traffic generated by smartphones and tablets, are known to be a potential source for network resource utilization inefficiencies [10]. As a result, there is concern regarding the impact that M2M systems could have on the regular operation of LTE networks, which, if not architected properly, may be overwhelmed by the surge in both traffic and signaling load [11]. Given the number of threat vectors against embedded devices, there is also great interest in the potential impact of botnets of compromised devices and malicious signaling storms [22].

As mobile networks evolve and transition toward 5G, the capacity and throughput of the wireless interface is scaled up to tackle the goals of *massive device connectivity* and *1000 times more capacity*. To do so, researchers are already prototyping advanced systems at high millimeter-wave frequencies and implementing massive multiple-input and multiple-output (MIMO) systems. However, a common topic of discussion at a major 5G industry forum was how it is not all about speed, but also about scalability [23]. The scalability of billions of embedded devices joining existing LTE and future 5G networks is one of the major availability challenges within the field of IoT security.

20.2.1 *LTE network operation*

LTE mobile networks were designed to provide IP connectivity between mobile devices and the Internet based on the architecture depicted in Figure 20.1. LTE mobile networks are divided into two separate sections: the RAN and the core network, referred to as the EPC.

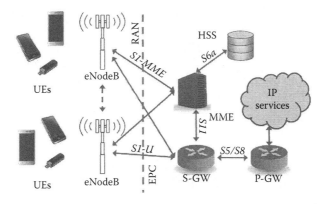

Figure 20.1: LTE network architecture.

A number of user equipment (UE) devices, or mobile terminals, and the eNodeBs, or LTE base stations, compose the RAN. This wireless access portion of an LTE network is in control of assigning radio resources to mobile terminals, managing their radio resource utilization, performing access control, and, in the case of the implementation of the X2 interface between eNodeBs, even managing mobility and handoffs independently of the EPC.

The EPC is the core in charge of establishing and managing the point-to-point IP connectivity between UEs and the Internet. Moreover, certain MAC (medium access control) operations at the RAN are triggered or controlled by the core network. The EPC is composed of the following network nodes. The serving gateway (SGW) and the PDN gateway (PGW) are the routing points that anchor a point-to-point connection, known as a *bearer*, between a UE and the Internet. The mobility management entity (MME) manages the control plane bearer logistics, mobility, and other network functions. To authenticate end users, the MME communicates with the home subscriber server (HSS), which stores the authentication parameters and secret keys of all the UEs.

To operate the network and provide connectivity, LTE networks execute a series of signaling processes, known as non access stratum (NAS) functions [24]. Such functions are coordinated and triggered by means of nonuser data messages among the LTE network nodes, known as the *control plane signaling* traffic.

After the device is switched on, a series of steps and algorithms are executed to reach the connected state. At this stage, an IP default bearer is set up between the UE and the PGW, and an IP address is assigned to the UE. The device executes the Cell Search procedure to acquire both time and frequency synchronization, and, by means of the random access procedure, radio resources are assigned to the UE, setting up a radio resource control (RRC) connection

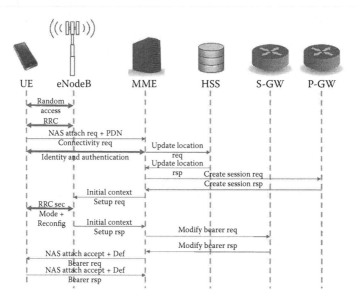

Figure 20.2: NAS attach signaling procedure.

between the device and the eNodeB. The NAS identity and authentication procedures are then executed between the UE and the MME, which in turn communicates with the HSS. At this point, the data traffic bearers through the SGW and PGW are set up, and the UE's RRC connection is reconfigured according to the type of IP service and quality of service (QoS) requested by the UE.

This entire NAS attach procedure is illustrated in Figure 20.2, which gives a clear visual intuition of the large number of messages exchanged among EPC elements to connect a mobile device [25]. Note that the random access procedure, the RRC connection establishment, and the NAS authentication and identity procedures involve a substantial number of messages not shown in the figure for simplicity.

Although all devices are assigned radio resources to communicate, there are not enough resources for a simultaneous connection from or to all the UEs. As a result, to efficiently assign and manage the spectrum, strict resource management and reutilization policies are implemented. Whenever a UE has been observed as idle by the eNodeB for more than a few seconds (often between 10 and 15 s), the RRC connection for this UE is released and its associated radio resources freed to be reused and assigned to another device, moving the UE to an idle state [26]. Although just one message from the eNodeB to the UE is sufficient to transition it down to an idle RRC state, this procedure still involves a series of messages among the EPC nodes to release the dedicated bearer. In parallel, a UE that is in idle state but needs to either transmit or receive data must be transitioned back to a connected state. To do so, a similar procedure to the NAS attach is executed.

The main differences are that certain authentication and bearer operations are not required. For example, a UE transitioning from idle to connected state is not required to have a new IP assigned. Note that each time a UE transitions from idle to connected, the procedure must always start with a random access and an RRC connection.

The functionality of a mobile network involves further signaling procedures not listed here, such as paging and handoffs. The aforementioned network functionalities are described because they involve NAS signaling procedures that have been discussed in standardization bodies as a potential trigger for signaling storms in LTE mobile networks.

20.2.2   *Control plane signaling storms*

The spreading of IoT applications over mobile networks brings security implications for the LTE packet core. The operation of mobile networks must be considered before designing wireless embedded devices that leverage cellular connectivity to prevent network utilization inefficiencies and, potentially, larger network availability threats. As has been described, each transaction or traffic flow between the IoT devices and other mobile devices or the Internet results in control plane signaling at the EPC. Unnecessary connection establishment and release signaling could potentially overburden the core network and reduce the QoS of other devices [27].

The concept of mobile core overloading due to control plane signaling was introduced in [28], which described a theoretical signaling overload threat against cellular networks. A low-volume attack, consisting of small data packets addressed to a large number of mobile devices, would theoretically induce a large number of RRC state transitions and, theoretically, overload the packet core of a mobile network. In this context, certain categories of M2M devices are characterized by small and frequent communication bursts [10]. This traffic pattern differs from the typical smartphone or tablet usage patterns that LTE was designed to support. Such frequent traffic bursts could cause a large amount of signaling in the network as devices transition between idle and connected states.

This negative impact of control plane overloads has already been observed in the wild over the last couple of years in a series of signaling storms summarized in Table 20.1.

All the aforementioned signaling storm instances were caused by misbehaving mobile applications. However, security researchers argue that a signaling storm could potentially be triggered from within the network by means of a malicious botnet of compromised M2M devices [36]. The authors of [37] discussed feasible techniques and platforms to build and operate such a botnet, including potential command and control channels.

Table 20.1   Sample of known signaling overload events

Cause	Event	Reference
Chatty app	IM app checking for new messages too frequently caused outage at U.S. carrier	[29]
Signaling spike	Outage for 3 million users at the sixth largest operator in the world	[30]
Smartphone native apps	Native apps from one of the main mobile operating systems causing signaling overloads to Japanese operator	[31]
Chatty apps	Operators at Open Mobile Summit discuss actions to mitigate signaling spikes from chatty apps	[32]
Adds in popular app	Signaling spikes caused by ads displayed in a popular mobile game	[33]
LTE-connected tablet	Connectivity from popular tablet increases control plane signaling substantially	[34]
Mobile cloud service	Frequent reconnect attempts to a cloud service under outage resulted in signaling spike	[35]

Additionally, certain M2M devices send or receive traffic at predefined time periods: for example, security cameras reporting a picture every few minutes or temperature sensors reporting a reading periodically. A large number of devices acting in this fashion could potentially generate peaks in both signaling and data traffic that could impact the core network. A similar situation could arise if an external event triggers a large number of devices to report or communicate, resulting in all the devices transitioning to an RRC connected state simultaneously. This is a specific use case of control plane signaling overload being discussed by standardization bodies [27].

In addition to M2M-related signaling overloads and network congestion, over the last few years, security researchers have theoretically proposed malicious ways to overload and congest LTE mobile networks. This congestion can occur in two different ways. RAN congestion is the result of many simultaneous M2M connection requests, modifications, and releases to the same eNodeB [38]. Congestion in the core network can affect the MME, SGW, and PGW when a large group of M2M devices attach to different cells, transition between RRC states, and move to different tracking areas [28]. Researchers have also theoreticized the potential impact of a signaling overload against the HSS node in legacy 2G and 3G networks, known as *home location registry* (HLR) [39].

Table 20.2 **Main threat scenarios and solutions proposed by 3GPP to mitigate control plane signaling overloads at the HSS**

Threat Scenario	Proposed Solutions
Overlaid RATs and failure of radio access technology (RAT) Node	Optimization of periodic tracking area update (TAU) signaling
Flood of registrations	NAS reject solution
Flood of RRC resource allocation	HLR/HSS overload notification
Flood of location information Reporting	Subscription data download optimization

20.2.3 Industry and security standardization work around M2M communications

Standardization bodies are actively working at proposing new security architectures to secure mobile M2M systems and the IoT. Certain industry forums are particularly engaged in proposing methods to alleviate potential signaling overloads and other security threats that could arise from the surge of IoT cellular systems.

The 3rd Generation Partnership Project (3GPP) is actively involved in defining a framework to mitigate control plane signaling spikes in cellular M2M systems [27], known as machine type communications (MTC) in the context of 3GPP. This effort focuses on a series of threat scenarios that range from a sudden flood of attach signaling load after a node failure to a flood of mobile terminated events.

The major threat scenarios and proposed solutions proposed by this 3GPP task force are summarized in Table 20.2. Most of the solutions proposed by this effort provide the HSS with means to filter and, potentially, block its incoming signaling load. For example, a new feature is defined such that the HSS can notify the MME in the event of a spike in traffic. The MME, in turn, can then reject attach attempts from mobile devices without requiring a prior handshake with the HSS. Moreover, optimization on various HSS operations is proposed.

Beyond security architectures for the cellular IoT, 3GPP is also actively involved in proposing enhancements for MTC traffic over cellular networks [40]. Finally, there is also ongoing discussion at 3GPP in areas that, though not directly related to IoT security, would make a big impact. For example, new enhancements to reduce the amount of control plane signaling load during RRC state transitions are proposed [41].

The oneM2M organization, closely related to the European Telecommunications Standards Institute (ETSI), also has active projects defining the security architecture for IoT cellular deployments in the framework of the oneM2M Release 1 specifications [42]. This project is focused on providing security at the application layer of IoT services provided over mobile networks. Some of the threat scenarios under analysis range from the deletion of service encryption keys stored on the memory of embedded devices to handling malicious or corrupted software in the M2M core service provider network.

A series of recommendations are issued to ensure the confidentiality and availability of cellular M2M systems. These recommendations, which had already been defined in the previous releases of these specifications in [43], ensure that strong encryption is applied at the application layer, with encryption keys stored in secure compartments. Note that, given the privacy threats of legacy 2G cellular links, it is not a good practice to rely solely on the wireless link encryption to secure M2M traffic.

This generalized interest in mitigating potential control plane traffic spikes in mobile networks is motivating certain industry players to develop appliances for mobile network infrastructure. These security solutions are designed as a control plane firewall, which sits between the RAN and the MME and monitors for signaling spikes, mitigating the impact on the mobile core [44]. Mobile infrastructure manufacturers are also increasing their efforts to supply new tools to assist the packet core in optimizing the control plane, minimizing the risk for overloads [45].

20.2.4 IoT security research

Extensive research is aiming to design new network mechanisms to efficiently handle the surge of cellular traffic originated from the IoT. In [46], new congestion control techniques for M2M traffic over LTE are proposed. Other techniques are suggested in [47]. The authors of [48] introduce new adaptive radio resource management to efficiently handle M2M traffic, and [49] proposes enhancements to the random access channel (RACH) of LTE systems to handle large numbers of embedded wireless devices.

20.3 Scalability of Large Deployments of Cellular IoT Systems

There is increasing interest among the telecommunications industry in understanding and, ideally, forecasting the scalability dynamics of IoT growth on LTE networks. Given the scale and device population expected, mobile network operators are expecting a large increase in both data and control plane traffic, to which network resources must adapt.

This necessity is driving a substantial increase of research work in this area. For example, the authors of [10] introduced the first detailed study of the traffic characteristics of emergent M2M applications. The authors highlight radio resource and network resource inefficiencies of these communication systems as a challenge for mobile infrastructures. Other research projects have analyzed the characteristics of IoT traffic over LTE mobile networks [50], reaching similar conclusions: certain M2M applications send periodic small bursts of traffic, which induce frequent RRC state transitions and, hence, are not efficient in terms of network resource utilization.

To be able to forecast and understand the scalability of the IoT over mobile networks, accurate modeling of the interaction of M2M systems with the cellular network is critical. The main goal of such modeling is to understand the non-linearity of control plane signaling traffic, as it scales with the number of connected devices. The heterogeneous traffic patterns from different M2M device categories result in a great diversity of signaling traffic load statistics. Certain device types, reporting measurements periodically, generate very low data loads (i.e., one 100 kb measurement per hour), but induce a large number of RRC state transitions and, thus, control plane signaling load. In parallel, devices sending a large daily summary of readings at the end of the day (i.e., 100 Mb) inject 1000 times more data traffic but have a marginal impact in terms of signaling load.

The study of the scalability of the IoT on LTE mobile networks requires a large-scale analysis. Laboratory testbeds are not sufficient to gauge the actual effects and implications of security threats involving the IoT. And, more importantly, laboratory based research does not provide means for rapid prototyping and test of M2M security technologies at scale. The potential risk of signaling overload as well as mobile botnets of compromised embedded devices requires a security analysis only possible on a simulation testbed.

For example, a fully LTE standards-compliant security research testbed can be used [51]. The testbed is designed and implemented to be fully standards compliant and can be scaled up over multiple virtual machines to simulate arbitrarily large scenarios. Moreover, the mobile devices simulated by this testbed run statistical traffic models derived from actual LTE mobile network traces, fully anonymized, from a tier-1 operator in the United States. Therefore, it can generate highly realistic results for smartphone traffic as well as several IoT device categories, such as smart grid, asset tracking, LTE-connected automobiles, telemedicine, remote alarming systems, and security cameras.

This testbed is leveraged to provide some insights into the scalability of IoT devices on mobile networks [52]. The experiments consist of deploying a simulated generic LTE network containing an instance of the EPC (MME, SGW, PGW, and HSS). IP communications occur between UEs and an external Internet server. The capacity of this server is assumed to be infinite so as not to interfere with the scalability impact at the EPC. A number of IoT devices, from different M2M categories (tele medicine, asset tracking, smart grid, etc.) are deployed on the network, with this device population being scaled up. As this occurs, multiple load metrics at the EPC are collected.

A generic experiment is summarized in Figure 20.3 a and b, examining MME-SGW link load and MME central processing unit (CPU) use to provide insight into the signaling impact of scaling M2M devices. It is intuitive that the M2M categories with the highest control plane signaling load impact (asset tracking and personal tracking devices in the case of the experiment in Figure 20.3)

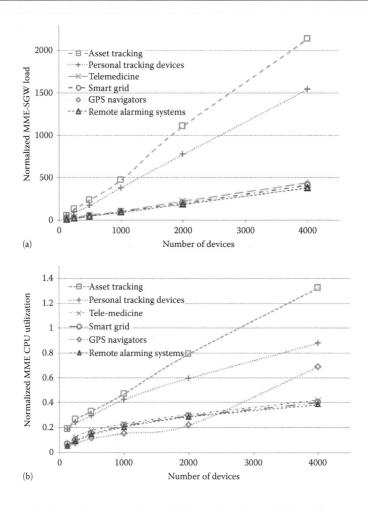

Figure 20.3: M2M scalability signaling impact: (a) normalized MME-SGW load and (b) normalized MME CPU use. GPS, global positioning system.

also result in the largest increase in MME CPU use. The load results in the figure are normalized.

The two device categories with the highest signaling load generate roughly the same load at the MME-SGW link for 125 devices. However, the load increases much more rapidly for asset tracking devices, with 2000 of these devices inducing 42.6% higher signaling load than the same number of personal tracking devices. A similar trend can be observed with the MME CPU use. Roughly the same MME CPU use is induced by 125 personal tracking devices and 125 asset tracking devices. However, 2000 asset tracking devices incur 32.5% higher load than the same number of personal tracking devices, and in the case of 4000 devices, the load is 50.5% higher.

These results present important insights into the scalability dynamics of the IoT on LTE mobile networks. As expected, the control plane signaling load spikes as the number of connected devices increases. This increase is linear, which indicates that, for example, an exponential increase in load should not be expected. However, the largely heterogeneous scalability characteristics for different IoT categories present a challenge for network operators. This is due to the fact that the connectivity and resource use must be optimized for very diverse traffic dynamics, as opposed to current mobile networks mainly optimized for smartphone traffic.

The challenge of deploying and scaling up IoT systems over mobile networks is one of the main areas of discussion and work in standardization bodies and will be one of the main challenges in the design of 5G mobile systems. Although advanced technologies at the physical (PHY) layer will provide orders of magnitude more capacity and throughput than current wireless links, 5G systems must be designed so that the core network is not a capacity bottleneck due to control plane signaling issues.

20.3.1 New network enhancements for mobile IoT systems

Proactive efforts are being established by cellular operators to encourage and ensure proper network resource use by M2M nodes [53]. Among other guidelines, recommendations to hardware and system manufacturers are provided to avoid applications that repeatedly check with servers or send sporadic data traffic with little or no flow control. In other words, M2M applications should not behave in cellular networks in the same way as they do in wireless local area networks (wLAN) or wired connections. It is essential that all network operators ensure proper implementation of such guidelines to minimize the impact of the surge of M2M appliances connected to mobility networks.

Some solutions to the described threats have been proposed by the industry and standardization community [27, 54]. Although there is a certain amount of improvement that can be made from the application/device itself, network-centric solutions are known to be the most effective. Mitigations for the above scenarios are challenging and difficult to implement and test to understand their potential benefits. For example, extending the idle timeout period of UEs has been proposed so that devices transition less frequently between RRC states [54]. This could potentially result in an increased cost, though, due to resources reserved for a longer time for an active device session, such as radio resources at the wireless link. The trade-off between the security benefit and this cost is very difficult to determine. Other proposed solutions envision techniques to filter signaling load at, for example, the MME to protect the HSS from an overload [27].

Figure 20.4: LTE random access procedure and capture from a real network.

In parallel, extensive work is ongoing in the research community and standardization groups to propose new techniques to provide data links over cellular networks for IoT devices with minimal impact on the mobile core. An interesting proposal introduces a connectionless protocol to communicate with IoT devices over LTE cellular links with zero control plane signaling [55]. This technique, aimed at M2M device categories with periodic small bursts of traffic, is designed within the framework of the 3GPP standards, requiring no standards modification.

IoT connectionless communications leverage the LTE PHY layer channels used for the RACH procedure, as shown in Figure 20.2. This handshake between the UE and the eNodeB is executed every time a mobile device requires to communicate and transition to the RRC connected state. The first message in this handshake is also used to achieve uplink (UL) synchronization with the eNodeB.

The transmission on the RACH channel is shared by all users within a cell or sector and follows the slotted ALOHA (S-ALOHA)/code division multiple access (CDMA) protocol, so collisions might occur. A signature is randomly chosen out of 64 possible signatures, and a preamble packet is sent over the RACH. Upon reception of a preamble, the eNodeB generates a reply message known as random access response (RAR). This message contains five fields: the ID of the time-frequency slot where the preamble was received, the selected signature, a time-alignment instruction, an initial UL resource grant, and a network temporary ID for the UE (radio network temporary identifier (RNTI)). A real lab capture of a standard RACH procedure is shown in Figure 20.4, including the handshake between a smartphone and a commercial laboratory eNodeB. This capture was obtained with an off-the-shelf LTE traffic sniffer [56].

A connectionless link for IoT traffic over LTE networks encodes uplink traffic in the RACH preambles. Six bits of information are encoded by the choice of a signature out of the 64 available. Given a number k of RACH resources per 10 ms LTE frame, the total available throughput in a given cell would be $\frac{k \cdot 6}{0.01}$ bits per second. Note that this throughput would be shared by all IoT devices within the cell and with the RACH traffic of regular smartphones and other mobile devices within the same cell. However, network measurements indicate that the RACH channel is substantially underused in very dense areas, leaving plenty of room for a connectionless link [55].

Given the possible configurations of the LTE RACH defined in the standards, k ranges from 1 to 10. The latter is the case of one RACH resource allocated in every slot [57]. Given a RACH collision probability of $p_{collision}^{UE} = 1\%$, including collisions and decoding errors, $k = 10$ and 64 signatures, a connectionless link can support an uplink load of up to $R_{RACH}^{max} = -10 \times 64 \times \ln(1 - p_{collision}^{UE}) = 6.432$ preambles per frame. This results in a maximum throughput of $R_{UL}^{max} = (6.432 \times 6)/0.01 = 3.86$ kbps.

In parallel, downlink traffic is encoded in the 16-bit RNTI field of the RAR, which is not necessary for a connectionless link. Assuming the same system configuration ($k = 10$), the maximum downlink throughput that can be delivered over a connectionless link is $(10 \cdot (16 + 11))/0.01 = 27$ kbps. This total raw capacity would also be shared by all the IoT devices within a given cell and the RAR messages sent to regular mobile devices and smartphones.

A mobile-initiated connectionless link is triggered by a series of preambles with a predefined pattern of signatures, while a network-initiated link is triggered by leveraging the three unused padding bits at the tail of the LTE paging message.

A system simulation of a connectionless link under a realistic load equivalent to that of a highly populated area is summarized in Figure 20.5. Results indicate that the RACH background load of one of a highly densely populated area would not impact the performance of the link. As a result, assuming a controlled M2M deployment and absence of adversarial UEs, the impact of the connectionless link on regular LTE communications would be almost null. This is a particularly positive result in the context of signaling overload threat mitigation and other large-scale security threats that could be triggered by a swarm of misbehaving or infected IoT embedded devices.

Connectionless links are a viable solution to provide connectivity to IoT devices over mobile networks without expensive control plane signaling impact against the core network. This particular characteristic is necessary to protect the network from potential saturation attacks against the cellular core that could be triggered by, for example, a malicious botnet of IoT devices. In parallel, industry consortiums and research laboratories are working on further implementations to mitigate the impact of large M2M deployments on mobile networks,

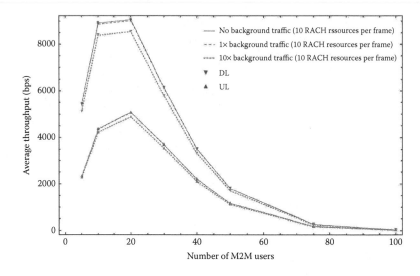

Figure 20.5: UL and downlink (DL) connectionless throughput with background LTE RACH load.

guaranteeing security and availability for both the IoT systems and the mobile network itself.

Bibliography

[1] A. Iera, C. Floerkemeier, J. Mitsugi, and G. Morabito, "Special issue on the Internet of Things," in *IEEE Wireless Communications*, vol. 17, 2010, pp. 8–9.

[2] K. Benedict, *M2M News Weekly*, Sys-Con Media, 2011, http://goo.gl/CI7T4D.

[3] T. Norman, "Machine-to-machine traffic worldwide: Forecasts and analysis 2011–2016," Analysys Mason, Technical Report, 2011.

[4] "The connected car: making cars smarter and safer," AT&T, 2014, http://goo.gl/PJYp3g.

[5] A. Berg, "Gadgets: New connected devices put smartphones in the middle," *Wireless Week*, 2012, http://www.wirelessweek.com/articles/2012/01/gadgets-new-connected-devices-put-smartphones-middle.

[6] "M2M News Weekly," *Connected World Magazine*, 2012, http://goo.gl/n6DB04.

[7] Ericsson, "More than 50 billion connected devices," Ericsson White Paper, 2011, http://goo.gl/7LGq4h.

[8] NTT Docomo, "5G radio access: Requirements, concept and technologies," NTT Docomo, 2014, http://goo.gl/L72689.

[9] D. Lewis, "Closing in on the future with 4G LTE and M2M," Verizon Wireless News Center, 2012, http://goo.gl/ZVf7Pd.

[10] M. Shafiq, L. Ji, A. Liu, J. Pang, and J. Wang, "Large-scale measurement and characterization of cellular machine-to-machine traffic," *Networking, IEEE/ACM Transactions on*, vol. 21, no. 6, pp. 1960–1973, 2013.

[11] A. Prasad, "3GPP SAE-LTE security," in *NIKSUN WWSMC*, Princeton, NJ July 25–27, 2011.

[12] A. Ukil, J. Sen, and S. Koilakonda, "Embedded security for Internet of Things," in *Emerging Trends and Applications in Computer Science (NCETACS), 2011 2nd National Conference on*, 2011, pp. 1–6.

[13] S. Agarwal, C. Peylo, R. Borgaonkar, and J. Seifert, "Operator-based over-the-air m2m wireless sensor network security," in *Intelligence in Next Generation Networks (ICIN), 2010 14th International Conference on*, 2010, pp. 1–5.

[14] A. Jara, M. Zamora, and A. Skarmeta, "An architecture based on Internet of Things to support mobility and security in medical environments," in *Consumer Communications and Networking Conference (CCNC), 2010 7th IEEE*, 2010, pp. 1–5.

[15] C. Miller and C. Valasek, "A survey of remote automotive attack surfaces," in *Blackhat USA*, 2014, http://goo.gl/k61KzN.

[16] C. Heres, A. Etemadieh, M. Baker, and H. Nielsen, "Hack all the things: 20 devices in 45 minutes," in *In DefCon 22*, 2014, http://goo.gl/hU7a8G.

[17] A. Cui and S. J. Stolfo, "A quantitative analysis of the insecurity of embedded network devices: Results of a wide-area scan," in *Proceedings of the 26th Annual Computer Security Applications Conference*. ACM, 2010, pp. 97–106. Austin, TX.

[18] K. Nohl and S. Munaut, "Wideband GSM sniffing," in *27th Chaos Communication Congress*, 2010, http://tinyurl.com/33ucl2g.

[19] D. Bailey, "War texting: Weaponizing machine to machine," in *BlackHat USA*, 2011, https://www.nccgroup.trust/globalassets/newsroom/us/news/documents/2011/isec_bh2011_war_texting.pdf.

[20] Hunz, "Machine-to-machine (M2M) security," in *Chaos Communication Conference Camp*, 2011, https://events.ccc.de/camp/2011/Fahrplan/attachments/1883_m2m.pdf.

[21] A. R. Prasad, "3GPP SAE/LTE security," in *NIKSUN WWSMC*, 2011, http://goo.gl/e0xAWQ.

[22] R. Piqueras Jover, "Security attacks against the availability of LTE mobility networks: overview and research directions," in *Wireless Personal Multimedia Communications (WPMC), 2013 16th International Symposium on*, Atlantic City, NJ, 2013, pp. 1–9.

[23] "2015 5G Brooklyn Summit," http://brooklyn5gsummit.com/.

[24] S. Sesia, M. Baker, and I. Toufik, *LTE, The UMTS Long Term Evolution: From Theory to Practice*. Published online. Wiley, 2009.

[25] S. Rao and G. Rambabu, "Protocol signaling procedures in LTE," Radisys, White Paper, 2011, http://goo.gl/eOObGs.

[26] 3rd Generation Partnership Project; Technical Specification Group Radio Access Network, "Evolved Universal Terrestrial Radio Access (E-UTRA) - Radio Resource Control (RRC) - protocol specification. 3GPP TS 36.331," vol. v8.20.0, 2012.

[27] 3rd Generation Partnership Project; Technical Specification Group Services and Systems Aspects, "Study on core network overload and solutions. 3GPP TR 23.843," vol. v0.7.0, 2012.

[28] P. Lee, T. Bu, and T. Woo, "On the detection of signaling DoS attacks on 3G wireless networks," in *INFOCOM 2007. 26th IEEE International Conference on Computer Communications. IEEE*, 2007.

[29] M. Dano, "The Android IM app that brought T-Mobile's network to its knees," *Fierce Wireless*, 2010, http://goo.gl/O3qsG.

[30] "Signal storm caused Telenor outages," *Norway News in English*, 2011, http://goo.gl/pQup8e.

[31] C. Gabriel, "DoCoMo demands Google's help with signalling storm," *Rethink Wireless*, 2012, http://goo.gl/dpLwyW.

[32] M. Donegan, "Operators urge action against chatty apps," *Light Reading*, 2011, http://goo.gl/FeQs4R.

[33] S. Corner, "Angry Birds + Android + ads = network overload," *iWire*, 2011, http://goo.gl/nCI0dX.

[34] E. Savitz, "How the new iPad creates 'signaling storm' for carriers," *Forbes*, 2012, http://goo.gl/TzsNmc.

[35] S. Decius, "OTT service blackouts trigger signaling overload in mobile networks," Nokia Networks, 2013, http://goo.gl/rAfs96.

[36] J. Jermyn, G. Salles-Loustau, and S. Zonouz, "An analysis of DOS attack strategies against the LTE RAN," *Journal of Cyber Security*, vol. 3, no. 2, pp. 159–180.

[37] C. Mulliner and J.-P. Seifert, "Rise of the iBots: Owning a telco network," in *Proceedings of the 5th IEEE International Conference on Malicious and Unwanted Software (Malware)*, Nancy, France October 2010 IEEE.

[38] M. Khosroshahy, D. Qiu, M. Ali, and K. Mustafa, "Botnets in 4G cellular networks: Platforms to launch DDoS attacks against the air interface," in *Mobile and Wireless Networking (MoWNeT), 2013 International Conference on Selected Topics in*. Montreal, Canada August 2013 IEEE, pp. 30–35.

[39] P. Traynor, M. Lin, M. Ongtang, V. Rao, T. Jaeger, P. McDaniel, and T. La Porta, "On cellular botnets: Measuring the impact of malicious devices on a cellular network core," in *Proceedings of the 16th ACM Conference on Computer and Communications Security*, ser. CCS '09. New York, NY, United States: ACM, 2009, pp. 223–234.

[40] 3rd Generation Partnership Project; Technical Specification Group Services and System Aspects, "Machine-type and other mobile data applications communications enhancements. 3GPP TR 23.887," vol. v12.0.0, 2013.

[41] 3GPP work item description, "Signalling reduction for idle-active transitions. 3GPP RP-150426," 2015.

[42] oneM2M R1, "Security solutions. ETSI TS 118 103," vol. v1.0.0, 2015.

[43] ETSI, "Machine-to-machine communications (M2M: Threat analysis and counter-measures to M2M service layer. ETSI TR 103 167," vol. v1.1.1, 2011.

[44] "Open channel traffic optimization," Seven Networks, Tech. Rep., 2015, http://goo.gl/uz5YOH.

[45] "9471 Wireless Mobility Manager," Alcatel Lucent, Tech. Rep., 2015, http://goo.gl/0n2v8F.

[46] S. Duan, "Congestion control for M2M communications in LTE networks," Technical Report, University of British Columbia, 2013.

[47] S.-Y. Lien and K.-C. Chen, "Massive access management for QoS guarantees in 3GPP machine-to-machine communications," *Communications Letters, IEEE*, vol. 15, no. 3, pp. 311–313, 2011.

[48] Y.-H. Hsu, K. Wang, and Y.-C. Tseng, "Enhanced cooperative access class barring and traffic adaptive radio resource management for M2M communications over LTE-A," in *Signal and Information Processing Association Annual Summit and Conference (APSIPA), 2013 Asia-Pacific.* Kaohsiung, Taiwan October 2013 IEEE, pp. 1–6.

[49] A. Laya, L. Alonso, and J. Alonso-Zarate, "Is the random access channel of LTE and LTE-A suitable for M2M communications? A survey of alternatives," *Communications Surveys Tutorials, IEEE*, vol. 16, no. 1, pp. 4–16, 2014.

[50] C. Ide, B. Dusza, M. Putzke, C. Muller, and C. Wietfeld, "Influence of M2M communication on the physical resource utilization of LTE," in *Wireless Telecommunications Symposium (WTS), 2012.* London, UK April 2012 IEEE, pp. 1–6.

[51] J. Jermyn, R.P. Jover, M. Istomin, and I. Murynets, "Firecycle: A scalable test bed for large-scale LTE security research," in *Communications (ICC), 2014 IEEE International Conference on.* Sydney, Australia June 2014 IEEE, pp. 907–913.

[52] J. Jermyn, R.P. Jover, I. Murynets, and M. Istomin, "Scalability of machine to machine systems and the Internet of Things on LTE mobile networks," in *IEEE International Symposium on a World of Wireless, Mobile and Multimedia Networks (WoWMoM).* Boston, MA June 2015 IEEE.

[53] L. Iyengar, Y. Zhang, J. Jun, and Y. Li, "AT&T network ready device development guidelines," AT&T Network Ready Laboratory, Tech. Rep., 2011, http://goo.gl/nmUrSl.

[54] "System improvements for machine-type communications (MTC). 3GPP TS 23.888," vol. v11.0.0.0, 2012.

[55] R. P. Jover and I. Murynets, "Connection-less communication of IoT devices over LTE mobile networks," in *IEEE International Conference on Sensing, Communication and Networking (SECON).* Seattle, WA June 2015 IEEE.

[56] Sanjole, "WaveJudge 4900A LTE analyzer," http://goo.gl/ZG6CCX.

[57] 3rd Generation Partnership Project; Technical Specification Group Radio Access Network, "Evolved Universal Terrestrial Radio Access Network (E-UTRAN); Physical channels and modulation. 3GPP TS 36.211," vol. v10.3.0, 2011.

Index